CW01281418

EAI/Springer Innovations in Communication and Computing

Series editor

Imrich Chlamtac, CreateNet, Trento, Italy

Editor's Note

The impact of information technologies is creating a new world yet not fully understood. The extent and speed of economic, life style and social changes already perceived in everyday life is hard to estimate without understanding the technological driving forces behind it. This series presents contributed volumes featuring the latest research and development in the various information engineering technologies that play a key role in this process.

The range of topics, focusing primarily on communications and computing engineering include, but are not limited to, wireless networks; mobile communication; design and learning; gaming; interaction; e-health and pervasive healthcare; energy management; smart grids; internet of things; cognitive radio networks; computation; cloud computing; ubiquitous connectivity, and in mode general smart living, smart cities, Internet of Things and more. The series publishes a combination of expanded papers selected from hosted and sponsored European Alliance for Innovation (EAI) conferences that present cutting edge, global research as well as provide new perspectives on traditional related engineering fields. This content, complemented with open calls for contribution of book titles and individual chapters, together maintain Springer's and EAI's high standards of academic excellence. The audience for the books consists of researchers, industry professionals, advanced level students as well as practitioners in related fields of activity include information and communication specialists, security experts, economists, urban planners, doctors, and in general representatives in all those walks of life affected ad contributing to the information revolution.

About EAI

EAI is a grassroots member organization initiated through cooperation between businesses, public, private and government organizations to address the global challenges of Europe's future competitiveness and link the European Research community with its counterparts around the globe. EAI reaches out to hundreds of thousands of individual subscribers on all continents and collaborates with an institutional member base including Fortune 500 companies, government organizations, and educational institutions, provide a free research and innovation platform.

Through its open free membership model EAI promotes a new research and innovation culture based on collaboration, connectivity and recognition of excellence by community.

More information about this series at http://www.springer.com/series/15427

Dagmar Cagáňová • Michal Balog
Lucia Knapčíková • Jakub Soviar
Serkan Mezarcıöz
Editors

Smart Technology Trends in Industrial and Business Management

Springer

EAI
RESEARCH MEETS INNOVATION

Editors
Dagmar Cagáňová
Slovak University of Technology
in Bratislava
Trnava, Slovakia

Lucia Knapčíková
Technical University of Košice
Department of Industrial Engineering
and Informatics
Prešov, Slovakia

Serkan Mezarcıöz
Temsa Ulaşım Araçları San. ve Tic. A.Ş.
Seyhan, Adana, Turkey

Michal Balog
Technical University of Košice
Department of Industrial Engineering
and Informatics
Prešov, Slovakia

Jakub Soviar
University of Žilina
Faculty of Management Science
and Informatics
Žilina, Slovak Republic

ISSN 2522-8595   ISSN 2522-8609 (electronic)
EAI/Springer Innovations in Communication and Computing
ISBN 978-3-319-76997-4   ISBN 978-3-319-76998-1 (eBook)
https://doi.org/10.1007/978-3-319-76998-1

Library of Congress Control Number: 2018949322

© Springer International Publishing AG, part of Springer Nature 2019
This work is subject to copyright. All rights are reserved by the Publisher, whether the whole or part of the material is concerned, specifically the rights of translation, reprinting, reuse of illustrations, recitation, broadcasting, reproduction on microfilms or in any other physical way, and transmission or information storage and retrieval, electronic adaptation, computer software, or by similar or dissimilar methodology now known or hereafter developed.
The use of general descriptive names, registered names, trademarks, service marks, etc. in this publication does not imply, even in the absence of a specific statement, that such names are exempt from the relevant protective laws and regulations and therefore free for general use.
The publisher, the authors, and the editors are safe to assume that the advice and information in this book are believed to be true and accurate at the date of publication. Neither the publisher nor the authors or the editors give a warranty, express or implied, with respect to the material contained herein or for any errors or omissions that may have been made. The publisher remains neutral with regard to jurisdictional claims in published maps and institutional affiliations.

This Springer imprint is published by the registered company Springer Nature Switzerland AG
The registered company address is: Gewerbestrasse 11, 6330 Cham, Switzerland

Preface

This publication is a collection of rigorous research projects that were presented at the *"Industry of Things and Future Technologies"* international conference. The event took place in Bratislava, Slovakia, during November 22–24, 2016, and was endorsed by the European Alliance for Innovation, a leading community-based organisation devoted to the advancement of innovation in the field of ICT. The conference was organised by the Slovak University of Technology in Bratislava, Faculty of Materials Science and Technology in Trnava (MTF STU) and the Technical University of Košice, Faculty of Manufacturing Technologies with a seat in Prešov (TUKE).

The Internet of Things (IoT) offers advanced connectivity of devices, systems, and services that reach beyond machine-to-machine communications and covers a variety of domains. Therefore, the main goal of the conference was to determine the next level technologies that will drive Industry 4.0 and IoT forward and to introduce industrial innovations, smarter key technologies, and novel approaches to data analysis, rethink strategic planning, develop new opportunities, address challenges, and explore its solutions. Additionally, the focus of the conference reflected the European Union thematic priorities for research and innovation to improve the quality of life for citizens and make cities more sustainable with decreasing impact on the environment.

The conference articles were presented in the four thematic areas:

- *Technology* (responsibility in IT business, creative technologies, intelligent transport systems, advances in robotics and machine vision as a key success factor in innovative companies, the role of the human factor in the performance and sustainability of manufacturing, challenges of IoT to cybersecurity, engineering secure IoT systems, IoT impact on critical infrastructures, and others)
- *Internet* (future Internet, Internet of Things, security and safety, smart cities, creative cities)
- *Innovation* (industrial networks and intelligent systems, social networks [social marketing] and innovation in social areas, new perspectives in transport innovation, green vehicles, fast track for transport innovation, socio-economic and

behavioural research for policy-making, novel trends in production devices and systems, and others)
- *Mobility* (mobility within Danube strategy, smart mobility, mobility and its consequences on health and well-being, urban mobility, E-mobility, congestion-free, and sustainable mobility)

This publication encompasses a total of 34 research articles with worldwide contributors. Among the project findings featured in this publication are those written by the conference keynote speakers, e.g. Dr.h.c. Mult. Prof. Ing. Juraj Sinay, DrSc., the Head of the Department of Safety and Quality in Mechanical Engineering at the Technical University in Košice and the President of Automotive Industry Association of the Slovak Republic. Prof. Dr. Dragan Perakovic, the Head of Department for Information and Communication Traffic and Head of Chair of Information Communication Systems and Services Management at the University of Zagreb, and the chief editor of the International Journal of Cyber-Security and Digital Forensics (IJCSDF), contributes with his research article on Smart Wristband System in Traffic Environment. His study is included in the Smart Transportation Applications and Vehicle Data Processing System for Smart City Buses section of this release.

Next, at the conference presented highly acclaimed scholar Prof. Pedrag Nikolič who serves as a Dean of the Faculty of Digital Production at Educons University, Serbia, together with Prof. Adrian David Cheok, the director of Imagineering Institute in Malaysia. Among others, we are proud to achieve the scholarly success of this edition which resulted from our cooperation with the Comenius University in Bratislava, the Faculty of Management, and their prestigious contributors Assoc. Prof. Ján Papula, Assoc. Prof. Zuzana Papulová with the research team, Assoc. Prof. Gabriela Bartáková-Pajtinková, and Assoc. Prof. Katarína Gubíniová. We were additionally delighted to welcome Assoc. Prof. Katarína Stachová and Assoc. Prof. Zdenko Stacho from The School of Economics and Management in Public Administration in Bratislava. The Department of Management at MTF STU and The Institute of Management from Slovak Technical University were represented by Prof. Tatiana Kluvánková, Prof. Maroš Finka, and Assoc. Prof. Daniela Gažová with their teams at the conference.

Moreover, thanks to an excellent collaboration between MTF STU and the Department of Marketing and Trade at the Slovak Agricultural University in Nitra. Prof. Ľudmila Nagyová and Dr. Mária Holienčinová shared their scholarly expertise in the Smart Technology Trends Business Management section of the publication. We are also thankful for the research article contribution under the leadership of the Assoc. Prof. Marian Králik from STU Faculty of Mechanical Engineering, Dr. Peter Pištek with his colleagues from Faculty of Informatics and Information Technologies, and Prof. Michal Cehlár, the Dean of Faculty of Mining, Ecology, Process Control and Geotechnologies (BERG) at the Technical University of Košice (TUKE) with his team whose contributions are included in the Smart Transportation Applications and Vehicle Data Processing System for Smart City Buses section of the publication. Furthermore, we are grateful to Assoc. Prof. Dagmar Petríková who leads the Department of Spatial Planning and Management at the MTF STU.

Additionally, we would like to show appreciation to the PhD candidates and other academics from the MTF STU Faculty of Materials Science and Technology, particularly to Dr. Natália Horňáková, MSc. M.A. Richard Jurenka, and MSc. Augustín Stareček who mastered the exemplary teamwork and for their rigorous contributions to this publication as well as many other outstanding contributions from various institutions. Other areas of expertise covered in this edition include Industry of Things and Future Technologies and Smart Technology Trends in Industrial Management and Materials.

As chairs of the conference, we were particularly impressed by the wide range of innovative research solutions presented at the symposium. In the light of the latest knowledge and findings from scientific projects, the authors present actual R&D trends in the given field. Therefore, the Scientific Committee members and organisers would like to express their sincere thanks to all the authors who attended the conference in Bratislava, Slovakia, and particularly to the authors, who contributed to the creation of this publication. This issue not only defines the state of the art in the field, but it additionally explores related topics for future research. Moreover, we would like to thank the audience members who actively interacted in the discussion on the topics mentioned above.

Trnava, Slovakia
                                                              Dagmar Cagáňová
                                                                 Dorota Horvath

Acknowledgment to the International Conference on Management of Manufacturing Systems (MMS 2016)

I would like to thank all the authors and reviewers, especially Prof. Juraj Sinay, Prof. Dragan Peraković, and Assoc. Prof. Juraj Pančík who are our keynote speakers. My thanks to the Dean of Faculty of Manufacturing Technologies with a seat in Prešov of Technical University of Košice Dr. h. c. Prof. Dr. Jozef Zajac, to the Head of the Department of Industrial Engineering and Informatics, to Assoc. Prof. Dr. Michal Balog, and to my colleague Dr. Jozef Husár.

Lucia Knapčíková

Contents

Part I Industry of Things and Future Technologies

1 **Application of AHP Method in Decision-Making Process** 3
Richard Jurenka, Dagmar Cagáňová, and Daniela Špirková

2 **An Integrative Spatial Perspective on Energy Transition: Renewable Energy Niches** 17
Filip Gulan, Maros Finka, and Michal Varga

3 **Can Concept of Smart Governance Help to Mitigate the Climate in the Cities?** 35
Alfréd Kaiser and Tatiána Kluvánková

4 **Operational Characteristics of Experimental Actuator with a Drive Based on the Antagonistic Pneumatic Artificial Muscles** 49
Miroslav Rimár, Peter Šmeringai, and Marcel Fedák

5 **Critical Values of Some Probability Distributions and Standard Numerical Methods** 61
Dušan Knežo, Jozef Zajac, and Peter Michalik

6 **Measuring Production Process Complexity** 71
Soltysova Zuzana, Bednar Slavomir, and Behunova Annamaria

7 **IoT Challenge: Older Test Machines Modernization in an Automotive Plant** 85
Juraj Pančík and Vladimír Beneš

Part II Smart Technology Trends in Industrial Management

8 **Industry 4.0: Preparation of Slovak Companies, the Comparative Study** 103
Ján Papula, Lucia Kohnová, Zuzana Papulová, and Michael Suchoba

9 Transformations of Urbanized Landscape Following
   the Smart Water Management Concept 115
   Matina Lazarová, Michal Varga, and Daniela Gažová

10 RFID Labels and Its Characteristics on Labeled Products 133
   Dušan Dorčák, Romana Hricová, and Peter Šebej

11 Basic Assumptions of Information Systems for Increasing
   Competitiveness of Production Companies within the EU
   and their Application of the CAPP System Design 145
   Katarina Monkova, Peter Monka, Helena Zidkova,
   Vladimir Duchek, and Milan Edl

12 Cooperation as a Key Element Between Universities
   and Factories 165
   B. Mičieta, J. Herčko, Ľ. Závodská, and M. Fusko

13 Zilina Intelligent Manufacturing System: Best Practice
   of Cooperation Between University and Research Center 183
   Milan Gregor, Jozef Hercko, Miroslav Fusko, and Lukas Durica

14 Improvement of the Production System Based
   on the Kanban Principle. 199
   Ľuboslav Dulina, Miroslav Rakyta, Ivana Sulírová,
   and Michala Šeligová

15 Quality Assurance in the Automotive Industry and Industry 4.0.... 217
   Štefan Markulik, Juraj Sinay, and Hana Pačaiová

Part III Smart Technology Trends Business Management

16 Potential of Human Resources as Key Factor of Success
   of Innovation in Organisations 229
   Stacho Zdenko, Stachová Katarína, and Cagáňová Dagmar

17 Environmental Policy as a Competitive Advantage
   in the Global Environment. 249
   Zuzana Tekulova, Zuzana Chodasova, and Marian Kralik

18 Sustainable Organization of Cooperation Activities
   in a Company: Slovak Republic Research Perspective 263
   Jakub Soviar, Viliam Lendel, Josef Vodák, and Jana Kundríková

19 Green Markets and Their Role in the Sustainable Marketing
   Management 281
   Katarína Gubíniová, Gabriela Pajtinková Bartáková,
   and Jarmila Brtková

20 A New Approach to Sustainable Reporting: Responsible Communication Between Company and Stakeholders in Conditions of Slovak Food Industry 291
Mária Holienčinová and Ľudmila Nagyová

21 Cooperative Relations and Activities in a Cluster in the Slovak and Czech Automotive Industry 303
J. Vodák, M. Varmus, P. Ferenc, and D. Zraková

Part IV Smart Technology Trends in Materials

22 Composites Manufacturing: A New Approaches to Simulation 319
Lucia Knapčíková, Michal Balog, Alessandro Ruggiero, and Jozef Husár

23 Study of the Cutting Zone of Wood-Plastic Composite Materials After Different Types of Cutting 327
Dusan Mital, M. Hatala, J. Zajac, P. Michalik, J. Duplak, J. Vybostek, L. Mroskova, and D. Knezo

24 Risk Analysis Causing Downtimes in Production Process of Hot Rolling Mill 337
Marcela Malindzakova, Dagmar Cagáňová, Andrea Rosova, and Dusan Malindzak

25 Evaluation of Roughness Parameters of Machined Surface of Selected Wood Plastic Composite 345
J. Zajac, F. Botko, S. Radchenko, P. Radič, A. Bernat, J. Roman, and B. Zajac

26 Evaluation of the Transverse Roughness of the Outer and Inner Surfaces of the Thin-Walled Components Produced by Milling 353
Peter Michalik, Jozef Zajac, Michal Hatala, Dusan Mital, and Łukasz Nowakovski

Part V Smart Transportation Applications and Vehicle Data Processing System for Smart City Buses

27 Designing Behavioral Changes in Smart Cities Using Interactive Smart Spaces 367
Predrag K. Nikolic and Adrian D. Cheok

28 Social Innovations in Context of Smart City 383
Richard Jurenka, Dagmar Cagáňová, Natália Horňáková, and Augustín Stareček

29 Towards Creating Place Attachment and Social Communities in the Smart Cities 401
Matej Jaššo and Dagmar Petríková

30 **Awareness of Malicious Behavior as a Part of Smart Transportation in Taxi Services**........................... 413
Peter Pistek and Martin Polak

31 **Alternative Lights for Public Transport in Smart Cities**........... 421
Michal Cehlár and Dušan Kudelas

32 **Smart Wristband System for Improving Quality of Life for Users in Traffic Environment**........................... 429
Dragan Peraković, Marko Periša, Rosana Elizabeta Sente, Petra Zorić, Boris Bucak, Andrej Ignjatić, Vlatka Mišić, Matea Vuletić, Nada Bijelica, Luka Brletić, and Ana Papac

33 **Smart Transportation Applications and Vehicle Data Processing System for Smart City Buses**..................... 451
Serkan Mezarcıöz, Enis Aytar, Murat Demizdüzen, Mert Özkaynak, and Kadir Aydın

34 **A Model Approach for the Formation of Synergy Effects in the Automotive Industry with Big Data Solutions: Application for Distribution and Transport Service Strategy**...... 467
Martin Holubčík, Gabriel Koman, Michal Varmus, and Milan Kubina

Index.. 489

Part I
Industry of Things and Future Technologies

Chapter 1
Application of AHP Method in Decision-Making Process

Richard Jurenka, Dagmar Cagáňová, and Daniela Špirková

Abstract The paper deals with the application of method – analytical hierarchy process in action of decision-making in the field of innovation management of industrial companies. Analytical hierarchy process is a tool for application of exact methods in process of decision-making. The paper contains theoretical description of analytical hierarchy process and subsequently also application of this method in the field of innovation management. Method of analytical hierarchy process brings into action of decision-making objectivity, exactness, and also the quality of the evaluation. Perhaps the biggest advantage of this method is that this method allows to evaluate comprehensively all the criteria of alternative solutions. This paper aims to highlight the widespread use of method – analytical hierarchy process in the decision-making process, including in the field of innovation management.

Introduction

In the modern business environment, organizations must respond to new challenges, constant changes, and opportunities and also must respond to the different requirements and various restrictions. Permanent transformation of the business environment is in contemporary world the necessity of continuous innovation and changes.

R. Jurenka (✉) · D. Cagáňová
Slovak University of Technology in Bratislava, Bratislava, Slovakia

Faculty of Materials Science and Technology in Trnava, Trnava, Slovakia

Institute of Industrial Engineering and Management, Trnava, Slovakia
e-mail: richard.jurenka@stuba.sk; dagmar.caganova@stuba.sk

D. Špirková
Institute of Management, Bratislava, Slovakia

Slovak University of Technology in Bratislava, Bratislava, Slovakia
e-mail: daniela.spirkova@stuba.sk

© Springer International Publishing AG, part of Springer Nature 2019
D. Cagáňová et al. (eds.), *Smart Technology Trends in Industrial and Business Management*, EAI/Springer Innovations in Communication and Computing,
https://doi.org/10.1007/978-3-319-76998-1_1

Nowadays from existing organizations, some flexibility, dynamism, and constant adaptation to changing conditions are expected. The current social conditions are reflected in organizational behavior of individual companies and also in their decision-making process.

Social environment in this century is very dynamic, changeable, unstable and hardly predictable. Result of these factors is instability and unpredictable development in social environment. Different changes in such an environment and conditions are becoming a necessity and everyday reality with which organizations must deal by their flexibility and ability to adapt to new and new conditions. In dynamic environment, in which organizations want to operate and develop, constantly bringing new enhancements, innovations, ideas, and thoughts is fundamental for companies. For selection the correct option of advancement is a necessary perfect decision-making process that will consider all the factors, risks, and opportunities in outside and inside environment.

Exact Methods in Managerial Decision-Making

Decision-making is an integral part of management processes in every company and intersects across all functions of management – planning, organizing, staffing and maintaining staff, leadership people, and controlling. All managers at all levels of directing make decisions, while the ultimate effect of these decisions can have a significant impact on a range of other activities. Some decisions are strategic in their nature and significantly affect company's survival on the market; other decisions can be seemingly pointless. However, all decisions have the certain impact on business performance; therefore it is very important to give adequate attention to decision-making issue [1].

Decision-making as a complex process involves a number of different phases that must be met in order to have a final decision. Final decision is selected from several variants of solutions. Managers and competent persons, respectively, have to make decisions even under changing conditions, under pressure, in a state of uncertainty or risk, and under the influence of certain restrictions, respectively. The decision-making process is therefore a nonrandom selection of one variant of the solution according to certain criteria in order to meet predefined objectives.

Decision-Making Process

Decision-making process constitutes a comprehensive system which is characterized by the following parts:

- Decision problem – is a reflection of the significant deviation between the desirable state (planned or established standards) of specific decision object and its real state
- Decision-making situations – is given by the object of decision (selected process or operation), by state of internal or external conditions which cause deviations

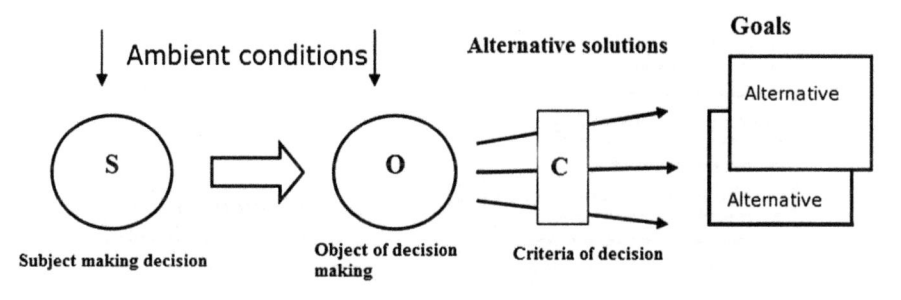

Fig. 1.1 Decision problem and decision-making situations [2]

from the desired state, by setting targets or its alternative aims, and by criteria, variants, and sometimes also consequences of possible decisions
- Decision-making process – the process of choosing between several possible solutions (search and selection of suitable alternatives solutions for the current problem)
- Variations of decisions – option of several possible and different solutions which ideally suits to specific criteria [2]

Decision-making process is affected among other things by ambient conditions, subject making decision, the object of decision-making, decision-making criteria, and setting of goals. The following picture shows how the abovementioned elements affect decision-making process (Fig. 1.1).

Necessary assumes for creating high-class decisions:

- Clear statement of the target goal which must be achieved by decision
- Adequate, verified, quality, timely, reliable information that are the basis for making any high-grade decision
- Adequate qualification and competence of decision-makers. Using appropriate methods, tools, and knowledge [3]

Decision-making process is divided by majority of authors into several main groups of steps (e.g., analytical part, project part, evaluation part). In books from many authors, it is possible to find several models of decision-making process. In context of this paper, we suggest to determine the content and steps of decision-making process in general as [2]:

1.	Discovering of the problem and acceptance of decision to resolve the existing problem
2.	Identification and analysis of the situation and analysis of the problem, causes of the problem, and possible consequences in the event of failure to resolve the whole problem
3.	Determining criteria for choosing the right solution (general requirements for possible solutions, which must be legal, ethical, economic, and feasible)
4.	Creating (generating) possible variants of solutions
5.	Assessment of possible variants of solutions according to established criteria
6.	Selecting the optimal solution
7.	Evaluation (testing) of the chosen decision – assessment of its positive and negative consequences
8.	Formulation of decision and determination of implementers, procedures, and forms

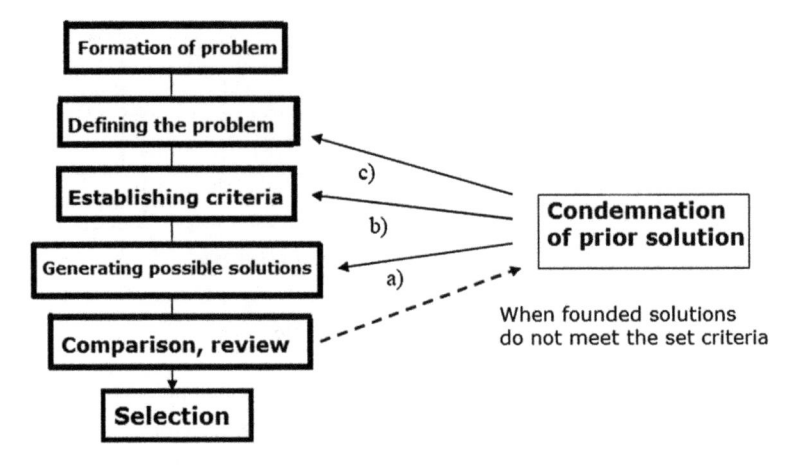

Fig. 1.2 Decision algorithm [2]

The following figure shows a decision algorithm:

In business practice the decision-making process is not always as objective, uninfluenced by each successive step, which is shown in Fig. 1.2, because a specific decision-making process can be affected by in-house arrangements, conventions, or stakeholder or under the influence of a supervisor person [4].

Exact Methods and Analytical Hierarchy Process (AHP)

Exact methods of operational analysis can be classified among the most advanced quantitative methods for decision-making. The main role of these methods is to find among the possible variants of solutions the best variant in accordance to existing problem or target goal. Among the exact methods, which are designed for resolving decision-making problems, are the relationships between the elements mainly expressed quantitatively which may include:

- Methods of mathematical statistics – theory of probability, correlation analysis, and time series analysis
- Methods of mathematical analysis and linear algebra – differential numbers, extrapolation, and matrix number
- Methods of operational analysis – economic-mathematical methods, structural analysis, network analysis, models of mass operation, etc.
- Multi-criteria decision-making methods [5]

The analytical hierarchy process is a systematic approach created in 1970 for structuring the experience, intuition, and heuristic-based decision-making into a properly defined methodology based on mathematical principles. The AHP method was created for the needs to return to quantitative assessment in decision-making process. This method provides a formalized approach for creating solutions [6].

The AHP method was created in the late 1960s by Thomas L. Saaty, an American professor who worked at the University of Pittsburgh. Saaty's analytical hierarchy process offers a methodology that allows to model complex decision situation and determine the appropriate choices.

This approach has been developed to help solve complex problems. Although this method was not originally intended for collective decision-making, nowadays thanks to its transparency and consistency is using also for group decision-making process [6].

The method of analytical hierarchy process is based on scientific analysis. This method, among other things, can be used also in the process of formulating the strategic objectives of stakeholders as well as in decision-making process in terms of crisis or risk management. The AHP method provides the framework for making effective decisions in situations, when the choosing of best decision is needed or fundamental. AHP method enables to prepare effective solutions in complex situations and simplifies the natural decision process [7].

The AHP method is one of the most exact objective methods of multi-criteria decision-making but still has several disadvantages. One of the major disadvantages of the method is burdening some steps in its application with a certain degree of subjectivity. For this reason it is necessary to objectify the allocation of specific weights for the individual criteria, which minimize subjectivity and ultimately lead to an overall objectification of the AHP method. Therefore it is necessary to create a tree structure as far as possible in an exact way that could minimize the subjective impacts of evaluators. Possible solution that partially removes such a problem is performing of an evaluation with the participation of a group of evaluators that are experts in the specific field. Objectivity could be increased by assignment of weights to individual evaluators; this process could reflect the importance of individual experts [8].

Structured hierarchy of AHP method represents a system for optimization, which consists of a primary objective, criteria, and alternatives, respectively, in other words variations. The criteria can be further spaced out to sub-criteria; this division can lead to as many levels as necessary for resolving the problem.

AHP is a method of decomposition of complex unstructured situation into simpler components. This method is first done by an expert method and then by mathematical method, which divides the main problem into smaller and more detailed elements [7].

Advantages and Disadvantages of the AHP Method

Advantages of the AHP method [7]:

- Pair comparison allows to make easier conclusion
- The method requires just one pair comparison between all criteria and between all variants of solutions
- The method is very simple and clear
- Exact weighting of individual criteria and clear results from quantitative assessments

Disadvantages of the AHP method [7]:

- The scale of the evaluation is created by the evaluator (evaluator could think that one criterion is more important than other)
- Assignment of weights by one evaluator may not be acceptable to other evaluators
- Formation of the tree structure is influenced by a certain degree of subjectivity
- In the case of additional adjustment of the decision matrix is coming to disruption of objectivity in process of evaluation;
- This method could be time-consuming in terms of the amount of comparisons that must be made
- Validity of the method is limited by consistency, because in practice is assessment sometimes inconsistent

Principles of AHP Method

The flexibility of the AHP method as a decision-making model helps to clearly identify the most optimal solution among the other possible variants of solutions. Decision-making process is influenced by three principles [9].

- Prioritization
- Priority
- Consistency

Prioritization – under this term it is possible to understand the linear structure that contains several levels, whereby each of the levels contains a number of elements. The arrangement of the individual levels of the hierarchical structure corresponds to the arrangement from general to specific. If the individual elements are more general in relation to the decision-making problem, then these elements have higher importance in the hierarchy and vice versa [7].

Priority – method is based on a pair comparison. On the other hand, the evaluation is based on the expert judgment. These expert judgments are made by experts from the field and after that compare mutual influences of the two factors. They rank criterions and variants between each other with the scale: same, weak, medium, strong, or very strong [7].

Consistency – expresses the level of credibility of the final result. In cases where it is necessary to compare a large number of criteria, the consistency gives us warranty or credibility that the final result will be objective and optimal [7].

Potential Uses of AHP

The AHP method can be applied to decision-making processes and evaluation of tasks in different sectors, but one condition must be always met that there must be at least two alternatives for decision-making process [8].

From the factors which make the AHP method a very popular decision-making method in the world, it is necessarily highlighted that this method quickly adapts to fixed data like price, speed of delivery, personal experience, and also intuition. This method allows mathematical deduction weights of the individual criteria instead of the subjective choice of weighting criteria. In the first phase, before the method is applied, the subject that makes evaluation must define all the criteria and sub-criteria on which will be evaluation based. The selection of individual criteria and sub-criteria is based on the previous knowledge and experience of each evaluator. If this is the first rating of a single subject, then this evaluator must sort criteria on their own intuition or in accordance with specific pattern [8].

AHP method can be applied to decision-making processes or to assessment of the goals in different business sectors, but one condition must be respected every time, that decision-making process must have at least two alternatives [8].

AHP method can be used in various fields as:

- Management
- Logistics
- Economy
- Health care
- Education
- Research
- Agriculture and forestry
- Energy
- Transportation
- Industry

Application of AHP Method in Innovation Management

AHP method was used to select the most innovative variants. For application method of analytical hierarchy process, the software Expert Choice was used, which comprises the following steps:

1. Goal setting
2. Determination of the individual criteria
3. Assessment of alternative solutions
4. The assignment of weights to each criterion
5. Evaluation of the alternatives by paired comparison
6. Reviewing the current state of the system of values

Goal setting represents stage of choosing the best innovation variant from six selected innovative activities that are in the field of innovation management very often used. List of individual innovation variants are as follows:

- Introduction of a new product
- Modifying of the product
- Change the size of the package
- Change the appearance of a product (design)

- Amendments of the product
- Introduction a new method of production

Criteria for evaluation of possible alternative solutions – for selection of the most appropriate innovative variants, seven criteria were selected. The criteria are actually kinds of parameters of individual innovation activities. The criteria are given below:

- Price
- Quality of manufacturing
- Material
- Appearance
- Utilization
- Durability
- Energy consumption

Application of AHP Method: Use and Application of Expert Choice

In finding solution for our target goal to select the most innovative variant in innovation management, the computer software Expert Choice was used. Expert Choice software offers multi-criteria decision-making process. Expert Choice offers the opportunity to split decisions in several hierarchical levels and sub-levels, respectively; thereby the entire decision-making process is divided into separate and simpler parts.

Such allocation is done through the tree structure which distributes and organizes the entire hierarchy. Criteria and options are stored into the decision matrix. Individual preferences are granted according to Saaty's nine-point scale. Preferences can be given by numerical method or by verbal expression. Expert Choice offers evaluation by numerical or graphical representations. Our goal was by the AHP method to find the most suitable alternative innovation activities. In order to find such a variant with computer software Expert Choice, at first we selected the main goal, at second we selected variants of innovation activities, and then we entered the decision-making criteria. Target goal can be defined by clicking on the icon Goal.

Defining Criteria

After setting the primary objective, the next step was defining the decision-making criteria that represent a kind of basic parameters in the selection process of the best innovative variant or activity. Defined criteria are shown in Fig. 1.3 (price, quality of manufacturing, material, appearance, utilization, durability, and energy consumption).

1 Application of AHP Method in Decision-Making Process

SELECTING CRITERIA

- Price
- Quality of manufacturing
- Material
- Appearance
- Utilization
- Durability
- Energy consumption

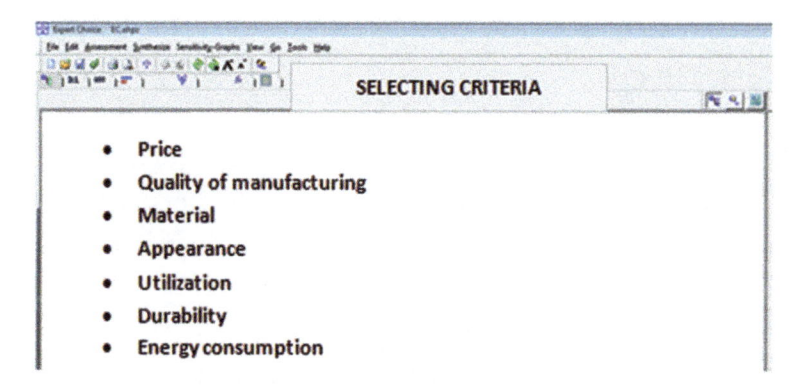

Fig. 1.3 The file of decision criteria

SELECTING INNOVATIVE VARIANTS

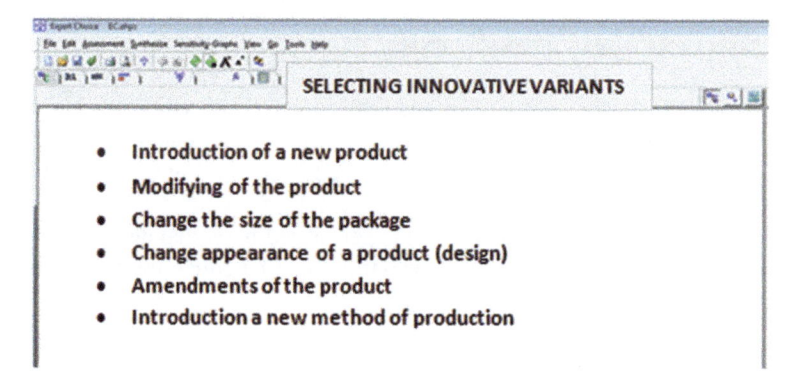

- Introduction of a new product
- Modifying of the product
- Change the size of the package
- Change appearance of a product (design)
- Amendments of the product
- Introduction a new method of production

Fig. 1.4 Determination of the individual variants

Assessment of Alternative Variants

The next step in the application was to determine the individual variants of innovation activities, which are used in the field of innovation management (Fig. 1.4).

Assignment of Weights to Variants and to Individual Criteria

The next task was to determine specific values for the selected variants in accordance with the established criteria. The allocation of weights to each variant and criterion is undoubtedly an important step in the whole process. For the objectivity and better results of calculation of the best innovative activities, the individual values in the table were determined by two experts in the field.

Consequently, it is necessary to compare each individual variant with others and do the same comparison between each criterion with each other. This comparison is

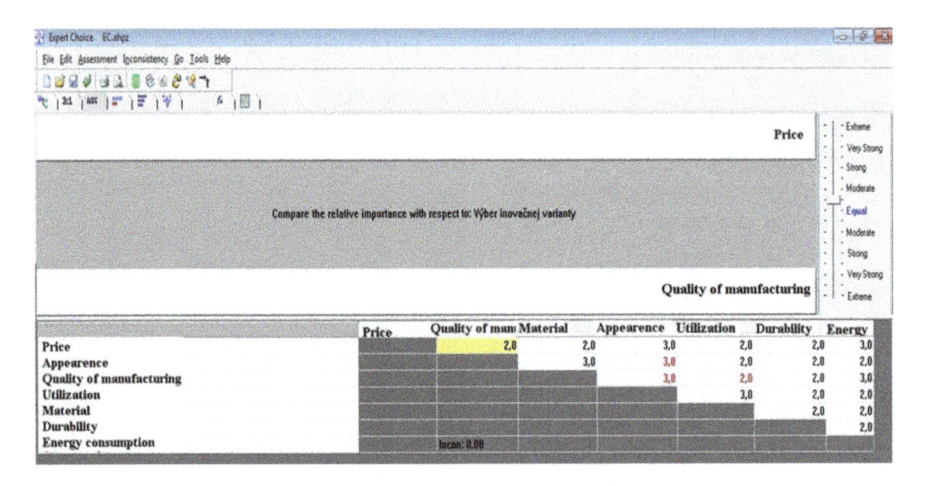

Fig. 1.5 Entering the values to individual criteria

performed separately for each variant and criterion. The procedure is identical in Expert Choice, and this software can provide for us verbal, numerical, and graphical, respectively, representation of a particular comparison. Figure 1.5 shows the assignment of values for them.

Word comparison of criteria (Fig. 1.5) is performed by verbal designation. The relationship between each criterion (or variation) can be equal, moderate, strong, very strong, or extreme.

Evaluation of Paired Comparison of All Criteria

After the successful assignment of particular values to compared criteria, it is possible to determine which criterion is the most important for us as well as which criterion is the least important in concrete decision-making process. Figure 1.6 is a displayed graph which tells us that price criterion is the most important criterion from all criteria in the process of choosing the best innovation activity. Consequently, it is possible to see the order of the other criteria importance.

Final Evaluation

The final evaluation is the ending phase. The paired comparison of variations and criteria with the software Expert Choice gave us a result, in the form of various graphs and diagrams. The final evaluation can be either numerical or graphical. In this project, the main goal was selecting the most innovative variant or activity. Evaluation of the whole process which was realized in Expert Choice can be seen in

1 Application of AHP Method in Decision-Making Process

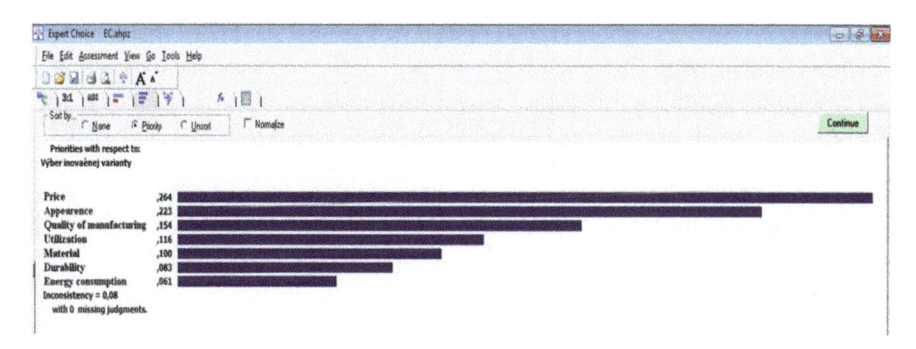

Fig. 1.6 Graphical comparison of the most important criteria in the process of choosing the best innovation activity

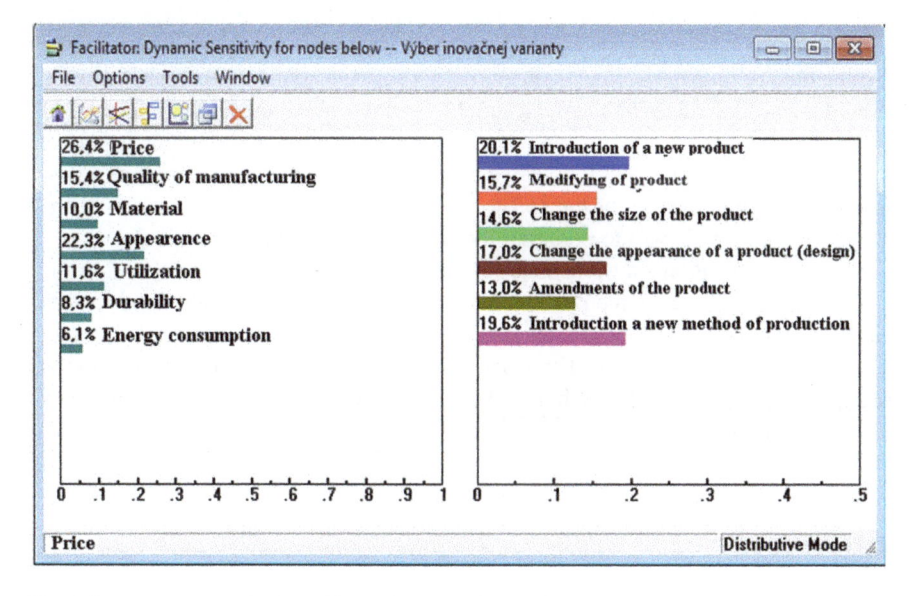

Fig. 1.7 Percentage evaluation of individual criteria and variation

the following Fig. 1.7 that gives us graphic and numerical information. From the figure it can be revealed that the most important criterion is price and the most important variant was the chosen introduction of a new product.

The abovementioned figure shows in detail the results of the percentage of the whole evaluation. As from the figure results, the most important criterion for choosing the best innovation activity is the price with 26.4%. The second most important criterion is the design, which is reflected by 22.3%. Next, the criteria quality of manufacturing with 15.4%, utilization with 11.6%, and material with 10.0% follow. Durability has 8.4%, and the least important criterion chosen was the energy consumption with 6.1%.

Introduction of a new product is determined by final evaluation of variants as the best innovation activity. This result is represented by 20.1% share. The most important innovation activity is followed by the introduction of a new process for the production, this activity was very close, and its importance is reflected with 19.6%. On third place is located the change of design with 17.0%. Next, modifying the product with 15,7%, change of the size of package with 14.6% and on the last place are amendments of the product with 13.0% share.

Conclusion

The article discusses and describes how the program Expert Choice works. Expert Choice applies the method of analytical hierarchy process in business practice. Application of AHP method was performed in decision-making processes falling into the field of innovation management; in this case the application of AHP method is an ideal tool for selecting the most suitable variants. AHP method as a method of multi-criteria decision-making can work with a number of criteria. In this paper we managed to choose the most innovative options, namely, the introduction of a new product with an assistance of the software program Expert Choice. The aim of the project and the article was to point out the large-scale use of AHP method.

The final evaluation shows us that the most important criterion for choosing the best innovation activity is the price with 26.4%. Introduction of a new product was determined by final evaluation of variants as the best innovation activity. This result is represented by 20.1% share. The work itself is served to the general public and in a simple way explains AHP method and the use of software Expert Choice in many sectors within the decision-making process.

References

1. Čambál, M., Holková, A., & Lenhardtová, Z. (2011). *Základy manažmentu*. Bratislava Alumni Press. ISBN 978-80-8096-138-1.
2. UNIZA. *Rozhodovacie procesy*. [online]. Accessible at: http://fsi.uniza.sk/kkm/old/publikacie/ma/ma_05.pdf
3. Máca, J., & Leitner, B. (2002). *Operačná analýza I.: Deterministické metódy operačnej analýzy*. [online]. Accessible at: http://fsi.uniza.sk/ktvi/publikacie/11_operanal1_u_2002.pdf
4. Baďo, R., & Vrablic, P. (2011). *Využitie metódy multikriteriálneho rozhodovania metódou AHP pri rozhodovaní v podnikateľskom prostredí. Transfer informácií 21/2011*. Bratislava: Strojnícka fakulta STU. [online]. Accessible at: http://www.sjf.tuke.sk/transferinovacii/pages/archiv/transfer/21-2011/pdf/082-085.pdf
5. Hrablik Chovanová, H., Sakál, P., Drieniková, K., & Naňo, T. (2012). *Operačná analýza : časť II*. Trnava: AlumniPress. ISBN 978-80-8096-165-7.
6. Ramík, J. (2010). *Analytický hierarchický proces (AHP) a jeho možnosti uplatenění při hodnocení a podpoře rozhodování.v Sborniku příspěvku z konference: Matematika v ekonomické praxi*. Jihlava. [online]. Accessible at: https://most.vspj.cz/files/11/Matematika-v-ekonomicke-praxi.pdf

7. Drieniková. K. et al. *VYUŽITIE AHP METÓDY PRI TVORBE UDRŽATEĽNEJ STRATÉGIE SZP – I. (ANALYTICKÝ HIERARCHICKÝ PROCES)* [online]. Accessible at: http://www.scss.sk/cd_apvv_lpp_0384_09_2011/V%C3%9DSTUPY%20Z%20VLASTNEJ%20VEDECKO-V%C3%9DSKUMNEJ%20A%20PEDAGOGICKEJ%20%C4%8CINNOSTI/PUBLIKA%C4%8CN%C3%81%20%C4%8CINNOS%C5%A4/KONFERENCIE/BRNO/I..pdf
8. Roháčová, I., & Marková, Z. (2009). *Analýza metódy AHP a jej potenciálne využitie v logistike*. [online]. Acta Montanistica Slovaca Ročník 14 (2009), číslo 1, 103–112.
9. Saaty, T. L., & Kearns, K. P. (1985). *Analytical planning*: the Organization of a System. International Series in Modern Applied Mathematics and Computer Science, Vol. 7, 208 pp. Pergamon Press, Oxford. ISBN 0-08-032599-8.

Chapter 2
An Integrative Spatial Perspective on Energy Transition: Renewable Energy Niches

Filip Gulan, Maros Finka, and Michal Varga

Abstract The shift towards sustainable use of renewable energy accompanied by noteworthy improvements in energy efficiency and reduced consumption is considered as a fundamental element of energy transition. However, framing the role of the sustainable energy development in a monofunctional way does not allow to use its multifunctional potential linked to sustainable development efficiently and may result in disconnection between energy and broader spatial and urban development agenda. This, in turn, can underplay the importance of potential synergy effects between renewable energy production and use and its localized context. This paper presents the narrative that single although innovative solutions are insufficient and poor integration of sustainable energy initiatives creates barriers to achieve sustainability and limits potential synergies with their spatial context. We discuss this phenomenon in relation to new demands on integrative and spatially sensitive approaches to the renewable energy development, often catalysed by ICT hand in hand with the implementation of smart grids – a backbone of the smart city concept. In our attempt was to better understand how renewable energy systems and initiatives emerge, how to maintain and enhance them, and under which conditions they co-evolve with their unique context in a more structured and productive way towards the vision of smart city and hence also recognizes the multifunctional potential of sustainable energy developments. In order to do so, we propose to learn more from both spatial planning and transition management (niche-based) perspectives. Finally, this paper outlines several areas for further research as well as reminds some of the related scientific challenges and disparities between energy and spatial planning.

F. Gulan · M. Finka (✉) · M. Varga
SPECTRA Centre of Excellence EU, Slovak University of Technology in Bratislava, Bratislava, Slovakia
e-mail: maros.finka@stuba.sk; michal.varga@stuba.sk

© Springer International Publishing AG, part of Springer Nature 2019
D. Cagáňová et al. (eds.), *Smart Technology Trends in Industrial and Business Management*, EAI/Springer Innovations in Communication and Computing, https://doi.org/10.1007/978-3-319-76998-1_2

Introduction

The development and implementation of the smart city concept seems to be one of the dominant issues in current urban agenda across Europe undergoing substantial shift from the high-tech-based approach to the comprehensive concept of smart management of the resources (natural and human resources, technologies, knowledge, etc.) of knowledge-based urban community. Energy, an inherent part of the smart city concept, is one of the most critical and cross-cutting issues facing the EU today, being a headline of climate change discussions as only few activities affect environment and economy as much and as continually as the production and use of energy. The more and more acute need of shifting towards more sustainable energy (SE) systems, manifested by climate change and ever-growing energy demand, is considered as one of the major societal goals at global level. Sustainable technologies and innovations (e.g. in the field of energy) accompanied by noteworthy improvements in energy efficiency, changes in the behaviour of consumers and market actors and promotion of "clean" sources of energy, particularly renewable energy sources (RES), may help to unlock accelerated transitions towards greater sustainability of energy systems. These aspects of energy transition can be considered as one of the essential pillars of societal climate change adaptation and mitigation efforts. On the other hand, sustainable (energy) development also drives and influences various developments across different areas and scales – such as development towards sustainable and liveable cities – often labelled as smart cities.

While frequently used, the notion of "energy transition" towards more sustainable energy systems lacks a widely accepted definition as well as common grounds in terms of its implementation are absent [3, 30]. Even though many aspects of future transitions are highly complex and uncertain, historical experiences can provide many useful lessons [16]. Clearly, past energy transitions suggest that sustainable energy transition will require a fundamental, structural change in the way how we use and think about energy, and it will take long time to unfold. This will most likely require reorganizing spatial structures, responsible institutions, governance structures and even much more, which in turn creates also new demands on novel, smart energy-conscious approaches offered by spatial planning and governance bodies.

Small-scale, decentralized and community-driven SE initiatives are widely acknowledged to be a desirable feature of low-carbon future; however, they face a range of challenges in the context of their integration into the physical and socio-economic landscape which goes far beyond the municipal or regional competences.

This approach can be considered as a part of current trend away from conventional top-down and centralized structures of energy provision towards more innovative, bottom-up and decentralized arrangements. The growing importance of regionalization, localization and decentralization of energy services coupled with changing role of local actors and smart networks that are transforming energy consumers towards active "prosumers" (they both produce and consume energy) demands new approaches for urban, suburban and rural areas [27, 33, 40, 41, 49]. Decentralized generation of energy is usually based on the diversity of actors that participate in "throughout" the energy value chain – from the development and construction of energy systems to the generation and delivery of energy services and their ownership and operation. Such value chain then involves institutions, local utilities, entrepreneurs and companies, associations and co-operations and, of course, municipalities and communities, who simultaneously provide and consume energy. This leads us to another phenomenon – which is in stark contrast with centralized energy systems, where major influence on energy policy and energy production and distribution is in the hands of large-scale geo-governance apparatuses and energy companies operating at larger (e.g. national or transnational) scales. We are witnesses to mushrooming of sustainable energy initiatives and energy cooperatives at the lower tiers of governance (e.g. regions, municipalities, neighbourhoods) particularly in western Europe that go beyond the "business as usual" in realizing the potential of local renewable energy resources. Provision and promotion of green electricity, improved social cohesion, empowered communities and optimized energy value chain associated with the investment of revenues in the local community are considered as some of the main motivations for these initiatives [7, 31, 42, 48]. In this context, the role of local and regional niches – initiatives, networks and "frontrunner" communities – with their own business models is often based on the use of localized territorial capital where (renewable) energy is well-integrated within larger supply chains within local economies.

Here, we argue that framing the role of the SE in a monofunctional way[1] does not allow efficient use of its multifunctional potential (social, economic, environmental, etc.) at multiple spatial scales (local, regional, national, etc.). Eventually, such a limited view can result into disconnection between energy and broader spatial development agenda, what might also underplay the role of potential synergy effects linked to implementation of SE (e.g. RES integration concepts). In fact, the development of sustainable energy with emphasis on renewable energy deployment raises many profound questions about its benefits, limitations and negative impacts, which are necessarily associated with such "energy transition". In this sense, only the development and implementation of SE in close connection with territorial concepts can react more properly on many profound questions about the benefits, limitations and potential negative impacts that are associated with deployment of RES and related energy transition in spatial contexts. On this basis, a sociotechnical transition towards sustainable energy can be envisioned as a complex, long-term process involving reconfiguration of dynamic spatial patterns of socio-economic and ecological landscape, what has often been neglected in relation to spatial development practice.

---

[1] For example, given the centrality of carbon reduction (e.g. of a spatial unit such as country/region/city/building) within policies, strategies and projects and emphasis on their economic feasibility over other aspects (such as impacts on sustainability, architecture and urbanism, landscape, biodiversity, etc.).

The emergence and development of more sustainable energy systems will inevitably imply a careful consideration of their complex relationships with the territory and their dynamics. Given this context, the notion of supralocal (regional) development is often under-appreciated when it comes to the development of SE and vice versa, so the integrative solutions and potential synergy between both domains are frequently omitted. However, cultural and political economic factors co-evolve with changes to the quality, location and environmental impact of energy resources, and extensive changes in the energy mix have often underpinned social and geographical change [23]. Therefore, one of the core assumptions guiding the reported research is that "weak" spatial-structural and socio-economic integration of energy-related initiatives creates barriers to their potential synergy with sustainable development of the territory. On this basis, the capacity of local actors, communities, municipalities and regions to create, maintain or adopt innovative SE practices and translate them into the sustainable development is perceived as once of the major spatially embedded "qualities" to be followed by energy-conscious spatial planning that serves as a "toolkit" for making our settlements smarter.

Make Energy Smarter, but in Time and Space

"Smart city" concept has gained its relevancy as a concept or a vision for a more sustainable development of cities and urban areas, yet has no universally accepted definition. Smart city is envisioned as having a high quality of life by excelling in multiple key areas – such as economy, mobility, environment, governance and, of course, energy – with improved efficiency of services to meet peoples' current and future needs. Considering that this concept inherently implies a transition (a widespreading and systemic innovation reshaping our cities), this paper suggests that it might be useful to learn from the transition management (TM) theory. In this context, TM provides a systemic view on change at all levels, its dynamics and complexity and thus exceeds "usual" smart city frameworks. TM provides a useful framework for considering the temporal aspects of (energy) transitions largely, but renders to be rather fuzzy on how it relates to other dimensions of space such as territorial or administrative spaces and their typologies [45]. Therefore, we attempt to take the perspective of spatial planning on the smart city energy transition continuum in order to explore more about renewable energy development at the interface between local and regional level.

Spatial planning can be described as largely public-sector-led repertoire of activities to influence the future spatial distribution of activities, to enhance the integration between different sectors, to create a more rational territorial organization of land uses including the linkages between them, to balance demands for development with the need to protect the environment and to achieve social and economic objectives. Spatial planning is regarded a key instrument for establishing long-term sustainable frameworks for social, environmental and economic development [14, 32] and thus directly relates to the concept of smart city. Arguably, spatial planning has for some time been ignorant of the complexity and traditionally has focused on well-defined and tame problems, operating within an atemporal framework [36]. Often, solutions for the past problems have been sought rather than looking for flexibility and adaptability enhancing solutions. Therefore, making the connection between spatial planning and TM seems to be logical to foster potential synergies as well as to address and at least minimize their shortcomings in relation to smart energy transition.

Potentially, this integrative approach can be the ground to provoke and challenge our understanding of the role of spatial planning in the context of smart city concepts and "smart energy transition", considered as a concept relying heavily on automation, information and communications technology (ICT), driven by both social and technological innovations. We need to place stronger emphasis on the relationship between spatial development and renewable energy development at the local and supralocal level (e.g. urban area, a city, its neighbourhoods and hinterland). In this attempt to bring complexity-based "transitions" science on boards, the idea of sociotechnical energy systems (e.g. including smart grids) considered as complex adaptive systems is adopted. Under such circumstances, "hard infrastructure" for energy should be equally important in planning as dealing with nontechnical aspects, which in turn involves building of soft infrastructure capacities. Crucial questions in this context are linked to the possibilities to rethink the nature and character of sustainable and low-carbon (urban) development processes (e.g. smart city initiative) in a more productive way forward in the context of spatial planning practice.

Theoretical Contexts

Changing nature and framing of spatial planning across all levels can be partially attributed to the recognition of the complexity, uncertainty and irreversibility, for example, demonstrated by the climate change awareness and its relationship to energy use and sustainable development [13]. In this case, observation that spatial implications of sustainable energy (RES in particular) are undeniably problematic has become an imperative for further research. Despite many limitations, the evidence gathered from academic literature and policy sources leave little doubt that the planning system has a major part to play in climate change policy agenda [12, 22].

The place-based approach towards the implementation of sustainable energy systems based on the broader use of indigenous renewable energy can be built upon their territoriality and ability to build on specialized supralocal (regional) assets – efficient use of territorial capital – implying careful utilization of available RES while considering their ubiquitous spatiality. The various forms of spatiality, territoriality and decentralized character of RES constitute the premise of (relational) proximity between people, settlements and renewable energy technologies. This consideration becomes increasingly relevant in order to inform the spatial planning interventions more comprehensively and efficiently [5, 50].

Indeed, the fact that RES have substantially different characteristics compared to fossil or nuclear fuels needs to be taken into an account. RES as one of the most common alternative energy sources are projected to increase in the long run.[2] This phenomenon suggests that energy supply based on these spatially dispersed and often variable and volatile sources of energy implies change in the land-use patterns and requires substantial land resources. Hence, RES development involves complex reorganization of the territory as well as profound changes in the energy systems. The consideration of the fundamental differences in the socio-spatial distribution and socio-economic properties related to RES in contrast to high-carbon energy resources used in incumbent energy systems and regimes (particularly fossil fuels) then seems crucial.

RES along with the concept of smart grids are deemed as a viable alternative to the conventional, rather centralized, energy systems based on the fossil fuels. However, incumbent high-carbon energy systems with their underlying infrastructure (understood in the broader sense – i.e. technical, social, institutional infrastructure) have evolved gradually over decades and thus are strongly established and can prove resilient to changes. And in order to meet the basic goals of a smart city, it is evident that traditional power systems will need to undergo a complex, gradual transition towards a more sustainable, more decentralized, sociotechnical networks that will be indeed more "smart".

One of the practical responds to current challenges within the sustainable energy research and practice that has been largely adopted also in the concept of smart city is the ongoing evolution towards smart grids.[3] Basically, smart grids are envisioned as sociotechnical energy networks, where consumers become more autonomous – they become prosumers (e.g. they produce energy from RES, supply the grid and consume electricity while possessing improved information, control and choice) and interact with the grid (e.g. via smart meters) – enabling a two-way flow of energy and information. The main idea behind smart grid solutions is that they are designed to counteract the natural disadvantages of RES through improved grid integration and interaction between actors (market, generation, distribution and transmission utilities, end-users, consumers, etc. are integrated particularly through ICT). The potential benefits associated with smart grids are in line with the idea of smart city: they are designed to bring improved grid stability, security, efficiency and reliable integration of RES that are largely variable in time and space at both large and smaller scales [21, 26]. However, smart grid development is not a purely technical matter. In this regard, the need for careful consideration of the integration of RES within the (smart) power systems as well as in their localized context needs to be taken into account and investigated in more detail [4, 46].

---

[2] Also in the short term, referring to the 2020 Climate and Energy Package which introduced three key objectives: a 20% reduction in greenhouse gas emissions, a 20% share of renewables in total energy consumption and a 20% improvement in the EU's energy efficiency by 2020 [9], Renewable energy Directive, Energy Efficiency Directive.

[3] For instance, see also [15, 21, 26].

A "technical" or "technocratic" perspective on development, which often dominates our thinking and planning approaches, proves to be seriously limited and rather ossified when it comes to dealing with the complex web of relationships between actors and networks in physical, socio-economic and institutional environment, i.e. in the case of energy systems. In fact, if we look at energy systems as being an integral part of our society (understood as sociotechnical system), we also might assume that it will be more feasible to frame the phenomena of renewable energy development in a way that its social and spatial embeddedness are considered in planning [8, 34, 47]. On this basis, we suggest to approach energy systems from the complex adaptive system's perspective. In this regard, complex systems theory proves to be highly relevant for understanding of how complex, sociotechnical energy systems evolve and continuously adapt to changing internal and external conditions as we are witnessing in everyday reality. Arguably, this perspective can promote interesting ways to frame the development of sustainable energy systems (i.e. renewable energy) and their (spatial) integration. Focus on the build-up processes of indigenous energy innovation capabilities for low-carbon development as part of "energy transition", where renewable energy and technology are perceived as active elements in the territory, offers potential for interventions that go far beyond business-as-usual planning methodologies.

Management of different activities to "set-up" processes related to the energy transition towards more sustainable energy can learn and draw inspiration from TM approach. Interestingly, TM is built upon the key notions of complex systems theory and governance based on complexity and uncertainty in order to guide and influence the course and pace of complex systems change. TM literature generally assumes that dealing with persistent, wicked problems such as climate change and energy transition is long term and requires transitions. Moreover, it points out rather consistently that directing of such complex changes can be supported substantially by insights into general patterns of complex systems dynamics. In other words, TM attempts to overcome the conflict between long-term imperatives and short-term concerns [25]. That being said, "transitions" can be understood as fundamental changes in the structure, culture and practices of societal systems that cannot be controlled but might be influenced and guided – also with the assistance of TM. Consideration that energy transition consists of different phases is in place so the fact that different transition phases will probably require different strategies needs to be taken into account [29, 37, 38]. At this point, we argue that bringing complexity-based, temporal perspective of TM together with "spatiality" of spatial planning can enrich spatial development practice and improve our understanding of how we can put our cities on a "smarter and sustainable" trajectory.

In this context, this paper intends to introduce the "multilevel perspective" on sustainability transitions and particularly "niche management" approach in order to improve our understanding of the processes of formation and upscaling of innovative sustainable energy initiatives and infrastructures in relation to the ever ongoing process of smart city development.

Hence, the theoretical premise is that bridging the gap between current planning and governance of energy transition (TM) in relation to the broader spatial-physical and socio-economic integration of RES into the territory can allow for more synergy and thus offer potential answers to these questions. This paper intends to argue that synergy stimulated by co-evolution between sustainable energy systems and their localized and supralocal context embodies the very idea of the smart city concept – a synergy between technology, environment and society. In this regard, effective spatial planning should help to avoid the duplication of efforts by actors [32]. This paper intends to argue that synergy stimulated by co-evolution between sustainable energy systems and their localized context embodies the very idea of the smart city concept – a synergy between technology, environment and society.

Adding a Spatial Dimension to Managing Energy Transition

The multilevel perspective has been used to describe and unfold structural innovations in sociotechnical systems or, in other words, transitions to sustainability in sociotechnical systems.[4] From the sociotechnical perspective, technology is seen as "being formed by, and embedded within, particular economic, social, cultural and institutional structures and systems of beliefs" [6]. In fact, the energy system can be described as "all actors and artefacts that together produce the societal function of energy" [43]. In practice, the sociotechnical perspective can be illustrated at the example of social barriers related to the development of renewable energy, which can take many forms. The impacts of RES deployment on sustainability, individuals, communities and landscapes have often been underestimated yet the main focus has been on "hardware" solutions and their economy. Such approach has often proved insufficient in terms of the integration of energy initiatives at multiple scales. Hence, a more inclusive approach which is prone to acceptance by the local society and less vulnerable for failure has been advocated elsewhere.[5] The emerging field of energy transitions presents us with interesting frameworks for guided change of sociotechnical systems. These insights are particularly relevant in terms of our focus on the development and (socio-spatial) integration of sustainable energy systems. Therefore, we treat basic principles of the multilevel perspective on transitions in order to enhance our understanding of the co-evolutionary behaviour of both energy systems and their localized context. In this context, the multilevel perspective on transitions takes into account that nonlinear processes of change result from the interplay of developments at three qualitatively different scale levels and attends to the dynamic relationship between them:

- Micro level of niches, where innovations, norms, practices, alternatives and novelties emerge and eventually can form the seed for systemic change. Niches can be considered as relatively protected "experimental settings" or "incubators" for

---

[4]As in [6, 9, 15, 17, 19, 35, 36].
[5]For example, in [34, 47, 49].

innovation. Niches can be part of the higher scale level (regime), located on the periphery or outside of the existing regime/system [24, 37].
- Meso-level, occupied by a (sociotechnical) regime. Regime can be described as the space of established practices and associated rules that form the stability of the sociotechnical system. At the regime level, patterns of institutions, culture, practices, lifestyles, artefacts, rules and norms are aligned in a coherent and self-reinforcing way in order to perform economic and social activities (Berkhout et al. 2004). Regime refers to the dominant culture, structure and practice embodied by physical and immaterial infrastructures patterns of institutions [28].
- Macro level or (sociotechnical) landscape. The exogenous level of landscape refers to overall societal setting (worldviews, paradigms) in which processes of change occur. Sociotechnical landscape represents a wider context in which a regime and niches are embedded. Sociotechnical landscape consists of autonomous trends, paradigms and slow changes (e.g. geopolitical dynamics, macroeconomic trends, etc.) that are beyond the direct influence of a regime or niches. Therefore, landscape typically develops autonomously and changes more slowly, however, influences the dynamics at the lower levels (regime and niches) (D. Loorbach 2007b). An example of a destabilizing element in the landscape is climate change.

The multilevel perspective as a leading theoretical framework on transitions maps the entire transition process, which is seen as a result of alignments between multiple developments at different levels, and brings innovative agency into play. The intermediate level of a regime is central in a way that the transition (systemic change) is viewed as a shift from one sociotechnical regime to another, whereas both niche and landscape levels are characterized by their interactions with the regime level. In this sense, both niches and landscape are similar because they interact with the intermediate regime scale.

Niches form the micro level where experimentation, novelties and innovations (e.g. promising technologies and their adaptation at the micro level such as small-scale renewable technologies, local energy cooperatives) may emerge and can be tested in a "sufficient distance" from the mainstream practices. In reality, niche practices are rather unstable configurations carried and developed by small networks of dedicated actors or frontrunners that often come from outside environment. In some cases, if niche practices are strong enough, they can be mobilized and scaled-up (e.g. the rate of the technology/practice application has increased significantly in the region). The interaction between niches can have different forms in terms of cooperation, learning, competition and so on. As niches extend and accumulate, they mature into larger assemblages of practices, technologies, skills or norms, which can eventually challenge or gradually disturb the incumbent regime. At the regime level, such tensions can be manifested in a way that niches are absorbed and appropriated by the dominant regime because the dominant regime is stabilized by many lock-in mechanisms (path dependency). In this regard, path dependency can be described as systemic and self-referential preference for fossil energy resources and reproduction of practices that fit within such fossil fuel based regimes. As a result, many lock-in mechanisms are active or can be activated. In such cases, we can refer to processes of maintenance, optimization and improvement (adaptive behaviour, self-organization). On rare occasions, however, niches can surpass the niche status more drastically and reach the take-off momentum (tipping point) when they can spark more fundamental, systemic changes (nonlinear yet gradual). After such alternative yet coherent constellations appear, incumbents can lose faith and legitimacy and thus lead to rapid changes of the existing regime (e.g. centralized energy system based on fossil fuels), which potentially might end up being replaced in the long run. Departure from the existing regime implies a systemic change. Systemic change or system innovation refers to gradual and long-term regime shift towards a new balance (dynamic equilibrium) in the system, where structures, relationships and patterns "drastically" change and new ones are created (emergence, self-organization, co-evolution). System innovation is in stark contrast to processes that tend to stabilize and reinforce the incumbent system, referring to processes that optimize or improve the performance of incumbent system.

On the other hand, macro-trends, deep cultural patterns and climate are examples of slowly changing phenomena beyond the direct influence of niche or regime actors. Together with the category of external shocks such as wars or natural disasters, they are characteristic for the macro level of sociotechnical landscape. Sociotechnical landscape can support as well as destabilize the dominant regime, which in turn exerts influence on the micro level. Therefore, windows of opportunity for niche innovations can be created due to the landscape pressure. Although the variables at the macro level are equally relevant (for transition management), they can be considered as inert or slowly changing, which means that they cannot be changed in the short-run and influenced directly. Following complex systems thinking, transitions represent particular type of system dynamics where the dominant regime self-organizes as well as co-evolves with its contextual environment (landscape level) towards a state of dynamic equilibrium. From the transition management perspective, the challenge at the macro level is mainly related to the broader societal setting that frames the transition processes. This would suggest that changing societal needs can put incumbent regime under pressure and eventually opens up for change. Therefore, "transition managers" should also direct their attention towards societal discourses, economic trends, governance and policy-making that shape the framing of sociotechnical transition in the long run. Independently from the scale and "form" of sustainability transitions that offer collective benefits, identification and steering of the breakthrough novelties ("sustainable developments" or "smart city") that put the dominant regime under the pressure becomes crucial.

Finally, we argue that energy-conscious spatial planning can draw inspiration from the following steps that underline the reflexive and cyclical process of searching, learning and experimenting in relation to different development phases at various scale levels, which is considered as an essential component of transition management framework [39]:

1. Stimulate niche development (emergence, variation) at the micro level and try to interconnect niches with the same direction. In the transition management framework, one does this by establishing and organizing a transition arena, a quasi-protected area for frontrunners (niche players and change-inclined regime players).
2. Try to find new attractors for the system by developing a sustainability vision and derived pathways at the macro level that can act as guidance for niche development.
3. Try to stimulate the formation of niche regimes by creating coalitions and new networks around the transition agenda and the different pathways.
4. Create diversity by setting out transition experiments that are related to specific pathways onto the vision.
5. Select the most promising ones that can be scaled up to a higher level as you learn from these experiments and develop an upscaling strategy.
6. Try to further modulation between the micro and macro levels (co-evolution) by adjusting the vision, agenda and coalitions, if necessary, by monitoring and evaluating (analysing patterns and mechanisms) the transition management process, after which the cycle starts again.

Given these context, we can observe that compared with more or less analytical perspective of complex systems thinking, transition management is also interventionist in its essence, concerning its desire to guide and influence the direction and pace of sociotechnical change. Seeing that decision-making is one of the most essential aspects of spatial planning, we argue that the basic tenets of transition management as introduced above should be followed, reflected and incorporated into our considerations and frameworks related to (energy-conscious) spatial planning. In contrast to conventional and atemporal "blueprint" planning, the emphasis on the process of redirecting and steering rather than on the end point can equip the domain of spatial planning with better understanding of the dynamic interactions inherent to sociotechnical change. Indeed, we propose that transition management can assist and inform spatial planning practice substantially and navigate its rationality towards the "becoming" aspects rather than focus on the "being".

Incumbent high-carbon energy systems with their underlying infrastructure have evolved gradually over decades and thus are strongly established and can prove resilient to changes. Hypothetically, a transition from such centralized energy systems (based on fossil fuels) to a more differentiated and decentralized energy systems (e.g. relying largely on RES and smart grid solutions) can be envisioned as one of the alternative visions within the TM framework.[6] In this context, the multilevel perspective can help to map, analyse and explain how the variety of sustainable energy initiatives may emerge, develop further and eventually scale up. That being said, TM framework holds the potential to provide spatial planning with deeper

---

[6] We can follow the trend of moving from centralized energy systems towards alternative practices involving nonconventional electricity generation systems characteristic for decentralized energy generation and supply, which are becoming increasingly attractive options [1, 2, 30, 49].

understanding on how can the build-up momentum connected to the niche level be facilitated, steered and influenced, which is in line with the focus of this research on the local-regional interface. The idea that (sustainable energy) niche development carried by local networks of actors can be strategically managed and translated into the sociotechnical regime is particularly compelling and offers a fertile ground for innovation in our spatial planning approaches. This corresponds well with the grounding perspective of this paper, which is determined by the niche-based approach and its emphasis on the local-regional interface – the scale of smart city. In other words, the local level resembles the micro level of niches, which are situated at the base of a multilevel system, beneath incumbent sociotechnical regimes and overarching landscapes ([44], p.1). In this regard, the recognition of energy as a spatially determined sociotechnical system is crucial. Embracing the niche-based perspective on RES development in the context of energy transition suggests to adopt a contextual, place-based approach in our governance and spatial planning approaches. This concerns particularly our desire to support mapping and building of indigenous innovation capabilities (through learning, networking, etc.) in the field of sustainable energy and to facilitate their spatial and socio-economic integration through the means of spatial planning. Indeed, we argue that "sustainable energy niches" become more meaningful, and hence their innovation potential is higher:

- When niches and other energy initiatives make use of their unique contexts and its special qualities (territorial capital).[7]
- When they promote integration and sustainability – they should be based on an understanding of local needs, conditions, dynamics and potentials, and that includes (local) residents and stakeholders in a collaborative planning process.
- When they activate area-based linkages from which both niches as well as the physical and socio-economic landscape in which they are located can benefit (synergy effects) – processes of co-evolution can be stimulated.
- When they have the ambition of system innovations at higher levels.

From this perspective, niches develop largely within the constraints endowed by existing regional assets; however, their broader integration into the territory can eventually press for "reforms" at the regime level. In this regard, "better" spatial and socio-economic integration of niches might also create a window of opportunity for optimization of renewable energy value chain and second-order learning between niches, empowering them (niche networking), and, in some cases, allows for their upscaling and diffusion. In addition to the focus of TM to attend to the dynamic patterns at and between different scale levels of sociotechnical system, the proposition to entail the dimension of space in relation to the area-specific conditions of niches is in place. Given the contexts sketched above, the area-based framing of niches allows to make use of indigenous local or regional assets and advantages and hence also recognizes the multifunctional potential of sustainable energy developments (e.g. in relation to regional development). In other words, the ability of niches

---

[7] Territorial capital can be described as the system of territorial assets of economic, cultural, social and environmental nature that ensures the development potential of places [10].

to "valorize" their unique context can be understood as one of the main drivers of their innovative capacity. The latter is a key point in relation to spatial planning, since the area-based understanding of niches reveals the potential to frame and conceptualize both innovation and integration of the energy initiatives in a more integrative manner – as spatially and socio-economically embedded phenomena. On this basis, the core assumption here is that "sustainable energy" niches can have manifold synergy effects towards sustainable spatial development if created and developed in line with the four basic principles that have been formulated above.

Several areas of special interest for further theoretical and empirical research can be outlined:

- Niche-space relationship. To explore potential ways on how sustainable energy niches can make use of their context (territorial capital) with the focus on rural spaces
- The role of various niche-internal processes (e.g. learning, networking, adapting, visioning) and their importance for energy-conscious spatial planning
- The role of various niche-external processes (e.g. diffusion, upscaling, co-evolution seen as alignment with ongoing processes outside of niches/at the regime or landscape level) and their importance for energy-conscious spatial planning
- The area-based conditions under which the value chain of renewable energy is improved and synergy effects between niches (in the field of renewable energy deployment) and their contextual environment (region) emerge
- Integration of sustainable energy systems based on spatially demanding renewable energy sources into the wider physical and socio-economic landscape

In this context, several disparities and scientific challenges can be identified:

- Current spatial planning and spatial development practice is quite well equipped to accommodate seemingly static and straightforward issues. However, it struggles to tackle complexity, uncertainty, dynamics and nonlinear change inherent to persistent and wicked problems such as climate change and related energy transition that encompass issues and timescales that lie beyond the traditional scope spatial planning.
- Energy (planning) and spatial development has been often treated as two separate domains. The mainstream framing of (sustainable) energy development is limited and underplays the multifunctional role of sustainable energy and its potential synergy effects, e.g. between the phenomena of renewable energy development and sustainable local/regional development. Arguably, the smart city concept has been opening the doors for integrated energy solutions in line with spatial development of urban areas.
- Although (sustainable) energy systems, considered as sociotechnical complex adaptive systems, are constituted spatially, spatial aspects of their development have been poorly understood and accommodated in the spatial planning practice. However, the spatial dimension becomes indispensable for smart and sustainable energy systems that co-evolve with their unique context and thus contribute to the sustainability.

- Sustainable energy and particularly the use of renewable energy sources have spatially demanding characteristics and thus require localized and spatially sensitive solutions for their socio-economic integration and sustainable deployment.
- Area-based understanding of niches (in local and regional contexts) in connection to innovative renewable energy developments can enhance our understanding of their role in local and regional development agenda. However, the research on transition management has tended to neglect spatial contexts of the transition.

Conclusions

Bluntly stated, the challenge for further research is better understanding of the relationship between innovation at the level of niches and regime with emphasis on the niche creation, niche development and their co-evolution and their local and supralocal context. Doing so can help to explore new ways to tackle and utilize the ongoing and still growing interest in bottom-up, community-led renewable energy developments in a more structured and strategic way and optimize their value chain. Although the niche-based approach is built upon the TM framework, it differs in its focus since the TM deals with the management of the whole system. In this regard, such bottom-up approach can prove to be limited in terms of its bias towards niche-driven processes, meaning that niche pressures on the regime are overly articulated. Therefore, attention needs to be paid also to ongoing processes at the regime and landscape level; however, subsequent reflection upon their geographies and spatial implications will be essential. Indeed, niches can be incorporated into the existing regime and thus support it or even improve it. This is related to one of the main critiques of early works on the multilevel perspective concerning their tendency to place too much emphasis on the role of niches. Niches have been frequently presented as building blocks for systemic changes towards sustainable development; however, they do not necessarily compete with the prevailing regime. Indeed, they can be incorporated into the existing regime and thus support it or even improve it. On the other hand, the possibility of innovations coming directly from the incumbent regime or other regimes should not be neglected. The role of both bottom-up and top-down influences that shape the regime needs to be incorporated in our considerations. Therefore, balance between top-down and bottom-up elements within our approach to transition management is needed. Since all societal actors exert influence on transitions but no single actor is not capable to control the pace and direction of transitions entirely [37], the demand for more informed and thus smart planning and governance is growing. TM remains a multifaceted research topic which has still a long way to go. Both theoretical framework and exercise, TM is still in its "beginnings", and in the sense, there are still many opportunities to interrogate, revise and validate its perspectives before it can be considered part of mainstream science and applied with greater confidence [18, 20]. Nevertheless, relatively robust foundations of the TM have been already laid and even empirically applied to several ongoing projects in different contexts. On this basis, a strong conclusion arises from the literature: the empirically driven methodology and integrative framework of TM can inform and assist a broad range of scientific disciplines. Moreover, TM can be promoted as a potential guidance for planners and policy makers to learn and understand how they can contribute to the transition towards sustainable energy systems and hence towards smart city initiatives in a more structured, coordinated and sophisticated way at multiple levels.

Acknowledgement This contribution is the result of the project implementation: SPECTRA+ No. 26240120002 "Centre of Excellence for the Development of Settlement Infrastructure of Knowledge Economy" supported by the Research and Development Operational Programme funded by the ERDF.

References

1. Ackermann, T., Andersson, G., & Söder, L. (2001). Distributed generation: A definition1. *Electric Power Systems Research, 57*(3), 195–204. Available at: http://www.sciencedirect.com/science/article/pii/S0378779601001018.
2. Alanne, K., & Saari, A. (2006). Distributed energy generation and sustainable development. *Renewable and Sustainable Energy Reviews, 10*(6), 539–558.
3. Araújo, K. (2014). The emerging field of energy transitions: Progress, challenges, and opportunities. *Energy Research and Social Science, 1*, 112–121. Available at: http://www.sciencedirect.com/science/article/pii/S2214629614000164.
4. Bagliani, M., Dansero, E., & Puttilli, M. (2010). Territory and energy sustainability: The challenge of renewable energy sources. *Journal of Environmental Planning and Management, 53*(4), 457–472. Available at: http://www.tandfonline.com/doi/abs/10.1080/09640561003694336 Accessed 15 Dec 2014.
5. Basta, C., van der Knaap, W., & Carsjens, G. J. (2012). Planning sustainable energy landscapes: From collaborative approaches to individuals' active planning. *Sustainable Energy Landscapes: Designing, Planning, and Development, 7*, 187.
6. Berkhout, F., Smith, A., & Stirling, A. (2004). Socio-technological regimes and transition contexts. 117 *System innovation and the transition to sustainability: theory, evidence and policy. Edward Elgar, Cheltenham, 44*(106), 48–75.
7. Bomberg, E., & McEwen, N. (2012). Mobilizing community energy. *Energy Policy, 51*, 435–444. Available at: http://www.sciencedirect.com/science/article/pii/S0301421512007276.
8. Bridge, G., et al. (2013). Geographies of energy transition: Space, place and the low-carbon economy. *Energy Policy, 53*, 331–340. Available at: https://doi.org/10.1016/j.enpol.2012.10.066.
9. Byrne, R., et al. (2011). Energy pathways in low-carbon development: From technology transfer to socio-technical transformation.
10. Camagni, R. (2008). Regional competitiveness: towards a concept of territorial capital. In *Modelling regional scenarios for the enlarged Europe* (pp. 33–48). Berlin: Springer Verlag.
11. Commission, C.F. the & the Commission, C.F. (2010). *Europe 2020: A strategy for smart, sustainable and inclusive growth*. Brussels: European Commission. Available at: http://scholar.google.com/scholar?hl=en&btnG=Search&q=intitle:A+strategy+for+smart,+sustainable+and+inclusive+growth#0.
12. Davoudi, S. (2009). Framing the role of spatial planning in climate change. *Electronic Working Paper, 43*(43), 1–44.
13. Davoudi, S., Crawford, J., & Mehmood, A. (2009). Climate change and spatial planning responses. In *Planning for climate change: strategies for mitigation and adaptation for spatial planners* (pp. 1–18). London: Earthscan.

14. European Commission. (1999). European spatial development perspective, Available at: http://ec.europa.eu/regional_policy/sources/docoffic/official/reports/pdf/sum_en.pdf.
15. European Innovation Partnership. (2013). European innovation partnership on smart cities and communities strategic implementation plan. *European innovation partnership on smart cities 2013*, (Strategic implementation plan).
16. Fouquet, R., & Pearson, P. J. G. (2012). Past and prospective energy transitions: Insights from history. *Energy Policy, 50*, 1–7.
17. Geels, F. W. (2002b). *Understanding the dynamics of technological transitions: a co-evolutionaryand socio-technical analysis*, Twente University Press Enschede.
18. Geels, F. W. (2011). The multi-level perspective on sustainability transitions: Responses to seven criticisms. *Environmental Innovation and Societal Transitions, 1*(1), 24–40.
19. Geels, F. W. & Kemp, R. (2000). *Transities vanuit sociotechnisch perspectief*, MERIT Maastricht.
20. Genus, A., & Coles, A.-M. (2008). Rethinking the multi-level perspective of technological transitions. *Research Policy, 37*(9), 1436–1445.
21. Giordano, V., et al. (2011). Smart Grid projects in Europe : Lessons learned and current developments,
22. IPCC. (2014). *Climate change 2014 synthesis report summary chapter for policymakers.* Ipcc, p. 31.
23. Jiusto, S. (2009). Energy transformations and geographic research. In *A companion to environmental geography* (pp. 533–551). Wiley-Blackwell. Available at: https://doi.org/10.1002/9781444305722.ch31.
24. Kemp, R., Schot, J., & Hoogma, R. (1998). Regime shifts to sustainability through processes of niche formation: the approach of strategic niche management. *Technology analysis & strategic management, 10*(2), 175–198.
25. Kemp, R., & Loorbach, D. (2006). 5. Transition management: a reflexive governance approach. In *Reflexive Governance for Sustainable Development* (pp. 103–130). *Cheltenham/Northampton*: Edward Elgar.
26. Kempener, R., Komor, P., & Hoke, A. (2013). Smart grids and renewables - a guide for effective deployment. *International Renewable Energy Agency*, (November), p. 47.
27. Kostevšek, A., et al. (2013). A novel concept for a renewable network within municipal energy systems. *Renewable Energy, 60*, 79–87. Available at: http://www.sciencedirect.com/science/article/pii/S0960148113002292.
28. Loorbach, D. (2007b). *Transition management: new mode of governance for sustainable development*, Dutch Research Institute for Transitions (DRIFT).
29. Loorbach, D. (2010). Transition Management for Sustainable Development: A prescriptive, complexity-based governance framework. *Governance, 23*(1), 161–183. Available at: http://doi.wiley.com/10.1111/j.1468-0491.2009.01471.x.
30. Loorbach, D., Van Der Brugge, R., & Taanman, M. (2008). Governance in the energy transition: Practice of transition management in the Netherlands. *International Journal of Environmental Technology and Management, 9*(2–3), 294–315.
31. Müller, M. O., et al. (2011). Energy autarky: A conceptual framework for sustainable regional development. *Energy Policy, 39*, 5800–5810.
32. Nations, U. (2008). SPATIAL PLANNING - Key Instrument for Development and Effective Governance with Special Reference to Countries in Transition. , pp.1–56. Available at: http://www.unece.org/fileadmin/DAM/hlm/documents/Publications/spatial_planning.e.pdf.
33. OECD. (2012). Linking Renewable Energy to Rural Development. *OECD Green Growth Studies*. Available at: http://www.oecd-ilibrary.org/content/book/9789264180444-en.
34. Pasqualetti, M. J. (2011). Social barriers to renewable energy landscapes*. *Geographical Review, 101*(2), 201–223.
35. Rip, A., & Kemp, R. (1998). *Technological change*, Battelle Press.
36. Roo, D., Hillier, J., & Van Wezemael, J. (2012). Complexity and spatial planning: introducing systems, assemblages and simulations. In *Complexity and Spatial Planning: Systems, Assemblages and Simulations* (pp. 1–32). Farnham: Ashgate Publishing.

37. Rotmans, J., Kemp, R., & Van Asselt, M. (2001). More evolution than revolution: Transition management in public policy. *Foresight, 3*(1), 15–31.
38. Rotmans, J., & Loorbach, D. (2008). *Transition management: Reflexive governance of societal complexity through searching, learning and experimenting*. Cheltenham: Edward Elgar.
39. Rotmans, J., & Loorbach, D. (2009a). Complexity and transition management. *Journal of Industrial Ecology, 13*(2), 184–196.
40. Sathaye, J., et al. (2011). *Renewable energy in the context of sustainable development*. IPCC Special Report on Renewable Energy Sources and Climate Change Mitigation, pp. 707–790.
41. Sharma, D. C. (2007). Transforming rural lives through decentralized green power. *Futures, 39*(5), 583–596.
42. Van Der Schoor, T., & Scholtens, B. (2015). Power to the people: Local community initiatives and the transition to sustainable energy. *Renewable and Sustainable Energy Reviews, 43*, 666–675. Available at: https://doi.org/10.1016/j.rser.2014.10.089.
43. Verbong, G., & Loorbach, D. (2012). *Governing the energy transition: reality, illusion or necessity?*, Routledge.
44. Smith, A. (2007). Translating sustainabilities between green niches and socio-technical regimes. *Technology Analysis & Strategic Management, 19*(4), 427–450.
45. Smith, A., Voß, J.-P., & Grin, J. (2010). Innovation studies and sustainability transitions: The allure of the multi-level perspective and its challenges. *Research Policy, 39*(4), 435–448.
46. Solomon, B. D., Pasqualetti, M. J. & Luchsinger, D. A. (2003). Energy geography. *Geography in America at the dawn of the 21st century*, (pp. 302–313). Oxford University Press, UK.
47. Stremke, S., van den Dobbelsteen, A. (2012). *Sustainable energy landscapes: designing, planning, and development*, CRC Press. ISBN: 978-1-4398-9404-0.
48. Walker, G. (2008). What are the barriers and incentives for community-owned means of energy production and use? *Energy Policy, 36*(12), 4401–4405. Available at: http://www.sciencedirect.com/science/article/pii/S0301421508004576.
49. Wolsink, M. (2014). Distributed generation of sustainable energy as a common pool resource: social acceptance in rural setting of smart (micro-) grid configurations. In B. Frantál & S. Martinát (Eds.), *New rural spaces: Towards renewable energies, multifunctional farming, and sustainable tourism* (pp. 36–47). Brno: ÚGN.
50. Wolsink, M. (2013). The next phase in social acceptance of renewable innovation. *EDI Quarterly, 5*(1), 10–13.

Chapter 3
Can Concept of Smart Governance Help to Mitigate the Climate in the Cities?

Alfréd Kaiser and Tatiána Kluvánková

Abstract Label of smart city is in recent years very fashionable and attractive. Because of this we are focusing on one of the main pillars of this concept; in our case it is smart governance. Smart governance is emerging concept that can be used at different scales and environments. In our paper, we address the potential of implementation of smart governance toward mitigation of climate change effects in cities. By these changes we have in our minds, heat island effect is felt especially in the hot summer time. We are about to provide a literature review of smart governance and how its implementation can improve the urban environment. The reason why the use of smart governance is inevitable is that the implementation of this concept may improve the communication of all stakeholders starting from local residents through nongovernmental organization to municipalities. Improvement of communication can also lead to better addressing of the requirements of local residents toward improvement of their lives and also to mitigate the effects of climate change in cities. When people can see profits from the actions that have been taken to manage the local environment, then they are likely to participate into the system.

Introduction

Nowadays, more and more people live in urban areas. This transition brings numerous challenges, providing the necessary utilities to large amounts of people living in the same area. Thanks to the revolution of the information technology and big data sets, the concept of the "smart city" has emerged, and nowadays a lot more information is available about different types of utilities. According to Caragliu et al. [1], the increasing urbanization is creating a need for the city planners to deal with

---

A. Kaiser (✉) · T. Kluvánková
SPECTRA Centre of Excellence EU, Slovak University of Technology in Bratislava, Bratislava, Slovakia

Institute of Forest Ecology, Slovak Academy of Sciences, Bratislava, Slovakia
e-mail: alfred.kaiser@stuba.sk; tatiana.kluvankova@stuba.sk

© Springer International Publishing AG, part of Springer Nature 2019
D. Cagáňová et al. (eds.), *Smart Technology Trends in Industrial and Business Management*, EAI/Springer Innovations in Communication and Computing,
https://doi.org/10.1007/978-3-319-76998-1_3

reinforced complexity regarding urban factors, such as food and water supply, traffic management and waste disposal, and climate change. Against this background, the concept of "smart cities" has been introduced, described as a device for dealing with these service problems in a common framework. When it comes to climate change, Mancarella [2] states further that reducing the energy footprint in cities is crucial. Integrated operation and planning of the urban system are described as essential tools for maximizing the environmental efficiency. It also means that the cities should strive to reduce their problem with heat islands especially cities where temperature in summer exceeds 30 degrees of Celsius and for day temperatures which stay over 30 degrees for more than 1 or 2 days.

Climate change in cities raises plenty of challenges for the community. These challenges are mitigation of the negative effects of climate change and adaptation to them. Heavy rainfall and connected flooding and extreme temperatures are commonly regarded as the effect of climate change [3]. These factors are already influencing the society especially in cities in cases of abnormal temperatures. In summer cities are struggling with residual heat, and they are not able to handle this problem or just ignoring it and ascribe this to climate change. Some cities use water showers as a solution, but we all know that it is not the solution and just patching of problem and definitely not solving it. These measures are only technical and are not sufficient enough to achieve mitigation of heat islands in the long term. In this paper, we will focus on improvement of understanding of concept of smart governance, and this concept subscribes the outline of the system for mitigation of heat islands in cities.

The main problem is prevailing struggle with extremely high temperatures in some parts of the city (heat island) and continuous denial of the attention of authorities to this problem. The following problem is that broad community weakly understands the problem of heat islands and is just complaining about the increasing temperatures during summer especially in time of the peaks of temperature. The concept of smart governance can bring more light into this problem and contribute to the mitigation of temperatures during extreme days.

A growing number of concepts of smart city consider a smart governance as one of the main pillars of smart city concept. The main focus of "smart city" concept seems to be concentrated on the role of ITC (information communication technologies), although the vast research was carried out on the role of human capital, social capital, and environmental capital as important drivers of growth of the cities [1].

Extensive research has been carried out in the area of heat islands in cities using modern satellite technologies and thermal-sensitive cameras to identify heat spots in the cities [4, 5]. Also the literature about a smart governance concept is focused on different levels and measures from global scale down to the local [6]. Research done in this field is considerable and can provide information and basics for new concepts of smart governance.

Academic research in the field of smart cities and smart governance is relatively young, and researchers are not united in defining what is "smart governance." Usually they vary in minor differences or the scale of the focus, but the main idea behind the theory is the same [7–9]. Mooji [8] and Willke [9] and Johnston and Hansen [7] agree that the information technologies play a significant role in the implementation of smart governance concepts. Some researchers pay more attention toward participants and involving people rather than in the technical background of concept and believe that they can positively contribute to the system of the local government. All concepts are developed from already working governance concepts, and researchers are trying to improve them. All these concepts are based on a decision-making system where multiple actors can contribute and defend their interests in different levels, and also across multiple levels [10], this concept is called multilevel governance. We were looking for more focused concepts; local scale is visible in the work of Janssen and Estevez [6] where they proposed new term "I-Government" with a focus on closeup local scale. Despite these works, little is known about the specific use of the concept of smart governance on climate change mitigation in local scale. In this scale residents can contribute to the system and therefore influence local climate. During table research we did not encounter any paper or work which pays attention toward our issue.

The purpose of this paper is to contribute to the theory about smart governance and to justify managerial abilities/possibilities of this theory. We see potential in the use of smart governance in managing of urban areas toward climate change mitigation. So the paper's research objective is: how can smart governance model contribute to the mitigation of heat islands in urban areas?

Smart Governance

Role of Governance

There is a difference between government (control/steering mechanism) and governance (decision-making/managing). Term governance is usually used for the description of the process of governing, and government is connected with public administration of municipalities or states. Therefore, government is broadly known as some kind of formal institution of government and control, while term governance is far broader and we can say that governance is an interaction of processes, relations between the state and other institutions (including private business and civil society), information structure, rules, etc. [11]. We argue that the classic understanding of governance seems as inefficient in the modern world. Nowadays we are adapting to multilevel governance where the nature of vertical relationships is redefined and overcoming existing structures. While the traditional way of understanding is connected to political control and centralization, multilevel governance puts in favor coordination of social relations while there is no authority. The multilevel governance is prevailing in the European Union as well as in other developed countries (Bache et al., 2010). The understanding of multilevel governance system is necessary for further understanding of smart governance, which adapts the principles of multilevel governance for environment of the new technology era where many actors are involved in different levels. In particular, multiple actors can interact at different levels and contribute to system, and also the system and environment can prosper from these contributions.

Governance is alternative to sustaining hierarchical control in policy making. This hierarchical form can be found in both the public and private sector. Smart governance can be defined as new form which has better conditions for cooperation and interaction between state (as authority) and civil society actors in the process of decision-making as a network. The local government needs to have a permanent process with continuous comparison and exchange between private sector and public sector rather than just copy the current state where single actor (state, municipality, etc.) operates with the support of bureaucracy according to its own vision of common public good [12]. This is what we are trying to promote by this paper.

Defining Smart Governance

The concept of smart governance can provide easy access to data and information to local residents as well as for local authorities and all other stakeholders and encourage them to participate on mitigation of climate change. Following the first definition of smart governance by Eger [13] who stated that in the time of post-industrial world full of the global economy which rapidly expands is the age of information. On this wave of change, all institutions, it does not matter if private or public, should be forced to adapt on this change. So the prevailing forms of governance are going to be replaced with new form which is called smart community, so sustaining governance is changing into "smart governance." There was no specification of what should smart governance consists of, so there was a need to set some basic factors for further specifications, and already aware of this need, Mooji [8] outlined elemental factors which define main attributes of smart governance. In this purpose smart should be moral and responsible for its actions, react to unforeseen circumstances, act according to moral rules, and be transparent. The first outline of specific factors is set, and it creates an environment for advancement of the theory, and researches took the opportunity. Willke [9] defines smart governance as a complex group of assumptions, aspects, and capacities that consists a structure of governance, which is capable to deal with the circumstances and needs of modern community based on knowledge, information technologies, and expertise. As he commented, new forms of governance should transform into smart governance and by this way transform prevailing system of governance. This new smart governance supposes to be sufficient for current needs of modern society. Further development of the concept proposed above inspired Johnston and Hansen [7] to the idea that the smart governance infrastructure should provide more clarity in the manner of public aspiration, promote the prosperity of culture, and further raise accountability. The assumption to be accountable is to be highly responsible. So in an ideal state, the one who is in governing position should be responsible for his actions. In this work authors determined a smart governance as a tool for participation and the way of collecting information from contributors because they are convinced that people want to contribute as an individual. They see a problem in creating the structure for system of smart governance but opportunity in meaningful contributions while individuals are still enabled to engage in the system. Despite that this development of the concept is far more narrowed down than Wilke [9] and Mooji [8] and is accurate for the purpose of local governance, academics, Janssen and Estevez [6] introduced the concept of smart government which they call "I-Government." Concept of I-Government is very interesting due to its innovation and more concentrated focus. This concept concentrates its focus on smaller groups in the small-scale environment and thanks to ICT (information and communication technology) better addresses problems and solutions between groups, communities, places, and so on. This is what they call "doing more with less" and it is an accurate name for the concept.

The concept of smart governance is nested in principles of multilevel governance (similar to earth system governance and global governance [14]), so it shares the same structure of multiple actors involved in different levels, and in some cases actors can be involved across multiple levels. Smart governance concentrates on the implementation of new information technologies into the existing concept of multilevel governance following the Eger's [13] definition "cyberspace and cyberplace" [13]. There is common agreement of scientists that a prevailing system needs to be changed and that implementation of new technologies in ICT (information and communication technologies) needs to be used while the system has to be moral and responsible, able to adapt to changes, and be transparent [8]. And we do agree that the change is inevitable for functioning systems of governance.

There are multiple theories about smart governance; however, Mooji [8] and Willke [9] and Johnston and Hansen [7] agree that information technologies are significantly important and innovation of governance is vital for further development. Some of them pay more attention toward people and their aspiration for participation in decision-making. Johnston and Hansen highlighted that participation of the common people can provide a positive contribution to society [7]. Janssen and Estevez went even further with their "I-Government" where the approach is more bottom-up than top-down, because they focus on small environments and not on the big-scale environment [6]. So the most significant difference is just on the scale of an environment where theory is implemented.

In our work we use the concept of smart governance to manage semi-public areas. We argue that these areas can be defined as common-pool resources because they are shared with local communities where every individual is an actor [15]. Common-pool resource includes natural and anthropogenic resources which are in our case semi-public areas (inner blocks). As we are about to encounter multiple actors, the managing of these semi-public areas is complex. Ostrom (1999) already identified that the vast number of actors creates difficulties in system, organization, and agreeing on rules and enforcement of those rules. We agree that especially in local scale individuals play a very important role. Usually in this kind of environment where individuals are facing choices, all others will be affected by "one man's choice" because they are interdependent. If the individual prefers short-term self-interest choice, that means for the others that this one individual leaves them worse [16]. This is where smart governance should step in, and use of new technologies can improve processes of common agreement on rules and enforcement. With smart governance, we create better, more accurate, and responsive system of local governance. Thus governance becomes more transparent and more beneficial.

So smart governance should provide individuals enough information to not take selfish decisions and focus on long-term benefit. From the past we already know about many cases such as case from Nepal, where irrigation systems are the example of well-managed common-pool resources which rely on rules and norms created and developed by local participants who are in this case local farmers. In the case where authorities created the system of irrigation, it was less efficient than one created by farmers themselves. It was caused mainly because of strict focus of authorities on modern engineering and ignoring of rules and norms that farmers had before [17]. It seems logic that the communication of participants is vital for livable system or concept. Only if local stakeholders and individuals are involved in creation of norms and rules, system have the chance to work with highest efficiency. It was proven by many studies, and in the case of Nepal irrigation systems, it is nicely described. Use of new information and communication technologies can improve the process of developing rules and norms, especially if it is in the interest of local residents. Inclusion of new technologies in local government does not immediately means that the city can promote itself as smart city, but it is a step toward it because smart governance is one of the pillars of smart city concept.

The fact is that the term smart governance is relatively young, and there is no common agreement on its definition, and there is still a considerable level of vagueness. Definitions are mostly the variety of the same idea and they differ slightly, but we can say that the main interest of all authors is concentrated in connecting of people to government through new ways of information technologies. Probably strict focus on information communication technologies is not the best way to define smart governance. Of course ICTs are important for the concept, but it is not the only important part. So we should provide further development of concept of smart governance and by this way contribute to this issue. In this paper, we will put a greater focus on the possibility of mitigation of heat islands in cities by using the concept of smart governance, so this concept is relevant for us. We will focus on lower levels of governance such as local governance of urban areas and discuss the possibilities that semi-public places offer opportunity for mitigation of the negative effects of global climate change in cities. By this way we should fulfil the research question and propose the possibilities of urban areas toward mitigation of heat islands which affect cities.

Stakeholders can play a role as sensors that can measure the quality of public green spaces, and the scientific measurements should be used as a supplement for the research. Data obtained from stakeholders in the form of interviews are crucial for research.

To sum up, smart governance should provide the environment for all individuals from selected location discussions on their needs, what they expect from the future, and places where they can contribute in any form, like knowledge, practical skills, social skills, and others. Our goal is to positively influence the residents to take more into account the aspect of global climate change and encourage them to take actions in the form of various adaptation measures toward climate change mitigation and heat island effect mitigation. Change of behavioral patterns of residents is one of the most important goals. We have to keep in mind that the decision that they will take will be under uncertainty and in multilevel governance conditions. With the use of smart governance concept, we create smart community fully aware of all circumstances of global climate change. Information communication technology is an integral part but more important is the social aspect.

Smart Governance as Tool for Local Climate Mitigation

Information and communication technology (ICT) is literally changing every aspect of our life. It can be found in our workplace, homes, sports, free time, and even in bars. All aspects are affected by the ICT, so it is no wonder that these technologies are now applied or are in the process of application in many countries in state governance and are used as a way of communication between authorities and common people.

Community is a term which represents correlation of three nexuses: the community of relationship, the community of place, and the community of interest. People are their activities and organizations, *community of relationship*; they have tended to communicate about questions of common interest with their neighbors, *community of* place; and also they usually have a common goal which makes them cooperate, *community of interest* [18].

There is a global trend which leads people to cooperate toward everyday happiness, and they do not wait for governance to build or serve anything for them. Today's condition may drive people to start cooperating and build a community on their own without intervention of the government. Thanks to advancing technologies, building communities is far easier, thanks to the fact that people still have the comfort of their homes but simultaneously they can communicate together. Smart community emerges, especially from the fast developing technologies. In the last few years, the development of mobile devices such as laptops and phones is making them more affordable and widespread. Now even elder people are using these devices. Also, many devices are equipped with computing technologies. Another recent trend is rapid the growth of social networks. Social networking can be described as a group of individuals connected through diverse social relations, like friends, family, coworkers, and school, to mention at least few. The spread of the Internet strategically contributed to the creation of social sites like, Facebook, LinkedIn, MySpace, and others. Also the Internet is the source of information used by a variety of people, like teenagers, managers, businessman, or researcher. So the smart community can be seen as a group of connected individuals that interact between each other on the network and deliver smart services or solutions. These individuals can be anyone or anything; a dog or tree can also interact by playing a significant role in the decision-making process [19]. According to Feng Xia and Jianhua Ma, smart community could by identified by these attributes:

- Smart communities are socially and physically aware systems.
- The scale of community varies with each case.
- Smart communities will be developed in time, and also the size of the community is changing during this development. That means that the smart community and its size are flexible.
- The Internet-based community is not the condition for the smart community; smart community may be functional in the local scale environment even without an Internet connection.
- In some cases lifetime of smart community can be long, while the life cycle of the other smart community can be short, depending on the supported application [19].

The use of ICT for development of a community at present is not a novelty. Now we are striving to add some higher value to the communities. So we do believe that only the implementation of ICT in communities does not make them instantly SMART. Smart community should be not just informed community, but also has to be concerned about environment and be able to communicate issues about the environment with authorities and care about future generations.

Semi-Public in Smart Governance

The inner blocks are always facing the problem of open access and lack of rules for managing them. Usually the municipalities manage these places, but in many cases provided maintenance by municipalities is not sufficient enough to fulfil the needs under the absence of participation of local residents.

Definition of public space as research object has been discussed for decades [15]. Our focus concentrates on management of semi-public spaces represented by inner blocks. Authors usually involve exterior and interior places in their understanding of what public space is.

The concept of semi-public spaces is a new concept and is missing the exact scientific definition [15]. WAUA [20], a blog about architecture, urbanism, and art, differentiates semi-private and semi-public spaces. We have to keep in our minds that both share the characteristics of private space. Although first group represents access-controlled environment which is accessible only for residents and their associates, the second represents private space which is publicly accessible and access is not controlled. It can be seen in the figure below the actual scheme of where semi-public spaces are situated. From this scheme is also clear that the semi-public spaces have the characteristics of public space, but from a legal point of view, they have private owners who share this place.

There are three basic attributes of semi-public spaces which are defined by Maco [15] in his thesis.

A: Physical environment – these are reflected as physical, visual, and functional qualities of the environment. In semi-public spaces these objects are usually used by smaller groups of people than things that are in open public space, so their function is more specific. Also the visual aspect is more specific and depends on the taste of local users.
B: Property relations – in the city we can say that the objects in open or public space are symbolically managed by the people, but actually these objects are managed by public authorities. In semi-public space it is different. Objects are in close range of residential living, so people tend to manage these spaces and objects by themselves as individuals or groups.
C: Institutional – in this attribute we can include regulation, rules, competences, and arrangements. Semi-public spaces do not require as much management from authorities because self-management is higher than in public space. Usually in public space, there is always a set of rules created by the authorities, but in the case of semi-public spaces, it is more common that this rule establishment process is bottom-up. People from local community tend to create their own set of rules, and then the self-control mechanism is more effective [15].

To distinguish between public and semi-public space, we have to define characteristics that will help us differentiate between different types of places. In Fig. 3.1 we can see difference between public, semi-public, and private space. In simplified perception, semi-public space is accessible for one who somehow participates in development of place or cultivates or contributes by any form that can be also some kind of payment. PPS [21] defined four main categories of quality: *accessibility* of place (a successful place which is easy to access and is also visible), *comfortability and good image* (a clean place with a safe environment and available places to choose from and rest in), *usability* (a place which provides reason for people to spend time in there and to return there), and *sociability* (a place which is good for meeting other people and create friendships). These elements propose four main conditions under which public space can be created.

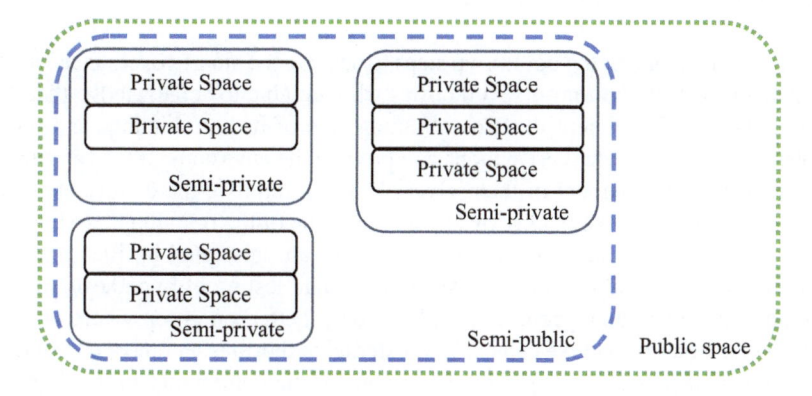

Fig. 3.1 Semi-public spaces. (Adopted from [20])

Difficulties of defining of semi-public space can be simplified by clear boundaries of discussed place. When we know the boundaries, it is easier to exclude elements or people who are not part of one exact semi-public space. This can lower the possibility of inequity in the system. Maco [15] also defined a clear set of qualities which indicates the quality of the space. These nine qualities are based on PPS [21] and specific characteristics of semi-public spaces.

(a) *Safety* – physical conditions of the space, which provides secure habitat for all users
(b) *Accessibility* – visual and physical orientation toward the local environment, allowing easy entrance with no barrier blocking the entrance
(c) *Sociability* – the ability to provide space for meetings and interactions within the community of users
(d) *Usability* – sufficient provision of functions for users, which reflect the character of space and needs of local community
(e) *Aesthetic value* – appearance of the space
(f) *Greenery* – quantitative and qualitative state of green areas within the space, sustainability of species
(g) *Scale* – appropriate size of the space toward the total amount of users (not too small or too big)
(h) *Cleanness* – a dynamic quality, which depends on the current degree of maintenance, but can also be measured in a long-term perspective
(i) *Collective service* – provision of external service, which is not directly performed by the community (e.g., waste disposal, parking, maintenance of sidewalks) [15]

Based on these qualities, we can qualify the state of place, but we still don't have the knowledge of whether the community within the place is functional or not. But the good condition can still indicate that the community exists within the place, and they are socially and environmentally aware.

Self-Governance in Semi-Public Space

Self-governing is one of the bottom-up steps to achieve the smart governance which can be adopted from bottom-up. Authorities who are able to acknowledge that the community is self-sufficient in terms of taking care of its surroundings and other problems should embrace these actions and provide the environment for communities for further development of themselves. This can improve the quality of semi-public spaces and also the quality of life in cities. One of the first things we have to do is to determine the scale of community that we are interested in. Best practice shows that initial focus should be on small scale, and testing of hypothesis should be initially done on small group of people or, even better, on group where we can easily set boundaries, and we can exclude external authorities and most of external actors who will potentially represent the threat for the community in the form of disrespecting of rules set by self-governing small group of stakeholders. According to Plichtová [22] any small social units like local community association can be more democratic than the big social unit like nation. She also proposed five reasons why small groups will provide higher quality in government of place:

1. Small communities are able to work together on principle of equity, openness, and reasonable arrangement better than bigger groups.
2. Small communities create place for active deliberation of citizens for their collaboration toward common goal.
3. Members of the community, bonded together with common interest and history, are highly motivated to achieve the deal or at least compromise.
4. Local community know its needs better than civil servants and is capable to solve the problems on its own.
5. Members of the community, according to fact that they will bear the consequences, are going to be highly motivated to take the power responsibly [22].

These reasons are encouraging for further development of self-governing smart communities based exactly on limited number of members with shared interest, place, and relationship like stated by Morse's nexuses above. We believe that the consensus can arise from small group faster and be better addressed than the solutions from external authorities represented by municipalities in environment of the cities. Also these communities do work inside their own environment and usually have the common interest and resource which leads us to commons which are discussed next.

Smart Governance and Common-Pool Resource

Common-pool resources (CPRs) are natural- or human-constructed systems that contain or generate limited amount of resources. Therefore, if one entity or person uses the resource, its use is subtracted from the abundance of resource units feasible to other entities [23]. The theory of commons copes with common-pool resources, with benefits from these resources and with a distribution of these benefits and with governing of these resources. In general, we can envision commons as all what we share with others or as a product of environment that belongs to everybody, and this resource should be taken care of for future generations [24]. The commons can be also perceived as shared resources within stakeholders with equal amount of interest [25].

Common-pool resources are generally distinguished into two groups, traditional and new commons. With traditional commons we understand forests, irrigation systems, fisheries, groundwater, grazing lands, and the air that we breathe [23]. The main objective of commons is setting of pair of rules that can prevent the overuse of shared resource, and therefore the resource can be beneficial for every stakeholder of the resource. CPRs are facing social dilemmas where short-term interest of an individual is in conflict with long-term interest of group or society; therefore the governance of the commons is a challenging field of research especially in terms of economics and policy [26]. In our case we are more concerned with the policy and setting of rules that will help to maintain the local environment for future generations. The conventional approach to this dilemma originates from the theory of property rights in resource management which is understood as right to sell and estrange the right to harvest the goods from resource [27].

New types of commons are developing rapidly, such as urban commons, knowledge commons, neighborhood commons, and so on. Regarding Hess [25] as one of main protagonists of commons today, she reminds that it is often necessary to limit the access of CPRs for those who are not part of the community. However, such measurements like limited access are difficult and expensive. Problem with excludability and non-excludability is highly complex in CPRs, and it results in confusion of setting boundaries in the environment even if these boundaries are conceptual or not. In this case these boundaries can be defined as borders of inner blocks at Legionárska and Vazovova Street in Bratislava.

There still occurs one question: what is in the way of implementation of CPR regime in common neighborhood? Probably the current institutional monoculture represents top-down approach in managing process where authority is represented by municipalities. Traditionally way of management of resources claim that the only way how to manage resource effectively is whether private or state-centralized management as it is most common example of management today. On the other side, Ostrom (2009–2012) and Poteete et al. [28] claim that local users in CPR regimes are able to create their own rules to manage CPRs and also these rules provide high level of acceptance among the users of CPRs. Due to the capability of self-organizing in group, they are able to solve problem on their own without intervention from external authorities.

Conclusion

In this paper we focus on the concept of smart governance and its possibilities to contribute in mitigation of global climate change in the form of the mitigation of heat islands. Heat islands appear to be a continuous problem, especially in dense cities. The increasing temperature in urban areas is caused by high concentration of soil sealing and surfaces like concrete, tarmac, or interlocking paving. For this reason, there is a need to create a concept which can be used in a wide range of environments like urban areas, inner blocks, and many more to face the heat island phenomenon. When we are talking about facing heat islands, we have in mind the smart decision-making on local level supported by authorities and also further support of local of local communities and creation of new communities on places where they do not exist.

There are usually many actors and stakeholders who tend to use and manage one specific semi-public space. The theory of the commons is concentrated on common-pool resources (CPRs), products which are coming from these resources and on the question on how these resources should be managed. The commons are acknowledged as a term for shared resource where each stakeholder has an equal share or interest [25]. CPRs are either natural or human made, and in the case of inner blocks, it is human made even if it has natural character, but it has been created by anthropologic activity. The traditional distinction of commons is on forest, irrigation system, groundwater, fisheries, agriculture, pasture, or even the air, which we share all around the world [23]. Along with the traditional commons, new type of commons has developed. The new commons are urban commons and digital commons [25]. Our research is relevant, especially urban commons because we will focus on developing collective regimes for semi-public spaces inside the inner blocks. Goods from CPR are supposed to benefit certain group of stakeholders who are related to clearly specified common-pool resource. We focus on inner block, and stakeholders are represented by local residents and close surrounding residents who are facing a social dilemma in the process of decision-making.

References

1. Caragliu, A., Del Bo, C., & Nijkamp, P. (2011). Smart cities in Europe. *Journal of Urban Technology, 18*(2), 65–82.
2. Mancarella, P. (2012). Distributed multi-generation options to increase environmental efficiency in smart cities. In *2012 IEEE power and energy society general meeting* (pp. 1–8). IEEE.
3. Shaw, K., & Theobald, K. (2011). Resilient local government and climate change interventions in the UK. *Local Environment, 16*(1), 1–15.
4. Chen, X. L., et al. (2006). Remote sensing image-based analysis of the relationship between urban heat island and land use/cover changes. *Remote Sensing of Environment, 104*(2), 133–146.
5. Stathopoulou, M., & Cartalis, C. (2007). Daytime urban heat islands from Landsat ETM+ and Corine land cover data: An application to major cities in Greece. *Solar Energy, 81*(3), 358–368.
6. Janssen, M., & Estevez, E. (2013). Lean government and platform-based governance—Doing more with less. *Government Information Quarterly, 30*, S1–S8.
7. Johnston, E. W., & Hansen, D. L. (2011). Design lessons for smart governance infrastructures. In *Transforming American governance: Rebooting the public square* (pp. 197–212). Armonk: M.E. Sharpe.
8. Mooij, J. E. (2003). *Smart governance?: Politics in the policy process in Andhra Pradesh, India,*
9. Willke, H. (2007). *Smart governance: Governing the global knowledge society.* Campus Verlag.
10. Bache, I., & Flinders, M. (2010). *Multi-level Governance.* New York: Oxford University Press.
11. Kluvánková-Oravská, T., et al. (2009). From government to governance for biodiversity: The perspective of central and eastern European transition countries. *Environmental Policy and Governance, 19*(3), 186–196.
12. van Staden, M., & Musco, F. (2010). *Local governments and climate change: Sustainable energy planning and implementation in small and medium sized communities.* Dordrecht/Heidelberg/London/New York: Springer. Dordrecht.
13. Eger, J. (1997). Cyberspace and cyberplace: building the smart communities of tomorrow. *San Diego Union-Tribune, Insight*, 2. San Diego
14. Biermann, F. (2007). "Earth system governance" as a crosscutting theme of global change research. *Global Environmental Change, 17*(3–4), 326–337. doi:10.1016/j.gloenvcha.2006.11.010.

15. Maco, M. (2015). *Theory of commons in urban governance : Application to semi-public spaces theory of commons in urban governance : Application to semi-public spaces.* Slovak University of Technology.
16. McGinnis, M. D. (2000). *Polycentric games and institutions: Readings from the workshop in political theory and policy analysis.* University of Michigan Press.
17. Ostrom, E., et al. (1999). Revisiting the commons: Local Lessions, global challenges. *Science, 284*(5412), 278–282.
18. Morse, S. W. (2004). *Smart communities - how citizens and local leaders can use strategic thinking to build a brighter future.* San Francisco: Jossey-Bass. Available at: http://scholar.google.com/scholar?hl=en&btnG=Search&q=intitle:smart+communities#9.
19. Xia, F., & Ma, J. (2011). Building smart communities with cyber-physical systems. Proceedings of 1st international symposium on From digital footprints to social and community intelligence - SCI '11, p.1. Available at: http://dl.acm.org/citation.cfm?id=2030066.2030068. New York, USA
20. WAUA. (2008, July). On the strange disappearance of semi-spaces in London. Writings on Architecture, Urbanism & Art.
21. PPS. (2009). What Makes a Successful Place? http://www.pps.org/reference/grplacefeat/
22. Plichtová, J. (2010). *Občianstvo, participácia a deliberácia na Slovensku: teória a realita*, Veda. Bratislava, ISBN 978-80-224-1173-8
23. Ostrom, E., Gardner, R., & Walker, J. (1994). *Rules, games, and common-pool resources.* University of Michigan Press.
24. Walljasper, J. (2010). *All that we share: How to save the economy, the environment, the Internet, democracy, our communities, and everything Else that belongs to all of us.* New York Press. Available at: https://books.google.com.au/books?id=DnKtQAAACAAJ.
25. Hess, C. (2008). Mapping the new commons mapping the new commons. Syracuse University: SURFACE, (July), pp. 14–18.
26. Kluvánková, T., & Gežík, V. (2016). Survival of commons? Institutions for robust forest social–ecological systems. *Journal of Forest Economics, 24*, 175–185, Cheltenham, UK..
27. Demsetz, H. (1967). Toward a theory of property rights. *American Economic Review, 57*(2), 347–359. v10.1126/science.151.3712.867-a.
28. Poteete, A. R., Janssen, M. A., & Ostrom, E. (2010). *Working together: Collective action, the commons, and multiple methods in practice.* Princeton University Press.

Chapter 4
Operational Characteristics of Experimental Actuator with a Drive Based on the Antagonistic Pneumatic Artificial Muscles

Miroslav Rimár, Peter Šmeringai, and Marcel Fedák

Abstract This chapter describes the experimental measurements carried out in order to describe the actions taking place in antagonistic involvement of pneumatic artificial muscles (PAMs). Operational capabilities of PAMs operation were recorded and described.

Introduction

In the field of manufacturing technique, there are applications requiring replacement or enhancement of human muscular system power, especially in materials handling applications; for manipulators, the use of the drives based on the pneumatic artificial muscles' power can be considered as very advantageous. Application of PAM as a drive is possible, for example, in an environment where the use of conventional drive is disadvantageous or not conceivable. Relevant examples are aggressive or explosive environments. The reason for continuously more intensive research into the characteristics of PAM and their application is their high proportion of the generated power to their weight. For PAM there is characteristic inherent strength and safety. PAM itself has several additional benefits such as clean operation, easy maintenance, low acquisition costs, maintenance costs, and high durability, and in condition for safe operation, sparking or ignition is not possible. These benefits are the reason why PAM can become an alternative in substitution for traditionally used drives in selected applications. This article deals with an experimental measurements made on experimental device with an installed propulsion drive consisting of two pneumatic artificial muscles located in opposition and connected by chain

---

M. Rimár · P. Šmeringai (✉) · M. Fedák
Department of Process Technique, Faculty of Manufacturing Technologies with a Seat in Prešov, Košice, Slovakia
e-mail: miroslav.rimar@tuke.sk; peter.smeringai@tuke.sk

transmission with shaft for the transmission of torque. Opposed involvement of two pneumatic artificial muscles is termed as antagonistic involvement of pneumatic artificial muscles (PAMs) [1–5].

A wider range of usability of these drives prevents poor cope with precise position control. PAMs require accurate positioning control system that would be capable of rapid response to changes taking place in the PAMs. In position control today, mathematical and statistical methods are often applied to create the most accurate mathematical model of PAMs with the estimated optimal control parameters. In the case of wider application of PAMs and unforeseen changes in workload, the task of the control system is a control of nonlinear system. Therefore, an effort of the measurements referred in this article describes the conditions of PAMs operation ensuring stable operation of the experimental device [1–5].

For potential of mass applications of PAMs manufacturing nodes, it is possible to also consider the effectiveness of PAMs use, as well as the effective use of the working media. Therefore it is necessary to fix the limit of the conditions of PAMs operation, for identifying the potential for operation of the device with the lowest consumption of process media [1, 6, 7].

Static Characteristics of Artificial Muscles

PAM works on the transformation of pneumatic energy into mechanical energy. PAM contraction is the result of the impact of pressure of medium (gas) to the inner layer of the membrane, as performing input labor W_{in}: [2, 8, 9].

$$dW_{in} = \int_{S_i} (P - P_0) dl_i . ds_i = P'dV \qquad (4.1)$$

where P is gas absolute pressure inside the PAM, P_0 is ambient gas absolute pressure, P' is relative pressure $(P-P_0)$, S_i is total inner surface of the muscle, ds_i is area differential, dl_i is inner surface differential, and dV is volume differential [8, 10].

Work output W_{out}, arising when PAM length is reducing, is defined as:

$$dW_{out} = -FdL \qquad (4.2)$$

where F is PAM axial tractive force and dL is displacement in axis direction [8].

If energy loss caused by the system will be neglected, then under the law of energy conservation, the work at the output is equal to input work:

$$dW_{out} = dW_{in} \qquad (4.3)$$

After substituting both Eqs. (4.1) and (4.2):

$$F = -P'\frac{dV}{dL} \tag{4.4}$$

Ratio of *dV/dL* is intended assuming idealized active part of PAM as an ideal cylinder, where L is cylinder height, θ is the braid fiber and the cylinder axis angle, D is cylinder diameter, n is fiber circles around the cylinder, and b is fiber length [8].

With constant parameters, n and b can be expressed L and D (Fig. 4.1) as a function of θ:

$$L = b\cos\theta \tag{4.5}$$

$$D = \frac{b\sin\theta}{n\pi} \tag{4.6}$$

The volume of cylinder:

$$V = \frac{1}{4}\pi D^2 L = \frac{b^3}{4\pi n^2}\sin^2\theta\cos\theta \tag{4.7}$$

The output force is defined as a P and θ function:

$$F = -P'\frac{dV}{dL} = -P'\frac{dV/d\theta}{dL/d\theta} = \frac{P'b^2\left(3\cos^2\theta - 1\right)}{4\pi n^2} \tag{4.8}$$

The output force, developed by pneumatic artificial muscle, is thus linearly dependent on the pressure of the working fluid inside the PAM. It is also the function of angle between the braid fiber and the cylinder axis θ [2, 8].

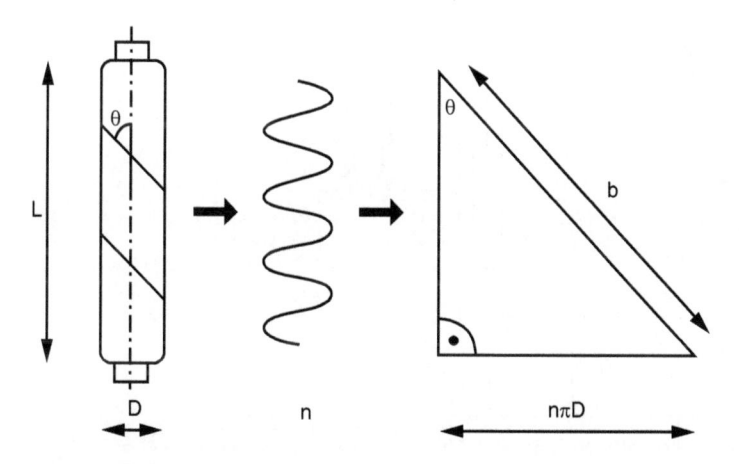

Fig. 4.1 n and b coefficient determination [8]

Experimental Apparatus

Experimental device consists of the antagonistic involvement of two pneumatic artificial muscles, compressed air supply system, and control system. The task of the control system is to measure, process, and record information about the static and dynamic characteristics during operation of actuators. On the basis of the data control, which is computing inside each iteration of the main control algorithm, the control system interferes with the operation of experimental assembly. PAMs are mounted on the support structure. Arm serving for attaching a weight is mounted to the drive shaft (Fig. 4.2) [7]. The figure shows an experimental assembly where PAM_L is left pneumatic artificial muscle, PAM_P is right pneumatic artificial muscle, LM is distance of the center of gravity of weight M from the shaft of the arm, S_P is potentiometric encoder, α is angle of the arm rotation from the neutral position 0, P_L is left PAM pressure sensor, P_P is right PAM pressure sensor, P is pressure, I is electric current, R is electrical resistance, P_K is pressure at the compressor outlet, V_{1L}/V_{1P} is left/right inflation electro valve, V_{2L}/V_{2P} is left/right drain electro valve, P_{VZD} is vessel pressure, and A/B/C is acceleration sensors. Highlighted signals from measuring devices are in orange; highlighted signals from vibration sensors are in red. Control signals for controlling electropneumatic valve are drawn in light green, and the blue color represents transportation routes for the compressed medium. Control of this facility is realized by means of four electropneumatic valves. Two electropneumatic valves are used to connect the PAM system and the compressed medium (Fig. 4.3). Other two electropneumatic valves are used for deflation of each PAM [7].

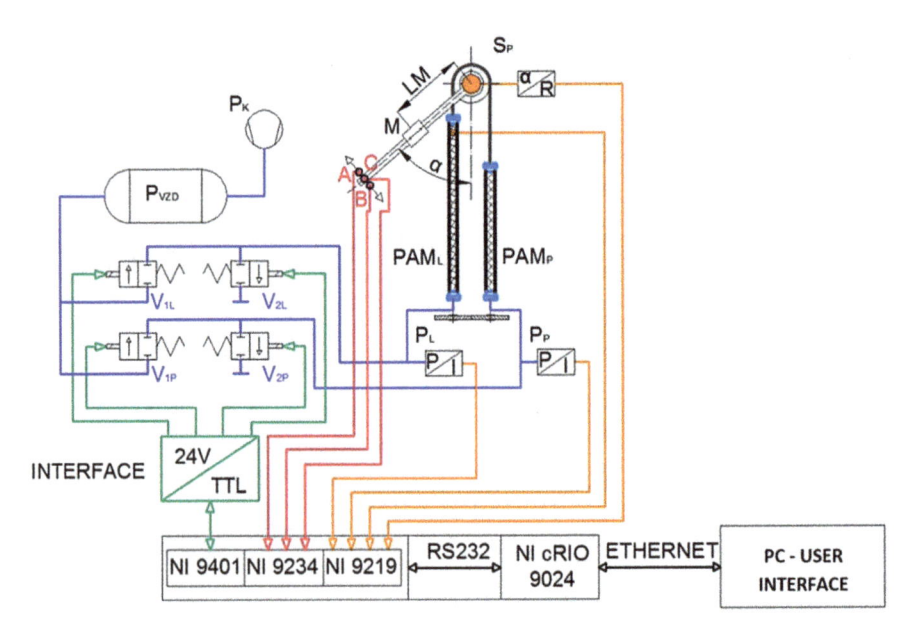

Fig. 4.2 Schematic depiction of experimental assembly with the NI cRIO 9024 [7]

4 Operational Characteristics of Experimental Actuator with a Drive Based... 53

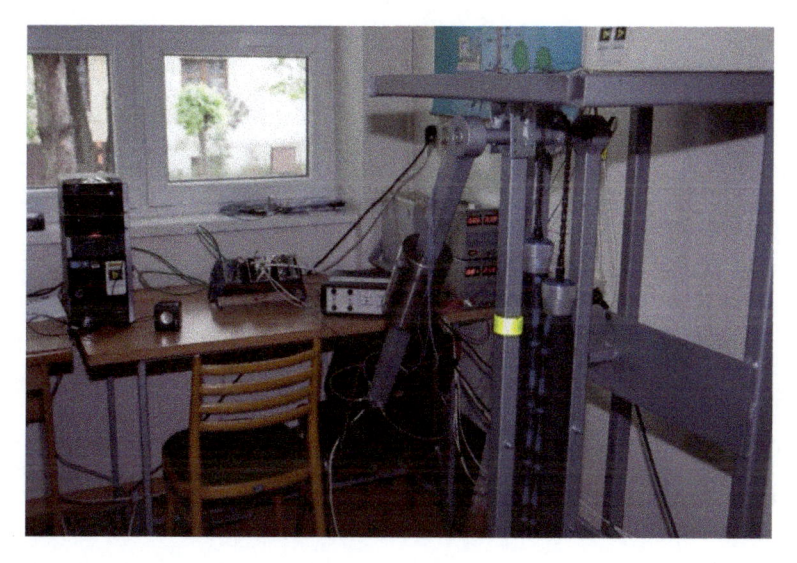

Fig. 4.3 Experimental device with drive based on PAM [7]

Control system of this experimental device provides more operation modes. For the purpose of this chapter, only the mode for controlling the position of the carrying arm was used. Position of the carrying arm is controlled by the angle of its rotation around the axis of the drive shaft. The zero, respectively, starting position is set to be identical with direct vertical direction.

The main purpose of these measurements is to describe static and dynamic characteristics of the experimental device, to minimize their impact to device operation, before it will be supplemented with another higher-order control subprogram, such as control based on neural networks or fuzzy logic [10].

Algorithm for position control is based on the elementary proportional regulation, where in the moment when pressure in one PAM is increased, the pressure in the second PAM is decreased. Only exceptions are in situations where one of the PAMs reached the maximum or minimum operational pressure. In this case only the pressure in other PAM can be changed.

Because the main active parts of experimental facility control and regulation system are proportional to electropneumatic valves, it was necessary to take into account the longitude of their lifecycle. That is the reason why the range of the usable pulse-width modulation was set to 20–89%. If the required PWM is lower than 20%, the zero value of PWM is used instead. If the required PWM is greater than 89%, the value of PWM equal to 100% is used instead. Control program also monitors whether the device is in a faulty state (Fig. 4.4).

In Table 4.1 are described basic states, which can happen in the experimental system, during experimental measurement. The main variables are $P_{L,P}$ which is pressure in the left and right PAM, respectively, $\Delta\alpha_{MIN}$ which is the difference between the required and measured angle of the carrier arm rotation, and PWM_{PROG} which is the programmed value of the pulse-width modulation. After detection of

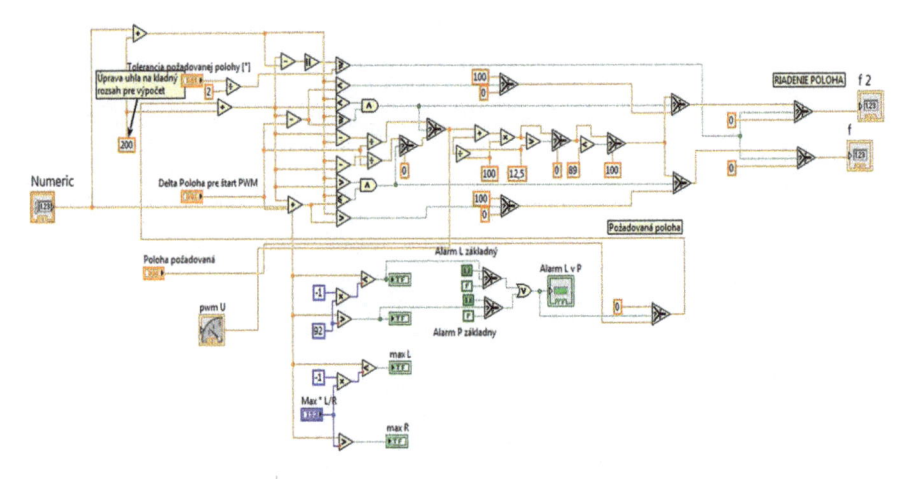

Fig. 4.4 Position control algorithm [7]

Table 4.1 Definition of operating conditions of the experimental system

	$PWM_{L,P}$	ALARM	System shut down
$P_{L,P}$-min	–	T	F
$P_{L,P}$-max	–	T	T
Δp_{MIN} ($\Delta\alpha_{MIN}$)	F	–	–
Δp_{MAX} ($\Delta\alpha_{MIN}$)	T	–	–
α_{PROG}-min	F	T	T
α_{PROG}-min-max	T	F	F
α_{PROG}-max	F	T	T
PWM_{PROG}-(0–19.99%)	0%	–	–
PWM_{PROG}-(89–100%)	100%	–	–

the faulty state, the control system will inform the system operator about this state by turning on the alarm. If critical operation conditions are detected, the control system will stop the system operation (Table 4.1).

Experimental Measurements

During experimental measurements, characteristics of the device were monitored during its operation. Arm was rotated into four positions (15°, 30°, 45°, 60°). For this experimental measurement, the size of the maximum PAM working pressure has been set to 5.1 bar and tolerance of accuracy of pressure setting in PAM to

0.05 bar. Tolerance of the arm rotational angle size has been set at ±1.5°. Size of the pulse-width modulation has been set to 20–100%.

Measured were pressures in each PAM (PS 016 V-504-LI2VPN8X H1141), value of arm rotational angle (potentiometric resistive divider), and arm acceleration (DeltaTron 4514-B).

The control system operates in cycles. During performance of one cycle, data corresponding to one group (cluster) of data from each sensor element and control intervention variable are measured and recorded. In this chapter the data about the size of rotation of the carrier arm and its acceleration and the data about size of pulse-width modulation for the opening of electropneumatic valves are taken into consideration. Position control is performed by using a simple algorithm providing a deflation for one PAM and simultaneous inflation for the second PAM.

In Figs. 4.5 and 4.6 only data about the value of PWM for the left PAM are drawn which represents the opening of the electropneumatic valve serving for inflation of PAM (plotted in the positive direction) and electropneumatic valve for PAM deflation (plotted in a negative direction). Waveforms are arranged one below the other, and they represent the values of the angle of the carrier arm rotation and support arm acceleration and information about opening of the electropneumatic valves used to control the left PAM.

Measured data are representing processes in PAMs during time when they are holding the desired position. There is a significant difference between the sizes of acceleration when positioning the arm to different positions. With an enlarged rotation of the carrier arm, there is significant decrease in the value of the measured acceleration. This is caused by increasing the rigidity of the system. Impact emerging in PAMs from the operation of electropneumatic valves is responding to fluctuations in the measured acceleration. Waveforms display a phenomenon occurring within the opening of electropneumatic valves on more than one iteration of the control program. If the electropneumatic valve is open for more than one iteration, glimpse of the support arm to the opposite side occurs, even in the smallest size of PWM.

From the measurement results, the following postulates and conclusions were made:

- In case when the life cycle of the electropneumatic valves is taken into account, pulse-width modulation must be used in the range of 20–100%.
- Impact emerging in PAMs from the operation of electropneumatic valves is responding to fluctuations in the measured acceleration and position.
- The rigidity of the system is increasing with increasing pressure in both pneumatic artificial muscles.

In the future, it is possible to establish limit values for the acceleration of the support arm in order to allow the device to achieve smoother operation at the start of movement of the carrier arm, stopping it and holding it to the desired position. In the future the experimental device could be supplemented by elements that would allow the control of acceleration of the carrier arm during the start of movement and during braking [11–13].

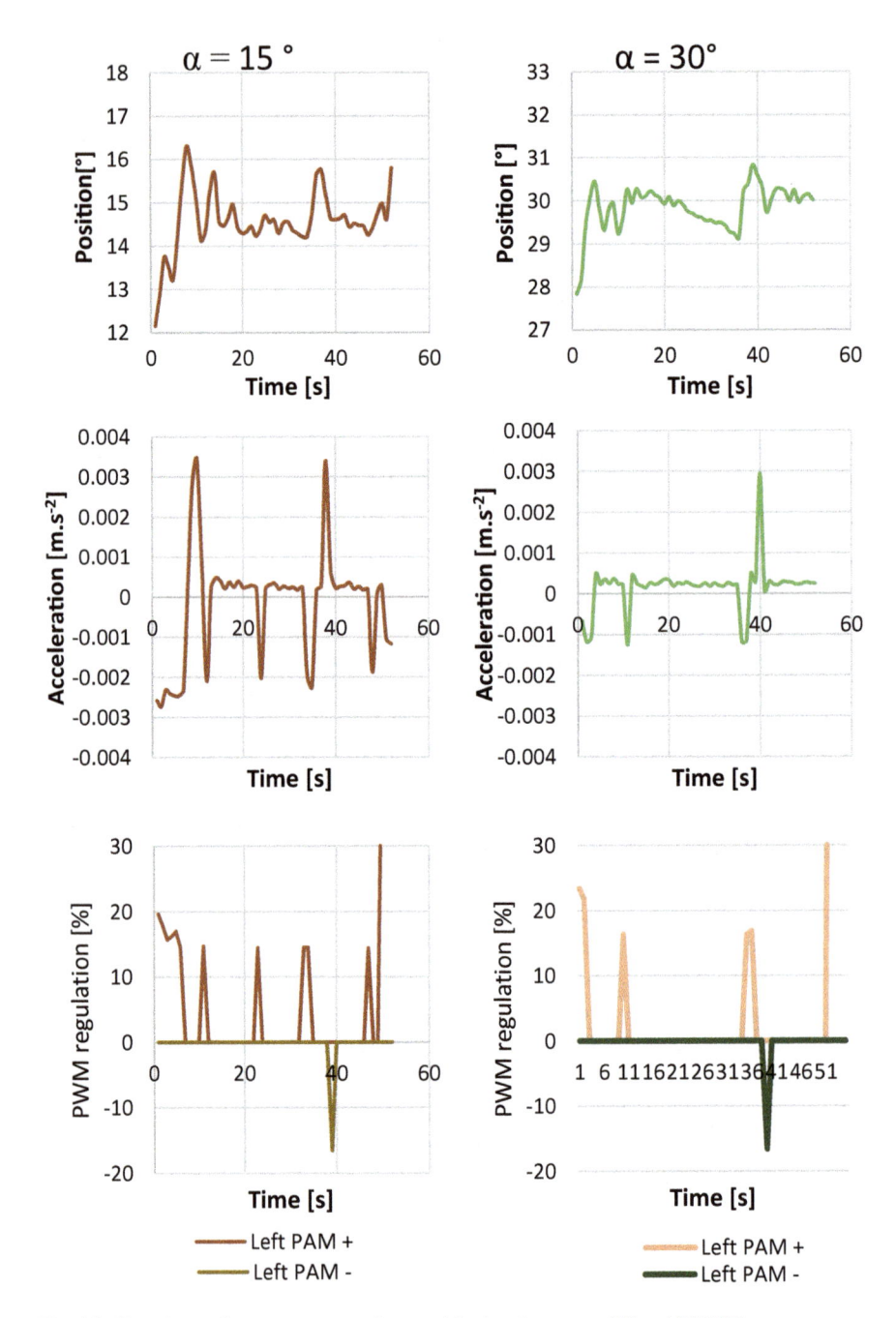

Fig. 4.5 Experimental measurements when positioning the arm to 15° and 30 ° [7]

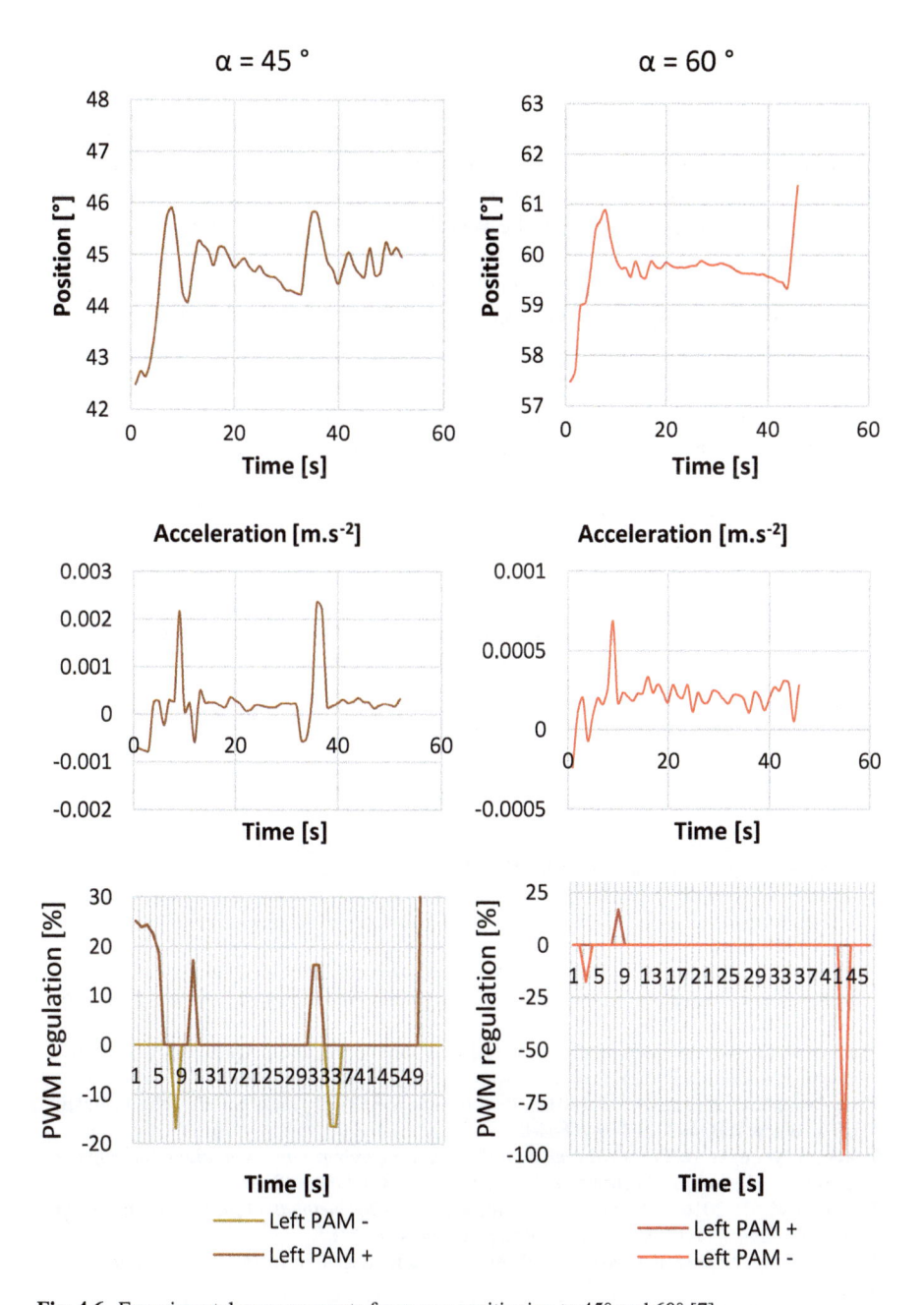

Fig. 4.6 Experimental measurements from arm positioning to 45° and 60° [7]

Conclusion

At the Department of Process Technique, Faculty of Manufacturing Technologies, with a seat in Prešov, an experimental device with propulsion consisting of two antagonistic pneumatic artificial muscles and relevant equipment was designed and assembled. This chapter describes the waveforms of selected static and dynamic variables (rotation of the carrier arm and its acceleration) during operation of the experimental actuator using simple proportional controller. The article describes the data obtained from the experimental measurement performed to describe the behavior of the mechanism during its operation. The described facts allow further determination of the limit values of the individual operating variables in which the experimental actuator does not exceed the selected static or dynamic quantities which are to be observed for the operating conditions of the particular application of antagonistic assembly of pneumatic artificial muscles. Once the device is assessed as stable as possible, mathematical model applications can be used in the control system of the experimental station.

Acknowledgments This chapter is supported by the project VEGA 1/0338/15 "Research of effective combinations of energy sources on the basis of renewable energies" and also by the Project of the Structural Funds of the EU, ITMS code: 26220220103.

References

1. Tondu, B. (2012). Modelling of the Mckibben artificial muscle. A review. *Journal of Intelligent Material Systems and Structures*. 255–268, USA: SAGE. ISSN: 1530-8138.
2. Dearden, F., Lefeber, D. (2002) Pneumatic artificial muscles: Actuators for robotics and automation. *European Journal of Mechanical and Environmental Engineering*. Belgium, 11–21. ISSN 1371–6980.
3. Klute, G. K., & Hannaford, B. (2000). Accounting for elastic energy storage in McKibben artificial muscle actuators. In *Journal of dynamic systems, measurement and control* (pp. 386–388). Washington: ASME.
4. Kuna, Š. (2015). Research methods for real time monitoring and diagnostics of production machines. Dizertačná práca. Prešov. TUKE.
5. Laplante, A. P., & Ovaska, J. S. (2012). *Real-time systems design and analysis* (4th ed., 560 p). USA: Wiley. ISBN 978-0-470-76864-8
6. Kuo, S. M., et al. (2006). *Real-time digital signal processing: implementations and applications* (2nd ed., 646 p). Chichester: Wiley. ISBN-13 978-0-470-01495-0.
7. Šmeringai, P. (2016). *The research of online monitoring methods of manufacturing equipment's with artificial muscles*. Dizertačná práca. Prešov, TUKE.
8. Kopečný, L., & Šolc, F. (2003). Mckibben pneumatic muscle in robotics. *AT&P Journal, 2*, 62–64.
9. Šmeringai, P., Rimár, M., Fedák, M., Kuna, Š. (2016). Real time pressure control in pneumatic actuators. in: key engineering materials: operation and diagnostics of machines and production systems operational states 3. (Vol. 669, pp. 335–344). ISBN 978-3-03835-629-5, ISSN 1662-9795.

10. Piteľ, J. a kol. (2015). *Pneumatické umelé svaly: modelovanie, simulácia, riadenie*. Košice: Technická univerzita v Košiciach. 275 s. ISBN 978-80-553-2164-6.
11. Rimár, M., Šmeringai, P., Fedák, M., Hatala, M., & Kulikov, A. (2017). Analysis of step responses in nonlinear dynamic systems consisting of antagonistic involvement of pneumatic artificial muscles, 2017. *Advances in Materials Science and Engineering, 2017*, 1–14. ISSN 1687-8434.
12. Šmeringaiová, A., Vojtko, I., & Monková, K. (2015). Experimentelle Analyse der Dynamik von Zahnradgetrieben – Teil 1. *TM-Technisches Messen, 82*(2), 57–64. ISSN 0171-8096.
13. Šmeringaiová, A., Vojtko, I., & Monková, K. (2015). Experimentelle Analyse der Dynamik von Zahnradgetrieben – Teil 2. *TM-Technisches Messen, 82*(4), 224–232. ISSN 0171-8096.

Chapter 5
Critical Values of Some Probability Distributions and Standard Numerical Methods

Dušan Knežo, Jozef Zajac, and Peter Michalik

Abstract Statistic processing of experimental data, part of which is use of interval estimation, statistic hypothesis tests, and other statistical methods, requires critical values of some probability distributions. To obtain critical values, critical values tables are usually used, which are part of literature focused on mentioned statistical methods. Significant disadvantage of mentioned approach is the limited number of discrete values. If it is required to process large amount of experimental data using own programs, then it is appropriate to possess suitable critical values calculation methods. Presented article is focused on critical values calculation methods for standardized normal distribution, Student's t-distribution, Fisher-Snedecor F-distribution, and χ^2-distribution using standard numerical methods, specifically approximate calculation of definite integrals and approximate solving of nonlinear equations.

D. Knežo (✉)
Technical University of Košice, Faculty of Manufacturing Technologies with a Seat in Prešov, Department of Mathematics, Informatics and Cybernetics, Prešov, Slovak Republic
e-mail: dusan.knezo@tuke.sk

J. Zajac
Technical University of Košice, Faculty of Manufacturing Technologies with a Seat in Prešov, Department of Computer Aided Manufacturing Technologies, Prešov, Slovak Republic
e-mail: jozef.zajac@tuke.sk

P. Michalik
Technical University of Košice, Faculty of Manufacturing Technologies with a Seat in Prešov, Department of Manufacturing Technologies, Prešov, Slovak Republic
e-mail: peter.michalik@tuke.sk

© Springer International Publishing AG, part of Springer Nature 2019
D. Cagáňová et al. (eds.), *Smart Technology Trends in Industrial and Business Management*, EAI/Springer Innovations in Communication and Computing, https://doi.org/10.1007/978-3-319-76998-1_5

Introduction

Statistical processing of experimental data is often performed using statistical methods, which require knowledge of some probability distribution critical values, especially standard normal distribution, Student's t-distribution, Fisher-Snedecor F-distribution, and χ^2-distribution. Let us assume that random variable X is subject to probability distribution with distribution function $F(x)$. Then critical value x_α x_α of this distribution on significance level α is such value of x_α, for which it is valid:

$$P(X > x_\alpha) = 1 - F(x_\alpha) = \alpha. \tag{5.1}$$

Determining critical values is usually made by using tables or specialized statistical software. This approach also has its disadvantages, but in some cases it is advantageous to have a method by which it is possible to calculate evaluated critical values. Calculation of critical values is also possible using standard numerical methods, specifically method of approximate calculation of definite integrals and methods of approximate calculation of nonlinear equations. Analogical uses of methods are presented in article according to [1–3].

Study [2] prescribes methods for determination critical values of standard normal and Student's t-distribution.

Equation (5.1) is used for determination of critical value x_α of standard normal distribution $N(0,1)$ in the level of significant α and implies

$$1 - \phi(x_\alpha) = \alpha, \tag{5.2}$$

where $\phi(x)$ is distribution function of standard normal distribution $N(0,1)$ and is defined by equation

$$\phi(x) = \int_{-\infty}^{x} \varphi(t)\,dt = \frac{1}{\sqrt{2\pi}} \int_{-\infty}^{x} e^{-\frac{t^2}{2}}\,dt. \tag{5.3}$$

From the properties of the distribution function, $\varphi(x)$ implies that

$$\phi(x) = \begin{cases} 0.5 - \dfrac{1}{\sqrt{2\pi}} \int_{0}^{|x|} e^{-\frac{t^2}{2}}\,dt, & x < 0, \\ 0.5, & x = 0, \\ 0.5 + \dfrac{1}{\sqrt{2\pi}} \int_{0}^{x} e^{-\frac{t^2}{2}}\,dt, & x > 0. \end{cases} \tag{5.4}$$

From Eq. (5.4) it can be stated that is sufficient to limit consideration of the integral [4].

5 Critical Values of Some Probability Distributions and Standard Numerical Methods

$$\int_0^x \varphi(t)\,dt = \frac{1}{\sqrt{2\pi}} \int_0^x e^{-\frac{t^2}{2}} dt \tag{5.5}$$

valid for $x > 0$. Calculating integral (5.5) can be used with Simpson's rule and can be deduced.

$$\left|\varphi^{(4)}(t)\right| \le \varphi^{(4)}(0) = \frac{3}{\sqrt{2\pi}}. \tag{5.6}$$

For integration step h of approximate calculation of integral (5.5), using Simpson's rule with accuracy ε implies [5]

$$h \le \sqrt[4]{\frac{60 \cdot \varepsilon \cdot \sqrt{2\pi}}{x}}. \tag{5.7}$$

Based on previous equations, it is possible with accuracy ε to calculate values of the distribution function; thus it is possible to set critical value x_α with required accuracy to evaluate equation

$$\phi(x_\alpha) + \alpha - 1 = 0. \tag{5.8}$$

Because function

$$G(x) = \phi(x) + \alpha - 1 \tag{5.9}$$

is increasing continuous function,

$$\lim_{x \to -\infty} G(x) = \alpha - 1 < 0 \tag{5.10}$$

and

$$\lim_{x \to \infty} G(x) = \alpha > 0, \tag{5.11}$$

then Eq. (5.8) has exactly one solution. Approximate solution of Eq. (5.8) can be done using bisection method.

Probability density according to Student's t-distribution with n degrees of freedom is evident; that distribution function can be defined as follows [6]:

$$F_n(x) = \begin{cases} 0.5 - c_n \cdot \int_0^{|x|} \left(1 + \frac{t^2}{n}\right)^{-\frac{n+1}{2}} dt, & x < 0, \\ 0.5, & x = 0, \\ 0.5 + c_n \cdot \int_0^{x} \left(1 + \frac{t^2}{n}\right)^{-\frac{n+1}{2}} dt, & x > 0. \end{cases} \quad (5.12)$$

where

$$c_n = \frac{\Gamma\left(\frac{n+1}{2}\right)}{\sqrt{n\pi} \cdot \Gamma\left(\frac{n}{2}\right)}. \quad (5.13)$$

For approximate integral calculation

$$\int_0^x g_n(t) dt = \int_0^x \left(1 + \frac{t^2}{n}\right)^{-\frac{n+1}{2}} dt \quad (5.14)$$

for $x > 0$ can be used as trapezoidal rule. It can be demonstrated by

$$\left| g_n''(t) \right| \leq g_n''(0) = \frac{n+1}{n}. \quad (5.15)$$

For the step of approximate integral calculation (5.14), using trapezoidal rule with accuracy ε is valid [7]:

$$h \leq \sqrt{\frac{12n\varepsilon}{(n+1) \cdot x}}. \quad (5.16)$$

The aim for use of determination of critical value x_α by Student's t-distribution is to fnd the equation root

$$1 - F_n(x) = \alpha. \quad (5.17)$$

Approximate solution of Eq. (5.17) can be realized using bisection method.

Critical Values of χ^2-Distribution

Probability density χ^2-distribution with n degrees of freedom is defined by the following equation:

$$f_n(x) = \begin{cases} \dfrac{1}{2^{\frac{n}{2}} \cdot \Gamma\left(\dfrac{n}{2}\right)} \cdot x^{\frac{n}{2}-1} \cdot e^{-\frac{x}{2}}, & x > 0, \\ 0, & x \leq 0. \end{cases} \qquad (5.18)$$

This is because $n \in N$ and $n \geq 2$ for calculation values of function Γ can be used in equation

$$\Gamma(k) = (k-1)! \qquad (5.19)$$

or

$$\Gamma\left(k + \frac{1}{2}\right) = \frac{1 \cdot 3 \cdot 5 \cdots (2k-1)}{2^k} \cdot \sqrt{\pi}, \qquad (5.20)$$

where $k \in N$.

Equation (5.1) for determination of critical value x_α of χ^2-distribution (Fig. 5.1) with n degrees of freedom on confidence level α is apparent:

$$1 - F_n(x) = \alpha, \qquad (5.21)$$

where $F_n(x)$ is distribution function defined by equation

$$F_n(x) = \int_0^x f_n(t)\,dt. \qquad (5.22)$$

Derivation properties of function $f_n(x)$ show that in general it is not possible to define the step of approximate integral calculation (5.22) to obtain a required accuracy. In that case which is approximate integral calculation (5.22), it would be

Fig. 5.1 Critical values of χ^2-distribution

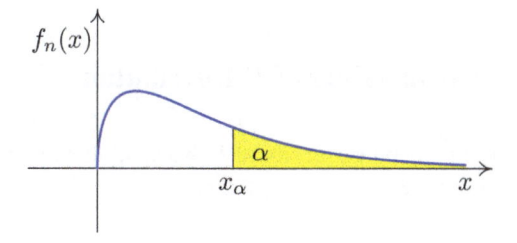

Table 5.1 Selected critical values of χ^2-distribution

N	$\alpha = 0.1$	$\alpha = 0.05$	$\alpha = 0.025$	$\alpha = 0.01$	$\alpha = 0.001$
3	6.251	7.815	9.348	11.345	16.267
4	7.779	9.488	11.143	13.277	18.467
5	9.236	11.070	12.832	15.086	20.515
6	10.645	12.592	14.449	16.812	22.458
7	12.017	14.067	16.013	18.475	24.322
8	13.362	15.507	17.535	20.090	26.125
9	14.684	16.919	19.023	21.666	27.877
10	15.987	18.307	20.483	23.209	29.588
11	17.275	19.675	21.920	24.725	31.264
12	18.549	21.026	23.337	26.217	32.910
13	19.812	22.362	24.736	27.688	34.528
14	21.064	23.685	26.119	29.141	36.123
15	22.307	24.996	27.488	30.578	37.697
16	23.542	26.296	28.845	32.000	39.252
17	24.769	27.587	30.191	33.409	40.791
18	25.989	28.869	31.526	34.805	42.312
19	27.204	30.144	32.852	36.191	43.820
20	28.412	31.410	34.170	37.566	45.315
21	29.615	32.671	35.479	38.932	46.797
22	30.813	33.924	36.781	40.289	48.268
23	32.007	35.172	38.076	41.638	49.729

appropriate to use Richardson extrapolation and Simpson's method. Calculation uncertainty/error E is estimated by equation

$$|E| \cong \frac{\left|I_h - I_{h/2}\right|}{15}, \qquad (5.23)$$

where I_s is approximate value of the integral calculated by Simpson's method using step s.

Equation (5.21) has exactly one solution, and its approximate values can be found using bisection method. Table 5.1 shows some critical values for χ^2-distribution with n degrees of freedom calculated using the described evaluation method.

Critical Values of F-Distribution

The probability density of F-distribution with m, n degrees of freedom is in interval $[0, \infty)$ prescribed by equation

5 Critical Values of Some Probability Distributions and Standard Numerical Methods

Fig. 5.2 Critical values of F-distribution

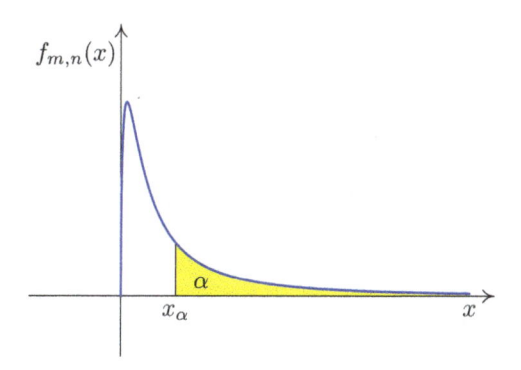

$$f_{m,n}(x) = c_{m,n} \cdot \frac{x^{\frac{m}{2}-1}}{(mx+n)^{\frac{m+n}{2}}}, \tag{5.24}$$

where

$$c_{m,n} = \frac{\Gamma\left(\frac{m+n}{2}\right) \cdot m^{\frac{m}{2}} \cdot n^{\frac{n}{2}}}{\Gamma\left(\frac{m}{2}\right) \cdot \Gamma\left(\frac{n}{2}\right)}. \tag{5.25}$$

This is because $m, n \in N$ values of function Γ can be calculated using Eqs. (5.19) and (5.20).

Equation (5.1) for determination of critical value x_α, subsequently F-distribution with m, n degrees of freedom on significant level α, is apparent (Fig. 5.2):

$$1 - F_{m,n}(x_\alpha) = \alpha, \tag{5.26}$$

where $F_{m,n}(x)$ is distribution function and is defined by equation

$$F_{m,n}(x) = \int_0^x f_{m,n}(t)\,dt. \tag{5.27}$$

In analogy with the previous section, in general, it is not possible to define the step of approximate integral calculation (5.27) to obtain a specified accuracy. In this case it is also appropriate to use Richardson extrapolation method and Simpson's method [4, 8].

Equation (5.26) has exactly one solution, and approximate value can be possible to find using bisection method. Tables 5.2 and 5.3 list some of critical values of F-distribution with m, n degrees of freedom calculated using prescribed method.

Table 5.2 Selected critical values of F-distribution for $\alpha = 0.1$

	$m = 3$	$m = 4$	$m = 5$	$m = 6$	$m = 7$
$n = 3$	5.391	5.343	5.309	5.285	5.266
$n = 4$	4.191	4.107	4.051	4.010	3.979
$n = 5$	3.620	3.520	3.453	3.404	3.368
$n = 6$	3.289	3.181	3.107	3.055	3.014
$n = 7$	3.074	2.961	2.883	2.827	2.785

Table 5.3 Selected critical values of F-distribution for $\alpha = 0.05$

	$m = 3$	$m = 4$	$m = 5$	$m = 6$	$m = 7$
$n = 3$	9.278	9.117	9.013	8.941	8.887
$n = 4$	6.592	6.388	6.256	6.163	6.094
$n = 5$	5.410	5.192	5.050	4.950	4.876
$n = 6$	4.757	4.534	4.387	4.284	4.207
$n = 7$	4.347	4.120	3.971	3.866	3.787

Conclusions

Described methods present possibilities of calculation critical values for the most common probability distributions used in practice. Advantage of these methods is that they can be used as part of automatic data processing, e.g., at interval estimates or statistical testing. They are advantageous in that they allow the calculation of the critical values for any significance level. Methods can also be used to create tables included in books and textbooks in the field of statistics. Program for each of described methods was designed in a language Free Pascal-Lazarus IDE. Algorithms of these programs are of course usable in other programming languages. Application of described methods is limited by technical possibilities of the computer, because at some values, the computer works at its boundary limits. Described methods are based on standard numerical methods of calculation definite integrals. Other methods such as Monte Carlo and others can be used as well.

Acknowledgment This work is a part of the research project VEGA 1/0619/15. The article was prepared within the invitation OPVaV-2012/2.2/08-RO named "University Science Park TECHNICOM for Innovation Applications with the Support of Knowledge-Based Technologies" code ITMS 26220220182 and the project VEGA 1/0594/12.

References

1. Knežo, D. (2014). Inverse transformation method for normal distribution and the standard numerical method. *International Journal of Interdisciplinarity in Theory and Practice, 5*, 6–10. ISSN 2344-2409.
2. Knežo, D. (2014). Application of the numerical methods for the teaching of the statistics. *International Journal of Interdisciplinarity in Theory and Practice, 5*, 58–61. ISSN 2344-2409.
3. Knežo, D. (2012). O metóde Monte Carlo a možnostiach jej aplikácií. *Transfer inovácií, 24*, 178–181. ISSN 1337-7094.
4. Bonnar, J. (2010). *The gamma function.* Seattle: CreateSpace Publishing. ISBN:978-1463694296.
5. Walck, C.H.. (1996). *Hand-book on statistical distributions for experimentalist/2007*, Internal report SUF-PFY/96–01 University of Stockholm. Last modification 2007. [Online] [17.3.2017] http://www.fysik.su.se/~walck/
6. McKinsey & Company [Online] [24.4.2017]. http://www.mckinsey.com/business-functions/operations/our-insights/manufacturings-next-act
7. Kokoska, S., & Nevison, C. H. (2012). *Statistical tables and formulae.* New York. 1989: Springer. ISBN:978-0-387-96873-5.
8. Neave, H. R. (2002). *Elementary statistics tables.* Routledge: Taylor and Francis Group. ISBN:978-0415084581.

Chapter 6
Measuring Production Process Complexity

Soltysova Zuzana, Bednar Slavomir, and Behunova Annamaria

Abstract This paper identifies and quantifies static complexity indices of two layout types: job shop and cellular layout. Subsequently, both layout types are compared in terms of static complexity by three static complexity metrics, and conclusions are made toward the research question stated. As it was demonstrated on the two production layout models, lower complexity in the case of the flow-shop production system has been confirmed.

Introduction

Mass customization (MC) has become an increasingly important product strategy bringing more freedom in choosing the "right" product with efficiency that almost equals that of mass production. Such customization has been around for a long time with number of popular brands, but the very first efforts have never really caught on. In recent decades, market has changed rapidly as well as minds of the customers. They want to differ from same "mass" produced product and miss an added value coming out of such individualized products. Customers may find themselves making decisions either on buying standard product of lesser preference fit or buying an MC product providing a better preference fit at a higher price they are actually willing to pay [1, 2].

A number of authors so far have performed measurements of complexity within manufacturing systems with the use of various methods. Their most frequent finding was that, in the case of job-shop layout, complexity of the production reaches higher values than in the case of flow-shop production layout [1–5]. Complexity is often considered as a negative aspect of all types of activities, as it generally negatively affects efficiency of a production system [6–8].

S. Zuzana (✉) · B. Slavomir · B. Annamaria
Technical University of Kosice, Faculty of Manufacturing Technologies, Kosice, Slovakia
e-mail: zuzana.soltysova@tuke.sk; slavomir.bednar@tuke.sk; annamaria.behunova@tuke.sk

© Springer International Publishing AG, part of Springer Nature 2019
D. Cagáňová et al. (eds.), *Smart Technology Trends in Industrial and Business Management*, EAI/Springer Innovations in Communication and Computing,
https://doi.org/10.1007/978-3-319-76998-1_6

Each production system can be described operational, dynamic complexity [9–12]. Operational complexity is linked with measurement of dynamical aspects of a production system during its operation. Therefore, this research will adopt principles of static (structural) complexity measurement as it appears as a result of various production system's layouts.

We will consider two individual layout cases, with technologically oriented production system (job shop), and with batch production system (flow shop). The aim of this research is to assess which production system layout has higher structural complexity. With respect to this research aim, the following research question (RQ) must be answered: Will the assumption of lower complexity of the flow-shop layout be confirmed also using selected metrics?

As it was stated in the RQ, we assume with lower complexity of the flow shop and with higher structural complexity of the job-shop machine layout, as it was anticipated in the works of other authors [13, 14] using various types of structural-static complexity metrics.

The research will be performed via the assessment of structural complexity of production systems using six individual complexity coefficients, applying specific methodologies [2]. Each of the measures refers to characteristics of each layout and its properties. These characteristic include: number of links between work places, paths, cycles, decision points, and redundant paths. All the mentioned characteristics influence the complexity of production systems. Next, these two layout types will be compared by another two static complexity metrics, where one of them is based on Shannon's information theory.

Approaches to Assessment of Production System Structural Complexity

At this stage, we aim to replace "more complex" production system design (layout) with a "less complex" system arrangement, while nodes with pre-defined inputs and outputs and material flow and its direction are retained. The initial node refers to the beginning of the production process and of the material flow, while the end node is the end of both production process and material flow. Each node within a diagram is a representation of a workstation, while each workstation has at least input and output flows (except for two – initial and end nodes).

The subsequent step was to create a matrix of relation (M). Such matrix shows important relationships between nodes based on the existence of a relation. If there is a link between nodes, the elementary matrix value is equal to "1," and if such a relationship does not exist in a matrix, then the value is equal to "0." Arrangement of the nodes in the matrix is the sequence of material flow from the start till the end node.

6 Measuring Production Process Complexity

An important step is the calculation of complexity indicators or of so-called complexity indices. The formulas of all complexity indices are as follows:

Density index (*D*)

$$D = \frac{k}{n(n-1)}, \tag{6.1}$$

where *k* is the number of links and *n* is the total number of nodes (representing workstations).

Path index (*P*)

$$P = 1 - \frac{p}{N}, \tag{6.2}$$

where *p* is the maximum theoretical number of paths and *N* is the number of all existing paths.

Cycle index (*C*)

$$C = \frac{c}{MC}, \tag{6.3}$$

where *c* is the number of valid cycles and MC is the maximum theoretical number of cycles.

$$MC = \sum_{i=2}^{n} C_{(n,i)}, \tag{6.4}$$

where *n* is the number of nodes and *i* equals 2.

Decision points index (DS)

$$DS = 1 - \frac{SP}{LP}, \tag{6.5}$$

where SP is the number of within the shortest possible path and LP is the number of nodes within the longest possible path.

Distribution redundancy index (RD)

$$RD = \frac{r}{a}, \tag{6.6}$$

where *r* is the number of redundancy states occurring between two adjacent nodes and *a* is the maximum theoretical number of redundancy state occurrence among all adjacent nodes.

Magnitude redundancy index (RM)

$$RM = \frac{pr}{w} = \frac{w-a}{w}, \quad (6.7)$$

where pr is the total number of redundant parallel arrows, w is the total number of assigned arrows, and a is the number of adjacencies.

Obtained values of complexity indices will be applied to determine the overall structural complexity of a manufacturing system using the following three approaches:

Average value (A)

$$A = \frac{1}{n}\sum_{i=1}^{n} a_i = \frac{a_1 + a_2 + a_3 + \ldots + a_n}{n}, \quad (6.8)$$

where n is the total number of complexity indices obtained, a_i is the value of individual complexity indices, and i equals $1,\ldots n$.

Aggregated complexity index (CI)

$$a = \frac{1}{2}\left[(C_t * C_1) + \sum_{i=1}^{i=t-1}(C_i * C_{i+1})\right]\sin\left(\frac{360}{t}\right), \quad (6.9)$$

where C_i is the value of individual indices, $i = 1,2,\ldots t$, and t is the total number of complexity indices obtained.

$$A = \frac{t}{2}\sin\frac{360}{t}, \quad (6.10)$$

where A is the "total plot area" and t is the total number of complexity indices obtained.

In order to obtain aggregated complexity index (CI), the following equation is used:

$$CI = \frac{a}{A}, \quad (6.11)$$

Vector method (V)

Vector size V can be determined using the Pythagorean theorem as follows:

$$V = \sqrt{(V_t)^2 + (V_u)^2 + (V_w)^2 + (V_x)^2 + (V_y)^2 + (V_z)^2}, \quad (6.12)$$

where V_{t-z} is, in this case, the value of individual complexity indicators.

The second part of this section is focused on the static complexity method based on Shannon's information theory proposed by Zhang [15]. He proposed to measure static complexity according probabilities of any machine j being in any state i. Accordingly, he used Shannon's information entropy calculation to express static complexity measurement as follows:

$$H_s = -\sum_{j=1}^{M}\sum_{i=1}^{s_j} p_{ij} \log_2 p_{ij}, \qquad (6.13)$$

where H_s is static complexity of manufacturing system, M is the number of machines, S_j represents the number of possible planned states the machine j can achieve plus one idle state, and p_{ij} is probability of any machine j being in state i.

The third metric, proposed by Deshmukh, who proposed to measure static complexity of manufacturing system by calculation of the maximum static complexity, is expressed as follows [16]:

$$H_s = \log m^2 nr. \qquad (6.14)$$

where m is the number of operations, n is the number of parts processed by production system, and r is the number of machines.

As an example, if we consider manufacturing system consisting of 20 machines, 100 operations, and 30 parts, then static complexity of manufacturing system using Eq. (6.14) equals 5.78 bits.

Subsequently, the following three metrics will be applied on two layout types to compare them and confirm the research question in the conclusion. The last step is generation of summary table to compare all complexity measures and their values against each other.

Application of the Methods for the Job-Shop System Layout

We will now proceed according to the methods presented above. First, we started to apply the first metric to measure static complexity. For this purpose, let us have theoretical model of job-shop production site with five machines in each group A–D representing four types of machines according to technology. Group A may represent lathes, group B as CNC machines, group C as drilling machines, and group D as grinding machines. Then, it is necessary to transform the initial scheme of the production site (Fig. 6.1) into a simplified diagram (Fig. 6.2).

After obtaining a simplified diagram of the job-shop production site arrangement, we get a model with 20 nodes (four groups with five nodes/machines) and with single input and output nodes. In this specific case, each node has five possible output paths to other technological cells (see Fig. 6.1). As this is technologically

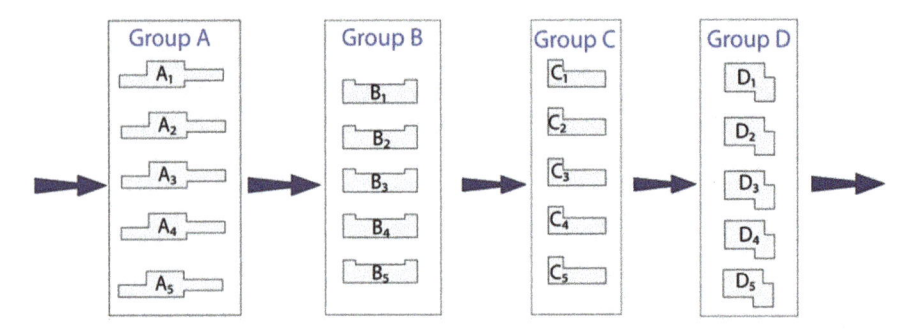

Fig. 6.1 Job-shop arrangement of the production facility

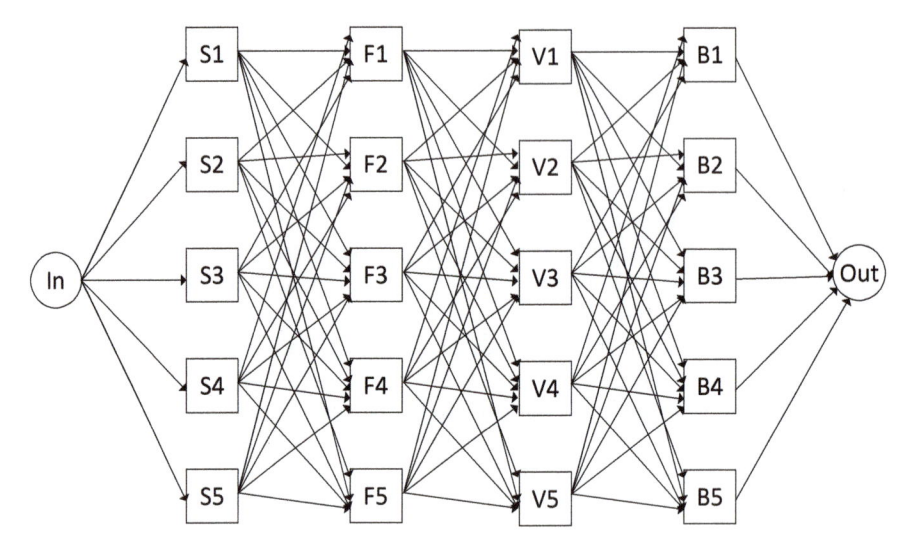

Fig. 6.2 A simplified diagram of the job-shop production layout (generated by a simulation tool TECNOMATIX)

based (job shop) production system, then the first five machines are very similar, or they are of the same type. These 5 machines have in total 25 possible output pathways to subsequent job-shop cell with different types of machines. From all 25 paths, only 5 paths are occupied/used, while the other 20 pathways are not used, and therefore, they are redundant. The first group of machines is denoted as A, and then the second group can be denoted by B, the third group by C, and the last group of similar machines by D (see Fig. 6.1).

6 Measuring Production Process Complexity

Generating the matrix of relations (*M*):

$$AM = \begin{array}{c} \\ \text{In} \\ \text{A} \\ \text{B} \\ \text{C} \\ \text{D} \\ \text{Out} \end{array} \begin{array}{c} \text{In} \quad \text{A} \quad \text{B} \quad \text{C} \quad \text{D} \quad \text{Out} \\ \begin{pmatrix} 0 & 5 & 0 & 0 & 0 & 0 \\ 0 & 0 & 25 & 0 & 0 & 0 \\ 0 & 0 & 0 & 25 & 0 & 0 \\ 0 & 0 & 0 & 0 & 25 & 0 \\ 0 & 0 & 0 & 0 & 0 & 5 \\ 0 & 0 & 0 & 0 & 0 & 0 \end{pmatrix} \end{array}$$

Calculating complexity indices:

Index D – we will take the total number of possible paths (75 for the whole job-shop layout except for 5 initial and 5 output connections) and the total number of nodes (20 for job-shop layout) into account. The number of possible paths among nodes is obtained as the sum of all arrows linking nodes (25 + 25 + 25 = 75). Substituting the value into Eq. (6.1), one would obtain 0.125.

Index P – considers the theoretical minimum number of paths, which is five. Each node results with five subsequent paths to group of five nodes. Each connection links single machine with five machines of subsequent group of machines, etc. The total number of possible existing paths is 15,625 = 25^3. Substituting the value into Eq. (6.2), one would obtain 0.9997.

Index C – reflects the actual number of cycles, which is zero (cycle is the path that starts and ends in the same point, and in this case, we are not talking about a cycle), and the maximum theoretical number of cycles is 33,554,406. Substituting the value into Eq. (6.3), one would obtain 0 (zero).

Index DS – takes the total number of nodes located on the shortest possible path (from the initial till the end node, while these nodes are not counted) into account. Then, the value for the shortest possible path is four, and the value for the longest possible path is also four, so the shortest and the longest paths are equal in this production layout. Substituting the value into Eq. (6.5), one would obtain 0 (zero).

Index RD – gives a ratio of path occurrence between two groups of nodes with redundant paths (three paths) and the theoretical number of redundant relations among the groups of nodes (again three paths). Substituting the value into Eq. (6.6), the result is one.

Index RM – takes the ratio of total number of redundant arrows (oriented connections), which is 60 among the nodes, and the total number of all oriented connections among nodes, which is 75. As there are 75 possible oriented connections among the nodes and 25 possible connections between each group of nodes, and five of them are fixed, then the other 20 are redundant (20 + 20 + 20 = 60). Substituting the value into Eq. (6.7), one would obtain 0.8 as a result.

Structural complexity and their values have been obtained by the complexity indices using the following three approaches:

Average value (A) or the average value of complexity can be calculated from the Eq. (6.8), with the result of 0.488.

Radar *chart (a)* value can be obtained by substitution of complexity indices into Eq. (6.9), and we get the result of 1.69. To get *aggregated complexity index* (CI), we must determine the relationship between *a* "radar area" and *A* "total plot area" from Eq. (6.10). Then, the calculated value is 2.504. Then, the relation CI from Eq. (6.11) results in a value of 0.68.

Determination of the complexity by the "vector method" by substitution into Eq. (6.12) returns the complexity as 1.63.

The resulting values are shown in the Table 6.1:

The next metric to measure static complexity of job-shop system layout by Zhang's metric is calculated as follows using Eq. (6.13):

$$H_s = -\sum_{i=1}^{20}\sum_{j=1}^{101} \frac{1}{101} \log_2 \frac{1}{101} = 133 \text{ bits}$$

Because the job shop is organized completely flexibly, the probability of any machine being in any state plus one idle state, the probability of manufacturing system p_{ij}, is 1/101, where the number of possible states on each machine is 100 plus 1 idle state (together 101 scheduled states).

The third metric proposed by Deshmukh, the static complexity, using Eq. (6.14) is calculated as follows:

$$H_s = \log(100^2 * 20 * 100) = 7.3 \text{ bits}$$

where the number of parts processed by job-shop manufacturing system equals to 100, the number of operations is 100 too, and the number of machines is 20.

Table 6.1 Summary of the complexity values per each index

Structural complexity indices	
Density index (D)	0.125
Path index (P)	0.9997
Cycle index (C)	0
Decision points index (DS)	0
Distribution redundancy index (RD)	1
Magnitude redundancy index (RM)	0.8
Average value (A)	0.488
Aggregated complexity index (CI)	0.68
Vector method (V)	1.63

Application of the Methods for the Flow-Shop System Layout

In order to proceed toward the application of the methods to measure complexity, the original five cell-based (flow shop) production layout in Fig. 6.3 had to be transformed into its simplified form, as seen in Fig. 6.4.

Considering that the scheme in Fig. 6.3 above is a flow system arrangement, production machines within each of the five cells are organized based on the sequence of technological operations. The system is divided into five individual parallel cells. Each of the cells contains four different machines arranged based on the sequence of operations needed to produce the final product. Therefore, each cell

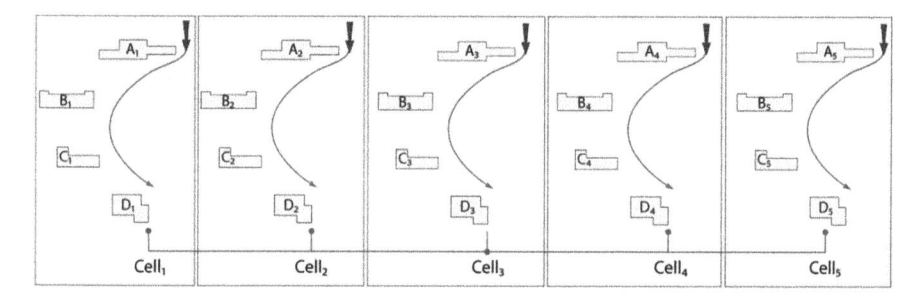

Fig. 6.3 Original scheme of the five cell (flow shop) production system arrangement

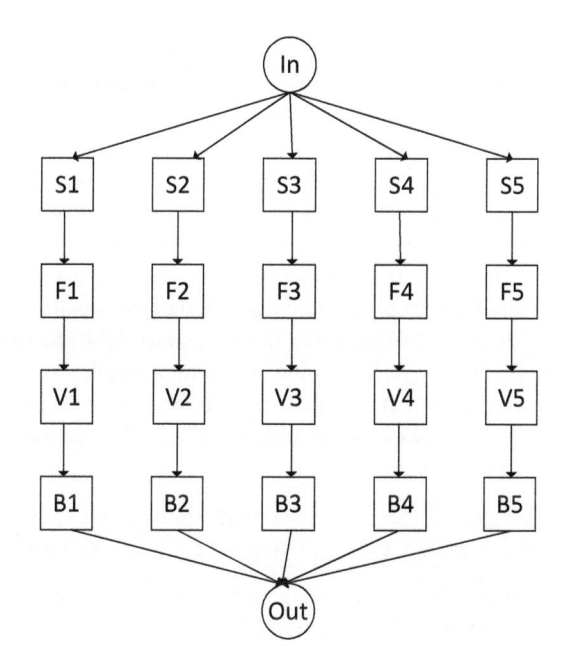

Fig. 6.4 A simplified diagram of the flow-shop production system using the simulation tool TECNOMATIX

contains four nodes, which is 25 nodes together for the whole workplace. In this special case, within each cell, there is only one path that the product can travel.

To create a matrix of relations, each individual cell can be divided into four nodes (A, B, C, D) with one input and one output. Then, such matrix of relations is the same for each of the five cells.

Matrix of relations (M) for the flow-shop layout:

$$AM = \begin{matrix} & \text{In} & A & B & C & D & \text{Out} \\ \text{In} & 0 & 1 & 0 & 0 & 0 & 0 \\ A & 0 & 0 & 1 & 0 & 0 & 0 \\ B & 0 & 0 & 0 & 1 & 0 & 0 \\ C & 0 & 0 & 0 & 0 & 1 & 0 \\ D & 0 & 0 & 0 & 0 & 0 & 1 \\ \text{Out} & 0 & 0 & 0 & 0 & 0 & 0 \end{matrix}$$

Calculating complexity indices:

Index D – takes the total number of possible paths/connections into account. In the case of the flow-shop system, we have 15 possible paths out of 25 nodes. Substituting the value into Eq. (6.1), one would obtain 0.025.

Index P – takes into account the minimum theoretical number of paths (five paths) and the total number of possible existing paths is also five; substituting the value into Eq. (6.2), one would obtain 0 (zero).

Index C – considers the actual number of cycles, which equals to zero, and the maximum calculated theoretical number of cycles, which equals 33,554,406; substituting the value into Eq. (6.3), one would obtain 0 (zero).

Index DS – takes into account the total number of nodes located on the shortest possible path from the entry till the output node, which in this equals four, and the number of nodes located on the longest possible path, which also equals four; substituting the value into Eq. (6.5), one would obtain 0 (zero).

Index RD – takes the ratio of the number of path occurrence between any two nodes, where also redundant paths are present (zero in this case), and the total theoretical number of locations with redundant connections among nodes (equals 15 in this case); substituting the value into Eq. (6.6), one would obtain 0 (zero).

Index RM – takes the ratio of the total number of redundant arrows (oriented connections), which equals zero in this case, and the total number of all connections among nodes, which equals 15; substituting the value into Eq. (6.7), one would obtain 0 (zero).

Again, in this case, structural complexity and their values have been obtained by the complexity indices, using the following three approaches:

Average value (A) or the average value of complexity can be calculated from the Eq. (6.8), with the result of 0.00417.

Radar chart (a) value can be obtained by substitution of complexity indices into the Eq. (6.9), and we get the result zero (0).

To get an *aggregated complexity index (CI)*, we must again determine the relationship between *a* "radar area" and *A* "total plot area" from the Eq. (6.10). The calculated value is 2.504. Then, the relation CI from Eq. (6.11) results in a value of zero (0).

Determination of the complexity by the "vector method" by substitution into Eq. (6.12) returns the complexity as 0.025.

The resulting values are shown in the following Table 6.2:

The next metric to measure static complexity of flow-shop system layout by Zhang's metric is calculated as follows using Eq. (6.13):

$$H_s = -\sum_{i=1}^{5}\sum_{j=1}^{21} \frac{1}{21}\log_2 \frac{1}{21} = 22 \text{ bits}$$

Cellular manufacturing is actualized in order to improve the efficiency of production in the job shop. All machines are composed of five production cells, and every cell contains four machines. So, the products are divided into 5 families of 20 products each. So, for each machine, probability p_{ij} equals to 1/21, where each machine has 20 production states and 1 idle state (together 21 scheduled states). So, we set the number of machines to 5.

The third metric proposed by Deshmukh, the static complexity, using Eq. (6.14) is calculated as follows:

$$H_s = \log(100^2 * 20 * 100) = 7.3 \text{ bits}$$

where the number of parts processed by flow-shop manufacturing system equals to 100, the number of operations is 100 too, and the number of machines is 20.

Table 6.2 Summary of the complexity values per each index of the flow-shop system arrangement

Structural complexity indices	
Density index (*D*)	0.025
Path index (*P*)	0
Cycle index (*C*)	0
Decision points index (DS)	0
Distribution redundancy index (RD)	0
Magnitude redundancy index (RM)	0
Average value (*A*)	0.00417
Aggregated complexity index (CI)	0
Vector method (*V*)	0.025

Results and Discussion

Mutual comparison of the structural complexity values as metric No. 1, metric proposed by Zhang as metric No.2, and metric proposed by Deshmukh as metric No. 3 for both job-shop and flow-shop production arrangement can be seen in the following Table 6.3:

The diversity of the approach used against other approaches lies in the fact that different structural characteristics of the production system have been considered, such as the number of paths and cycles and redundancy of mutual paths. The higher the frequency of, e.g., paths, the higher the value of structural complexity.

By comparison of the results, lower complexity in the case of the flow-shop production system arrangement has been demonstrated. The research question stated in the introduction of this paper can therefore be answered in the affirmative: Even using this metric, lower structural complexity of the flow-shop system arrangement has been confirmed. The lower complexity in the case of flow-shop production system was also confirmed by metric No.2, but the same result by metric No.3 was achieved in both cases, because both layout types have the same number of machines, parts, and operations.

Conclusion

The assessment of the structural complexity is only a partial view on the complexity of the production process [15–17]. For this reason, it is not scientifically correct to state that the general complexity of the flow production systems is lower. In order to be clear about such statement, it is needful to examine the dynamic complexity aspect of the production system, for which a methodology is not yet united.

Table 6.3 Mutual comparison of structural complexity indices

Metrics	Indices	Job-shop arrangement	Flow-shop arrangement
No. 1	5.1.1.1.1.1.1.1.1. D	0.125	0.025
	5.1.1.1.1.1.1.1.2. P	0.9997	0
	5.1.1.1.1.1.1.1.3. C	0	0
	DS	0	0
	RD	1	0
	RM	0.8	0
	5.1.1.1.1.1.1.1.4. A	0.488	0.00417
	CI	0.68	0
	5.1.1.1.1.1.1.1.5. V	1.63	0.025
No. 2		131 bits	22 bits
No. 3		7.3 bits	7.3 bits

Acknowledgment This paper has been partially supported by VEGA project no. V-16-013-00 granted by the Ministry of Education of the Slovak Republic and is part of actual research activities in the project SME 4.0 (Industry 4.0 for SMEs) with funding received from the European Union's Horizon 2020 research and innovation program under the Marie Skłodowska-Curie grant agreement no. 734713.

References

1. Dima, I. C. (2010). Location and importance of logistics in the company's organisational structure. *Polish Journal of Management Studies, 1*, 36–43.
2. Modrak, V., Marton, D., & Bednar, S. (2015). The influence of mass customization strategy on configuration complexity of assembly systems. *Procedia CIRP, 33*, 539–544.
3. Modrak, V., Bednar, S., & Marton, D. (2015). Generating product variations in terms of mass customization. In: *SAMI 2015 – IEEE 13th international symposium on applied machine intelligence and informatics, proceedings*, pp. 187–192 Herlany.
4. Modrak, V., Marton, D., & Bednar, S. (2014). The impact of customized variety on configuration complexity of assembly process. *Applied Mechanics and Materials, 474*, 135–140.
5. Matt, D. T. (2012). Application of axiomatic design principles to control complexity dynamics in a mixed-model assembly system: A case analysis. *International Journal of Production Research, 50*(7), 1850–1861.
6. Espinoza Vega, V. B. (2012). *Structural complexity of manufacturing systems layout*. Electronic Theses and Dissertations, 2012. Paper 5351. http://scholar.uwindsor.ca/etd/5351
7. Jančík, M., Panda, A., & Behún, M. (2012). Production management. *Studiaimaterialy, 31*(1), 57–59. ISSN 0860-7761.
8. Modrak, V., Bednar, S., & Semanco, P. (2016). Decision-making approach to selecting optimal platform of service variants. *Mathematical Problems in Engineering*. https://doi.org/10.1155/2016/9840679. Frizelle, G., & Suhov, Y. M. (2001). An entropic measurement of queueing behavior in a class of manufacturing operations. *Proceedings of Royal Society Series, A 457*, 1579–1601.
9. Frizelle, G., & Woodcock, E. (1995). Measuring complexity as an aid to developing operational complexity. *International Journal of Operations and Production Management, 15*(5), 26–39.
10. Guimaraes, T., Martensson, N., Stahre, J., & Igbaria, M. (1999). Empirically testing the impact of manufacturing system complexity on performance. *International Journal of Operations & Production Management, 19*(12), 1254–1269.
11. Kováč, J., & Rudy, V. (2014). Innovation production structures of small engineering production. *Procedia Engineering, 96*, 252–256.
12. Suzic, N., Stevanov, B., Cosic, I., Anisic, Z., & Sremcev, N. (2012). Customizing products through application of group technology: A case study of furniture manufacturing. *Strojniskivestnik/Journal Of Mechanical Engineering, 58*(12), 724–731.
13. Ostertagová, E., Kováč, J., Ostertag, O., & Malega, P. (2012). Application of morphological analysis in the design of production systems. *Procedia Engineering, 48*, 507–512.
14. Papakostas, N., Efthymiou, K., Mourtzis, D., & Chryssolouris, G. (2009). Modelling the complexity of manufacturing systems using nonlinear dynamics approaches. *CIRP Annals-Manufacturing Technology, 58*(1), 437–440.
15. Zhang, Z. (2011). Modeling complexity of cellular manufacturing systems. *Applied Mathematical Modelling, 35*(9), 4189–4195. ISSN 0307-904X. https://doi.org/10.1016/j.apm.2011.02.044.
16. Deshmukh, A. -V. (1993). Complexity and chaos in manufacturing systems, Purdue University, PhD Thesis.
17. Modrak, V. (Ed.). (2016). *Mass customized manufacturing: Theoretical concepts and practical approaches*. Boca Raton: CRC Press.

Chapter 7
IoT Challenge: Older Test Machines Modernization in an Automotive Plant

Juraj Pančík and Vladimír Beneš

Abstract In the presented paper we describe engineering design and implementation of an information system based on an Internet of things (IoT) approach for remote collection, storing, and processing of information on the number of carried out mechanical cycles for a group of older existing fatigue test machines in an automotive components production. In the modernization of existing test machines infrastructure, we built the electronic counter modules with developed Ethernet interface. We conducted three ways to verify the accuracy of this manner of counting and collecting of mechanical cycles, and we reached results comparable with the manually collected data. Thanks to automatic detection of ends of fatigue tests, we reached more effective utilization of test machines (more than 25%).

Introduction

What is modernizing in our case? Simply stated, it is the addition of existing machines designed for fatigue tests of components in an automotive plant. With an IoT approach we developed an information system for collection, storage, and analysis of data. The goal was not just to get the data but also to transform data into information. The task seems simple, but transforming data into valuable information is no small task. As a starting point, we asked ourselves what is the value of the obtained data and derived information for procedural role before and after implementation of the information system. The most important thing is to maximize utilization of test machines. Fatigue tests have no clearly defined end time. So the knowledge of when to stop the test machines is key information for efficient test machine utilization. Another objective was to facilitate the automatic processing of

---

J. Pančík (✉) · V. Beneš
Bankovní institut vysoká škola, a.s., Prague, Department of Informatics and Quantitative Methods, Nárožní, Praha 5, Czech Republic
e-mail: jpancik@bivs.cz; vbenes@bivs.cz

© Springer International Publishing AG, part of Springer Nature 2019
D. Cagáňová et al. (eds.), *Smart Technology Trends in Industrial and Business Management*, EAI/Springer Innovations in Communication and Computing, https://doi.org/10.1007/978-3-319-76998-1_7

data in the VBA language (Visual Basic for Applications is a script language for Microsoft Excel spreadsheet in order to monitor and visualize data).

Existing Fatigue Test Machines

A fatigue test machine may be classified from different viewpoints such as purpose of the test, type of stressing, means of producing the load, operation characteristics, type of load, etc. [1]. Based on the purpose of the test, a fatigue test machine can be divided into the following [2]: (1) general purpose fatigue test machine, (2) special purpose fatigue test machine, and (3) equipment for testing parts and assemblies (this was our case).

Cycle Counting

Cycle counting is used to summarize (often lengthy) irregular load-versus-time histories by providing the number of times cycles of various sizes occur. The definition of a cycle varies with the method of cycle counting. These practices cover the procedures used to obtain cycle counts by various methods, including level-crossing counting, peak counting, simple-range counting, range-pair counting, and rainflow counting. Cycle counts can be made for time histories of force, stress, strain, torque, acceleration, deflection, or other loading parameters of interest [3].

Specifications of the Mechanical Cycle

According to [3] we applied simple-range counting method. For this method, a range is defined as the difference between two successive reversals, the range being positive when a valley is followed by a peak and negative when a peak is followed by a valley. Positive ranges, negative ranges, or both may be counted with this method. If only positive or only negative ranges are counted, then each is counted as one cycle. In this text we used the term mechanical cycle as synonymous with one cycle.

Electromechanical Digital Counter

Electromechanical counters are digital counters that serve as a part of test machines. They give the equivalent value of the number of mechanical cycles. In this case the electromechanical counter is synonymous with the totalizing counter. A totalizing counter displays a total of events, parts, products, strokes, revolutions, etc. Electronic and electromechanical counters accept a variety of inputs, from sensors, switches, encoders, and relays that increment the counter each time a pulse is received. In our case an original digital totalizing counter accepts inputs of 24 V pulses from a test machine at the end of each mechanical cycle.

IoT Modernization Challenge: Information System Design Objectives

Our goal was to create the information system for collecting, storing, processing, and visualization of data from a set of test machines with digital electromechanical totalizing counters. The design of this information system has to take into account the existing information system with electromechanical counter of mechanical cycles and with manual data collection. The new information system must enable remote data collection and must take advantage of the existing plant network infrastructure at the plant (local area network, LAN). The secondary objectives are the price of solutions using modern technology based on the IoT (Internet of Things), ease of implementation, scalability, web-based user interface, and use of modern tools for data processing and their analyzing. The manager's objective is to obtain an overview of the utilization of the test machines, reducing of downtime of test machines, and the developing of a visualization management dashboard based on web technologies.

System Requirements

- The need to maintain existing electromechanical totalizing counters.
- Increasing of the utilization efficiency of test machines by reducing downtime (note: tested parts will be suddenly destroyed and test engineers should be informed about it in the shortest time).
- Automated data collection will replace the manual collection of data from electromechanical counters which is in connection with manual reading and writing data into the paper sheets.
- As legacy there is the need of ensuring reading data from electromechanical counters (manual reading) and simultaneously from electronic modules (automatic reading).
- The utilization of existing network infrastructure so that the existing counter of mechanical cycles will be resolved with Internet protocol (IP)-based electronic counter module which will have LAN connectivity.
- The need to centralize data on accumulated mechanical cycles to the database.
- The need for data evaluation and visualization in one place using the application server.
- To take advantage of client-server architecture, where the client's IP counter module and server will consist of web server, application server. and database.

- To ensure long-term reliability and accuracy of mechanical cycles counting.
- To build application servers on the basis of free available open-source tools/components.
- Good price of a programming and hardware solutions.

Electronic Counter IP Module Requirements

- The counting inputs are electrical pulses from the test machine with positive amplitude 0–24 V (this is also parallel input to the existing electromechanical counter (totalizing counter)) with 2 Hz input frequency (therefore two mechanical cycles per second).
- In the COUNT mode the electronic counter divides the input pulses from the test machine in any ratio (the grouping of mechanical cycles). It will generate output information for the application server based on a defined number of input pulses.
- Electronic counter IP module has the possibility of operating in a simulated input pulses counting mode (PHANTOM mode). In this mode it automatically generates output information in the application server
- The IP module supports common Internet protocol.

Application Server Requirements

- Possibility of parallel information process entry from at least eight IP modules (at a frequency of 2 Hz input pulses for each IP module, such that the overall frequency can reach 16 Hz).
- Free-of-charge Linux operating system.
- Graphical visualizing web page (document) generation with application server databases as sources of data.
- Common software framework will serve as base for application server building.

Design and Implementation

Design Scope

The proposal for our information system is based on the future use of the technology IoT (Internet of Things). We decided at this moment for Simple Network Management Protocol (SNMP) – a well-known Internet protocol [4]. Next we decided to take on the market available IP hardware module with SNMP and its known hardware and software solution. This decision can be changed later and we can select a much modern IoT protocol such as MQTT, DDS, etc. Another base for system design can be the using of cheap operating system and database server. We utilised Linux operating system and MySQL database server. Very important for subsequent processing of collected data is the ability to export data (CSV file extension) and directly connect the appropriate analytical server to the MySQL database.

Architecture of the Information System

The system architecture is described in Fig. 7.1. We consider that this figure clearly illustrates all components of the information system. Explanation is needed for the embedded web server, which serves to collect data on temperature and humidity from a sensor. Like the IP module for an electronic counter, this is another type of manufactured IP module [5]. This IP module communicates with its surrounding via HTTP protocol. Our application server respects this and manages provided data as the client. The client parses received messages via HTTP.

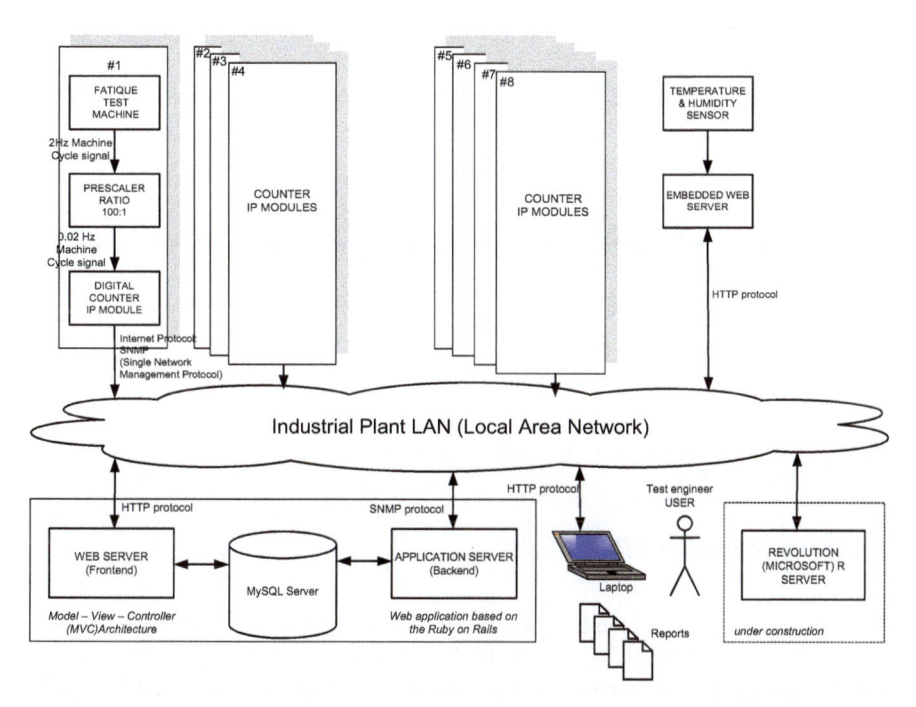

Fig. 7.1 Architecture of the system for collection data from fatigue test machines in automotive industry plant

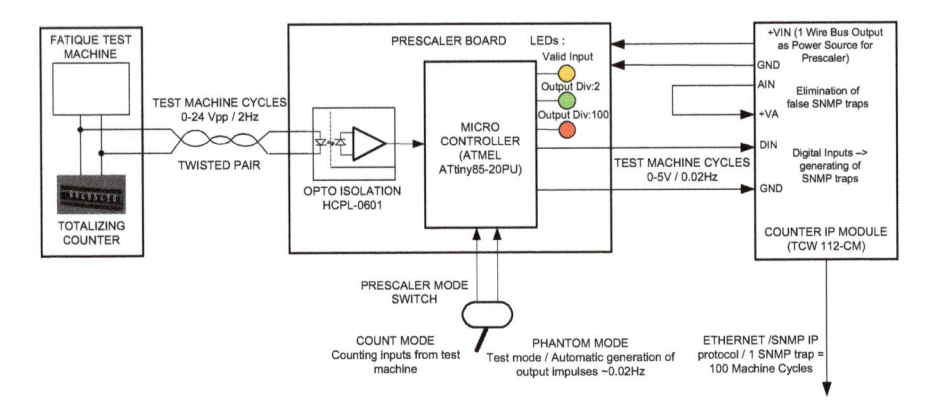

Fig. 7.2 Electronic counter IP module – signal processing diagram

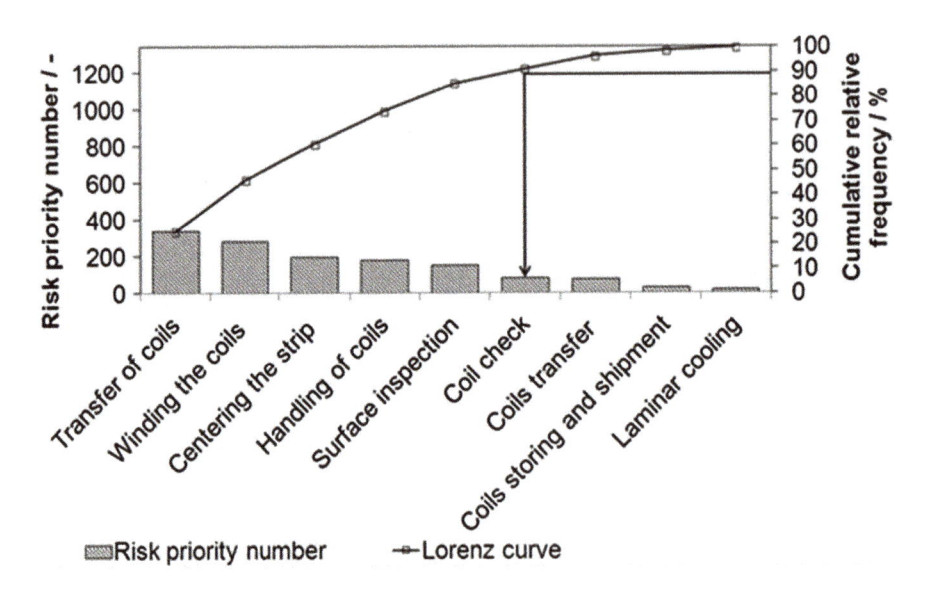

Fig. 7.3 Photography of the electronic counter IP module

Electronic Counter IP Module

The core of an electronic counter is a frequency divider with Atmel ATtiny85 and IP module TCW112-CM [6] (Figs. 7.2 and 7.3). Optically isolated input signals ensure fault tolerance. The frequency divider mode switch ensures the COUNT and PHANTOM modes mentioned in the requirement part. The frequency divider board is powered from the IP module. The frequency divider divides the frequency of the input pulses with split ratio 1:100. Output of the frequency divider ensures a signal for digital input of the module TCW112-CM. On this base the IP module creates

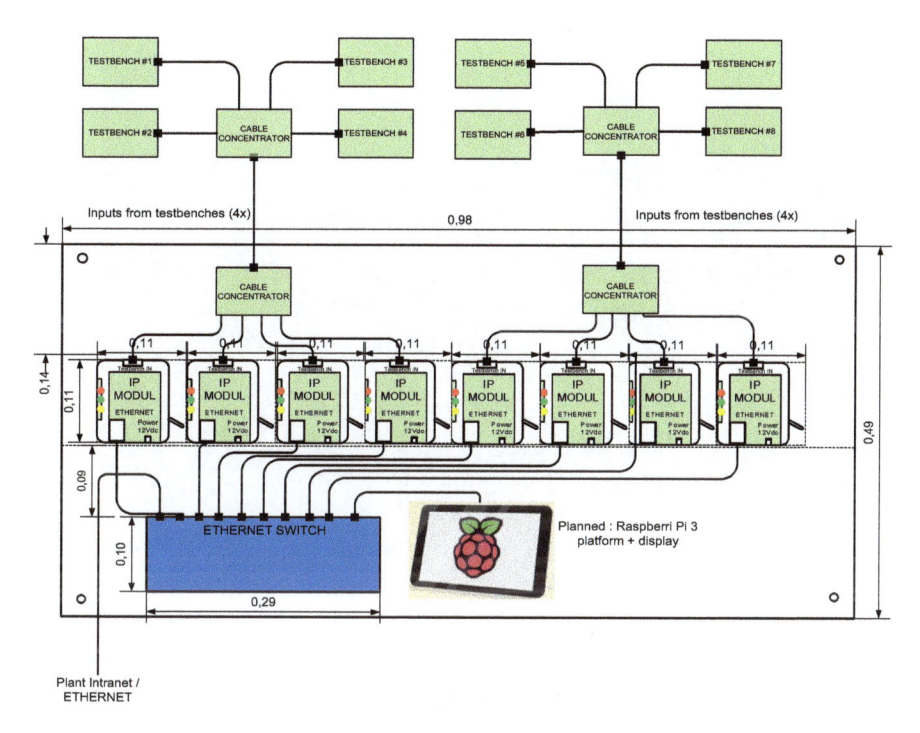

Fig. 7.4 Wall-mounted panel with eight electronic counter IP modules – construction view

one SNMP event per each 100 input pulses (mechanical cycles) which is processed by the application server. To avoid generating SNMP events that belongs to analog input, we joined analog input with analog power (Fig. 7.2). We placed all eight electronic counter IP modules on one wall-mounted panel for presentation purposes for our projects sponsors (Figs. 7.4 and 7.5).

SNMP

Simple Network Management Protocol (SNMP) is an Internet-standard protocol for collecting and organizing information about managed devices on IP networks and for modifying that information to change device behavior [4]. Devices that typically support SNMP include routers, switches, servers, workstations, printers, modem racks, and more. SNMP traps enable an agent (therefore in our case, the electronic counter IP module) to notify the management station (application server) of significant events by way of an unsolicited SNMP message. The trap includes current `sysUpTime` value, `OID` numbers identifying the type of trap, and optional variable bindings. Destination addressing for traps is determined in an application-specific manner typically through trap configuration variables in the management information base.

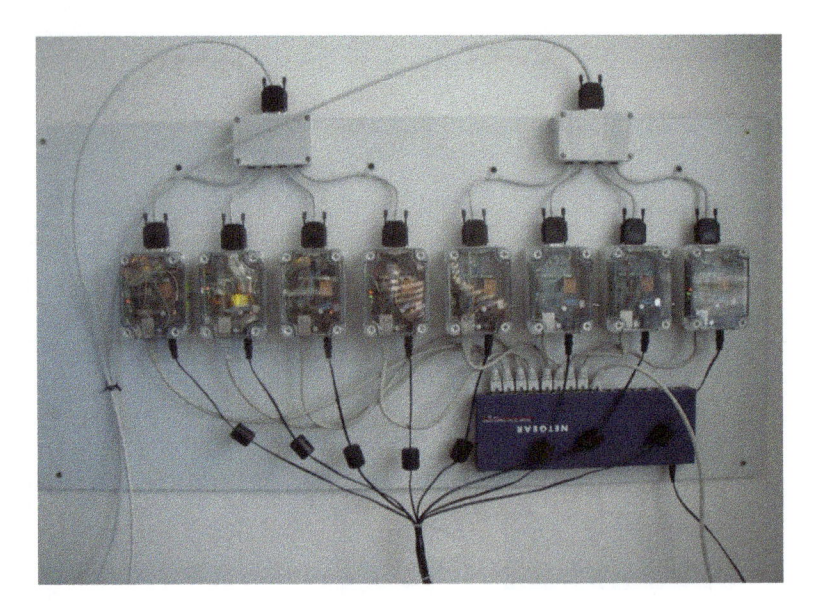

Fig. 7.5 Photo of wall-mounted panel with eight electronic counter IP modules

Application Server Side Description

On the server side we developed the application with the name 'Monitoring Dashboard'. This web application was written on the framework Ruby on Rails, or simply Rails. Rails is a Model–View–Controller (MVC) architecture framework, providing default structures for a database, a web service, and web pages. It encourages and facilitates the use of web standards such as JSON or XML for data transfer and HTML, CSS, and JavaScript for display and user interfacing [7]. The application server can be divided into three parts.

Web Server (Front-End)

It consists of a web-based user interface, which provides the user output (collected data from database), as well as user input (system configuration or data export). The main advantage of web applications is that the user requires no installation and they are independent of platforms (OS) and device (PC, tablet, smart phone) when users access the system. The only requirement is a modern web browser. The web server ensures the user interface for front-end. The application `'Monitoring Dashboard'` uses the web server Puma.

Database System

Relational database management system MySQL records mechanical cycles and also stores system configuration. It consists of several tables with one-to-many relations. Non-relational database system Redis is used by task queues system with the name Net-SNMP. Net-SNMP is a packet of tools for work with SNMP. It ensures the receiving of SNMP traps, the notification coming from the electronic counter IP module, processing ones, and its subsequent forwarding to the application server (backend).

Application Server (Backend)

It consists of a set of components that are invisible to ordinary users. Backend application is designed for the Linux platform (ideal for Ubuntu LTS). Backend ensures the data stream designed for processing each received SNMP messages (each of them represents 100 input mechanical cycles):

1. *Electronic counter IP module: sends SNMP trap notification*
2. *Net-SNMP: SNMP trap notification acceptance (snmpd Linux process), processing notifications (snmptt Linux process), transmission of relevant data to the web application*
3. *Backend: receiving data, building job, insertion of the job to the task queue (task queue system is used to prevent the system bottlenecks)*
4. *Backend: job worker processes tasks in the task queue and writes data to the relational database*

SNMP Traps Processing in the Application Server

To show incoming SNMP messages, we can use Linux console command: `tcpdump -i eth0 port 162`. An example of the SNMP message dump is in Fig. 7.7. The number of displayed messages should be equal to the number of messages recorded in the database MySQL. A description of the template for filtering SNMP traps is in Fig. 7.6. The template works with OID string '.1.3.6.1.4.1.38783.0.1'. The processing of the SNMP trap notification begins its assignment to the template and launches the system command that is defined in section EXEC. For each assigned trap notification, start running the following code in the web application (`Resque.enqueue (CounterlabsWorker," $aA", "$x $X")`). This code creates a new task in the task queue 'counterlabs' where $aA is the IP address of the sender trap notification and $x $X is the time of receipt of the notification. The task queue is processed by a queue worker process. It checks in an infinite loop the status of the task queue 'counterlabs'. If it finds a new task it carries out the method

SNMP message output :
```
192.168.1.2.65534 > 192.168.1.1.snmp-trap:  { SNMPv2c { V2Trap(80) R=1
system.sysUpTime.0=647815 S:1.1.4.1.0=E:38783.0.1 E:38783.3.1.0=1 } }
15:31:47.164107 IP (tos 0x0, ttl 100, id 1142, offset 0, flags [none],
                    proto UDP (17), length 127)
```
Template for SNMP filter trap :
```
moni_dashboard / config / snmptt.conf
#
EVENT clabs .1.3.6.1.4.1.38783.0.1 "Status Events" Normal
FORMAT counterlabs
EXEC cd /var/www/moni_dashboard && bin/rails runner -e production
     'Resque.enqueue(CounterlabsWorker, "$aA", "$x $X")'
SDESC
     counterlabs
EDESC
```

Fig. 7.6 Linux console command response as one SNMP message and the template for SNMP filter trap

'perform' with parameters IP address and time of receipt, which both were recorded during the creation of this task. The method 'perform' moves these parameters down to the method 'ClabsRecord.receive'. Method 'receive' moves parameters to method 'commit_record' which creates a new database record, and if check of integrity is successful, the record will be stored in the database.

Verification of Accuracy of Mechanical Cycles Counting

Counting of mechanical cycles with frequency of 2 Hz is a relatively long process. During this time, the number of mechanical cycles lies in orders of hundreds of thousands or millions (one million mechanical cycles is equivalent to 140 h or 6 days of testing). It is very important to be sure that the electronic counter IP module counts mechanical cycles correctly and without errors. We carried out three ways to verify the accuracy of counting of the mechanical cycles:

Verification of the Operation of IP Module Counter with Prescaler

We used Nanoline programmable logic controller (PLC) from Phoenix Contact (USA), which generates the exact numbers of 24 V pulses for input of electronic prescaler of the IP module counter. The number of pulses generated by the Nanoline PLC was compared with the number of received SNMP traps. Here, we have gained knowledge of the electrical interferences at unconnected inputs of the IP module (they generate false SNMP traps) that led to the grounding of the analog input of TCW112CM (see Fig. 7.2; see the shortcut between AIN and +VA inputs of the IP module).

Verification of the Application Server Data Throughput

The aim of this was to verify the application server data throughput in order to determine its ability of capturing SNMP traps in real time. SNMP traps came in to the application server with some frequency rate. Real-time verification was done by eight IP modules working in PHANTOM mode. Each frequency divider (prescaler) generated in this mode two pulses per minute (see Fig. 7.2), which means that an application server receives 16 random SNMP traps per minute. We compared the numbers of recorded SNMP traps by application servers with numbers of PHANTOM generated pulses (time of generation × frequency of input PHANTOM pulses (2 Hz) × number of PHANTOM generators).

Comparison Method

The last method uses comparison of recorded values of mechanical cycles in the application server database with the values recorded by totalizing counter. This method was performed during real mechanical tests. The test operator recorded both values regularly during 1 week. For access to the data in the application server database, he exploited the web interface (Figs. 7.7 and 7.8).

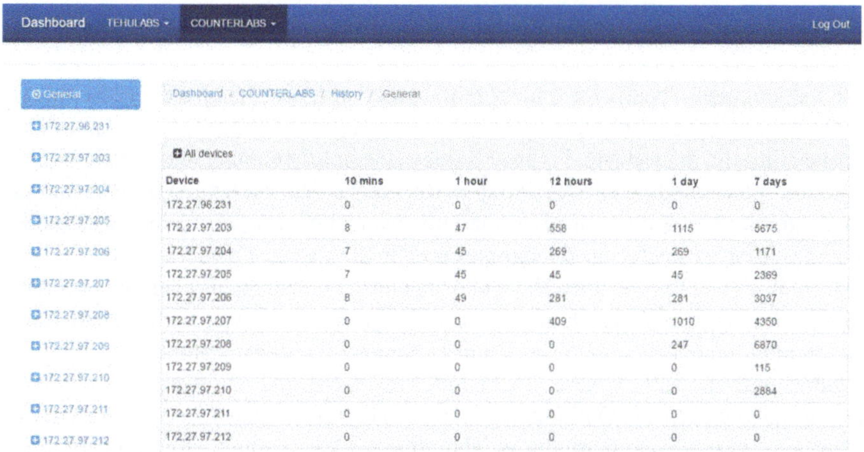
Dashboard
TEHULABS
COUNTERLABS
Log Out
General
172.27.96.231
172.27.97.203
172.27.97.204
172.27.97.205
172.27.97.206
172.27.97.207
172.27.97.208
172.27.97.209
172.27.97.210
172.27.97.211
172.27.97.212
Dashboard / COUNTERLABS / History / General
All devices
Device	10 mins	1 hour	12 hours	1 day	7 days
172.27.96.231	0	0	0	0	0
172.27.97.203	8	47	558	1115	5675
172.27.97.204	7	45	269	269	1171
172.27.97.205	7	45	45	45	2369
172.27.97.206	8	49	281	281	3037
172.27.97.207	0	0	409	1010	4350
172.27.97.208	0	0	0	247	6870
172.27.97.209	0	0	0	0	115
172.27.97.210	0	0	0	0	2884
172.27.97.211	0	0	0	0	0
172.27.97.212	0	0	0	0	0

Fig. 7.7 Example of administrator's web interface – information about performed mechanical cycles for all test machines

Fig. 7.8 Web graph visualization of performed mechanical cycles (×100) for 12 h – displayed for one electronic counter IP module

Developed Tools for Process Control, Data Visualization, Data Processing

Information System Administrator's Role

The administrator can set necessary settings like IP address of the IP counter module and his name and another via the dedicated web interface generated by the web server. As an example of a web user interface, Fig. 7.7 shows information about performed cycles for all IP counters.

Test Engineer's Role

The test engineer checks the end of fatigue test with two independent interfaces: Excel sheet and web interface. The basic display of measured mechanical cycles is reached through the website. Figure 7.8 shows the hourly number of reported SNMP traps for one IP module (1 SNMP trap = 100 mechanical cycles). Figure 7.8 shows a web graph visualization of performed mechanical cycles (×100) for 12 h, and it provides information for one electronic counter IP module. From this kind of information, it is hard to determine the end of the test. The graphical interface program module "Semaphores" developed in Excel VBA serves for visualizing of running and stopped tests (Fig. 7.9). The email is sent by the client computer to the test engineer automatically in case of test end. This is ensured with small VBA macro and installed Microsoft Outlook application. The test engineer needs information about temperatures and humidity in laboratories. One example of graphical representation of measured temperature and humidity in the test laboratory is displayed in Fig. 7.10.

7 IoT Challenge: Older Test Machines Modernization in an Automotive Plant   97

	A	B	C	D	E	F
1		Testmachine status monitor				
2						
3		●		RUN		
4						
5	LP Station Name	SEMAPHOR	Cycle No.	First Record Time	No. IP count module	
6	L1	○	0		3	
7	L2	○	2500	14.9.2016 15:19	4	
8	L3	○	2400	14.9.2016 15:19	5	
9	L4	○	2400	14.9.2016 15:19	6	
10	L5	○	2700	14.9.2016 15:18	7	
11	L6	○	0	8.9.2016 8:00	8	
12	L7	○	0	8.9.2016 8:00	9	
13	L8	○	0	7.9.2016 16:35	10	
14	L9	○	0	8.9.2016 8:00	11	
15						

Fig. 7.9 Graphical interface program module "Semaphores" serves for visualizing of running and stopped tests

Fig. 7.10 Web graph visualization of measured temperature and humidity in the test laboratory

Role of Manager's Excel VBA-Based Dashboard

The manager can reach information about utilization of the test machines in 1 day or 1 month with an Excel sheet powered by VBA program. His goal is also to ensure the maximum utilization of test machines also during weekends and nights. The graphical interface program module "History – Month" and "History – Day" developed in Excel VBA serves for visualizing the history of tests. The manager's dashboards are shown in Figs. 7.11 and 7.12. The test manager can choose arbitrarily the

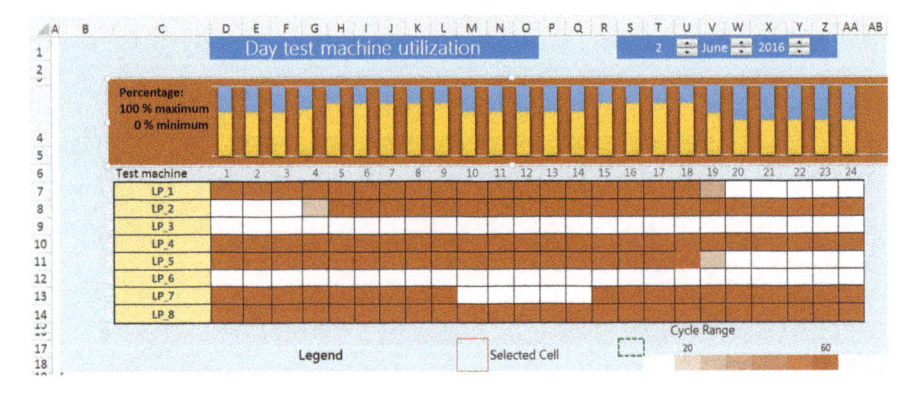

Fig. 7.11 Excel VBA dashboard visualization of a 1-day mechanical cycles history and test machines utilization

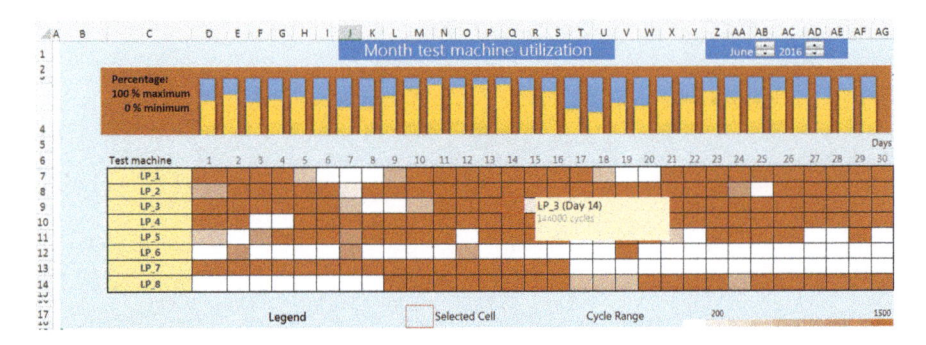

Fig. 7.12 The Excel VBA dashboard visualization of 1-month test machines history and test machines utilization

date (year, month, and day), and he can observe the KPI of utilization of tests machines very well.

Role of Future Data Analytics

For processing capabilities and data analysis, we are planning to extend with the R language server based on the free-of-charge Microsoft (previous Revolution) R Server. Revolution R open-source software allows R users to process, visualize, and model terabyte-sized data sets at a fraction of the time of legacy products without requiring expensive or specialized hardware. The data analyst works with R language and he can reach data from exported CSV files or he can read data direct from MySQL server with the "R MySQL" package [8].

Test Processes Optimization

More test processes were influenced with the modernized architecture of cycle counting. Thanks to automatic detection of ends of fatigue tests, we reached more effective utilization of test machines (more than 25%). Test machine automatically stops test after destruction of component. The software application recognizes mechanical cycle delay and sends email to the responsible persons (technician and test engineers) with information about the end of test. Further optimization of processes results from the use of databases and data processing using software tools such as Excel and R language.

Conclusions

In the presented paper we describe engineering design and implementation of an information system for remote collection, storing, and processing of information on a number of carried out mechanical cycles for a group of older existing fatigue test machines in an automotive components production. In the modernization of existing test machine infrastructure, we built the electronic counter modules with Ethernet interface developed by us. Electronic counter modules in the role of clients replaced the existing electromechanical totalizing counters. Web and application servers were built on Linux operating system. The MVC architecture (Model–View–Controller) serves as the base for web application, and it was written together with backend application server around the Ruby on Rails web framework. The database is built on MySQL server.

In addition to the web interface, a program for Excel VBA was developed for the purpose of accessing data in the MySQL database. For programming in Excel VBA programming language, we have decided thanks to fact that Excel VBA scripts are widely available in many organizations. The VBA script uses the ODBC database driver (manufactured by Oracle) to access the MySQL database.

We have done three ways to verify the accuracy of this counting method a collection of mechanical cycles, and we have achieved results comparable to hand-gathered data. Thanks to automatic detection of ends of fatigue tests, we reached more effective utilization of test machines (increasing of utilization is more than 25%).

Our possible contribution is the low-cost upgrade of older test machines with embedded platform and free-of-charge Linux server and web application framework without the need of buying expensive industrial-grade communication systems. We are planning to build the data analytical services around the free-of-charge Microsoft (Revolution) R open-source software. There is possibility to prepare reports and manager's dashboards with the common spreadsheet also (Microsoft Excel). Our solution of upgrading existing industrial infrastructure for fatigue tests is very reasonably priced and is suitable for older industrial facilities.

References

1. Waloddi, W. (1961). *Fatigue testing and analysis of results*. Oxford/London/New York/Paris: Pergamon Press. ISBN:1483154165.
2. Gbasouzor, A. I., Okeke, O. C., Chima, L. O. (2013). Design and characterization of a fatigue testing machine. San Francisco, USA, 2013. *Proceedings of the World Congress on Engineering and Computer Science* 2013 (Vol 1).
3. ASTM International. (2005) Standard practices for cycle counting in fatigue analysis, designation: E 1049-85 (Reapproved 2005). ASTM International, 100 Barr Harbor Drive, PO Box C700, West Conshohocken, PA 19428–2959, United States.
4. Mauro, D. R., & Schmidt, K. J. (2005). *Essential SNMP*. Sebastopol: O'Reilly Media. ISBN: 0-596-00840-6.
5. Teracom Ltd. *TCW122B-CM - Remote environmental monitoring*. [Online] Teracom Ltd. [30.9.2016] http://www.teracom.cc/products/tcw122b-cm-remote-environmental-monitoring/
6. *TCW112-CM – Environmental IP monitoring board*. [Online] [30.9.2016.] http://www.teracom.cc/products/tcw112-cm-environmental-ip-monitoring-board/
7. Hartl, M. (2015). *Ruby on rails tutorial: Learn web development with rails*. New York/Boston: Addison-Wesley. ISBN: 0134077709.
8. *Revolution R open the world's most popular data analysis software, enhanced*. [Online] Revolution Analytics. [Access: 30. 9. 2016] http://www.revolutionanalytics.com/revolution-r-open

Part II
Smart Technology Trends in Industrial Management

Chapter 8
Industry 4.0: Preparation of Slovak Companies, the Comparative Study

Ján Papula, Lucia Kohnová, Zuzana Papulová, and Michael Suchoba

Abstract Industry 4.0 is an initiative supporting the 4th Industrial Revolution. It comes with new technologies and concepts such as automation, system integration, autonomous robotics, digitization, Internet of Things, artificial intelligence systems, and many others. New technologies are rapidly changing the face of the economy, the way of living, and fundamentally also the field of industrial production. This movement brings a series of challenges that creates opportunities for many companies to develop a long-term competitive advantage and to ensure the competitiveness in a global environment. In our research, we studied the preparation and response of the companies from production industries to changes coming with Industry 4.0. We also made a comparative study comparing companies from Slovak Republic to companies from Germany, Austria, and Switzerland. The particular findings of the research are presented in this paper.

Introduction

Industry and the whole economy are undergoing radical changes due to the implementation of advanced ICT and artificial intelligence to manufacturing, services, and all sectors of the economy. The impact of these changes is so crucial that they are referred to as the new movement called the 4th Industrial Revolution. Industry 4.0 is a revolution built on the digitalization, on change of production processes, on change on business models with aim to accelerate the production and make it more effective, and also on the interconnection of systems from customer requirements to final product through digitalization.

The movement to Industry 4.0 has emerged in Germany as a response to the decline in industrial production after the release of production capacities into

---

J. Papula (✉) · L. Kohnová · Z. Papulová · M. Suchoba
Department of Strategy and Entrepreneurship, Faculty of Management,
Comenius University in Bratislava, Bratislava, Slovakia
e-mail: jan.papula@fm.uniba.sk

© Springer International Publishing AG, part of Springer Nature 2019
D. Cagáňová et al. (eds.), *Smart Technology Trends in Industrial and Business Management*, EAI/Springer Innovations in Communication and Computing,
https://doi.org/10.1007/978-3-319-76998-1_8

cheaper countries. It aimed to reindustrialize Germany with state-of-the-art technologies able to compete with even the cheapest workforce. Other countries, such as Slovakia, follow this movement. In 2016, the Ministry of Economy has introduced a state concept of Slovakia on Industry 4.0, which has been called "Smart Industry for Slovakia." The effort of the state strategy for Industry 4.0 is to promote the concept in Slovak production industries. We can see it especially in the automobile industry that has already started to apply many elements of Industry 4.0. The central idea of the concept is the transformation of production industries into a new type of industry that uses the knowledge in terms of digitization, the Internet economy, robotization, and the interconnection of industries with scientific research institutions and education.

Industry 4.0

Industry 4.0 or the 4th Industrial Revolution is representing the massive changes that are rapidly entering the current environment and also all industrial sectors. In the center of the Industrial Revolution is the connection of the virtual cyber world with the world of physical reality [5]. The world has already become digitally connected to the point of no return [10].

The main ideas of Industry 4.0 have been firstly published in 2011 [4] and have built the foundation for the Industry 4.0 manifesto published in 2013 by the German National Academy of Science and Engineering. This movement provides immense opportunities for realizing sustainable manufacturing using the ubiquitous information and communication technology infrastructure [12]. Automation and more flexible processes as well as horizontal and vertical integration are becoming more important features in a modern, competitive production structure [3]. One of the key features of Industry 4.0 is to create a smart factory, in which integration of various components inside a factory enables to implement a flexible and reconfigurable manufacturing system [14]. Intelligent and flexible processes characterize the manufacturing companies of the future [13]. New approaches will help to transform production from separate automated units into fully integrated and continuously optimized production environments and enable to combine several hi-tech technologies such as CAD systems, ERP systems, 3D scanning, 3D printing, and so on. Many of the advances in technology are already used in manufacturing, but with Industry 4.0, they will transform production to a fully integrated, automated, and optimized production flow, leading to greater efficiencies and changing traditional production relationships among suppliers, producers, and customers [9].

While the first three Industrial Revolutions resulted from mechanization, electricity, and IT, now it is mainly about the consequences of the introduction of new services and technologies related to the context of IoT (Internet of Things). Industry 4.0 assumes the maximum use of the Internet of Things and Internet services, which are based on mutual communication and cooperation. Examples of advanced technologies and prognoses for Industry 4.0 are [1, 2, 5, 9, 11, 15]:

- *System integration* (IT systems are not fully integrated yet, but with Industry 4.0 companies, departments, functions, and capabilities will be more cohesive.)
- *Additive manufacturing* (Additive manufacturing, such as 3-D printing, will be more used in production.)
- *Autonomous robots* (Robots will become more autonomous, flexible, and cooperative, and eventually, they will interact with one another and work safely side by side with humans and learn from them.) *and next-level automation* (There is still a lot of potential in increasing the use of automation in both blue-collar and white-collar work. In terms of blue-collar work, it can be expected that adoption of robotics will grow significantly in the next 5–10 years.)
- *Internet of Things* (More devices will be enriched with embedded computing and connected. It will allow field devices to communicate and interact both with one another and with more centralized controllers, as necessary.)
- *Simulation* (3-D simulations of products, materials, and production processes are already used, e.g., while developing new products, but in the future, simulations will be used more extensively in plant operations as well to leverage real-time data to mirror the physical world in a virtual models.)
- *Big data and analytics* (The collection and comprehensive evaluation of data from many different sources will become standard to support real-time decision making.)
- *Smart products* (Smart products are cyber-physical systems providing new features and functions based on connectivity with the ability for machine-to-machine communication, and embedded interfaces enable interaction with human users. Sensors in production will be able to provide information about their environment and, e.g., about their current use and status, and the data will be linked to an actuator able of triggering autonomous reactions to changes.)
- *Cloud computing* (Companies are already using cloud-based software for some applications, but more production-related undertakings will require increased data sharing, and the performance of cloud technologies will improve.)
- *Cyber security* (The need to protect critical industrial systems and manufacturing lines from cybersecurity threats increases dramatically; thus the secure, reliable communications as well as sophisticated identity and access management of machines and users are essential.)

Adaptation to new technologies and possibilities that come with Industry 4.0 can bring many benefits to companies, especially in the area of the flexibility, speed, productivity, and quality of the production process. These possibilities represent tremendous opportunities for innovative producers, system suppliers, or entire regions but also create severe threat to laggards [9]. According to the results of the recent McKinsey survey, there are also barriers mentioned by manufacturers that were still struggling with how to get started with Industry 4.0 implementation [15]: difficulty in coordinating actions across different organizational units; lack of courage to push through radical transformation; lack of necessary talent, e.g., data and scientists; concerns about cybersecurity when working with third-party providers; lack of a clear business case that justifies investments in the underlying IT architecture; concerns about data ownership when working with third-party providers; uncertainty about which Industry 4.0 applications to source internally and which to source from third-party providers as well as a lack of knowledge about suitable providers; and challenges with integrating data from disparate sources to enable Industry 4.0 applications.

Methodology

The aim of this research was to analyze the behavior of firms in response to changes coming with Industry 4.0. The research was focused on the comparison of Slovak enterprises (SK) with enterprises in Austria (AT), Germany (DE), and Switzerland (CH). Group of AT, DE, and CH companies is representing companies from more developed countries, which achieved higher ranking in different innovation evaluations. Research was also focused on companies operating in the industry sector, i.e., companies that are highly affected by changes coming with the 4th Industrial Revolution. Specifically, we compared the automotive, electrotechnical, machine, and construction industries.

Sample Data and Collection

The research was conducted during years 2015 and 2016 on a sample of 489 Slovak companies and 574 companies from Austria, Germany, and Switzerland. The data were collected using electronic questionnaire, which was distributed directly to respondents. Respondents were representatives of companies from different sectors and mostly held management and senior management positions. For the purposes of this research, we analyzed companies from the automotive industry, electrotechnical industry, machine industry, and construction industry. Ee analyzed 186 companies in total. The following diagram (Chart 8.1) shows the numbers of firms in each of the analyzed groups.

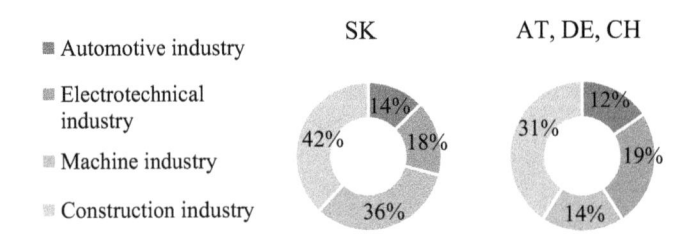

Chart 8.1 Sample of companies from different industries from Slovakia (SK) and from Austria, Germany, and Switzerland (AT, DE, CH) (Source: Authors)

In the survey, respondents were asked: "What kind of innovation has your business done over the last two years?" This is where options of "new technologies to support or automate processes" and "information technology software" were analyzed. Respondents had the possibility of yes/no answer, depending on if the innovation has been introduced in years 2014–2015.

Further we analyzed the question: "Which of the following areas do you consider essential for long-term financial performance of the company (= long-term sustainability performance objectives, such as earnings, ROA, …)?" For the purposes of this paper, we compared the responses to these answers:

- Implementation of IT/ICT for process support and increase of process automation
- Investment in human resource development
- Investing in own research, development, and innovation processes

The perception of the importance of respondents was rated on a scale from 1 to 5, where 1 is "not considered essential" and 5 "are considered very crucial." Analyzed groups were divided into group which considers selected area as essential, meaning companies that selected 4 or 5 on the scale, and group which does not consider selected area as essential, meaning companies that selected 1, 2, or 3 on the scale.

Then we subsequently analyzed questions focused on *measures implemented by Slovak enterprises within their organizations during years 2009, 2011, and 2014–2015*. Respondents could select all answers that are satisfactory. We focused on selected measures: informatization, automatization, and process innovation.

For evaluation we used basic descriptive statistics such as average and percentage comparison. To analyze the significance of differences between SK and AT, DE, and CH companies, we used statistical analysis using chi-square test.

Results

As we can see, the greatest need for the application of IT/ICT in Slovak companies is in the automotive and electrotechnical industry (Chart 8.2). In these sectors, it is also the biggest difference between the perception of Slovak and group of AT, DE, and CH companies. Analysis of the chi-square test showed that difference in each of both cases has a high level of significance. Similarly, the average value of all the responses to this question (Table 8.1) was higher among Slovak businesses compared to the group of AT, DE, and CH companies.

Most of the innovations in introducing new technologies to support and automate processes in 2014–2015 were introduced in the electrotechnical and machine industry, where more than 40% of the organizations surveyed introduced this type of innovation. Except for the construction industry, in each of the groups surveyed, more SK companies have introduced this innovation than the group of AT, DE, and CH companies (Chart 8.3).

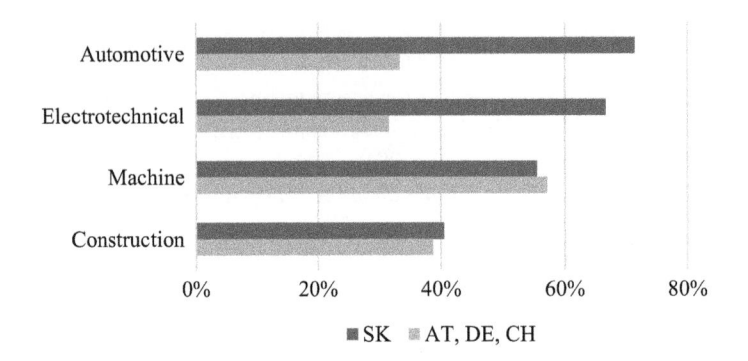

	Automotive	Electrotechnical	Machine	Construction
Chí SK-AT, DE, CH	0.052079	0.032818	0.919134	0.878823

Chart 8.2 Percentage comparison of SK and AT, DE, and CH companies in selected industries, which consider implementation of IT/ICT for process support and increase of process automation as key to long-term financial performance of the company (Source: Authors)

Table 8.1 Comparison of average assessment of SK and AT, DE, and CH companies in the question of considering implementation of IT/ICT for process support and increase of process automation as key to long-term financial performance of the company

	SK	AT, DE, CH
Average	3.31	3.29

Source: Authors

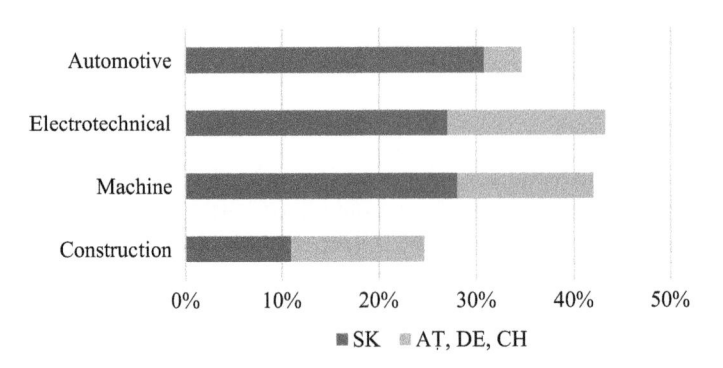

Chart 8.3 Percentage comparison of companies from selected industries which have introduced new technologies to support or automate processes in 2014–2015, with comparison of SK and AT, DE, and CH companies' ratio (Source: Authors)

In a deeper analysis, the percentages were compared in the individual sectors within the two countries surveyed, which were then compared in Chart 8.4. Among SK enterprises, the introduction of new technologies for process support and automation was observed most in the automotive (57%) and electrotechnical industries

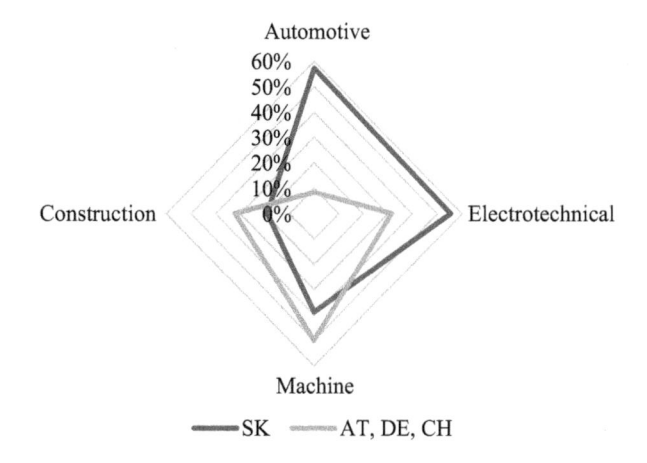

Chart 8.4 Percentage comparison of SK and AT, DE, and CH companies from selected industries which have introduced new technologies to support or automate processes (Source: Authors)

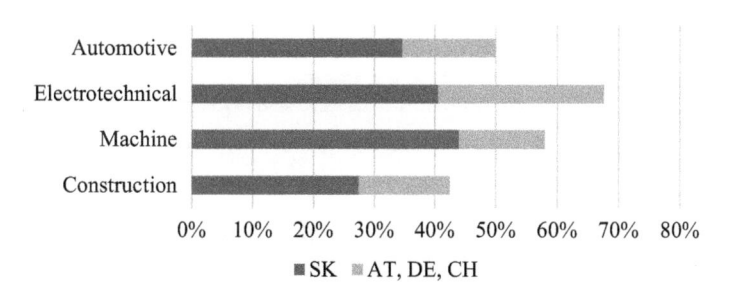

Chart 8.5 Percentage comparison of companies from selected industries which have introduced new information technology software in 2014–2015, with comparison of SK and AT, DE, and CH companies' ratio (Source: Authors)

(56%). On the contrary, it was observed least in the construction industry (19%). AT, DE, and CH companies have made this innovation on average less (but more than SK enterprises) in the construction industry (32%) and also in the machine industry (50%). At least the group of AT, DE, and CH companies introduced this innovation in the automotive industry (8%). Significant statistical difference was found in the automotive industry between SK and AT, DE, and CH companies, with p-value = 0.05 and chi = 0.009108481 (Chart 8.5).

New information technology software was most introduced in 2014–2015 in the electrotechnical industry (68%) and at least in the construction industry (42%). In each of the studied sectors, Slovak enterprises prevailed. On average, information technology software introduced in 2014–2015 was more common in the surveyed group than the process automation and automation technologies shown in Chart 8.6.

Within the Slovak enterprise group, information technology software in the years 2014–2015 was most introduced in the electrotechnical industry (83%) and then in

Chart 8.6 Percentage comparison of SK and AT, DE, and CH companies from selected industries which have introduced new technologies to support or automate processes (Source: Authors)

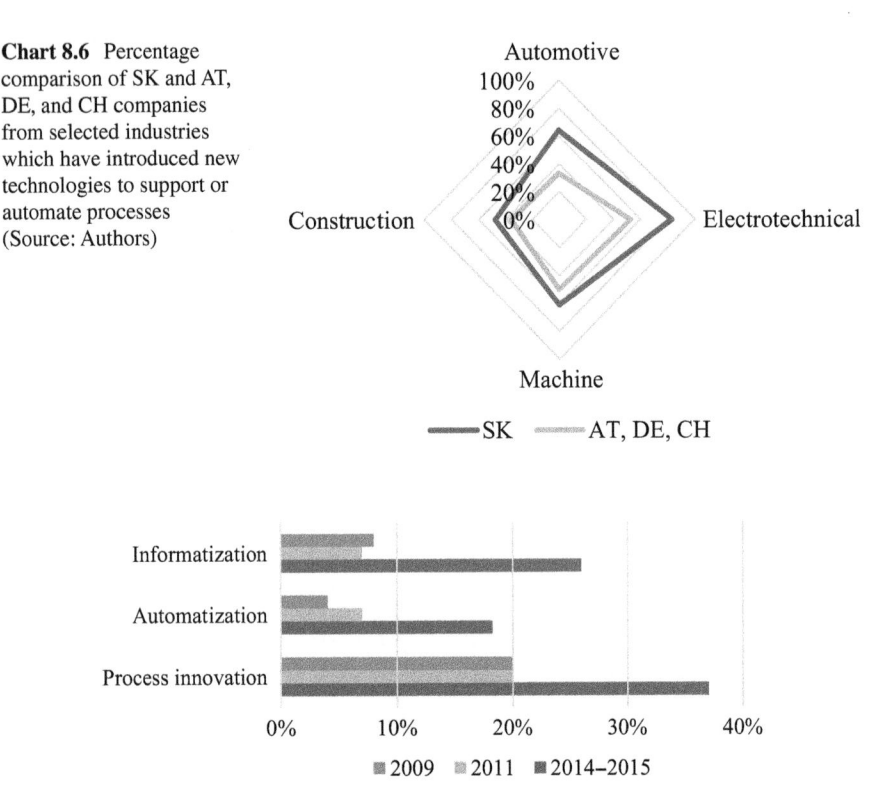

Chart 8.7 Annual comparison of the percentages of SK enterprises that implemented selected measurements within the company (Source: Authors)

the automotive industry (64%), while in the construction industry it was 48%. Within the AT, DE, and CH group, information technology software was most introduced also in the electrotechnical industry; however it was 30% less (53%) compared to Slovak companies, with the difference being statistically significant at p-value = 0.05 and chi = 0.046156407 (Chart 8.7).

In the evaluation of long-term research of Slovak companies, it is obvious that in recent years information engineering, automation, and process innovation have gained expressively increased interest and importance.

Investing in human resource development is highly perceived mainly in the automotive industry where there is no significant difference between the Slovak enterprises and groups of AT, DE, and CH companies. The significant difference is in the machine industry with p-value = 0.05 and chi = 0.046063, where higher interest was expressed by companies in the group of AT, DE, and CH companies (Chart 8.8).

Machine industry is also an area where companies from the group of AT, DE, and CH companies perceive a greater need for their own research and development. This difference is statistically significant with p-value = 0.05 and chi = 0.013284. Low interest is in automotive and construction industries (Chart 8.9).

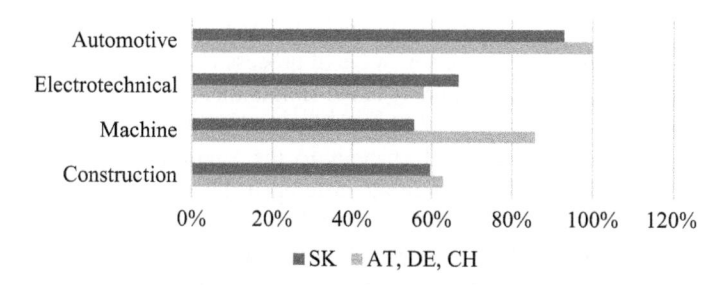

Chart 8.8 Percentage comparison of SK and AT, DE, and CH companies in selected industries, which consider investment in human resource development as key to long-term financial performance of the company (Source: Authors)

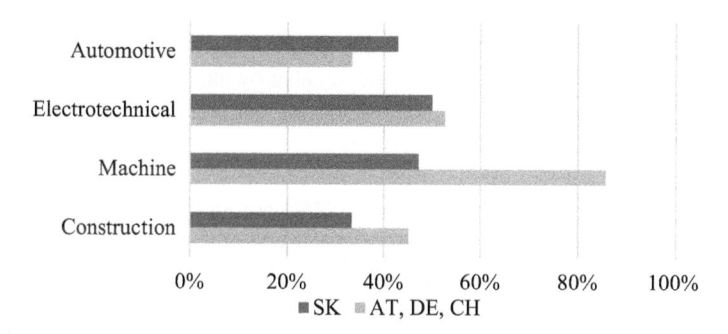

Chart 8.9 Percentage comparison of SK and AT, DE, and CH companies in selected industries, which consider investment in own research, development, and innovation processes as key to long-term financial performance of the company (Source: Authors)

Conclusion

The challenges that come with Industry 4.0, the challenges to boot competitiveness, to create new competitive advantages and to be able to compete on the global market, should be taken as key strategic challenges. These challenges require timely preparation, analysis of opportunities, identification of barriers and risks, and activation of hidden potential inside the companies. Process automation, investment in human resource development, investment to own research, collaboration with research institutions and clusters, and development and innovation of processes are some of the key areas to respond to new movements.

In the near future, we can expect a large number of technical devices and technologies that will support the Industry 4.0 mindset and visions. Managers will thus face a strong marketing pressure from providers of these technologies. The need to review existing competitive strategies or review existing business model of the organization will rise. The strategic transformation of companies will be necessary to get the most of emerging opportunities, as well as change the patterns of partner and customer behavior. At the corporate level, it will be necessary to thoroughly map the business processes and to create a virtual model, in which it will be able to realize the "to-be" simulations and calculate the impacts of these changes. Therefore, it is possible to expect a new wave of process management development focused on process measurement and dynamic simulations.

Rapid improvement and the availability of advanced technologies and services in the ICT field have already brought changes to the environment of companies. Significantly, it is seen in evaluation of the differences between Slovak companies in the years 2009–2011 and 2014–2015. Certainly, to a certain matter, the recovery from the crisis period contributed to it which is also accompanied by recoveries on the markets and at the same time by reduction of interest rates on loans.

High focus on the processes and on support of automation through the implementation of IT/ICT in Slovak companies is also seen in our comparative research. High interest especially in the automotive industry can be explained by the fact that in Slovakia it is one of the key industries, and its importance was reinforced by the arrival of major new investments in Slovakia. On the other hand, the low interest in the automotive industry was found in Germany also by analytical report [7], where 60% of respondents do not perceive the development of production toward Industry 4.0 as important.

Similarly the high level of perception, however, was expressed by companies in the machine industry. In the assessment of comparison of the overall results of the machine industry, where there is a very great interdependence with automotive industry, we see a lag of Slovak companies. It is in human resource development as well as in own research and development.

As for the discussion, we are evaluating the need for a comprehensive development of enterprises in Slovakia. The implementation of technology and automation on the level of the company itself are only a part of the concept of Industry 4.0. From our point of view, the main challenges are in corporate culture and in people in relation to the necessary innovation.

The positive news for Slovakia is the trend of interest in the topic Smart Industry, not only at the level of discussions but also in the activities of companies. Strong population of highly educated labor force, at the age of 25–45 years in Slovakia [8], now representing the majority of working population, provide an opportunity for innovative behavior and support from the management of enterprises.

As we can see on OECD data, shown in Chart 8.10, Slovakia has also the strongest rise of industrial production among the group of researched countries in this comparative study.

Industrial companies identified the geographical area and the economic environment in Slovakia as attractive, as they increase the volume of investments not only in the production volume but also obviously into modernization and automation. Suppliers follow these trends and invest in development as well. The rise of investments in automotive and electrotechnical industry, together with strong tradition of machine and construction industry and well-educated and technically prepared labor force, gives strong opportunities for the implementation of the Industry 4.0 concept in Slovakia.

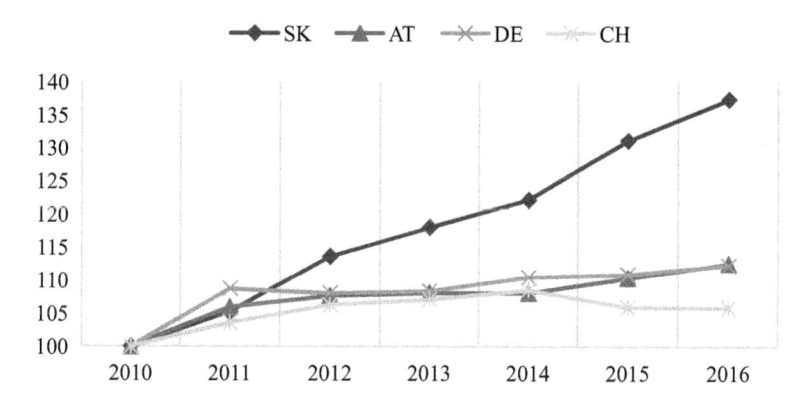

Chart 8.10 Industrial production between years 2010–2016, index based on a reference period of year 2010 = 100 (OECD Data) (Source: OECD [6])

References

1. Bechtold, J., Lauenstein, Ch., Kern, A., & Bernhofer, L. (2015). Industry 4.0 – The Capgemini Consulting View, Sharpening the Picture beyond the Hype. Available via Capgemini Consulting. https://www.capgemini-consulting.com/resource-file-access/resource/pdf/capgemini-consulting-industrie-4.0_0_0.pdf. Accessed Dec 2016.
2. Caganov, D., Bawa, M., Szilva, I., & Spirkova, D. (2016). Importance of internet of things and big data in building smart city and what would be its challenges. In *Lecture notes of the Institute for Computer Sciences, Social Informatics and Telecommunications Engineering* (pp. 605–616). Berlin, Switzerland: Springer.
3. Heng, S. (2014). Industry 4.0: Upgrading of Germany's Industrial Capabilities on the Horizon. Available via SSRN.. https://ssrn.com/abstract=2656608. Accessed Dec 2016.
4. Kagermann, H., Lukas, W., & Wahlster, W. (2011). Industrie 4.0 – Mit dem Internet der Dinge auf dem Weg zur 4. industriellen Revolution. VDI Nachrichten, 13:2.
5. Marik, V. (2016). *Prumysl 4.0: Vyzva pro Ceskou Republiku*. Praha: Management Press.
6. OECD (2017). Industrial production (indicator). doi:https://doi.org/10.1787/39121c55-en. Available via. https://data.oecd.org/industry/industrial-production.htm. Accessed May 2017.
7. PAC-IT (2013). Trends in der Automobil – IT-Investitionspläne in Deutschland. Available via PAC.. https://www.pac-online.com/industrie-40-der-automobilindustrie-aktuell-noch-geringe-bedeutung. Accessed May 2016.
8. Population Pyramid (2016). Available at Population Pyramid.. https://www.populationpyramid.net/slovakia/2016/. Accessed May 2017.
9. Rüssmann, M., Lorenz, M., Gerbert, P., Waldner, M., Justus, J., Engel, P., & Harnisch, M. (2015). Industry 4.0: The Future of Productivity and Growth in Manufacturing Industries. Available via ZVW.. http://www.zvw.de/media.media.72e472fb-1698-4a15-8858-344351c8902f.original.pdf. Accessed Sep 2016.
10. Schaeffer, E. (2017). *Industry X.0: Realizing digital value in industrial sectors*. Munchen: Redline Verlag.
11. Stanek, P., & Pauhofova, I. (2016). Adaptačné procesy a pulzujúca ekonomika v cylke paradigmy zmien v 21. storočnî. EU SAV, Bratislava.
12. Stock, T., & Seliger, G. (2016). Opportunities of sustainable manufacturing in industry 4.0. *Procedia, CIRP, 40*, 536–541.
13. Tomek, G., & Vavrova, V. (2017). *Prumysl 4.0 aneb nikdo sam nevyhraje*. Pruhonice: Professional Publishing.

14. Wang, S., Wan, J., Zhangb, D., Lia, D., & Zhanga, C. (2016). Towards smart factory for industry 4.0: A self-organized multi-agent system with big data based feedback and coordination. *Computer Networks, 101*, 158–168.
15. Wee, D., Breunig, M., Kelly, R., & Mathis, R. (2016). Industry 4.0 after the initial hype: Where manufacturers are finding value and how they can best capture it. Available via McKinsey Digital.. https://www.mckinsey.de/files/mckinsey_industry_40_2016.pdf. Accessed May 2017.

Chapter 9
Transformations of Urbanized Landscape Following the Smart Water Management Concept

Matina Lazarová, Michal Varga, and Daniela Gažová

Abstract In the global water arena, a consensus has emerged that urban water management urgently calls for smart solutions in order to adapt to climate change. The ways our society is managing water resources are clearly in need of innovation and experimentation but, on the other hand, call for reinstatement of traditional knowledge based on locally developed practices of water use. This paper describes smart water system as a system that implements meaningful data and transforms it into actionable intelligence, but at the same time built upon traditional knowledge. Desk research, transect coding, and multi-scale modeling are used as research methods to answer the question on how to smartly manage urban water systems at different scales and in different types of urbanized landscape.

Introduction

In recent decades, the world has experienced unprecedented urban growth. According to the United Nations predictions, by 2050, more than 65% of the world population will be living in cities [47]. Europe is, for example, an increasingly urbanizing continent, where currently roughly 75% of the total EU population lives in cities, towns, and suburbs [12], and predictions estimate more than 80% of total urban population [47]. With such urban population growth, it is inevitable that the demand for water increases and pressure on finite water resources intensifies [4, 29, 32, 47]. Another starting point for this research is the assumption that the global climate is changing, as stated in many scientific records [8, 15, 21], and, according to several publications, has an increasing trend [34, 38, 41, 44, 47]. Climate change express itself through the increase of disruptive events that puts the urban water

---

M. Lazarová (✉) · M. Varga · D. Gažová
SPECTRA Centre of Excellence EU, Slovak University of Technology in Bratislava, Bratislava, Slovakia
e-mail: martina.lazarova@stuba.sk

© Springer International Publishing AG, part of Springer Nature 2019
D. Cagáňová et al. (eds.), *Smart Technology Trends in Industrial and Business Management*, EAI/Springer Innovations in Communication and Computing, https://doi.org/10.1007/978-3-319-76998-1_9

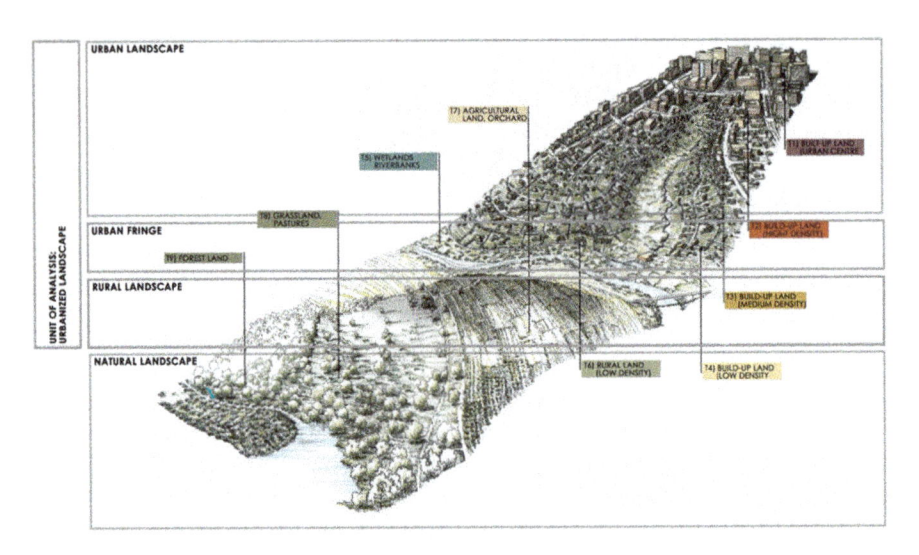

Picture 9.1 Typology of an urbanized landscape described by transect method. The concept represents a visible break in the continuity of the traditional urban-to-rural transect. Today's changes in society and environment had resulted to the creation of a landscape affected by urbanization processes. People become increasingly mobile, and the ecological footprint of the urbanites now stretches far beyond the cities. Resource: Authors according to [1, 40]

systems that include *water supply*, *sewage system*, *rainfall and evaporation*, *groundwater resources*, and *water in aquatic ecosystems* under high risks [14, 19]. Consequently, the need to ensure that water can be managed sustainably, operated efficiently, and maintained in a high-quality standard has been raised. Using big data techniques from all urban water components has, therefore, the potential to enable smart water management. The research understood the water system in all its complexity, which means all water forms and all its spatial appearances. Urbanization, accessibility, and globalization are the driving forces that cause the polarization between more intensive and more extensive land uses. In general, the speed of changes in landscape patterns, their frequency, and scale have an increasing trend [3, 5, 9, 24]. These landscape changes and emergence of new stretches far beyond the urban landscape justify why the paper's unit of analysis is the urbanized landscape (Picture 9.1).

Today the smart city concept is a widely spread concept, but often used without critical reflection to its proper meaning or specifics. The concept of the smart city has recently been introduced as "a strategic device to links legacy systems with new communication chains in order to achieve a common goal of human welfare without compromising the sustainability of dependent ecosystems" [22]. Further, the research defines the term smart water management that is built upon the broader smart city concept. From water management point of view, the concept of *smart water management* highlights the importance of information and communication technologies in the last 20 years for enhancing water sustainability. The concept was originally developed by large IT companies that focus on analyzing "big data" using software-centric, top-down approach. Nevertheless, when it comes to the modernization of hundred-year-old water system, advanced software and networking capabilities are rarely broad enough to make self-adaptive systems. Over the last decades, there has been increasing interest to participate in the decision process and to provide their locally based knowledge. Simply put, urban water systems need to be built upon time-tested practices and methods, which contribute to the variability of urban landscape patches [42]. Thanks to the local knowledge that many cultures were successful in sustaining their water resources for centuries. Therefore, smart water management concept needs to reclaim local knowledge of water management practices, thought the challenge of climate change to present day.

Problem Statement

Over the past decade, the concerns about extreme events and water scarcity have grown also within the EU [36, 47]. And why it is so? Simply put, water scarcity is often a by-product of climate change, current water management practice, and unsustainable water policies. Moreover, current spatial planning practice does not enhance smart water management of urban landscape with an adaptive capacity to water-related problems. Hence, cities do not easily adapt to the unpredictable events. As a result, the current academic debates are mostly talking about global water crisis that is heavy on problems and light on solutions. But what can be understood by the term – crisis? The word crisis, from its Greek roots *krisis*, refers to a time of great danger, difficulty, or confusion when problems must be solved or important decisions must be made [10]. Crisis signifies a time of decisive action, to a turning point, that may make things worse or better. A crisis also implies opportunity and not necessarily a disaster. Therefore, the paper defines the water crises as a highly needed wake-up call to action and as a problem that has to be solved; otherwise it will become irreversible [35].

In addition, the European Conference of Ministers responsible for Regional Planning [7] defined spatial planning as the "geographical expression to the economic, social, cultural and ecological policies of society" and "an interdisciplinary and comprehensive approach directed toward a balanced regional development and the physical organization of space according to an overall strategy." They directly spoke of the different management methods used in spaces of various scales [7]. In water planning practice, however, the application of resilience thinking is often limited in scope of institutional structure in relation to spatial scale [43, 48]. Multiple levels of governance typically do not fit all water issues, resulting in inefficiencies and management gaps. Tension exists also between the traditional rooted hierarchies of national systems and trends toward both the upscaling of governance in the form of multinational agreements or the growing influence of the European Union and the downscaling in the form of decentralization of decision-making involving a local context. It is important to realize that due to increasing uncertainties caused by climate and global change, there is a need to also change management boundary conditions.

Research Objectives

Although extensive academic research is focused on new water technologies [13, 20, 27], little attention has been given to the attributes of local knowledge and cultural values of historical water practices [23, 31, 35, 45] within the concept of smart city. Through analysis of recent scientific articles and research programs, the paper states that the literature on smart city concept does not describe smart water management as its sub-concept [39]. Thus, the first objective of the paper is to develop an adjusted framework of smart water management, moving beyond technical solutions and communication channels. In this paper, the research presents an analysis of three discourses in water management. One is the established discourse "fight the water" by water engineering approach, the other is the new discourse "the room for water" by water landscape approach, and the third is "living with water" by cultural approach. The second objective of the research is to design a cross-scale theoretical model that provides a proper insight on localizing smart water management solutions and governance options.

Research Question

Given the contexts sketched before, the basic theoretical question of the research is: *What scale is the most appropriate for addressing the principles of smart water system management?* The research question is focusing on localizing smart water management strategies on different levels.

Research Methodology

The proposed theoretical framework of water system resilience introduces vertical and horizontal axes, each with their own dynamic and spatial appearance. The theoretical framework uses tools as *desk research*, *transect* coding, and *multi-scale modeling*. In this phase, the spatial elements derived from the literature study were aggregated to different spatial levels (framework's vertical axis) and heterogeneity of urbanized landscape (framework's horizontal axis).

Framework's Vertical Axis

Cross-scale comparisons are made through an examination of the water management tools that differ across the scale. Scale is important in dealing with complex systems; therefore the research follows the concepts of systems thinking and hierarchy theory. A complex system is one in which many subsystems can be discerned.

Many complex systems are hierarchic – each subsystem is nested in a larger subsystem, and so on. Therefore, transect coding and multi-scale modeling are used as methods to addressing multi-scale relationships of water system resilience.

Framework's Horizontal Axis

A range of different types of urbanized landscape is expressed by transect coding method. Transect units are distinguished by their level of intensity of their natural, built, and social components, rate of urbanization, and disturbance regimes. Transect units may be coordinated to all scales of planning, from the national, regional, and municipal through the community scale down to the individual lot and building. The zoning used in the research is applied at the municipal scale.

Framework's Nodes

The research reviews policy documents and literature to define the water governance model in Slovakia under the directives, strategies, and plans in the national, regional, and local levels. To compare the basin and watershed boundaries with administrative ones, the framework analyzes data from official reports on the national water plan, river basin plans, flood risk plans, master plans, and statistics.

By assuming that the smart management of water is a messy and contested concept, the research takes different schools of thought that describe how water management can be influenced by various underlying perspectives. In the context of global water crises, it has become increasingly clear that water problems have traditionally been approached from three different perspectives. One is the established discourse "fight the water" by water engineering approach, the other is the new discourse "the room for water" by landscape-oriented approach, and the third is "living with water" by taking into account human factor.

Engineering Perspective

Until just a few decades ago, the materialization of great possibilities caused by industrial revolution brought the massive hydraulic development with large-scale, centralized water infrastructure systems for flood control, irrigation, water treatment, water distribution, and sewage systems (known as "hard path"). On the one hand, there was the precondition that technologies generate wealth and development. Under this view, the ancient wish to transport water from where it is abundant to where it is scarce was satisfied and applied. In general, we know that water is both a key to socioeconomic growth and quality of life [35]. But expanding research about engineering approach raises more and more questions. Concerns have been raised that "one-type-fits-all" solutions are not effective, and for the sake of water engineering, the urban landscape is losing its capacity to adapt to unexpected changes.

Moreover technical solutions often do not solve the problem but just transfer it elsewhere. Critics of engineering approach to water management also place spotlights on the phenomenon that have been mostly seen within urban landscape, where small water cycles are often disturbed. As a consequence, the urban landscape has no adaptive capacity to deal with internal or external disturbances. On the one side, following a century of blinding technicism, the current adaptation projects call for a new perspective that reintegrates water into spatial perception and water-sensitive design, but also gives a rise to new approaches that restore the ethics toward water and its cultural trails. However, in practice the right to decide is often in the hand of separate specialists (mostly engineers) that operate according to their own engineering techniques. Therefore the significant challenge of current water management is the implementation of adaptive water governance.

On the other side, according to UN-Water agency [46], the technology refers not only to physical equipment (as infrastructures and installations), but in a broader perspective, it also supports the innovations. Moreover, as stated by the United Nations, "there are a number of on-going international initiatives aimed at accelerating the development, diffusion and transfer of appropriate, especially environmentally sound, water technologies" [47]. It is important to realize that although several technological achievements have been reached in the last decade, there is still a gap in this research field [33]. Environmental technology adoption enables societies to reduce their environmental impacts on water circulation. These green technical solutions however require a data with complex scope (data from public and private sector) and proper understanding of the context in which they are applied. Only access to reliable information can help to overcome their failure. Furthermore, water systems evolve gradually and organically that cannot be fully improved or "smartened up." They are almost never designed on "blank slate." It is not the aim of this paper to contribute to scientific debate regarding new water technologies based on environmental ground, but the point of the paper is to show how the progress in technological innovation research is needed for smart water management.

Landscape-Oriented Perspective

In light of describing environmental concerns (global climate change, water pollution, landscape dewatering, etc.), the industrialized world faces a massive decrease in biodiversity that is reflected in political debates and new approaches [34, 38, 41, 44]. Taking into consideration the scope of water crises, ecologist and landscape architects warn that the current water management practices in the urban landscape are no longer resilient. A review of the way in which the natural hydrological cycle and adaptive capacity of landscape are in the core gives the first priority to the natural ecosystems. Under this perspective, ecosystems have similar rights to people and should be treated with the same sensibility. Moreover, in the context of water management, the main idea of this approach is to support natural water movement in two stages. The first is renewability of small water cycles by promoting natural water processes as filtration, infiltration, and self-purification. The second is preserving the large water cycles by resilient management of cultural landscape what means to sustain the regional diversity of cultural landscapes. Critics of this school of thought state that this approach has high space requirements and therefore cannot be applied in a wide variety of context. In a highly dense urban landscape, landscape-oriented solutions, retention basins, rainwater gardens, or irrigational belts, are not possible. However, there is a chance to replace them with less effective but meaningful solutions as green roofs or green building walls, etc. ([25, 26] Catalogue of hard and soft water solutions). Moreover, these "green" approaches, by contrast, have been relatively quick to address innovative on-site water solutions. However, most of these green projects focus exclusively on site-scale water management. Therefore, there is a great need to bring together the site-scale innovation being driven by the water landscape movement with the watershed management and integrated infrastructure planning being increasingly promoted and implemented by communities. Next spot on landscape-oriented approach is the concern, from the side of water engineers, that these "soft" solutions have small capacity to absorb the water. Another problematic issue is that in areas with high water table, this approach could cause soil instability (as a result of waterlogging). The paper simplifies that especially in the urban landscape, it is highly needed to combine the landscape solutions with technical ones. Following Lazarová [25, 26], the key areas suitable for implementation of landscape-oriented solutions with higher space requirements are the rural-urban fringe. These zones are seen as edge boundary between the settlement and open landscape, where the urban density is allowing usage of the ecosystem-oriented solutions.

Cultural-Oriented Perspective

What is important for the following perspective of cultural approach to water management is the fact that basic and collective perceptions about the world (such knowledge, attitudes, values, ethics, etc.) are stored within every culture. These long-time preserved patterns influence our behavior and management practice [23]. Under this view, the water culture refers to a certain stage of knowledge that is a result of mutual interactions between people and natural resources – as water. This knowledge is rooted locally in management practices, values, religions, and ethics that have been preserved in customary laws. Over the millennia people shaped their traditions in response to the distinct environment they inhabit. Along the way, the current water crises have been partially generated by techno-scientific cultures. Scientists who follow this cultural approach argue that current management models are still dominated by a paradigm of "expertise." In other words, mainly engineers, politicians, and water managers are deciding. Therefore the research calls for a broad-based cultural expertise that will take into account also the local knowledge and time-tested solutions. It is necessary to realize that local knowledge, cultural stewardship, and traditional practice developed by different communities preserve the cultural diversity. A better understanding of these cultural water beliefs and practices may lead to new concepts in understanding the water resilience – from flood management to water supply, sanitation, and irrigation management. For this reason the challenge that the present-day society has to face is to match up to its ancestors: to give the adequate response to the moment in which that society is living [6].

Defining Smart Water Management Concept

The paper focuses on three discourses of smart water management that are deeply described above. Firstly, engineering perspective has in fact too narrow focus on maintaining efficiency, constancy of the system, and predictable future. It aims to conserve what we have and to "fight the water." To do so, it offers progressive replacement of time-tested strategies by "one-type-fits-all" solutions. The invention of electrical pump was to water management as what the elevator was to high-rise building or the car to transport. They all include a paradigm shift with socioeconomic, environmental, and also spatial consequences. Many of the technical inventions are thus being critically reconsidered, but meanwhile there is still a need for new and next level technologies. Secondly, the landscape-oriented approach is strongly linked to ecosystems. This discourse is focused on building "the room for water." The third discourse is "living with water" by traditional cultural approach based on implementation of local knowledge. More specifically, the following research questions were stated: *How to smartly manage water systems at different scales and in different types of urbanized landscape?*

The paper defines the smart water management concept as an overarching, interdisciplinary framework in which insights from different approaches (engineering, landscape-oriented, and cultural) will fit and will result to the development of a water-resilient city. The paper considers mentioned approaches to be complementary and both useful at different territories and different scales (Picture 9.2). The paper appoints that smart water management should implement relevant data and transforms it into systematic and intelligent decision-making process at all levels (water governance). Conflict between these scales sometimes leads to conflicting management decisions, and subsequently an erosion of resilience.

Theoretical Framework of Smart Water System

Based on the aforementioned theoretical discussion, the research defines the concept of water system resilience. The focus of the thesis is to study the resilience of water systems and to meet the *new* challenges of water governance. Through this chapter, the research summarizes the following key principles for water system resilience framework:

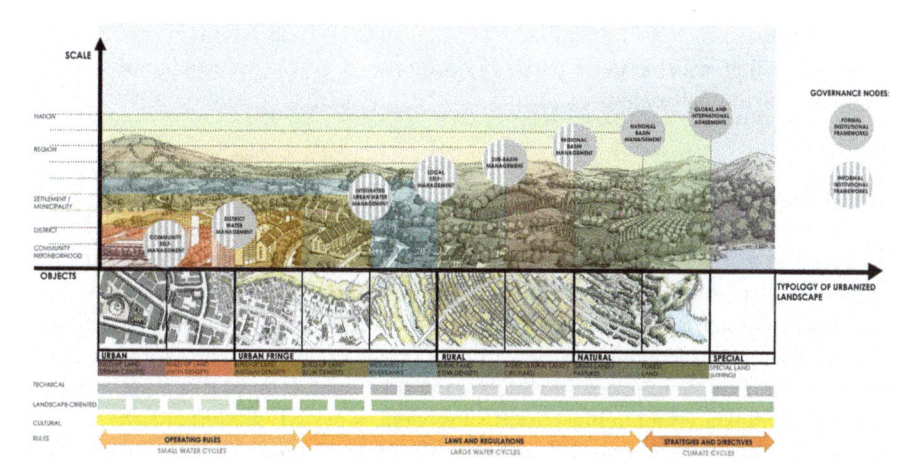

Picture 9.2 Model of smart water management as a cross-scale theoretical model that provides a proper insight on localizing water management solutions (model's horizontal axis) and governance options (model's nodes). Resource: Author according AECOM [1]

1. The increasing vulnerability of water resources calls for holistic strategies and diverse approaches to water management to be recognized and supported.
2. A move to technical management to true integration of the human and environmental dimensions, water system resilience is a combination of water landscape perspective, water engineering perspective, and culture-sensitive perspective.
3. Adaptive water management and adaptive governance are the key tools to address water system resilience. Make management more adaptive and flexible to make it operational under fast changing socioeconomic boundary conditions and climate change. At the same time, water issues must be tackled at various levels. Adaptive water governance requires a "dance between levels." Potential synergies between Regional Basin Management and Local Self-Management should be strengthened. Development and implementation of both of these management options have potential for:

   – Bringing a long-term, strategic focus covering large areas by using Regional Basin Management
   – Translation of local knowledge and traditional water use into Local Self-Management action

4. Link different vertical interplays (national, regional, local) with different types of landscape management.
5. Adaptive water management refers thus to a systematic process for continually improving management policies and practices by learning from the outcomes of implemented management strategies.
6. Preserving biodiversity of urbanized landscapes and complexity of water catchments need focus on the regional level to protect regional diversity and to overcome system fragmentation. Various landscape types have adequate buffer habitat and are connected by dispersal routes. On the other hand, socio-ecological

complexity needs to be researched by spatial analysis tools based on concepts of landscape that are able to be partially decomposed over a wide range of spatial and organizational scales.

Vertical Perspective: Scale

Safeguarding resilience requires appropriate management decisions by people using their society's cultural norms and institutions at different scales. Conflict between these scales sometimes leads to conflicting management decisions, and subsequently an erosion of resilience. Scale is important in dealing with complex systems. A complex system is one in which many subsystems can be discerned. Many complex systems are hierarchic – each subsystem is nested in a larger subsystem, and so on [2, 28]. For example, a small watershed may be considered an ecosystem, but it is part of a larger watershed that can also be considered an ecosystem and a larger one that encompasses all the smaller watersheds. Similarly, institutions may be considered hierarchically, as a nested set of systems from the local level, through regional and national levels, to the international level. Phenomena at each level of the scale tend to have their own management principles [17]. Therefore, complex systems should be analyzed, managed, and governed simultaneously at different scales. The need to use a multiplicity of perspectives follows from complex systems thinking. Because of a multiplicity of scales, there is no one "correct" and all-including perspective on a system. In complex systems, time flows in one direction, time's arrow is not reversible. Especially with social systems, it is difficult or impossible to understand a system without considering its history, as well as its social and political contexts. For example, each large-scale management system [18] or each local level [30] will have its unique history and context.

A broad range of hierarchical scales can be found within the water governance, ranging from global to local. The theoretical framework identifies the following scales and their governance nodes:

(A) *Large scales* (global, international, national) consist of stabilizing factors that constitute the context for water management. These scales reflect environmental variability, legal frameworks, and deeply rooted societal norms. The large-scale levels must not be independent from the micro- and meso-level since feedback processes can operate bottom-up (e.g., diffusion of innovation) and top-down (e.g., choice of strategies). These large scales are characterized by implementation of strategies and directives. The goal of large scales is to establish an ecological security pattern at the EU and national levels, which would protect the most water-sensitive landscape and strategy for both wise conservation and wise development.

(B) *Mesoscales* (basin, catchment, and sub-catchment areas) with stabilizing interdependencies between large and small scales are typical with implementation of laws and regulations. These scales establish a regional water security pattern to develop flood security at the basin and catchment level. The landscape is set up to provide maximum water-retaining capacity. Storm water management and flood protection depend on interconnected networks of wetlands, low-lying grounds, waterways, dikes, dams, etc. Special concerns are given to the most permeable lands that allow rainwater to infiltrate and create large water cycles.

(C) *Microscales* where innovative approaches can develop in a locally protected environment (e.g., subsidized pilot studies) and in new areas of application such as the restoration of riverine landscapes have started to become an integral part of water management. Operating rules are applied on these scales. A fundamental idea of these scales is to bring the settlements back to the water, to revive local knowledge and traditions of water use and management. The aim is to integrate and connect natural, engineering, and cultural management techniques.

Horizontal Perspective: Landscape Heterogeneity

The horizontal axis of the *theoretical framework of water system resilience* clearly to distinguishes the urbanized landscape from natural landscape that is further divided to spatially bounded patterns – *transect units*. The research unit of analysis is the urbanized landscape that is defined as the extension to urban landscape. The research defines urbanized landscape as landscape modified or influenced by urbanization activities. According to this understanding, such landscape describes a certain category of land that has been shaped by urbanization processes. In general, the speed of changes in landscape patterns, their frequency, and scale have an increasing trend [3, 5, 9, 24]. For example, increasing mobility, globalization, and urbanization are important driving forces of landscape changes and emergence of new stretches far beyond the urban landscape. The research classifies all these new stretches, landscape structures, and spatial appearances in the concept of urbanized landscape. The urbanized landscape therefore includes the following transect segments: urban landscape, urban fringe, rural landscape, and special types of landscape.

In order to preserve integrity, connectivity, and viability of natural habitats and compactness of urban zones the *theoretical framework* in its horizontal axis uses the transect method. The transect coding as a methodological tool defines the urbanized landscape and organizes the pattern of urban-to-natural elements. This alterative system of zoning has been developed by the New Urbanism Association in a compendium of principles named *The Lexicon of New Urbanism*, published by Duany et al. [11]. Duany's transect code consists of six basic transect units: *rural preserve, rural reserve, suburban, general urban, urban center,* and *urban core*. The research builds upon this classification and expands the original transect with detailed landscape classification and with hierarchical structure. The framework's transect segments consist of *urban core, urban fringe, rural areas, natural areas, special/other areas*. This main transect units are then further divided to subunits: *T1 Built-up land (urban center), T2 Built-up lands (height density), T3 Built-up land (medium density), T4 Built-up land (low density), T5 Wetlands/riverbanks, T6 Rural land (low density), T7 Agricultural land/orchard, T8 Grassland/pastures, T9 Forest land,* and *T10 Special types of land*, which are deeply defined in Picture 9.3. The transect

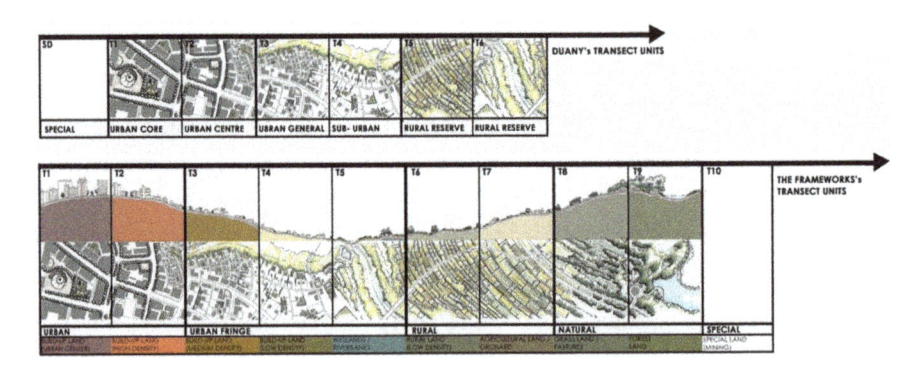

Picture 9.3 The research compares the original Duany's transect (T1–T6) and expands the original transect with landscape units (T1–T10). Resource: Authors according [1, 11]

identifies a range of habitats from the most urban to the most natural that have distinguishable structure and varying levels of urban intensity and disturbance regimes. Patterns are similar in some key elements and processes, but at the same time could be consider as heterogeneous. It is also important to realize that there are different ways in which transect zones can be fitted together. For instance, we have cases where high-density build-up land has common borders with protected natural sites of forest (see Picture 9.4, polycentric transect pattern). The transect as a conceptual framework relies on interdependencies in order to work. One zone is defined by its relation to others – it is either more or less urban, more or less rural, identified by its position along the rural-to-urban continuum. This chapter is a broad review of urban–natural continuum that ranges from the urban core through inner and outer urban fringe, a zone of urban shadow, and out to the rural and natural landscape. It is curtailing to realize that this continuum includes several levels (dimensions) of urbanization that create a complex spatial pattern.

It is important to realize that each of the transect units has an extent (area occupied or influenced) and grain (resolution). This means that different hierarchical levels target different landscape types. For example, the national level focuses mainly on the management of natural landscape through long-range comprehensive strategies. The regional level primary targets the rural landscape and urban fringe with mid-range plans, and the local level focuses on the urban landscape through short-range zoning codes. Picture 9.4 shows that each of the transect units is part of a larger system. The upper picture shows a simplified transect expression of a monocentric continuum defined by transect coding method. The lower picture shows different types of transect expression with more polycentric pattern with a complex boundary. In this case, urban areas are surrounded by a rural and natural landscape with a complex boundary. In the polycentric expression, the urbanized landscape becomes a geographical territory expressed as an area influence by urbanization processes. For example, large cities are often shaped by patterns of urban sprawl [3] along the main access roads or mobility nodes (metro stations, etc.) and by the phenomenon of urban shadow and urban implosion. Such urban networks cause that cities shape their surrounding areas, where the major city tends to expand its urban fringe and at the same time to influence the surrounding villages by functional urbanization.

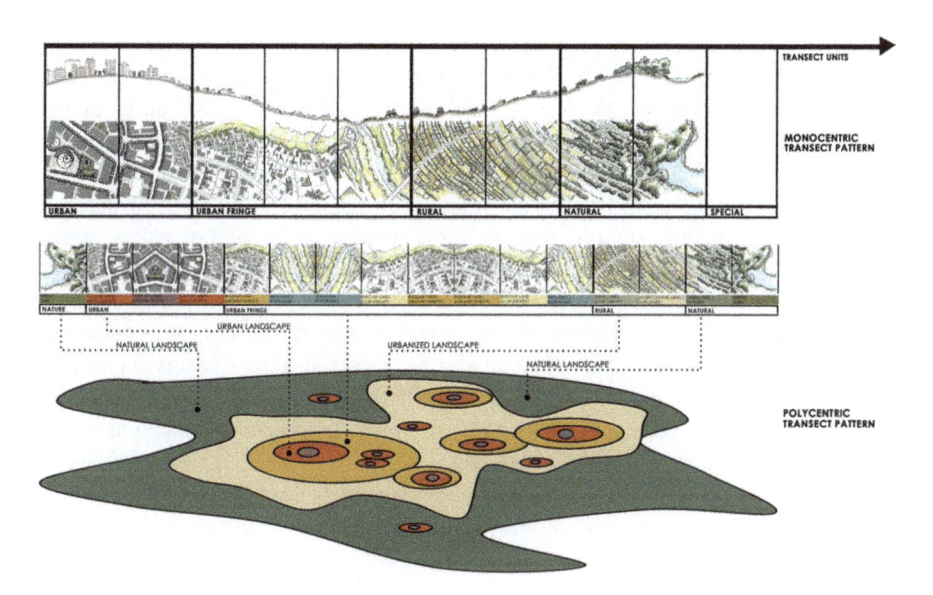

Picture 9.4 Graphically expressed transect of different landscape patterns and scales. (A) Monocentric transect pattern, as a method, is used to illustrate different types of urbanized landscape from urban to natural. This graphical interpretation shows how the aspects of urbanization, such as land uses, and density respond to research definition of urbanized landscape. (B) Using polycentric spatial patterns to link transect units with different hierarchical scales. This understanding is crucial for the theoretical framework interpretation in the next chapter. Resource: Author according to [1, 37, 40]

The transect method supports the view that the crisis of metropolitan form is not one of too high density or too low density, but it is caused by inappropriate mixing of elements that need to be reintegrated in a more suitable way. To sum it up, urbanization and its associated accessibility define the relationship between urban and rural landscape. Therefore, the research use in the *theoretical framework* is the transect method, which combines both (monocentric and polycentric) patterns and merges it with different ranges of hierarchical scales (national, regional, and local). For analyzing the urban continuum, these measures need to be considered: *density, continuity, concentration, centrality, mix of uses*, and *proximity* [16].

Framework's Nodes

The research starts by outlining the relevance of spatial and scale mismatches as a unifying concept that created water system resilience. The final piece in the theoretical framework concerns the responses of governance system. These come with the aim of solving problems stated in Problem statement. One of the most dramatic water issues besides (climate change, destruction of large and small water cycles, etc.) remain institutional mismatches and a gap between formal and informal institutional frameworks. To solve the water crisis, integrative strategies are needed that will link basin and catchment management with administrative boundaries. The key of this chapter is to build a theoretical framework with water spatial governance nodes processed across scales. These nodes can be qualitatively assessed by comparing the type of urbanized landscape (Picture 9.4 – horizontal axis) with a particular vertical scope (Picture 9.4 – vertical axis). It is important to realize that scale mismatch is more likely to be the rule rather than the exception for most natural resource problems. For example, river systems cover a wide range of scales, from international or national (such as the Danube River Basin) to multiple regions (such as the Vah River sub-basin). Importantly, even though many of these managed ecosystems are defined by fixed geographical or spatial scales, they are always subject to influence by ecological processes operating across different scales including changing climate or water-related problems. On a regional scale (basin management, catchment and sub-catchment levels), water resources are managed at naturally formed boundaries. On the other hand, on more local scales (municipality, district, community levels), the boundaries were socially defined as a reflection of history, culture, economics, and politics or a myriad of other driving forces. More importantly, despite connectivity to the water source, problems that arise in water governance may range from local to basin wide. Rather than a one-size-fits-all approach to water governance, a mechanism to adapt response scale to the problem is needed. Both formal and informal institutional frameworks across existing governing institutions may be one aspect of that mechanism. The role of law in network formation should be to provide authority for exchange of information and collaboration or to step aside when it creates barriers to such exchange.

Conclusions

Following an increasing series of unpredictable events (such as floods or droughts), more and more collective actions and initiatives calling for change are emerging. Furthermore, the current uncertainties pose special challenges, because planning processes based on uncertain predictions provide only an unclear approximation of the future and are a weak basis for smart water management. Therefore, the purpose of the research is not to find the "best solution" but outline the strategy to accept the unexpected as expected and plan ahead to fight current environmental changes. To overcome the limitations of problem-solving methods, the paper requires assessing the root causes of water problems. In such approaches, indigenous and local communities are recognized as invaluable partners that offer a wide variety of time-tested solutions. However, the local water problems are mostly managed and governed by regional, national, or sometimes even international organizational mechanisms (see Picture 9.2). As a result, water resources are brought under centralized, bureaucratic control, and the resilience of local water forms is strongly weakened [23]. Therefore, a better understanding of cultural values together with implementation of new and next level technologies is essential to catalyze change for smart water management regimes.

Acknowledgments This contribution is the result of the project implementation:

SPECTRA+ No. 26240120002 "Centre of Excellence for the Development of Settlement Infrastructure of Knowledge Economy" supported by the Research and Development Operational Programme funded by the ERDF

VEGA No. 1/0652/16 "Vplyv územného umiestnenia a odvetvového zamerania na výkonnosť podnikateľských subjektov a ich konkurencieschopnosť na globálnom trhu."

References

1. AECOM. (2010). *Kigali conceptual master plan*. Denver: Oz Architecture.
2. Allen, T. F., & Starr, T. B. (1982). *Hierarchy perspectives for ecological complexity*. Chicago, IL: The University of Chicago Press.
3. Antrop, M. (2000). Background concepts for integrated landscape analysis. *Agriculture, Ecosystems and Environment, 77*, 17–28.
4. Arup. (2011). *Water resilience*. s.l.: Arup Urban Life.
5. Brandt, J. (2000). Demands for future landscapes research on multifunctional landscapes. In *Proceedings of the conference on multifunctional landscapes—interdisciplinary approaches to landscape research and management*. Roskilde, October 18–21.
6. Cabrera, E. (2010). *Water engineering and management through time, learning from history*. Valencia: CRC Press. ISBN 9780415480024.
7. CEMAT (1983). *European regional/spatial planning charter*. European Conference of Ministers responsible for Regional Planning (CEMAT), Torremolinos, Spain.
8. COM. (2015). *The Paris protocol – blueprint for tackling global climate change beyond 2020*. Brussels: European Commision.
9. Council of Europe. (2000). *The European landscape convention*. Strasbourg: Council of Europe.
10. Dictionary, Oxford English (2004). *Oxford English dictionary online*. Mount Royal College Lib., Calgary, 14.
11. Duany Plater-Zyberk & Co. (2000). *The lexicon of the new urbanism, version 2.1*. Miami: Duany Plater-Zyberk & Co.
12. Eurostat (2016). *A statistical portrait of cities, towns and suburbs across the Europen Union*, STAT/16/2981, 7 September. Brussels: Eurostat Press Office.
13. Fishman, C. (2012). *The big thirst: The secret life and turbulent future of water*. New York: Simon and Schuster.
14. Fletcher, T., Mitchell, V. G., Deleti, A., Maksimovic, Č. (2008). Urban water system components. In *Management, data requirements for integrated urban water*. Leiden: Taylor & Francis.
15. Friedlingstein, P., et al. (2014). Persistent growth of CO_2 emissions and implications for reaching climate targets. *Nature Geoscience, 7*(10), 709–715.
16. Galster, G., Hanson, R., Ratcliffe, M. R., Wolman, H., Coleman, S., & Freihage, J. (2001). Wrestling sprawl to the ground: defining and measuring an elusive concept. *Housing policy debate, 12*(4), 681–717.
17. Gunderson, L. H., & Holling, C. S. (1995). *Barriers and bridges to the renewal of ecosystems and institutions*. New York: Columbia University Press.
18. Gunderson, L. H., Holling, C. S., & Peterson, G. D. (2002). Surprises and sustainability: Cycles of renewal in the Everglades. In *Panarchy: understanding transformations in human and natural systems* (pp. 315–332). Washington, DC: Island Press.
19. Howe, B., Cole, G., Souroush, E., Koutris, P., Key, A., Khoussainova, N., & Battle, L. (2011). Database-as-a-service for long-tail science. In *International Conference on Scientific and Statistical Database Management* (pp. 480–489). Berlin, Heidelberg: Springer.

20. Ingildsen, P., & Olsson, G. (2016). Smart water utilities: Complexity made simple. *Water Intelligence Online, 15*, 9781780407586.
21. IPCC. (2007). *The physical science basis, working group I contribution to the intergovernmental panel on climate change fourth assessment report*. New York: Cambridge University Press.
22. ITU. (2015). *Focus group on smart water management*. Reading: ITU Committed to Connecting the World.
23. Johnston, B. R. (2012). *Water, cultural diversity, and global environmental changes*. Paris: Springer. ISBN 978-94-007-1773-2.
24. Klijn, J., & Vos, W. (2000). A new identity for landscape ecology in Europe: A research strategy for next decade. In J. Klijn & W. Vos (Eds.), *From landscape ecology to landscape science* (pp. 149–162). Dordrecht: Kluwer Academic Publishers.
25. Lazarová, M. (2015a). *New water culture in cultural landscape*. Bratislava: Slovak University of Technology Press. ISBN 978-80-227-4443-0.
26. Lazarová, M. (2015b). New water culture under fuzziness. In České vysoké učení technické (Ed.), *Fuzzy responsibility, multi-actors decision making under uncertainty and global changes: Book of contributions* (pp. 104–116.) ISBN 978-80-01-05763-6.
27. Lu, F., Ocampo-Raeder, C., & Crow, B. (2014). Equitable water governance: Future directions in the understanding and analysis of water inequities in the global south. *Water International, 39*(2), 129–142.
28. Marceau, D. J., & Hay, G. J. (1999). Remote sensing contributions to the scale issue. *Canadian Journal of Remote Sensing, 25*(4), 357–366.
29. Nilsson, D. (2006). *Water for a few: a history of urban water and sanitation in East Africa*. Stockholm: KTH.
30. Ostrom, C. W. (1990). *Time series analysis: Regression techniques* (Vol. 9). Newbury Park, CA: Sage.
31. Ovink, H. (2015). Reform by design. *Journal of Extreme Events, 2*(01), 1502001.
32. PAI. (2011). *Why population matters to water resources*. Washington, DC: Population Action International.
33. Parodi, O. (2010). Towards resilient water landscapes. In *Proceedings of the international symposium on water landscapes at the university*. New South Wales: KIT Scientific Publishing. ISBN-13: 978-3866444980.
34. PBL, KNMI, WUR. (2009). *News in climate science and exploring boundaries, a policy brief on developments since the IPCC AR4 report in 2007*. Bilthoven: PBL. ISSN 500114013.
35. Priscoli, J. D. (1998). Water and civilization: Using history to reframe water policy debates and to build a new ecological realism. *Water Policy, 1*(August), 623–636.
36. Rahaman, M. M., Varis, O., & Kajander, T. (2004). EU water framework directive vs. integrated water resources management: The seven mismatches. *International Journal of Water Resources Development, 20*(4), 565–575.
37. Ravetz, J., Fertner, C., & Nielsen, T. S. (2013). The dynamics of Peri-urbanization. In *Peri-urban futures: Scenario and models for land use change in Europe* (p. 453). Berlin, Heidelberg: Springer. ISBN 978-3-642-30528-3.
38. Richardson, K., Steffen, W., Schellnhuber, H. J., Alcamo, J., Barker, T., & Kammen, D. M. (2009). *Climate change—synthesis report: Global risks, challenges and decisions, Copenhagen 2009*. Copenhagen: University of Copenhagen.
39. Rockström, J., et al. (2014). *Water resilience for human prosperity*. Cambridge: Cambridge University Press.
40. Soediono, B. (1989). Resilience and the cultural landscape, *Journal of Chemical Information and Modeling* (Vol. 53), Available at: https://doi.org/10.1017/CBO9781107415324.004.
41. Sommerkorn, M., & Hassol, S. J. (2009). *Arctic climate feedbacks: Global implications*. Oslo: WWF International Arctic Programme.
42. Sporrong, U. (1998). Dalecarlia in Central Sweden before 1800, a society of social and ecological resilience. In F. Berkes & C. Folke (Eds.), *Linking social and ecological systems:*

Management practices and social mechanisms for building resilience (pp. 67–94). Cambridge: Cambridge University Press.
43. Swyngedouw, E. (1997). Neither global nor local: "Glocalization" and the politics of scale. In *Spaces of globalization: Reasserting the power of the local* (Vol. 1, pp. 137–166). New York/London: Guilford/Longman.
44. Tin, T. (2008). *Climate change: Faster, stronger, sooner, an overview of the climate science published since the UN IPCC Fourth Assessment Report*. Brussels: WWF European Policy Office.
45. Tinoco, M., Cortobius, M., Doughty Grajales, M., & Kjellén, M. (2014). Water co-operation between cultures: Partnerships with indigenous peoples for sustainable water and sanitation services. *Aquatic Procedia, 2*, 55–66.
46. UNDP (2015). Transforming our world: The 2030 agenda for sustainable development. Available at: http://www.naturalcapital.vn/wp-content/uploads/2017/02/UNDP-Viet-Nam.pdf.
47. United Nations (2014). *World urbanization prospects: The 2014 revision, highlights*. Department of Economic and Social Affairs. Working Paper No. ST/ESA/SER.A/352.. ISBN 978-92-1-151517-6.
48. Young, O. R. (2002). *The institutional dimensions of environmental change: Fit, interplay, and scale*. Cambridge, MA: MIT Press.

Chapter 10
RFID Labels and Its Characteristics on Labeled Products

Dušan Dorčák, Romana Hricová, and Peter Šebej

Abstract Modern approaches to tracking, recording, and determining units in production technology are typically burdened by physical, chemical, and other influences that are characteristic of the current type of technological processes. This load causes changes in the properties of information carriers, stickers, labels, responders, tags what brings attrition, aging, reduced durability and loss of function. This article provides a preview of the principles, methods and options for designing impact tests on elements of the information process chain (RFID, IoT and others). Supplemented by the theoretical example of effective analyzes of the presented task.

Introduction

Virtually, every element in the production process carries the designation and information (IoT) of the unit itself and also the operations that have already been performed and their quality of execution. Another function of information carriers is the fulfillment of control, quantitative, and qualitative indicators. By implementing them, we will achieve better monitoring and control of the production process, while at the same time, there is a potential damage and a decline in their functionality. A serious problem occurs when products lose their function during their lifetime.

Its knowledge of fact, but confirms the properties and an exact system view, the current understanding of labeling (marking is creating and retaining the corresponding,

---

D. Dorčák (✉)
Development and Realization Workplace of Raw Materials Extracting and Treatment, Technical University of Kosice, Kosice, Slovakia

R. Hricová · P. Šebej
Department of Manufacturing Management, Faculty of Manufacturing Technologies with a seat in Presov, Technical University of Kosice, Presov, Slovakia
e-mail: romana.hricova@tuke.sk

© Springer International Publishing AG, part of Springer Nature 2019
D. Cagáňová et al. (eds.), *Smart Technology Trends in Industrial and Business Management*, EAI/Springer Innovations in Communication and Computing, https://doi.org/10.1007/978-3-319-76998-1_10

readable value for redefining the original attributes, with the need to present an entity creation genesis with the possibility of reapplicability throughout the entity's lifetime) does not confirm the ordinary equivalence (picture) with its knowledge of fact, but confirms the properties and possibilities of unambiguous attribution, determination to the original what means to the next subject either directly or by record, and also to the validity or correctness of the interpretation of the information being carried. (Other skills, abilities, human habits, information archiving, information recording, information transfer, and the need for communication, font and speech, etc. in this position are linked to the designation [1].)

All techniques need to be taken with their properties to compare under current conditions. Labeling by RFID tags is shown in case of large series as well as piece production in comparison with other technologies mostly as a better technology especially against bar code or manual entry [2, 3].

Basic Terms: Descriptions of Categories

The following categories could be included among the basic categories.

Reliability
It is the property of the product which determines the ability to guarantee the necessary functions to maintain the specified operating parameters within the required range and time, according to the technical conditions, respectively, according to contractual requirements. These are the following features of the unit: durability, reliability, sustainability, storage, and so on.

Sustainability
The product's dignity is the ability to remedy and prevent the causes of malfunction and subsequent revitalization of the resulting damage and faults, also in the ability to correct them with the possibility of achieving and repeatedly achieving the original functional state.

No-Failure Operation
Product attributes performance to function permanently during the specified service life and under the required conditions. Numerically, it is expressed, for example, by probability of trouble-free operation up to the specified time, fault intensity, and average trouble-free operation.

Testing of Reliability
Testing of reliability is the conscious process we determine by measuring and controlling the value of a real, objective size of reliability.

Assumptions for Shortening and Speeding Up Reliability, Lifetime, and Functionality Tests
Measurement, testing, and loading of selected parts are performed under customary, standardized conditions so that measured, gained product reliability information is

available in shorter times than under normal operating conditions and also under the conditions defined by the product's technical specifications. The selected special conditions must not alter the mechanisms of failure. For electronic products, it is proved that the acceleration mechanism is the temperature at which accelerated tests are carried out at a higher than normal operating temperature.

Fault Intensity Value, Mean Value–λ_s

Determining the time between failures, the system reliability time, especially in the case of indicative calculations, is the mean value of the failure–λ_s. It is the average magnitude of the fault intensity (point estimate) of the failure rate of the analyzed, examined class of the same type applied in the devices of the current technical and technological design. The quantity λ_s obtained by a credible procedure allows realistically determining system reliability in situations where all the influences necessary for an exact calculation are not known. The magnitude of this value is expressed by the ratio of the number of failures over a given time, for example, λ[FIT] is 1 fault/109 h.

Mean Time Between Failures (MTBF)

It is the ratio between the service life of the repaired product and the expected number of its faults during this operating time (or lifetime). This parameter belongs to repairable products among the most widespread. The value of this variable is expressed in units of time [h].

Categorization of Failures

Faults divided into:

- Catastrophic, in which a component or product ceases to perform its function (irreparably).
- Catastrophic, in which a component or product ceases to perform its function; but there is a possibility of repair and revitalization (repairable).
- Degradative, in which component or product changes parameters but can continue to operate with changed parameters [4, 5].

Theoretical Basics of Testing: Brief Description

Time accelerated (shortened) test functionality (reliability) to obtain reliable and comparable reliability parameters for a significantly shorter period of time with a lower number of products than would be measured in a standard way in normal operation (long-term and costly tests with a large number of products and virtually unguaranteed term end) [6].

Most of the measured entities and integrated assemblies of the test product have a very long service life under normal operating conditions (typically 108–1012 operating hours, 105–1014 repeated mechanical cycles). From the technical life expectancy considered, the expected component failure rate is in the range 8–10 to 10–12 per hour. Under the conditions considered, the total average time between label failures can be expected to be in the order of one million hours of operation. For the intensity of component or device faults at T_2, an empirical relationship exists:

$$\lambda(T) = A.e^{Ea_1.z} + (1-A).e^{Ea_2.z} \tag{10.1}$$

at which

$$z = \frac{1}{k}\left(\frac{1}{T_{amb,ref}} - \frac{1}{T_2}\right) \tag{10.2}$$

where:

A is a component's parameter.
E_{a1}, E_{a2} are the activation energy [eV].
k is Boltzmann constant $k = 8618.10{-}5$ eV°K.
$T_{amb,ref} = 313$°K is the reference ambient temperature (273° to +40 °C).
T_2 is temperature [°K].

To estimate such a value of reliability by classical procedures, an enormous amount of units (electronic products (several hundred pieces)) is needed and tests would take several thousand hours.

One of the possible acceptable procedures for verifying the reliability of the final products is to make abridged laboratory reliability tests. With shortened reliability tests, it is a good idea to determine the acceleration mechanism and to determine the acceleration factor, thus finding an acceleration magnitude for the individual components that make up the final product and its designation.

For electronic products, it is clearly confirmed that the acceleration factor is an elevated temperature (e.g., 100 °C) at which the acceleration tests are performed over a normal operating temperature (e.g., 23 °C). Possible additional acceleration factors are dependent on operating voltage, and the influence of operating currents does not have a dominant position for properly designed electronic circuits.

These accelerating influences range from 1 (for mechanical parts for which the increased operating temperature has virtually no effect (connectors, soldering points, etc.)) up to 100 (for elements very sensitive to operating temperature (electrolytic capacitors, optical elements, etc.)). When we put in the expression (1) $A = 1$, $E_{a2} = 0$, this example is consistent with Arrhenius theory. To determine the acceleration factor πT depending on the operating temperature (100 ° C) and normal operating temperatures (23 °C), it is best to use the Arrhenius equation to determine the intensity of the failure:

$$\lambda(T) = e^{Ea_1.z} \tag{10.3}$$

An accelerator factor is defined as follows:

$$\pi_T = \frac{\lambda}{\lambda_{\text{ref}}} \tag{10.4}$$

where λ_{ref} is the intensity of failure under normal, reference conditions, and π_T is a temperature dependency factor (acceleration factor, acceleration).

This approach is described by international standards in [7].

Applications of the RFID Tag Unit

An important indicator is the amount of load cycles to the normal pressure that the label holds. Based on the basic type of life of the designated piece:

- The unit is either in stock.
- A worker with other pieces for the next process.
- The normal life of the unit in the production process, for example, 2–30 days.
- Stay in store and in a technology worker 20–80%.
- Amount of units per worker 5–15.
- The stack height of the products in the workplace is 5–30 cm.

Determination of plain product pressure, of the same format size, understood as the mean pressure caused by a 1 mm column height is 0.000764 g/mm² pressure per 1 mm column height.

The pressure at the lower edge of the column is 0.038216–0.229297 g/mm² (we have assumed a value at the upper limit of 0.209775 g/mm² in our test), so we expect the unit to be present at each stack position and also at the bottom of the column. So we are considering the following pressures:

0.038216 g/mm² pressure at a depth of 50 mm from the top of the column
0.191081 g/mm² pressure at a depth of 250 mm from the top of the column
0.229297 g/mm² pressure at a depth of 300 mm from the top of the column

The operator stores the parts always on the top of the stack, but the actual position of the unit is random, and the central position is usually in the middle of the column, so the average pressure that acts on average on the label is the same.

If we usually work with equivalent parts, we consider one to two subscriptions and return the part to the column or row, and in the intensive process, this number of iterations may increase (estimated size for about 14 cycles). In the comparative multiunit process, these numbers increase significantly.

Abovementioned values (Table 10.1) show that few active live parts for life occur in the cycle stacks of units in the pressure range of 0.038–0.299 g/mm² with the number around 1e3 cycles. The parts in the active zone (intensive parts) assume a number of cycles over the lifetime of more than 2,600,000, so at least the number of cycles of stress tests is required to perform the tests. In order to confirm this feature (resistance to the label load by nominal pressure), it is advisable to perform these tests more with the estimated number of approx. 1e6–1e7 load cycles [2, 3].

Table 10.1 Frequency and load of RFID tags required for authentication (authors)

	Minimum	Maximum	Unit
Lifetime of the item	2	30	Days
Working time ratio	0,2	0,8	%/100
Number of units per workplace	5	15	Piece
Stack height at the workplace	50	300	mm
Brand pressure (RFID tag)	0,038	0,229	g/mm^2
Frequency of storage at the workplace	3	18	c/l
Estimated number of cycles over the life of the unit	1095,75	262,980	c/l

Selected Ways of Loading Test Labels

In this analysis, we considered multiple storage options, installing labels on a piece:

1. Parallel
2. Cross
3. From the outside
4. From the inside

From these options, the first tested version was the 3rd position. This selection has the appropriate properties for verifying good results. Then, it was necessary to select the surface size of the label to be loaded during the tests. The choice comprised three options:

- Pressed entire overlapping area
- 2/3 of the label area (the pressure interface passes over the part of the antenna)
- 1/2½ of the label area (the pressure limit passes through the label chip storing location and the contact between the chip and the antenna)

Other types of testing include:

1. Special load alternatives
2. The situation that parts are uneven
3. Attached chips, grains, or fragments causing significant local decrease/increase in pressure in the label area and surrounding area

Figure 10.1 shows tested label-type Alien 9540 Squiggle with the image of the antenna and the chip.

The situation in Fig. 10.2 shows if the units have the same format (size) and when deposited, the operator gives them pretty virtually all around. In this case, the label is pushed almost uniformly across the surface.

When placing parts into a column of different sizes and format, the label is not pressed on the entire surface. The same state of partial load on the surface of the label also occurs in the inaccurate or alternate orientation of the deposition of the same large parts in the partial overlap of the loading surfaces.

For these conditions, we perform tests for repeat load of the label area in 2/3 area and 1/2 (Figs. 10.3 and 10.4).

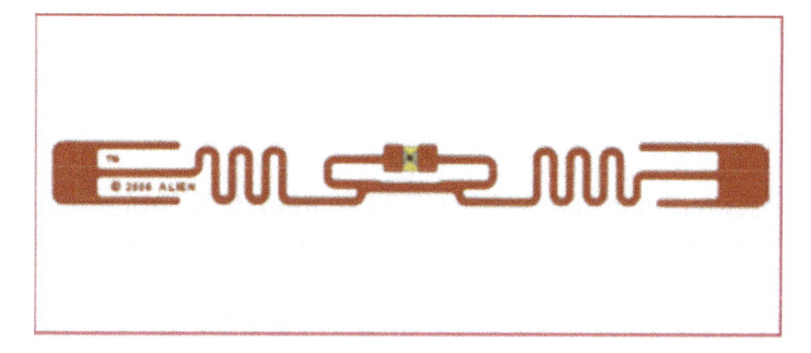

Fig. 10.1 Tested Alien 9540 Squiggle label types, with chip and antenna images

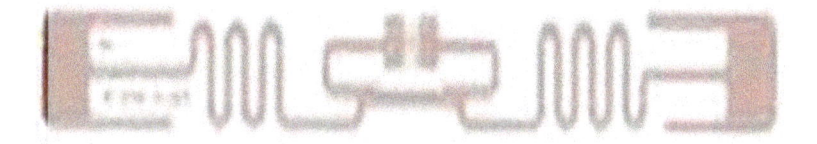

Fig. 10.2 Full load labels with nominal pressure edges (authors)

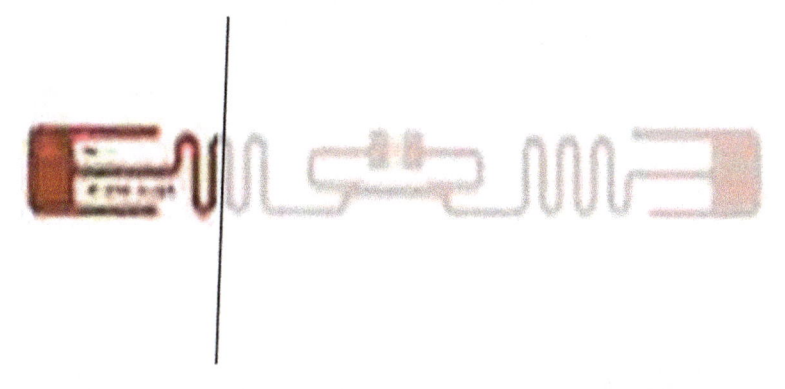

Fig. 10.3 Demonstration of 2/3 of the surface of the pressure-labeled label (authors)

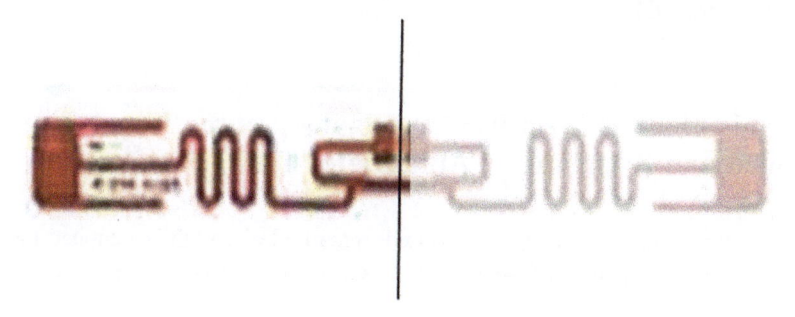

Fig. 10.4 Crush and contact cut off test, in ½ area of nominal pressure (authors)

The beginning of nominal pressure load was performed with features similar to normal handpiece stacking (contact speed is 0.3 m/s) (similar to manual storage of rigid parts) in the range proposed above.

Implementation of the Tests and Calculation of the Magnitude of the Acceleration Change Factor π_T to the Temperature

Testing is carried out in such a way that the actual conditions are taken into account as closely as possible. The test tag was cyclically loaded by storing the unit column, in the time range of about one working shift (6–9 h, following staff capability); after this stage of cyclic loading, the break occurred; the status of the column is on the monitored unit. This time can be considered relaxing.

For now, the pressure dynamic load is completed, and it can be assumed that there are some deviations on the label, so these will "show"

A weekend relaxation phase followed after 5 days of testing. This model was used until the end of the (required, estimated, necessary) number of cycles [4, 7].

To increase accuracy, π_T-empirical pattern describing the dependence of failure intensity on temperature [8] can be applied in the calculation. The two activation energies are used to include those cases where a pair or more mechanisms are affected by the failure process.

$$\pi_T = \frac{A.e^{Ea_1 \cdot z} + (1-A).e^{Ea_2 \cdot z}}{A.e^{Ea_1 \cdot z_{ref}} + (1-A).e^{Ea_2 \cdot z_{ref}}} \tag{10.5a}$$

$$z = \frac{1}{k}\left(\frac{1}{T_{amb,ref}} - \frac{1}{T_2}\right) a\, z_{ref} = \frac{1}{k}\left(\frac{1}{T_{amb,ref}} - \frac{1}{T_1}\right) \tag{10.5b}$$

where:

T_1 reference temperature [°K]
T_2 temperature [°K]
t_1 reference temperature [°C]
t_2 temperature [°C]

Determination the Intensity of the Faults λ_s of the Tested Products and Evaluation of Abbreviated Tests

Based on the initial (calculated or measured) magnitude λ and the calculated factor of temperature π_T dependence, we determine λ_{ref} according to the dependence mentioned above:

$$\lambda_{\text{ref}} = \frac{\lambda}{\pi_T} \tag{10.6}$$

We finish the test in accordance with STN IEC 60605-4. The constant intensity of the failure is expected.

Used designation:

T^* [h] accumulated valid test time to the test point
t [h] duration of the test
$\chi_p^2(n)$ theoretical magnitude of the p-quantile of distribution χ^2 with ν degrees of freedom
r the frequency of failures
n number of products
$\chi_{0,95}^2(2r_o)$ p-quantile distribution at 90% significance at $2r_o$ degrees of freedom

(a) Accumulated valid test time

$$T^* = \sum_{m=1}^{n} t^m \tag{10.7}$$

where t^m is the measured test time of the m-th product at the decisive time.

(b) Point estimate of mean time between MTBF failures

$$\text{MTBF} = \frac{T^*}{r}, \quad \text{for } r \geq 1,$$
$$\text{MTBF} = 3 \cdot T^* \quad \text{for } r \geq 0 \tag{10.8}$$

(c) Upper m_h and lower m_d of the significance level of the mean fail-safe period at a significance level of 90%

$$m_d = \frac{2 \cdot T^*}{\chi_{0,95}^2(2r+2)} \quad m_h = \frac{2 \cdot T^*}{\chi_{0,05}^2(2r)} \tag{10.9}$$

If a failure is not occurring during the test (also suitable for a small number of faults), only the lower limit of significance levels is determined as:

$$m_d = \frac{2 \cdot T^*}{\chi_{0,90}^2(2r+2)} \tag{10.10}$$

Quantities $\chi^2 p (\nu)$ are determined from STN IEC 60605-4 [7].

(d) Validity test of assumption of constant failure intensity

The test can be performed if the failure rate is greater than 3. For the number of failures less than 3, the assumption of constant failure rate is assumed without testing.

- Determine the cumulative valid test time T_k, for $k = 1,2, ..., r$

$$T_k = \sum_{m=1}^{n} t_{k,m} \qquad (10.11)$$

where $t_{k,m}$ is the valid test time of the product number m to the k-th disorder.
- The size (quantification) of the test statistics is determined

$$\chi^2 = 2 \times \sum_{k=1}^{r} \ln \left[\frac{T^*}{T_k} \right] \qquad (10.12)$$

- Table 10.1 of STN IEC 60605-6 gives the values $\chi^2_{0,05}(2r)$ and $\chi^2_{0,95}(2r)$,
- The assumption of a constant occurrence of faults is confirmed if it is valid:

$$\chi^2_{0,05}(2r) \leq \chi^2 \leq \chi^2_{0,95}(2r) \qquad (10.13)$$

Performing the Reliability Tests in a Shortened Procedure and Determination of Product λ_s Disturbances

A separate part of the tests consists of the practical measurement of product reliability. Indeed, it is time-consuming and financially demanding, but thanks to the necessary activities after the practical verification of the results, it is possible with the probability to determine the reliability of the products.

For this issue, the general view prevails that, to ensure objectivity or acceptance of the results of accelerated customer reliability testing, it is advisable to become an independent organization as the guarantor of accelerated tests [9].

Based on the above documentation and valid international standards, the following failure intensities can be determined for laboratory, shortened reliability tests (for each piece the value of the last digit after a comma).

```
                        date   time
ID=01 A5 7D FD 8A; 4. 9.; 8:34:14, 94
ID=01 A5 7D FD 8A; 4. 9.; 8:37:04, 94
ID=01 A5 7D FD 8A; 4. 9.; 8:37:06, 78
ID=01 A5 7D FD 8A; 4. 9.; 8:40:09, 79
ID=01 A5 7D FD 8A; 4. 9.; 8:40:11, 63
ID=01 A5 7D FD 8A; 4. 9.; 8:40:13, 78
ID=01 A5 7D FD 8A; 26. 10.; 7:43:00
ID=01 A5 7D FD 8A; 26. 10.; 7:43:03
```

ID=01 A5 7D FD 8A; 26. 10.; 7:43:06
ID=01 A5 7D FD 8A; 26. 10.; 7:43:09
ID=01 A5 7D FD 8A; 26. 10.; 7:43:12
ID=01 A5 7D FD 8A; 26. 10.; 7:43:16
ID=01 A5 7D FD 8A; 26. 10.; 7:43:18,

Measurement in the RFID lab in a number of 234,241 load cycles without loss of function of the labels subjected to the storing test. The measurement is now ongoing until there are no loss of function at least half of the measured elements. At this time, the calculated value will be calculated.

Conclusions

Using shortened tests in standard manufacturing practice, we receive, at reasonable times and costs, unnecessary sources of product information from the monitored plant, which show a number of direct and indirect benefits in the areas of tracking, managing, and enhancing product quality as well as technology and manufacturing processes.

From a given set of data, it does not include information about actual extrapolated functionality and also about their reliability, quality, and efficiency of involvement in the production process. The properties of shortened measurements and their benefits are well suited to each production. During the testing was shown that RFID technology is long-lived. Tags lifecycle is long what means that risk of the reduction or loss of functional properties of the technology, which can be critical to traceability, is not critical. It also has the potential to improve the overall qualitative as well as the economic effect.

References

1. Knuth, P., & Šebej, P. (2010). Overovanie vlastností RFID štítkov na označovaných predmetoch (2010). In *DoNT 2010: Day of New Technologies: zborník príspevkov a prednášok z vedeckej konferencie s medzinárodnou účasťou: 19.11.2010* (pp. 74–87). Žilina: EDIS Vydavateľstvo ŽU. ISBN 978-80-554-0279-6.
2. http://193.87.95.91/phpbb3/download/file.php?id=3045
3. Šebej, P. (1987). *Проверка надежности, (долговеренности, долговечности) електронных оборудовании, The Test of Reliability, Zborník: BAV, ROBCON-4*. Sofia: Svante August Arrhenius.
4. Klementev, I., & Kyška, R. (1990). *Elektrické meranie mechanických veličín*. Vydavateľstvo Alfa: Bratislava.
5. Balog, M., Szilagyi, E., & Mindas, M. (2015). Utilization of RFID technology for monitoring technical state of chosen elements of cargo railway wagon. *International Journal of Engineering Research in Africa, 18*, 167–174. ISSN 1663-3571.
6. Knuth, P., Šebej, P., & Zagora, M. (2010). Strata efektívnosti identifikačných a evidenčných technológií (2010). In *DoNT 2010: Day of New Technologies: zborník príspevkov a prednášok*

z vedeckej konferencie s medzinárodnou účasťou: 19.11.2010 (pp. 88–95). Žilina: EDIS Vydavateľstvo ŽU. ISBN 978-80-554-0279-6.
7. Klas, A. (2005). *Identifikácia predpokladov a možností prekonávania technologickej a inovačnej medzery SR*. Bratislava: Ústav slovenskej a svetovej ekonomiky SAV.
8. Strelec, S. (2009). *Vplyv fyzikálnych faktorov na spoľahlivosť RFID technológie*. Prešov: FVT TUKE.
9. Want, R. (2006). *RFID explained, USA*. Morgan and Claypool Publishers. ISBN 978-1-59829-108-7.

Chapter 11
Basic Assumptions of Information Systems for Increasing Competitiveness of Production Companies within the EU and their Application of the CAPP System Design

Katarina Monkova, Peter Monka, Helena Zidkova, Vladimir Duchek, and Milan Edl

Abstract This chapter describes the basic assumptions of information systems that must be satisfied to increase the competitiveness of a production company in the European Union. The application of these assumptions within a new computer-aided process planning (CAPP) system design is also discussed. Analyses of the cost structures of small- and medium-scale manufacturers indicate the significance of CAPP in production cost composition. Thus, considerable attention should be paid to the CAPP area because it can influence the output costs of a product and its quality in great measure. Thus, the first part of this chapter describes the basic requirements of information systems that must be satisfied in order for a company to be effective and successful in its activities. The second part of this chapter describes a new flexible CAPP system that was built on a modular structure; the functionality of this system was verified by implementing the CAPP system into the

K. Monkova (✉)
Department of Computer Aid of Manufacturing Technologies, Faculty of Manufacturing Technologies with the seat in Presov, Technical University of Kosice, Kosice, Slovakia
e-mail: katarina.monkova@tuke.sk

P. Monka
Department of Automobile and Manufacturing Technologies, Technical University of Kosice, Kosice, Slovakia

Department of Machining Technology, Faculty of Mechanical Engineering, University of West Bohemia, Pilsen, Czech Republic
e-mail: peter.monka@tuke.sk

H. Zidkova · V. Duchek · M. Edl
Department of Machining Technology, Faculty of Mechanical Engineering, University of West Bohemia, Pilsen, Czech Republic
e-mail: zidkova@kto.zcu.cz; duchekv@kto.zcu.cz; edl@kpv.zcu.cz

actual information structure of a regional company. The principles of this newly designed CAPP system could form the basis of a company's increasing competitiveness, as well as its integration into "Industry 4.0".

Introduction

In 2010, the European Commission (EC) proposed the Europe 2020 strategy as a means to focus the European Union (EU) and its Member States on the important task of improving the EU's competitiveness. The goal of this strategy is to transform the EU into "a smart, sustainable and inclusive economy, delivering high levels of employment, productivity and social cohesion." Today in Europe, the acute phase of the economic and financial crisis has passed. Signs of moderate but uneven growth and sluggish job recovery exist amid a number of risks and fragilities. However, renewed momentum is critical to achieve the long-term structural shifts required to meet these goals. At the heart of competitiveness is the level of productivity of an economy. As such, competitive economies are able to provide high and increasing standards of living, which allow all members of a society to contribute to and benefit from these levels of prosperity. In addition, competitive economies also have to be sustainable – meeting the needs of the present generation while not compromising the ability of future generations to meet their own needs [1].

Addressing the competitiveness divide will require differentiated strategies that take national and regional characteristics into account. While a concerted and united effort is desired from all EU Member States to improve Europe's knowledge driven economy, it is clear from the large regional disparities that paths towards this goal, and priorities for improvement, will differ across countries. For instance, innovation strategies for countries higher on the knowledge ladder will differ from strategies appropriate for countries lower down. However, for all European economies, investments in knowledge-generating assets will translate into important drivers for future productivity growth – those drivers being a common focus on education, information and communication technologies, the digital agenda and reforms to improve the overall enterprise environment across the region [2, 3].

Many East European (EE) post communism countries were forced to transform their production schedule, quantity, and types of products after political transformations. But many plants in these countries are working with unchanged philosophies to this day. This is one of the reasons why many are still not in competition compared to Western plants despite cheaper manpower. Basic problems in EE plants are [4–6]:

- Low level of information technology (live data, information availability and uniformity of structures)
- Computer aided systems have variable capacity; they use only subprograms, mostly on difficult, often incompatible, computing systems in Slovak plants

- Absence of tools that make it possible to analyze dynamic system properties important for the planning of high automation systems in a short time
- Lack of tools for alternative solutions, including options for testing and optimization
- No interconnection among systems in phase of design, with dominantly routine labour

The transformation or development of ESE production enterprises to new European environs is a complex process. Producers are permanently pressed by increasing domestic and foreign competition. One efficient way of competitiveness enhancement is related to advanced application of the computer aided (CA) systems. CA systems are sophisticated solutions which cover all areas of company activities. They integrate requisite information and provide live data that are necessary for efficient labour-saving work and for making sufficient decisions.

Computer Aided Process Planning (CAPP) uses a set of activities for the creation of all types of production documentation by means of computer assistance. Analyses of the cost structure in companies with small and medium series production indicate a significant ratio of CAPP in the production costs composition. Thus, it is very important to pay considerable attention to CAPP which can, as a result, greatly influence the output costs and quality of a product.

Impact of European Market Composition on CAPP Design

Basic present-day problems of production companies from the view of production information systems (IS) can are defined by their requirements: availability for use in a wide variety of production approaches, simple implementation in entrepreneurial surroundings, a modular concept for covering all necessary areas, reliable and secure data formats and structures, potential for flexible bilateral data sharing, potential for a trouble-free extension of IS, a relatively fast transfer to a higher level of IS at a reasonable price.

Generally, for selection of production software, companies can use all variations between two extremes: complex systems or independent solutions for every application field of enterprise activities.

For many small enterprises, complex systems are inaccessible by reason of system complexity, fixed structure, expensive price, large and complicated adaptation, time-consuming maintenance, etc. Second, they are generally characterized by flat interconnection possibilities to other information systems.

Basic problems dealt with by developers of manufacturing information systems are:

- Autonomous reasoning for a wide variety of technological approaches
- Flexible data structure for optimizing procedures
- Compromises for obtaining advantage of both extremes: complex systems vs. independent solutions

- Integration of manufacturing information systems in environs with specialized systems (CAD/CAM, salaries, financing, materials, accounting, etc.)
- High potential for data sharing by external applications and co-operators

A great number of Eastern European companies have different process planners which make different plans for the same parts, resulting in inconsistencies and extra paper work, applied in a heterogeneous data environ [7]. Computer Aided Process Planning (CAPP) systems can help in overcoming these inconsistencies. CAPP aids in creation of process plans for manufacturing and increases the flexibility of manufacturing. Process planning is a task which requires a significant amount of both time and experience. Computer support or computerized process planning systems can help reduce process planning time and increase plan consistency and efficiency.

In Computer Aided Process Planning (CAPP), there are two different ways of obtaining the process plan [8]:

- Variant process planning
- Generative process planning

Generally, process planning systems are oriented to one of these approaches, but each technique has its own advantages. Exploitation benefits of both can be obtained by using a combination of them.

When applying a technological approach, it is advantageous to subdivide processes into problem-orientated system areas, which represent a limited area of activity. The formulated scheme of plants shown in Fig. 11.1 [6] represents the first assignment stage for developing the concept of a production information system. The specific tasks and activities of the production process are focused on:

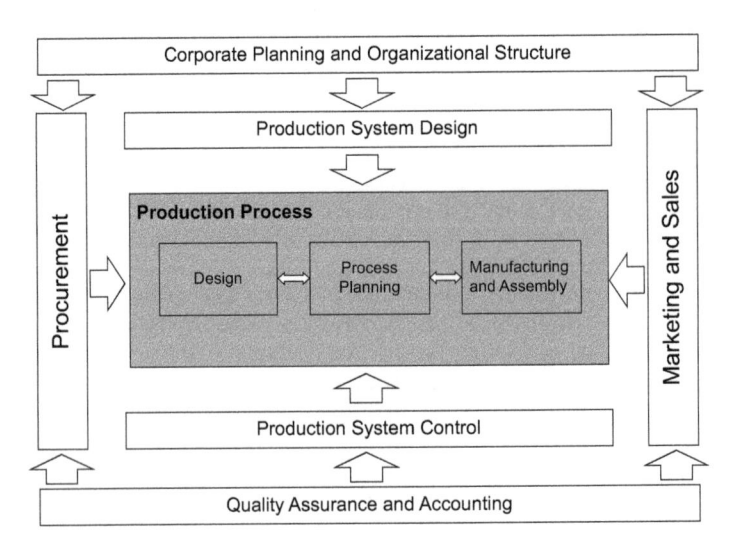

Fig. 11.1 Principal scheme of a plant [6]

- Design
- Process planning
- Manufacturing and assembly

These tasks form the conceptual basis of the corporate objective of design of this information system. The direct system areas, which are subdivided into the fields of Production System Design and Production System Control, are grouped around this conceptual level of production processes [9, 10].

The indirect system areas cover the external influential factors of the corporate process [11]:

- Corporate planning and organizational structure deal with the mid- and long-term planning of corporate aim and the resultant structuring of the corporation to reach the planned objective
- Production system design attunes the methods and processes, which contribute to the rational realization of the technical and economic overall objective in design, process planning and manufacturing
- Procurement deals with optimum material arrangement with respect to time/costs
- Marketing and sales cover the comprehensive promotional systems
- Quality assurance and accounting systems are auxiliary aids for the control function of the technical and economic corporate function of the technical and economic corporate objectives
- Production system control is to optimize the throughput times of the orders

The objective of product design is to raise or assure the technical value of a product not only in the systemization of the design process, but also in the build-up of work aids for the rational elaboration of drawings and planning data. Thus, the term product design describes the comprehensive application of methods and work aids used during the design process of a product to the release of the basic data for process planning. The objective of manufacturing design is to systemize the planning and preparation processes for the manufacture and assembly and to structure the planning data for rational realization of the different partial assignments within the framework of the concept. The aim of work study and wage structure is to optimize the working conditions at the work place and to increase output in accordance with scientific methods and processes [12].

On the basis of these objectives, a structural concept according to Fig. 11.2 was developed for the information system of production system design in the field of mechanical engineering. They constitute the basic form for further database structure building and for creation of a mathematical model of the manufacturing system.

In this system, it is possible to formulate the manufacturing system by the relation:

$$MS = S \cup SO \cup E \tag{11.1}$$

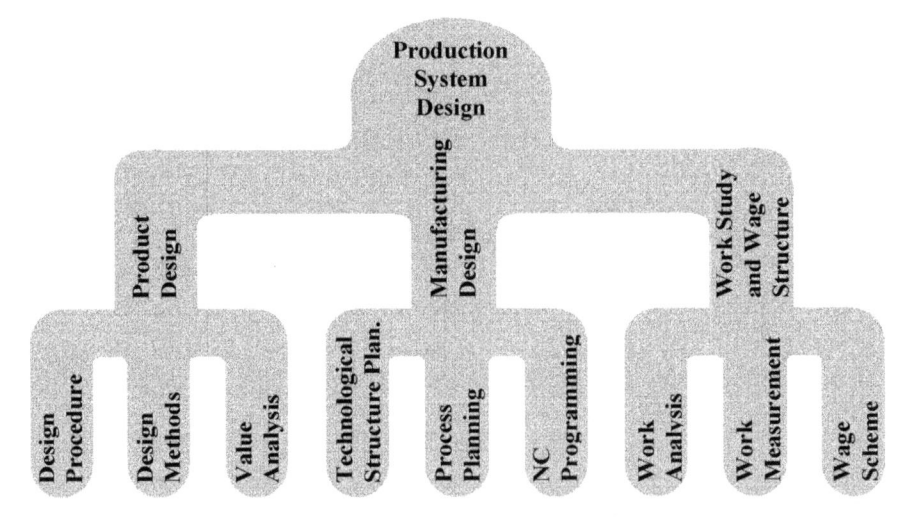

Fig. 11.2 Structural concept of an information system

where:

MS = Manufacturing System
S = Segment
SO = Structure of Operation
E = Equipment

Every set from relation (11.1) can be split into subsets and each subset can be divided into sub-subsets of lower and lower layers. Depth of the set definition (and circumstantials of storage data details) is dependent on required optimization and variation rules. In projects it can be used to describe the layer of set depth for the basic properties of every registered object in the database. This concept enables very good conditions for application of mathematical tools for manufacturing planning and registration of all relevant data.

Basic research about the composition of the European market (Fig. 11.3) [13] has come to some very interesting conclusions – the typical European enterprise unit is the micro enterprise. These types of companies constitute a substantial part of the European market as they comprise 92% of the overall number of companies and employ 39% of the employees. Small and medium-sized companies together comprise 7.5% of the overall number and employ 30.3% of the employees. Large companies comprise the rest (0.2% production units and 30.2% employees). Other results of this same study show that micro companies have the disposal of a free potential of 20% of the productivity and 15% profitability.

These characteristics, which describe a distinct ability of the dynamic growth production and the possibility of effective evaluation of micro company instruments, are very important.

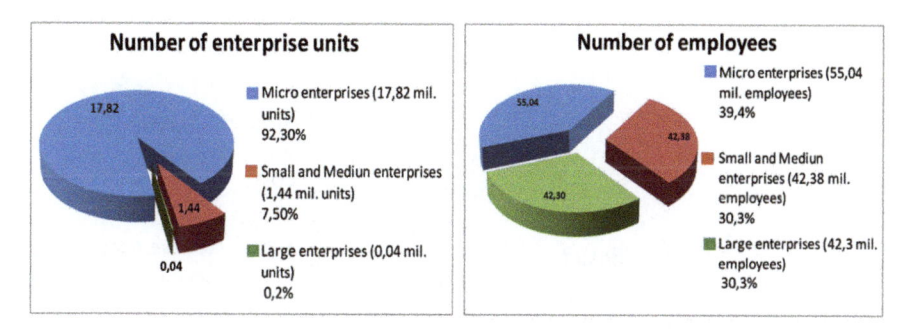

Fig. 11.3 Structure of business units in the European Union [13]

From the analysis of potential system users taking advantage of computer aided process planning, it can be said that it is the micro companies that constitute a significant part of the enterprise subjects.

Design of CAPP System

When designing a new product, the aim is to secure or increase its technical value not only by systematization of the production process but also by increasing the level of the supporting tools for the rational processing of the production documentation and data needed for planning [14]. Figure 11.4 shows the production system design logically divided into parts [15].

Specifications of the enterprise unit's structure imply diametrically different demands on information systems of micro and small-sized companies compared to medium or large-sized companies, including:

- Simple implementation
- Potential for a modular concept covering all necessary areas
- Reliable and secure data formats and structures
- Potential for flexible data sharing with IS (Information Systems) of the purchaser and supplier
- Potential for a trouble-free extension to the needed modules
- Potential for a relatively fast transfer into IS of a higher level when necessary
- Security
- Reasonable price

On the other hand, from the point of view of individual companies, the following conditions should be assessed when proposing IS:

- The system has to provide a user view on the production process from several angles.
- The enterprise subject should be limited as little as possible when launching new products to the production process.

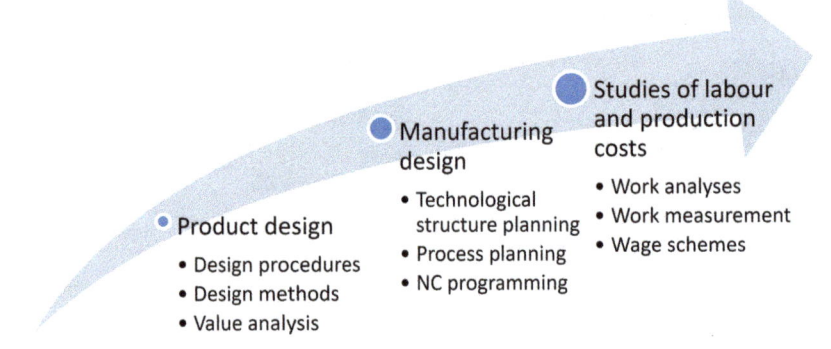

- Product design
  - Design procedures
  - Design methods
  - Value analysis
- Manufacturing design
  - Technological structure planning
  - Process planning
  - NC programming
- Studies of labour and production costs
  - Work analyses
  - Work measurement
  - Wage schemes

Fig. 11.4 Structural concept of information system

- It should be applicable for a wide range of business.
- It should be modular.

Another requirement for selection of a suitable CAPP system is its flexibility and the ability to quickly adapt to user requirements. For example, when purchasing a comprehensive program package from a general contractor, the user is usually not able to make adjustments to the structure of the system. If necessary, such adjustments require the supplier to make contact again, which is usually time consuming and cost-intensive. Appropriate software to support the creation of technological documentation can be obtained from these viewpoints:

(a) Complete delivery of the software application by the general contractor
(b) Purchase of software application from the general contractor and subsequent implementation by the user
(c) Purchasing all software application components from different vendors
(d) Development of the software application and purchase of other components from different suppliers, but self-integration
(e) Development of software application by the company itself

Tables 11.1, 11.2 and 11.3 display examples of advantages and disadvantages used for the decision analysis when the business management considers a CAPP purchase.

Newly designed computer aided systems for process planning can be built on the basis of a combination of the following technological approaches [16–18]:

- Individual technology (IT)
- Type technology (TT)
- Group technology (GT)

A designer can choose the best approach for process planning of a production object.

The *Individual approach* includes the creation of manufacturing documentation for each component individually without the possibility of using the same repeated

Table 11.1 Advantages and disadvantages of software delivered completely by general contractor

Advantages	Disadvantages
Fastest implementation; Low cost of ownership; Professional solution for each component and whole software application; It is possible to choose the best solutions for each part of the information system; Software application contains so-called „The best experience of a wide range of users" ; Software application is parametric and it is possible to solve a series of new requirements only by other parameter settings instead of reprogramming, and the integration of all components is guaranteed by the supplier; Supplier can also be guaranteed stability of software application development; Risk distribution between enterprise and system vendor; Lowering the entitlement to operating staff, an enterprise can better focus on its core business; Easier customization of IT capabilities according to business needs; In the case of a very strong software application system supplier for several customers, the cost of individual customers may be even lower than in the whole system purchase option from the general contractor; The possibility of using the most advanced technologies	Processes in the company must adapt options for software application; High dependence on the general contractor and his abilities, seriousness and stability; The risk of out-of-business confidential information leakage - the increase in dependence on the supplier as opposed to the purchase of software application alone

Table 11.2 Advantages and disadvantages of purchasing all parts of software from different vendors

Advantages	Disadvantages
Quick implementation; Lowest cost; Option to choose the best solutions for each part of the IS; Software application contains the so-called "best experience of a wide range of users"; Software application is parametric and it is possible to solve a series of new requirements by other parameter settings instead of reprogramming	Processes in the company must adapt options for software application; If software application parts are purchased from different vendors, it is difficult to integrate different applications into one information system (IS); High demands on the team; Problems with maintenance of bindings between applications and thus relatively low IS stability

Table 11.3 Advantages and disadvantages of software development by the company itself

Advantages	Disadvantages
Software application developed directly for the company needs; Incremental extension according to the business needs; Detailed knowledge of the software operation is directly inside the enterprise; The competing companies will not know the strengths and weaknesses of the software; Easier response to immediate user needs	High cost; World-class workflows are not always built into the software application; Lengthy solution time; Lower software application quality and problems with the integration of the complex software application caused by the relatively low qualification of home-based researchers; Low modularity of software application, if application development is derived from immediate user requirements and the requirements are not generalized enough, the solution is not implemented as parametric and thus prolongs and outweighs future maintenance; Significant risks of inconsistency in the system arise due to fluctuation of investigators

operations for certain sets of manufacturing objects (from parts through subassemblies and assemblies to final products). It can be said that this approach is not connected with standardization of technological processes and with the activities linked with them.

The term *Type technological process* represents the specific technological process for a group of parts with the equivalent technological characteristics. This process is suitable for a specific group of parts and defines the type and sequence of the main technological operations. The important term for 'Type technology' is the 'Type representative', namely, a real or abstract manufacturing object existing in a group of parts, for which the technological process contains all basic and auxiliary operations. The typification of technological processes can be achieved by two methods that vary in their use and in the objects of classification. The following procedures are typical in the technological processes typification [19]:

1. Classification of parts (or the elementary surfaces)
2. Projection of the Type technological process (operation)
3. Specification of individual technological process phases
4. Development of technological process for the Type representative
5. Transmission of Type technological instruction to a specific part

Another type of technological processes standardization is the *Group technology*. This is a manufacturing philosophy and strategy that assists a company in understanding what it manufactures and how those products are then manufactured. In manufacturing engineering, 'Group technology' focuses on similar machining operations, similar tooling, machine setup procedures and similar methods for transporting and storing materials [20].

By identifying similarities in manufacturing (machines, tooling, process sequences, etc.), similar work piece parts (geometric shape and size) can be grouped into distinct families and processed together in a dedicated work cell. Some parts may look similar to each other, but because of differences in materials, tolerances or other production requirements, they have different manufacturing conditions and so don't create a "manufacturing family of parts". In contrast to 'Type technological' processes, the 'Group process' is always specific and serves as a technical instruction to accomplish individual operations. Today, the approaches to 'Group technology' are based on the fact that all technical and organizational evolutions inside a specific manufacturing unit contain activities or data with some degree of similarity. So they can be combined into groups for which common solving methods are used. The methodological tools for the sorting of parts are different classification and coding systems.

Analysis of requirements for appropriate software application and the potential of developing an information system that is adapted to the East European market conditions initiated the idea to a create new system. For this new system, two basic requirements were specified: it has to be flexible and it should not restrict new products put into the manufacturing process.

Multi-variant Process Planning

The essential ambitions of designers is to build on a basic structure for a CAPP system which is very flexible from the viewpoints:

- Creating and modifying versions of process plans
- Approach of the design processes
- Sharing data among members of the manufacturing chain
- Modularity of components required by the user
- Possibility of cooperation with a wide variety of external software

According to these basic claims the theory of multi-variant process planning was designed. The theory deals with the production process (during its project phase, as well as during the production) as a homogenous whole, including technological and labour processes organized via various possible parallel phases in a way that the final product could be optimally processed for the set conditions whilst fulfilling all the demands required by a consumer. On the basis of this theory it is possible to create a combination of possibilities of various techniques used in individual process planning based on the strategy aimed at achieving the specific goal of the production unit. The main objective of this theory is:

- The creation of a unified definition environment for all the factors immediately influencing the result of the production process
- The flexible interface which enables bidirectional exchange of the required information with all surrounding systems

Via the unified definition environment, the philosophical and conceptual unity is secured within the whole issue falling into the formation area of multi-variable process planning, a distinct classification product constituent and the laws of production sequence for operation projection which allow the use of several possibilities designed by an information system based on this theory.

Flexible interface of the system must be effective in the production environment in the way that all the individual systems creating a heterogeneous information system (CAD/CAM application, wage records, accounting, material management) have inter-connection secured via suitable interfaces in order to prevent errors caused by data redundancy and human factors, but also to reduce the response time to a minimum.

The following steps were created by means of a database system: users rules for definition of rights, queries for selecting required data, forms for data editing by operators, reports for exploring relevant data, procedures for operation with data.

For correct database operation, it is necessary to fill in all relevant information to the interface for storing properties and characteristics of a production segment. For this information system, the term "segment" means all manufacturing objects from part, through subassembly, and assembly groups to final product. This interface requires basic information about production segments and further indications (see Fig. 11.5).

Procedures related to this system have to be worked out by operators with knowledge about advantages and disadvantages of every strategy, which can be used for processing a production segment. In the frame of this phase it is possible to prepare the classification of this segment for future handling as well as what is possible in manufacturing conditions.

After imputing all relevant information (production segment data and classification), an information system is ready for definition of manufacturing characteristics.

It is possible for operators to create more process plans suitable for actual production segments. For example, for every future hypothetic event, the operator creates one process plan with equivalent strategy, or in cases of a new unpredictable state, it can be approached with a new strategy. Examples of basic varieties of prepared process plants are:

1. Maximal efficiency
2. Minimal cost
3. Change of products flow (when in full capacity of machine tools)

It is possible to define more phases inside every process plan. Every phase has a relative independent line of operations. For example, the first phase may be the casting, the second – machining, and the last one – surface treatment.

In the phase frame, the operator may approach the manufacturing sequence as follows:

1. Handle writing of technological operation cycles
2. NC program – direct writing by operator or established for group in frame of GT or downloaded from NC program creator
3. Sequence of operation pictograms
4. Simulation sequence (video, animation, etc.)

The most profitable manner of NC program creation from the view of labour content is automatic generation of the NC program in CAD/CAM system conditions.

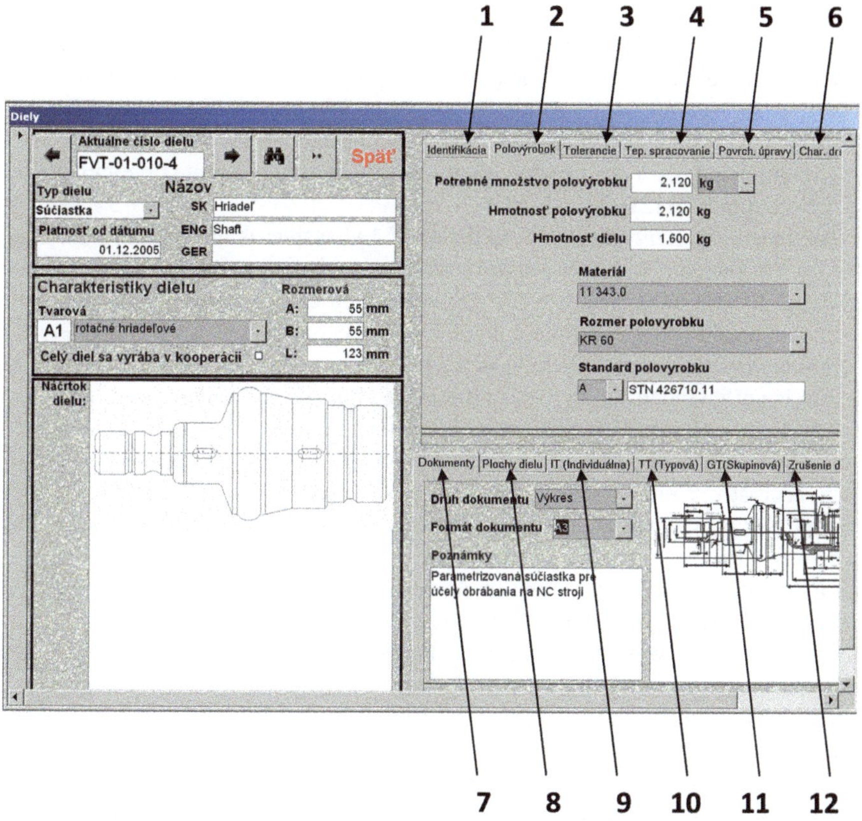

Fig. 11.5 Interface for definition of properties and characteristics of a production segment (1 Identification of segment by basic information. 2 Raw product identification. 3 Information about prescribed tolerances. 4 Heat treatment information. 5 Surface treatment information. 6 Surface roughness information. 7 Documents (definitions and full electronic form) related to segment of production. 8 Surfaces generating volume of production segment. 9 Indications for individual technology. 10 Indications for type of technology. 11 Indications for group technology. 12 Indications for case cancelling of a production segment)

For the majority of NC users, NC is about productivity and flexibility – making a lot of parts, and many different parts, on one machine tool. This was true even before computer numerical control (CNC) superseded an earlier generation of machine tools that did not have the benefit of microprocessor-based control technology. With the level of automation being used in CNC machining, the level of consistency and quality increased. CNC automation eliminated errors and provided CNC operators with time to perform more tasks. The CNC automation also allowed for more flexibility in set-up and job changes, and today's CNC machines are productive, capable and flexible, too.

How to create those new and different programs has taken various approaches. Many CNC machines can be programmed on the shop floor, with the operator entering data at the control panel. This method has been very popular, especially for simpler workpieces. Programs can also be prepared "off-line," away from the machine tool, using computer-aided manufacturing (CAM) software. This method is most often used for more complex workpieces. The latest CAM software for the PC (personal computer) provides many automated features that make NC programming largely a push-button affair, regardless of how simple or complex the workpiece might be [21].

The most important file produced by any CAM system is the NC program that will run the machine tool. General-purpose CAM systems produce programs to run a variety of different types of NC machines. There can be virtually an unlimited number of machine and control combinations. To accommodate this large variety of output requirements, most CAM systems produce a neutral tool path file. That is, the file is not intended for any specific machine but simply contains generic commands to do such things as change a tool or turn on the coolant. This neutral file traditionally has been referred to as the CL (cutter location) file. A separate program, generally referred to as the postprocessor, is used to format the neutral CL file into an NC program that is suitable for specific machine/control combination. Generally, there are some basic problems of post processing. The first problem is simply getting correct code output in the desired order at critical places in the NC program. The critical areas normally come at the start of the program, at a tool change, and at the end of the program. Getting the correct cutter compensation codes (diameter and length) output at the appropriate place is also a difficult task. As previously stated, individual companies and even departments within the same company have different requirements and frequently adopt different methods for employing such things as tool changes and cutter diameter compensation. Thus, a postprocessor configured for one company may not be suitable for another. The CAM user is then faced with three undesirable choices [22]:

- To accept the output as generated
- To have the NC programmer edit and modify each individual NC program before it is sent to the machine tool
- To modify the postprocessor configuration (requires personnel with appropriate expertise or aid from the CAM vendor)

The second problem is that an organization will purchase a machine tool, which their current CAM system cannot support. Many of the low to moderately priced CAM systems do not fully address the requirements of multi-axis machine tools. Limitations include the inability to control multiple rotary axes, such as two or more rotary tables or heads. Even in cases where the rotary axes can be controlled, the CAM software's postprocessor may not be able to accurately calculate the feed rates necessary to keep the tool tip moving at the programmed feed rate because the rotary axes are moving simultaneously. Problems such as these are frequently overlooked until after the machine tool is purchased and is on the floor waiting for a program. Changing or upgrading the CAM system is sometimes the only way to overcome such problems [23, 24].

The third problem frustrating CAM users is that postprocessors frequently do not support special features contained in their machine control unit. Examples include advanced probing cycles, support for variables within the NC program, and the naming of some sub programs. In most cases the staff will be trained on the advanced features of their control only to find out later that their CAM system cannot take advantage of these capabilities.

Some of these problems mentioned above could be removed using principles of 'Group technology' implemented inside the 'Information System'.

Processing of Parts Assigned to One Group Representative

Complex group combinations can be represented by 3D models (Fig. 11.6) with plenty of shapes and geometrical features (hole, round, chamfer, pin lock groove, slots, etc.). These are prepared as representatives for shaft parts to depict automatic NC program generation inside an information system.

On the basis of this component, the CAD/CAM systems can create other similar parts within 'Group technology'. Within software PTC Creo it is accomplished by means of the Family table module through the features variation using commands Yes/No or by dimensional change. The environment for new parts definition on the basis of already existing ones can be seen in Fig. 11.7.

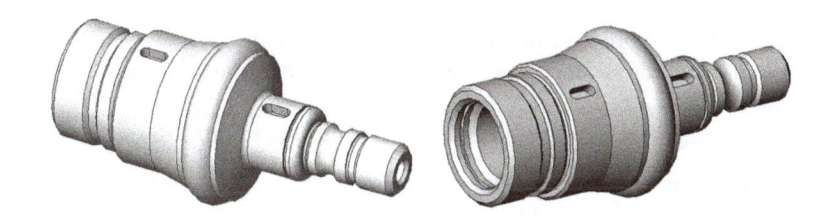

Fig. 11.6 Two views on complex-group representatives

Fig. 11.7 Preparation of a new component within group technology

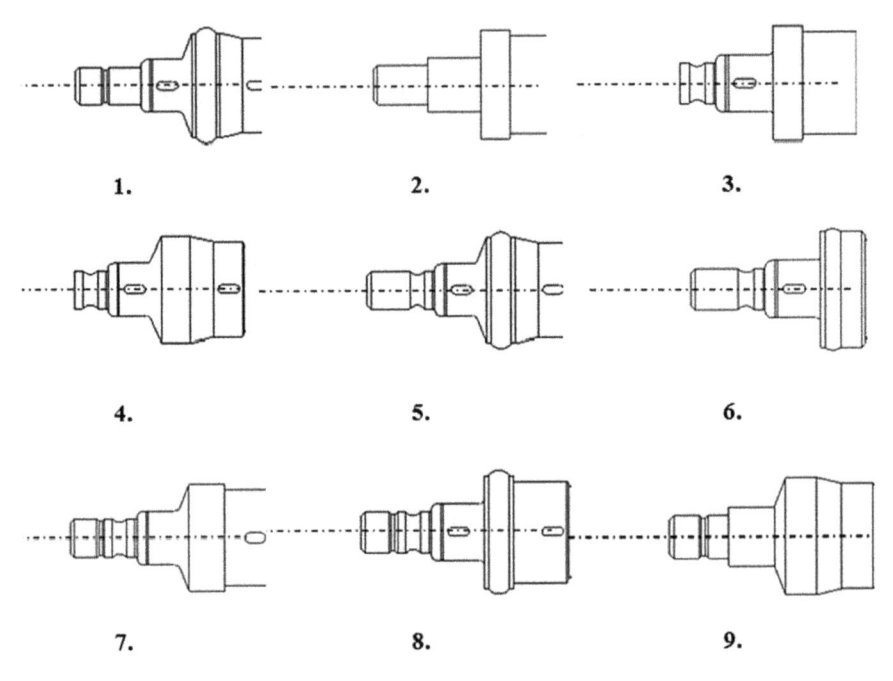

Fig. 11.8 Parts generated on the basis of group representative

Some parts generated by means of the Family table following one group representative are shown in Fig. 11.8. The parts from n.1 to 9 in this figure originate by varying some features (surfaces with certain shape, lock grooves, slots, chamfers, rounds, etc.).

Other advantages of Family table utilization inside the information system are:

- Creation and storage of many objects in a simple and compact way
- Time and labour savings that allow standardizing generated objects
- Potential for generating model variants without the necessity of creating a new model
- Possibility of creating object tables that can be stored as a file and used in an object register

Data verification was made on the part displayed in Fig. 11.9. It was prepared on the basis of a group representative (shown in Fig. 11.6) by means of the Family table in the CAD/CAM system PTC Creo.

The model listed above was an inside database assigned to a group representative and its characteristics were uploaded in the IS interface that is shown in Fig. 11.10.

CL data generated in a CAM system were transformed by means of a postprocessor to the NC program for the selected control system.

Verification of IS in relation to the object manufacturing was also made for the part when it was not possible to include it in some groups.

Fig. 11.9 3D model of parts for the NC program verification

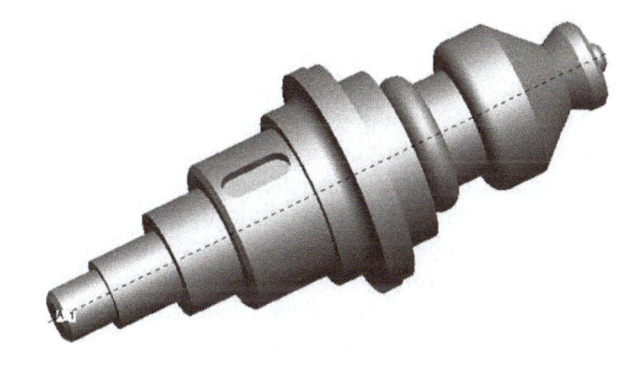

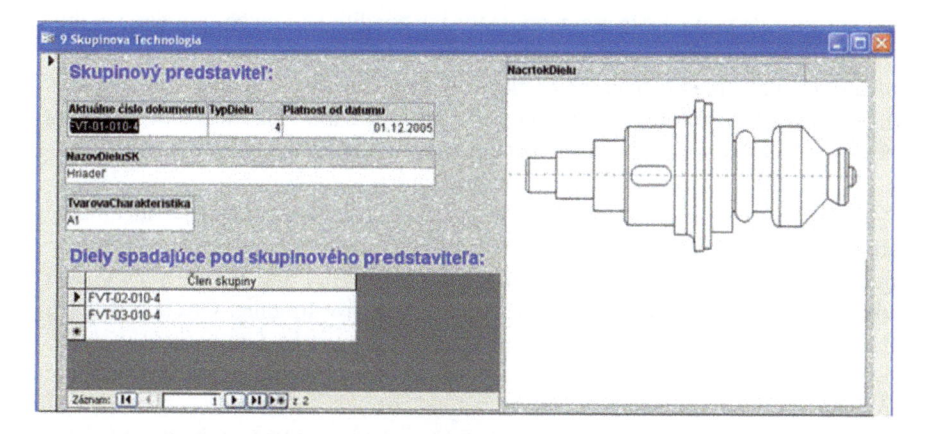

Fig. 11.10 The object assignment to the group representative

A tested information system was designed based on a multi-variant process plan strategy such that it is possible to connect it with a wide variety of CAD/CAM systems (models, CL data & NC programs, etc.) and also to be possible to set it for various approaches to the technological process plan design that corresponds with requirements of the European plants.

Conclusion

On the basis of the aforementioned theory characteristics, the information system was created and applied into real production conditions in computer aided process planning. A product designed with CAPP consists of approximately 6000 components.

The product was a result of co-operation between a German company, which provided investments and co-operation of the activities, and Slovak companies which provided technical process planning and production of the final product.

From the very beginning of the project individual real database objects (components, substructures, structures, finished product) were suitably analysed by the established IS and new analytical tools were created when required.

An established solution serves the purpose of easier and faster assignment of the process parameters, shortening of the computer aided process planning documentation time in real production conditions, and it also supports the effective utilization of the production plant based on the mathematical model description of object variation of the computer aided process planning, fulfilling the combination of the required characteristics within the given production conditions. Output system data can be used for processing of the details for the warehouse, economic and wage records for their control and optimization.

The main contributions of assigning IS, elaborated on the basis of the multivariable process planning in real manufacturing conditions, can be summarized into reduction of the variability of warehouse stock (at first application by nearly 30%); immediate information about the product elaboration; fast acquisition of the details via interfaces for the wage records and accounting; flexible analytical tools enabling the adoption of better decisions and acquisition of the statistical values of parameters applicable to plan production in the future.

The software tool is created in such a way to be easily implemented to an already existing information company structure via flexibly adjustable interfaces. It is also user-friendly, developed with the characteristics of GUI and typical for OS MS Windows, so that operation does not require expensive training. Of course, if the maintenance of this system is to be productive, it must be familiarized with the given philosophy and options of tactic and strategy planning, through which the production can be optimized.

The presented manufacturing information system is unique in its potential for cooperation with the CAD/CAM system (practically any type known) and connectivity to other systems (accounting, stock, wages, etc.). This concept brings advantages manly for micro companies. They are related to the modular conception, flexible interconnections to partners, opportunities for cooperation with a wide variety of external software, and affordable price.

Practically all know-how of the system described shows new tasks for future research of authors in the following scopes:

- Investigation of a system for comparison of 3D data to find the objects' similarity
- Looking for better interfaces for CL data creation, NC program sharing, alerts about 3D model changing, etc.
- Seeking certain data formats for communications between cooperated plants
- Research of a general format of process plan data
- Investigation of production environs in other European countries

Acknowledgments The present contribution has been prepared with direct support of Ministry of Education of Slovak Republic by the projects KEGA 007TUKE-4/2018 and APVV-17-0380.

References

1. Schwab, K. et al. (2014). The Europe 2020 competitiveness report: Building a more competitive Europe, world economic forum insight report, Geneva.
2. Dobransky, J., et al. (2016). Optimization of the production and logistics processes based on computer simulation tools. *Key Engineering Materials, 669*, 532–540.
3. Ungureanu, M., et al. (2016). Innovation and technology transfer for business development. *Procedia Engineering, 149*, 495–500.
4. Hloch, S., et al. (2008). Experimental study of surface topography created by abrasive waterjet cutting. *Strojarstvo, 49*(4), 303–309.
5. Panda, A., et al. (2014). Progressive technology – Diagnostic and factors affecting to machinability. *Applied Mechanics and Materials, 616*, 183–190.
6. Monkova, K., & Monka, P. (2014). Newly developed software application for multiple access process planning. *Advances in Mechanical Engineering, 2014*. article number 539071.
7. Krolczyk, G. M., et al. (2014). Influence of technological cutting parameters on surface texture of austenitic stainless steel. *Applied Mechanics and Materials, 693*, 430–435.
8. Kuric, I., et al. (1999). *Computer process planning in machinery industry.* Bielsko-Biala: Technical University Lodz.
9. Stoicovici, D. I., et al. (2008). An experimental approach to optimize the screening in the real operating conditions. Manufacturing Engineering, issue 2, 2008. *Technical University of Kosice, 2008*, 75–78.
10. Krehel, R., et al. (2016). Diagnostic analysis of cutting tools using a temperature sensor. *Key Engineering Materials, 669*, 382–390.
11. Jurko, J. (2011). Verification of cutting zone machinability during the turning of a new austenitic stainless steel. *Advances in Computer Science and Education Application, 202*(2), 338–345.
12. Krehel, R., et al. (2009). Mathematical model of technological processes with prediction of operating determining value. *Acta Technica Corviniensis: Bulletin of Engineering, 2*(4), 39–42.
13. Ackerman, J. (2007). *Ps structures and production plants in competence cell-based networks.* Warsaw: Advances in Production Engineering.
14. Crow, B. (1992) Rural livelihoods: Action from above, *Rural Livelihoods: Crises and Responses*, Oxford: Oxford University Press.
15. Arn, E. A. (1975) Group technology, *Springer Verlag*, Berlin, ISBN 3-540-07505-4
16. Suresh, N. C., et al. (1998). *Group technology and cellular manufacturing: State-of-the-art synthesis of research and practice* (pp. 568–572). Massachusetts: Kluwer Academic Publishers.
17. Hanzl, P., et al. (2017). Optimization of the pressure porous sample and its manufacturability by selective laser melting. *Manufacturing Technology, 17*(1), 34–38.
18. Baron, P., et al. (2016). Proposal of the knowledge application environment of calculating operational parameters for conventional machining technology. *Key Engineering Materials, 669*, 95–102.
19. Kadarova, J., et al. (2013). Proposal of performance assessment by integration of two management tools. *Quality Innovation Prosperity, 17*(1), 88–103.
20. Cep, R., et al. (2011). Testing of greenleaf ceramic cutting tools with an interrupted cutting. *Technical Gazette, 18*(3), 327–332.
21. Stojadinovic, S. M., & Majstorovic, V. D. (2014). Developing engineering ontology for domain coordinate metrology. *FME Transactions, 42*(3), 249–255.
22. Beno, P., Kozak, D., & Konjatic, P. (2014). Optimization of thin-walled constructions in CAE system ANSYS. *Tehnicki Vjesnik, 21*(5), 1051–1055.
23. Bozdech, J., Rehor, J., & Bruzek, P. (2012). *Streamlining tool management in an engineering company* (pp. 70–75). Slovakia: Proceedings - New Ways in Manufacturing Technologies.
24. Markovic, J., & Mihok, J. (2016). Legal metrology and system for calibration and verification of the radar level sensors. *Quality Innovation Prosperity, 20*(1), 95–103.

Chapter 12
Cooperation as a Key Element Between Universities and Factories

B. Mičieta, J. Herčko, Ľ. Závodská, and M. Fusko

Abstract This article deals with the necessity of education of employees and young people for the needs of Factories of the Future. People should be creating the core of enterprises, and ICT should be only means for acceleration of enterprise's development. Below, the best practices are described for the development of knowledge base of employees, students, postgraduate students and young scientific workers that are being trained for the needs of Slovak industry and for Factories of the Future. The article too describes best practices from Faculty of Mechanical Engineering and Department of Industrial Engineering from University of Žilina. In the end of this article are some statistics information.

Introduction

The changing customer demands put great emphasis on the flexibility of production, innovation and flexibility to respond to changes. It is needed to quickly modify the structure and organization of production, which is often time consuming and costly [7]. A key topic of today is the growth of technologies supporting so-called digital humanism. It is very important, even necessary, to realize that in the period of growing manifestation of digital businesses and digital workplaces, there are people and their potential being in the centre of interest of the progress. The overall development is heading towards further facilitation and efficiency improvement of human activities, easing them of their burden, their collaboration with machines, robots and so on. Many enterprises have already got on the imaginary way of digitalization,

---

B. Mičieta
Central European Institute of Technology, Zilina, Slovakia
e-mail: branislav.micieta@ceitgroup.eu

J. Herčko · Ľ. Závodská · M. Fusko (✉)
University of Zilina, Faculty of Mechanical Engineering, Zilina, Slovakia
e-mail: jozef.hercko@fstroj.uniza.sk; ludmila.zavodska@fstroj.uniza.sk; miroslav.fusko@fstroj.uniza.sk

© Springer International Publishing AG, part of Springer Nature 2019
D. Cagáňová et al. (eds.), *Smart Technology Trends in Industrial and Business Management*, EAI/Springer Innovations in Communication and Computing,
https://doi.org/10.1007/978-3-319-76998-1_12

many other still hesitate, and other will have to face it in the future. This way requires implementation of the right technologies in the right time and having the digital talents needed that understand the new vision of enterprises and that behave differently, so-called Generation Y.

The term Generation Y marketing describes techniques that are used to build and promote product brands among the group of consumers. Generation Y is also described as echo boomers, net generation, web generation or millennium generation. People belonging to this group are primarily characterized by their positive attitude towards information technology and Internet which provides them a lot of information about products and services [1, 11].

Objective and Methodology

The twenty-first century will be characteristic for the development and implementation of "intelligent solutions" in all fields of human life, not excluding the economy. Manufacture and technologies are becoming intelligent. Nowadays, intelligent production systems are mainly designed as so-called agent-based systems (systems with distributed intelligence). To optimize the behaviour of a production system, the methods of artificial intelligence are being used (expert systems, neural networks, genetic algorithms, etc.). These systems will require also new competencies and education of employees and specially prepared staff [1]. These systems will also use new business models, providing supplies of products in extremely short period of time for the whole world. Factories of the Future will no longer need employees with low qualifications. Sophisticated technologies, computers, various applications and so on will be used in these enterprises that have to be operated by employees with university education. As Fig. 12.1 shows, their salaries have been rising since 1980 at more significant pace when compared with salaries of less educated workers [6].

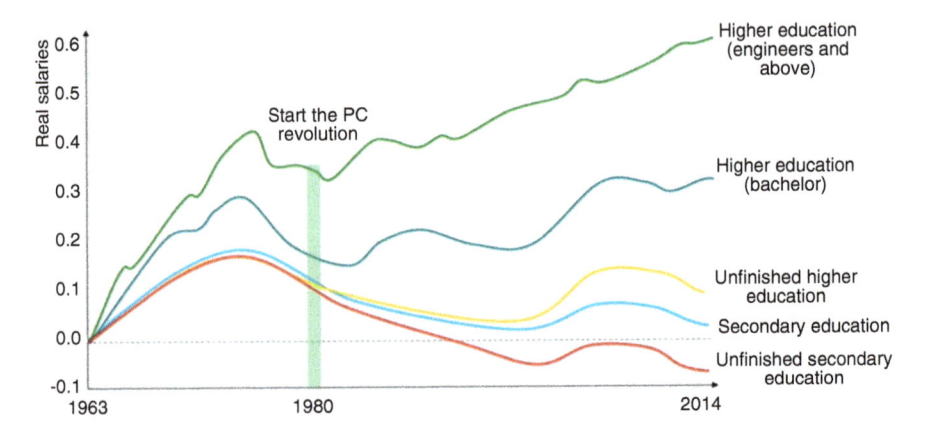

Fig. 12.1 Development of the real salary [6]

The demand for educated workers will rise enormously. These days, mainly human resources officers from companies are complaining about the shortage of qualified workers. Some people even have university education, but their focus is not corresponding with the needs of the market.

All this represents problems that will have to be solved in the future, if the mankind and individual states want to achieve sustainable development. In the University of Žilina, we are dealing with these challenges of the future. Within our region, several significant projects have been successfully organized, such as building the CEIT (Central European Institute of Technology) and ZIMS (Žilina Intelligent Manufacturing System), joining the national project Universities as Engines of Development of Knowledge Society, joining various international projects for the development of level of education of the new generation of workers, etc. [6, 13].

The main objective of the ZIMS initiative is to create research facility of European level. Future development of the ZIMS is oriented on deciding factors of productive and efficient manufacture, which supports competitiveness of production base of the Slovak Republic. New intelligent production systems will require also new competencies of workers. Therefore, there is, as a part of the ZIMS, also the module of innovative training of professionals and young research workers being created in collaboration of the CEIT, the Faculty of Mechanical Engineering and the Institute of Competitiveness and Innovations of the University of Žilina. The approach named the Learning University (Figs. 12.2 and 12.3) uses the latest educational technologies and supports process of selection and long-term training of new research workers and professionals for business practice.

Figure 12.3 represents the principle of building a knowledge environment that will encourage learning systems of active processes. This approach is validated in research area of technology for the industrial production of large optical single crystals.

The research in the field of intelligent production systems and the training of new generation of workers touch also the research and innovative priorities (Fig. 12.4)

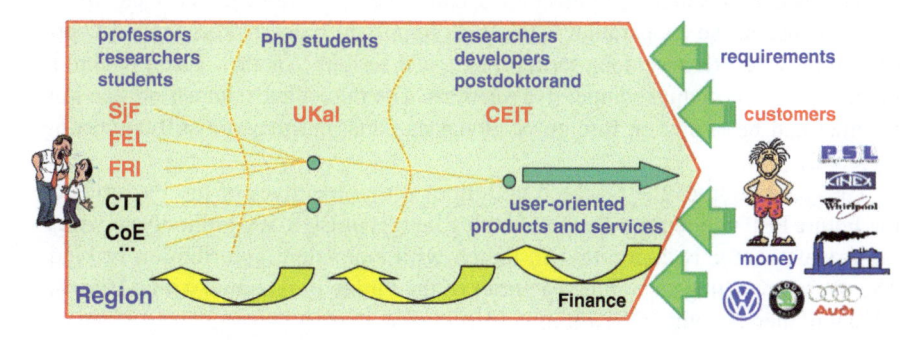

Fig. 12.2 Model of learning of university [6]

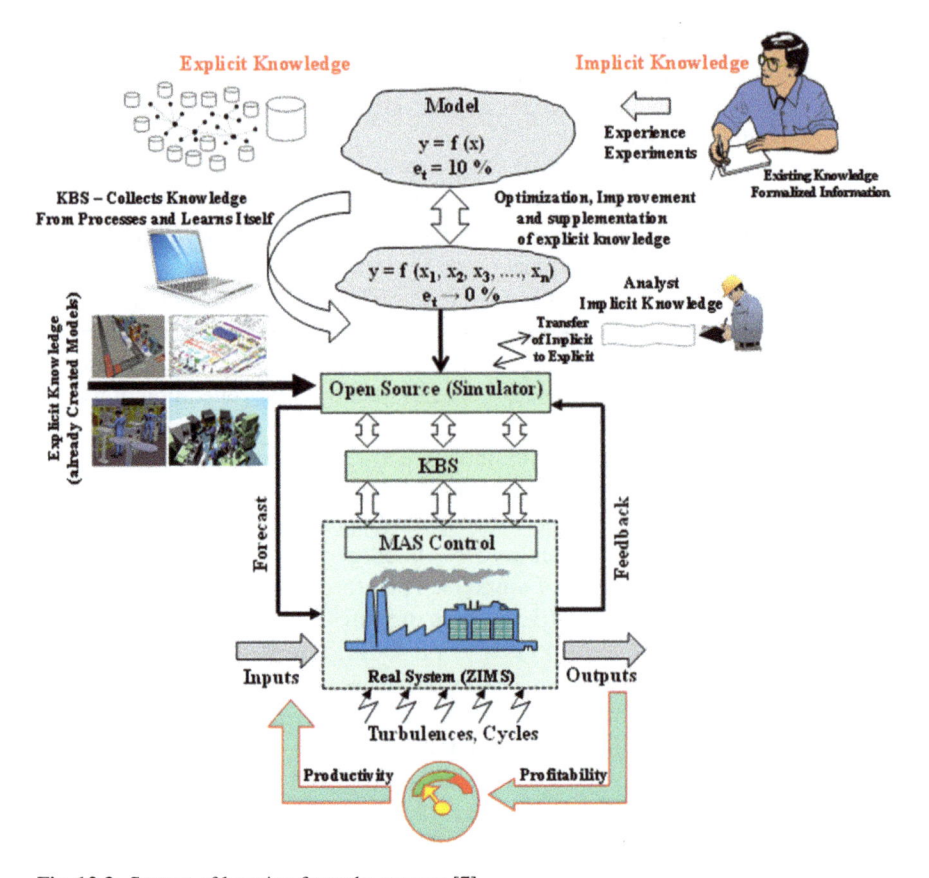

Fig. 12.3 System of learning from the process [7]

that have been defined by EFFRA – European Factories of the Future Research Association for the Factories of the Future.

Clusters are effective connection between the companies to each other and also with the wider surroundings: universities, banks, self-government, R&D, etc. In the cluster concept, the cooperation is the most important process. This kind of cooperation must be effective, long-term running and serious. Another cluster feature is the repression of mutual competitive relations. The immediate competitors can join together and be strong on foreign or global markets, but they are still competing regionally [15].

The realization of the research and innovation objectives of the Factories of the Future PPP will require a public funding budget of EUR 500 million/year which the private sector is committed to match with equivalent contribution in kind. The overall resulting size of the Factories of the Future programme within Horizon 2020 will then become EUR 7 billion [3].

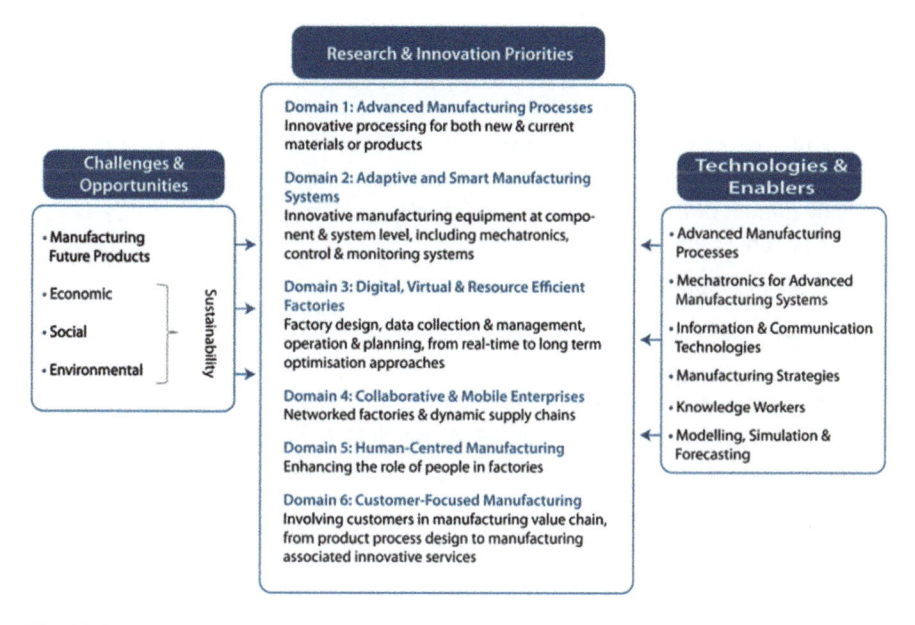

Fig. 12.4 The Factories of the Future roadmap framework (EFFRA) [3]

The Main Goal Is the Collaboration as a Necessity of Education of Young Talent

Knowledge creation, circulation and exploitation are the key elements of modern research and development (R&D) and innovation systems and underpin the evolution of so-called knowledge-based economies and societies [4].

Žilina Innovation Policy (ZIP) was the first step to the region potential in-depth analysis. It's a multidisciplinary team built with members of University of Žilina and partner organizations. Funding mechanism was provided by the EU and also from Self-Governing Region. It's the Regional Innovation Strategy creation process – these activities start in every Slovak region. Project starts in June 2005 within the Sixth Framework Programme of the EU. ZIP is focused on long-lasting activities bringing together R&D environment, business sector and the potential for innovation in Žilina region for continuous development. "Key objective of ZIP is to set up basis for regional institutional structures for innovation support, based on collaborative networks between existing institutions and organizations, and to implement a strategic innovation framework that will enable existing firms to introduce more innovation at all levels and create a positive culture for new entrepreneurs [9].

Project partner organizations include the following: University of Žilina, Žilina Self-Governing Region, Lower Austria Region, Region Södermanland, Sweden, BIC Group, Bratislava. Partnership and cooperation bring the ZIP project much further. Partners' skills and experience were significant help for project activities.

Žilina Self-Governing Region adopted the project results as a strategy for development in the field of innovations and clustering [9].

Main regional problematic areas

- Lack of access to qualified workforce
- Underdeveloped infrastructure (services, roads, railways, etc.)
- Small amount of making use of the EU funds
- Lack of cooperative activities

Main regional positive areas

- High credit of University of Žilina
- Demand for cooperation
- High credit of R&D facilities
- Development in sectors of ICT, automotive industry, finance sector and tourism industry
- Clustering potential

A survey of foreign literature and examples of good practice from abroad show the importance of various institutions involved in promoting brain circulation and reintegration of researchers as well as the existence of a unified reintegration system, which would be supported by the government of the country. However, the research conducted in Slovak enterprises showed that even enterprises themselves are not systematically prepared for adoption of reintegrated researchers. Therefore, the main identified problems and recommendations are focused on improving the situation directly in enterprises and on national and regional level [16].

Nowadays are developed varied programmes for young talent, for example, PhD. students or young researcher. Good example is Israel. The Ministry of Aliyah and Immigrant Absorption offers new-immigrant and returning-resident scientists and researchers diverse forms of assistance for promoting integration into the R&D sector in Israel. The Ministry offers you, scientists and researchers, professional counselling and guidance, along with financial help, in order to facilitate and improve your chances of integrating into research and development in Israel's public and private sectors. The Center for Absorption in Science operates a variety of programmes for promoting your vocational integration and provides research grants, participation in salaries and other forms of assistance. We believe that suitable professional support will enable you to integrate into the cutting edge of Israeli R&D, helping to lead the country to excellence in science and research [14].

Best Practice from Slovak Republic

Current requirements for continuous reduction of products, processes and systems life cycles increase the need of rapid design of "lean" and "flexible" production systems. This means that classical approaches of production systems design have to be extended by the application of advanced technologies and methods, such as digital factory, virtual and augmented reality, computer simulation, reverse engineering, etc. [10, 12]. Therefore within the Slovak Republic, under the auspices of the Ministry of Education, Science, Research and Sport of the Slovak Republic and the Centre of Scientific and Technical Information, the national project University Students for Practice – Universities as Engines of Development of Knowledge Society was realized. Its objective was to adapt university education to the needs of knowledge society. It should be done with the development of innovative forms of education and development of active cooperation of universities with private sector. And that while creating new fields of study and study programmes, and also while rationalizing and making the existing fields of study and study programmes of universities to be of higher quality. The cooperation should also be present in the process of education (Fig. 12.5). Also the engagement of universities in international cooperation should be higher.

The national project Universities as Engines of Development of Knowledge Society was created as a reaction to the need of better interconnection of university education with the needs of labour market and to identify and support those fields of study that are the most desirable on the labour market in the business sphere (Fig. 12.6). And that should be done especially in fields with high added value for the growth of the Slovak Republic. The national project was co-funded by resources of the European Union within Operational Programme Education. It was focused on students studying at universities in the whole area of Slovakia with exception for self-governing region of Bratislava.

Objectives set to fulfil the project were these [8]:

1. Adapt university education for the needs of knowledge society, i.e. the development of innovative forms of education, rationalization of study programmes of universities while making them of higher quality including the support of career consultancy.

Fig. 12.5 Students on excursion in factory

Fig. 12.6 Students on practice and excursion in factories

2. Adapt university education for the needs of knowledge society, i.e. support of development of human resources in research and development.
3. Adapt university education for the needs of knowledge society, i.e. support of active cooperation of universities and private sector in creation of new fields of study and study programmes and in the process of education.
4. Adapt university education for the needs of knowledge society, i.e. increase in engagement of universities and other organizations of research and development in international cooperation and networks of development and innovations.

Projects of similar orientation are known in democratic world for a long time. An example can be found in Austria, which despite "crisis" managed to maintain the lowest unemployment rate in Euro area. Austria and also Switzerland have highly developed system of practical education of young people already from the apprenticeship. For example, the state supports practical education in family companies. These create the core of healthy economy. Many of them invest also in Slovakia. The highest numbers of students are required by big enterprises, but there are also many small companies among applicants. According to the document of Ranking of Collaborations, the total number of offered positions was 3024. It was a great opportunity not only for students and employees of registered enterprises, but also for teachers, because there works the triangle: student – teacher – lector form practice, in the project.

National Project

Organizations are looking for opportunities to increase their efficiency and competitiveness in world markets [2]. The national project Universities as Engines of Development of Knowledge Society was created as a reaction to the need of better interconnection of university education with the needs of labour market and of identification and support of those study programmes that are most desired by labour market in business sphere, especially in fields with high added value for economic growth of the Slovak Republic. It is co-funded from resources of the European Union, and EUR 17.072 mil. was allocated on its realization within Operational Programme Education and its part Reform of Education System and Professional Training.

The project was intended for all kinds of university studies – Bachelor, Master, Master of Science, and also for Postgraduate studies. Students could join the project in forms varying from short-term excursions to long-term internships in conditions of business practice. Students were given the task to process specific professional topics during the set period. Alongside, educational bases were created at universities that will bring the environment of universities closer to the conditions of practice. The project was implemented in the Region of Trnava, Trenčín, Nitra, Žilina, Banská Bystrica, Prešov and Košice.

The project was implemented in following four activities [8]:

Activity 1.1 Evaluation of efficiency of study programmes of universities from the perspective of current and possible future needs of labour market and cooperation with business sphere.

The objective of this activity was to create methodology and to evaluate all relevant study programmes within eligible universities from the perspective of priority needs of practice and predictions of development of labour market. Based on this, optimization of procedures of adaptation of university education to the needs of employers and prospects of economic growth of the country was designed. The need of practice is understood especially as those segments of labour market that nowadays and in the predicted development contribute to the growth of GDP most significantly and correspond with the priorities of the National Plan of Building of Infrastructure of Research and Development 2012. Besides implementation of created methodology for selected study programmes, the importance will lie on designs for adaptation of new contents and forms of studies in 100 highly promising study programmes. It will create the basis for improvement of training of graduates and for increase of their successfulness in employment on the labour market, as well as for the increase of their actual value for the subject of labour market. Overall outcome of this activity should bring system recommendations in the subject area.

Activity 1.2 Active creating of networks of cooperation of universities and business sphere.

This activity is focused on creating and supporting the relations and cooperation of universities and private sector in the process of education. Activity 1.2 is the largest one from the whole national project. Its basic operation is creating prerequisites of practical education of university students and their engagement in recognition of real needs and real solutions of problems and innovations in practice.

Cooperation of universities and private sector will support creation of contents and forms of university education for real needs of labour market, as well as for the requirements of business sphere. There will be a space open for education of university students in real conditions of business practice. The intent is to create a pilot network of cooperation between universities (in case of already existing relationships, these will be supported and strengthened) and enterprises. An efficient system will be made, in which students will be educated in pre-agreed conditions and will acquire practical competencies directly in business practice or in established educational centres at university facilities of schools participating in the project.

That will create the prerequisite of their efficient employment on the labour market after finishing their studies.

Activity 1.3 Improvement of education content and support of innovative forms of education for the needs of labour market in selected promising fields of study.

The purpose of implementation of this activity is to improve university education through implementation of innovations into education content, as well as to support innovative forms of education for the needs of requirements of labour market in selected promising study programmers.

Activities will bring the content of education closer to the needs of business practice and to the requirements of labour market. Because of providing educational contact centres with modern educational tools according to the definitions of needs by business practice, the improved educational content and form will be made for students of promising study programmes in compliance with rules of the Operational Programme Education and with European Social Fund. These methods will correspond directly with requirements and needs of labour market. Alongside it will be possible to educate students of universities engaged in the national project in mentioned contact centres using innovative forms of education and also by direct engagement of students in solving of tasks at the workplace during their educational stays in the enterprise.

Activity 1.4 Popularization of studies in promising fields of study and of cooperation between universities and business sphere.

The essential element in this part of the project is, by using standard communication procedures, to create prerequisites of increased society-wide interest in the national project and in its individual intentions in a targeted way and to increase awareness of needs of interconnection of education with the needs of business practice. Another objective is to increase the interest of universities and enterprises themselves in mutual interaction and creating of prerequisites for cooperation oriented especially on utilization of development potential of young people for the economic growth and increase of competitiveness. Following that, objective is also to motivate students of first years of universities, studying in promising or more challenging study programmes, to actively use possibilities of the national project. So that they, through using practical education, find already as fresh graduates fast and promising employment on labour market and in business. Within individual media activities also opinions of experts on positive system changes in university education will be presented.

Experience from Universities

Study programmes integrated into technical sciences are the ones dominating in cooperation with business sphere and that especially study programmes from:
- Technical University of Košice
- University of Žilina

- Faculty of Materials Science and Technology in Trnava (Slovak University of Technology)
- Slovak University of Agriculture in Nitra

The highest number of collaborations with enterprises is reached by study programmes focused on technical sciences – 75%, and the second place is taken by study programmes focused on economic sciences – 11%. Top 50 places in the Ranking of Collaborations are represented by these companies (cooperation between company and university was mutually confirmed):

- Falck Záchranná, a.s.
- Volkswagen Slovakia, a.s.
- Vysokoškolský poľnohospodársky podnik SPU, s.r.o.
- INA Kysuce, spol. s r.o.
- PSA Peugeot Citroen, Slovakia, s.r.o.
- Zlatý ónyx Levice, s.r.o.
- Prvá zváračská, a.s., Bratislava
- Swedwood Slovakia spol. s r.o., o.z. Spartan
- Spinea Technologies, s.r.o.
- CEIT Consulting, s.r.o.
- Železnice Slovenskej republiky (Tables 12.1 and 12.2)

Table 12.1 Number and percentage of identified collaborations from the perspective of orientation of study programme

Orientation of study programme	Number of collaborations	Percentage
Technical sciences	2076	75%
Economic sciences	316	11%
Natural sciences	80	3%
Agricultural-forestry and veterinary sciences	80	3%
Military and security sciences	72	3%
Health care	57	2%
Social sciences	46	2%
Other	28	1%
Sciences in culture and art	0	0%

Table 12.2 Number and percentage of identified collaborations of students

Degree of study	Enterprises (%)[a]	Enterprises (number)
II. Degree	85	305
I. Degree	49	174
III. Degree	27	96

[a]As enterprises had a possibility to choose multiple answers simultaneously, cumulative percentage exceeds 100%.

Experience from Faculty of Mechanical Engineering, University of Žilina

The topic of managing cooperation activities is currently highly up to date. In present, cooperation as such represents for a company an important tool for increasing its competitiveness [17]. Practice in particular enterprises is always a valuable school for university students, but also their inspiring ideas can be a contribution for companies. That is how postgraduate student Ľudmila Závodská summed up the significance of interconnection of academic community and industry. She has tried already during her studies work in particular companies in Slovakia and also abroad, and she guides other students to it too. Young resident of Žilina, after finishing her studies of management, wanted to focus more on manufacturing and logistical processes, therefore she continued at the Department of Industrial Engineering of the Faculty of Mechanical Engineering at the University of Žilina. After she attended practice as an intern in CEIT in Žilina, she had the opportunity to try out the optimization of material flow also in German city of Mönchengladbach.

She was studying management as a field of study at the Faculty of Management Science and Informatics, and already in her third year there, operations management, manufacturing processes and especially logistics have attracted her attention. She wanted her diploma thesis to be "tailored" for a particular enterprise. Therefore she addressed Scheidt&Bachmann Slovensko, a company being in a long-term cooperation with the University of Žilina. It is a daughter company of German manufacturer of systems for car parks, railway security systems, petrol stations and equipment for passengers (Fig. 12.7).

Fig. 12.7 Ľudmila Závodská in front of the seat of German company

After starting her postgraduate studies, she has decided for "logistical" topic, for her doctoral thesis she is studying progressive approaches to the design of logistical strategy of an enterprise. After she had acquired valuable practical experience in the CEIT, which focuses among other things on optimization of material flows, Ľudmila used the opportunity to take a look into a foreign enterprise.

She was working at the department of production planning of railway security machinery, specifically signalling devices, where kanban was being implemented. The manufacture of signalling lights is not complex. The more complex issue lies in the fact that multiple components for these lights are produced by the enterprise itself. So, multiple kanban circles are being used among storehouses and various workplaces. The principle of how the system works is that kanban activates movement, manufacture or supplying. The objective is to make the flow of material in the manufacture clearer and to eliminate storehouses gradually. However she added that during the seven-week-long practice in German enterprise, among other tasks she was also dealing with designs of material flow between daughter company in Žilina and parent company. It was not just the logistics inside the enterprise. She was also designing the system of communication between German enterprise and the one in Žilina. In Žilina, cable ties are being produced that are used in the manufacture of signalling devices in Germany, and it is necessary to transport them between enterprises somehow. Ľudmila has created several designs using kanban system that were approved by both sides. Based on advantages and disadvantages, managers now have to pick the most suitable one.

Before starting the practice, after consultation with the contact person in the German enterprise, the schedule of the practice was set, which included the following activities:

- Familiarizing with the logistical strategy of the enterprise
- Analysis of material flows
- Working with the SAP system
- Implementing the kanban system
- Designing the improvement of organization of workplace from the perspective of material supplying

The set goals were met from the student's point of view. While processing the given tasks, she could use her experience and knowledge gained before, but on the other hand she had a sufficient support from employees in cases, when professional advice was needed.

Before the practice, student expected that she would get to know how logistics work in a big foreign enterprise. These expectations were fulfilled. Besides activities she had scheduled, she has learned much more about the business logistics. Employees of the enterprise have explained her processes of material receiving, dispatching, movement of material through production and so on. They have also explained her how information flows go, how material is planned and how these activities look in information system SAP. She also had an opportunity to see what technologies are being used in storehouses, e.g. automated storage systems (Figs. 12.7 and 12.8).

Fig. 12.8 Ľudmila Závodská in production system in company

Alongside working on specific tasks, it was interesting for the student to observe some particularities she has not come across in another enterprise, for example, glass door on offices, or monitors turned the way that anyone could see what was on the screen. She was communicating in English, so except professional knowledge she sees improvement in the language as a big benefit too. She recommends this kind of opportunity to every student because the time spent directly in the manufacturing enterprise gives fully new dimension to the knowledge acquired in the school. And the contribution can be mutual, not only students will gain practical experience, but also their inspiring ideas can, on the other side, represent enrichment for the enterprise.

After completing the practice, it was needed to fill the final report from the educational stay abroad. This report included, except verbal description of the completed stay, also questionnaire which was a feedback for CVTI. Each student, completing internship in Slovak or foreign enterprise, could express his opinion on how it went. The questionnaire included questions like whether the practice fulfilled expectations, what the level of competencies before and after the stay was, what competencies were being developed during educational stay abroad and with what result, what the biggest benefit of the educational stay is and so on. Based on results of these questionnaires, it is possible to see the satisfaction of students with the educational stay.

Experience from Department of Industrial Engineering, University of Žilina

Department of Industrial Engineering cooperates with leading companies (references in webpage: www.priemyselneinzinierstvo.sk). Into these companies have students the opportunity to go on excursions and practice. From the subjects Teamwork and Branch of practice were students at Schaeffler Kysuce, where they solved the real problems from factory environment.

On the first day, the company and the specific focus of the practice in the factory were presented to the students. The students were divided into 5 teams in which they jointly performed the assigned tasks. During the practice, students recorded all significant workplace deficiencies, also checked the real status of OEE and compared it with the established standard. During the day, a meeting was also available to discuss the findings and discuss their further intentions in their activities.

At the end of each day, a feedback was made, in which students presented their results during the day and plans for the next day. The penultimate day and the last day of the practice were mainly intended to complete their findings and prepare the presentation. Students presented their results on the flip chart, where they had to defend their claims to a wide audience composed of section leaders and pedagogical supervision. All five teams successfully defended their results, successfully identifying workplace problems and suggesting many improvements.

The practice was completed by granting a certificate of practice (Fig. 12.9). In addition to successfully completing the Prax course, however, students have gained much more and this practical experience of the industrial engineer's work in one of the best factories in the field of bearing production.

Fig. 12.9 Students with certificates

Conclusion and Statistics Information

The national project Universities as Engines of Development of Knowledge Society during its implementation fulfilled its objective and met the set indicators and outputs. It was successful also from the perspective of its acceptance by all engaged groups, as demonstrated by the active participation of target group of students, as well as the support of representatives of universities, of most significant enterprises and employers' associations in the Slovak Republic, representatives of MERDaS SR and experts from MLSAaF SR or professionals and laic public.

Continual effort to build new and sustainable research-development ecosystem, which will be able to prepare experts for designing and operation of Factories of the Future, is being produced in cooperation with the University of Žilina and CEIT. First results of the system approach to the creative environment of the future can already be seen in Žilina in these days. The new initiative gives a real form to creative ideas, in which first-class results of own innovations, or also of the training of new generation of workers for Factories of the Future, are being integrated. The options of using some mobile applications, Internet of Things, data gathering and their real-time evaluation and their sharing [5].

The following pictures (Fig. 12.10) show criteria for choosing. As can be seen, businesses are interested in students and positively evaluate the cooperation.

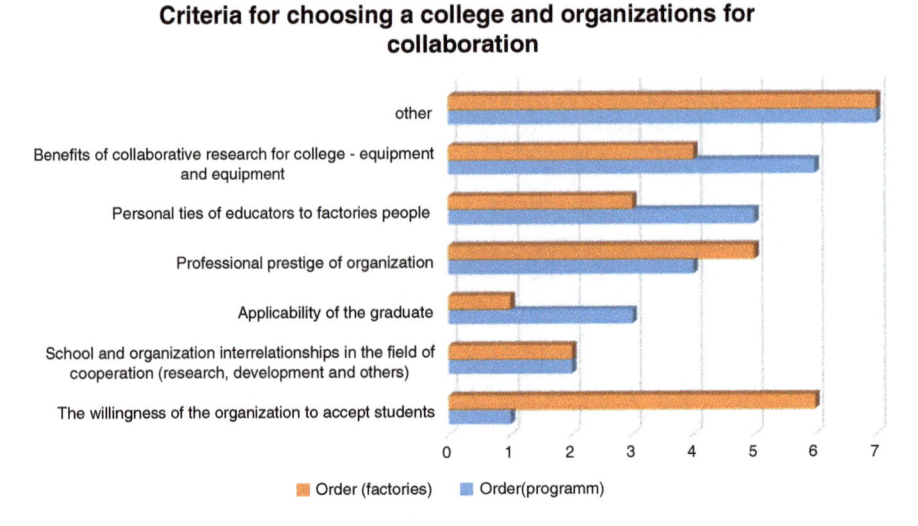

Fig. 12.10 Criteria for choosing

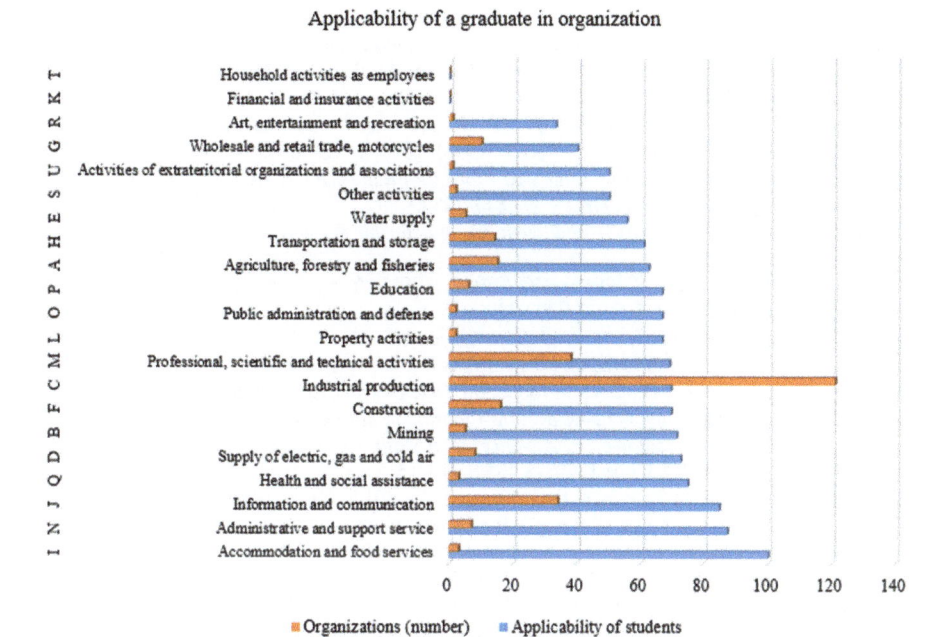

Fig. 12.11 Criteria for choosing

The following pictures (Fig. 12.11) show applicability of a graduate in organization. The greatest applicability of students is in accommodation and food services, but most of the enterprises were from industrial production.

Acknowledgements This paper was made about research work support: VEGA 1/0559/15.

References

1. Bubeník, P., Rakyta, M., & Biňasová, V. (2016). Proposal for the implementation of innovative approaches in teaching system based on interactive training applications. In *Metody i techniki kształtowania procesów producyjnych: monografia* (pp. 25–30). Bielsko-Biała: Wydawnictwo Akademii Techniczno-Humanistycznej. ISBN 978-83-65182-37-1.
2. Dulina, Ľ., & Bartánusová, M. (2015). CAVE design using in digital factory. In: *Procedia Engineering [elektronický zdroj]: International symposium on intelligent manufacturing and automation, DAAAM 2014*; Vienna, Austria; 26 November 2014 through 29 November 2014. ISSN 1877-7058. Vol. 100, online, s. 291–298.
3. European Commission. (2013). *Factories of the future – Multi-annual roadmap for the contractual PPP under horizon 2020*. Brussels: European Commission). ISBN 978-92-79-31238-0. https://doi.org/10.2777/29815.

4. Fernández-Zubieta, A., & Guy, K. (2010) *Developing the european research area: improving knowledge flows via researcher mobility*. In: JRC Scientific and Technical reports, European Commission. http://library.certh.gr/libfiles/PDF/MOBIL-103-DEVELOPING-ERAby-ZUBIETA-in-JRC58917-Y-2010.pdf.
5. Gašová, M., Gašo, M., & Štefánik, A. (2017). Advanced industrial tools of ergonomics based on industry 4.0 concept. In: TRANSCOM 2017: International scientific conference on sustainable, modern and safe transport. *Procedia Engineering, 192*, 219–224. https://doi.org/10.1016/j.proeng.2017.06.038. ISSN 1877-7058.
6. Gregor, M., Magvaši, P. (2013) *Intelligent Manufacturing Systems – Žilina's model*. www.researchgate.net.
7. Gregor, M., & Medvecký, Š. (2015). *CEIT 2030, CEIT – Technologické trendy do roku 2030* (p. 103). Žilina: CEIT, 2015, CEIT-Š002–03-2015.
8. https://vysokoskolacidopraxe.cvtisr.sk/sk/. Retrieved on 7 Oct 2016.
9. http://www.zip.utc.sk/. Retrieved on 5 Sept 2016.
10. Krajčovič, M., & Plinta, D. (2015). Production system designing with the use of digital factory and augmented reality technologies. In *Progress in automation, robotics and measuring techniques: Control and automation, Advances in intelligent systems and computing* (Vol. 350, pp. 187–196.) ISBN 978-3-319-15795-5. ISSN 2194-5357.
11. Lendel, V., Siantová, E., Závodská, A., & Šramová, V. (2017). Generation Y marketing — The path to achievement of successful marketing results among the young generation. *Springer Proceedings in Business and Economics*. https://doi.org/10.1007/978-3-319-33865-1_2.
12. Mičieta, B., Gašo, M., & Krajčovič, M. (2014). Innovation performance of organization. *Communications: Scientific Letters of the University of Žilina, 16*(3A), 112–118.
13. Mičieta, B., Závodská, Ľ., Rakyta, M., & Biňasová, V. (2015). Sustainable concept for green logistics and energy efficiency in manufacturing. In *DAAAM international scientific book 2015* (pp. 391–400). Vienna: DAAAM International Vienna. ISBN 978-3-902734-05-1. (Scientific book. ISSN 1726-9687).
14. Ministry of Aliyah and Immigrant Absorption: Scientists and researchers. In: *Ministry's Website, Returning to Israel* (2016) http://www.moia.gov.il/English/RETURNING.RESIDENTS/Pages/MadaUmechkar.aspx.
15. Soviar, J. (2016). Cluster initiatives in Žilina region (Slovak republic). *Economics and Management: 2009, 14*, 528–534. ISSN 1822-6515.
16. Šramová, V., Závodská, A., & Lendel, V. (2015). Reintegration of Slovak researchers returning to Slovak companies. In *Knowledge management in organizations: 10th international conference, KMO 2015: Maribor, Slovenia, August 24–28, 2015: proceedings*. Cham: Springer, 2015. ISBN 978-3-319-21008-7. - S. 353–364.
17. Vodak, J., Soviar, J., Lendel, V., & Varmus, M. (2016). Proposal of model for effective management of cooperation activities in Slovak companies. *Communications - Scientific Letters of the University of Zilina, 17*(4), 53–59. ISSN: 1335-4205.

Chapter 13
Zilina Intelligent Manufacturing System: Best Practice of Cooperation Between University and Research Center

Milan Gregor, Jozef Hercko, Miroslav Fusko, and Lukas Durica

Abstract This paper presents best practice of cooperation between university and research institute realized in Zilina, Slovakia. Creation of common workplace named ZIMS – Zilina Intelligent Manufacturing System – offers possibilities to realize common research activities with special focus on industry needs. As the name implies, key focus of workplace is on intelligent manufacturing system. The concept of workplace is based on a holonic approach that means all systems and subsystems are independent. At the end of the paper is presented one of the results of common research activities, which has been very successfully commercialized.

Introduction

Mass customization is a relatively new paradigm which principle is based on offering of many extending variants of existing product manufactured with costs of mass production. Such production of goods of the same type is different as by classical mass production. It presents production of high amount of product variants of the same product family by competitive costs of mass production and economy of amount. Following that companies must constantly react to changing requirements of customers through innovations of processes to be able to fulfill the expectations of customers in time and with appropriate quality. Factories are pushed to increase the amount of new technologies which nowadays exist as separate elements but without any cooperation within them. Therefore it is necessary for companies to cooperate with research institutions and universities on joint researches and

---

M. Gregor
Central European Institute of Technology, Zilina, Slovakia
e-mail: milan.gregor@ceitgroup.eu

J. Hercko (✉) · M. Fusko · L. Durica
University of Zilina, Faculty of Mechanical Engineering, Zilina, Slovakia
e-mail: jozef.hercko@fstroj.uniza.sk; miroslav.fusko@fstroj.uniza.sk; lukas.durica@fstroj.uniza.sk

innovation projects. In present, cooperation as such represents for a company an important tool for increasing its competitiveness [15]. Based on this, industry, universities, and research organizations need to cooperate. It is necessary not to forget the existence of common partnerships and mutual cooperation within supplier and customer relationships. The quality of such relationships is a prerequisite of the future success of a business on a market [14].

Cooperation to Reach Factories of the Future Challenges and Opportunities

In 2008 the European Commission created the European Recovery Plan which goal is to support the economy of the European Union after lingering economic crisis. One of the taken actions was the establishment of initiatives to support Factories of the Future development. The goal of this initiation is to support the growth of technologies in industry, so the industry will be competitive in global markets. Next after the creation of initiative, EFFRA – European Factories of the Future Research Association – was established. The main goal of this association is to determine direction in research and development focused on Factories of the Future. Publication Factories of the Future – multi-annual road map for the contractual PPP under Horizon 2020 published by EFFRA – defined research and development priorities (Fig. 13.1 shows an example) [3].

Research & Innovation Priorities

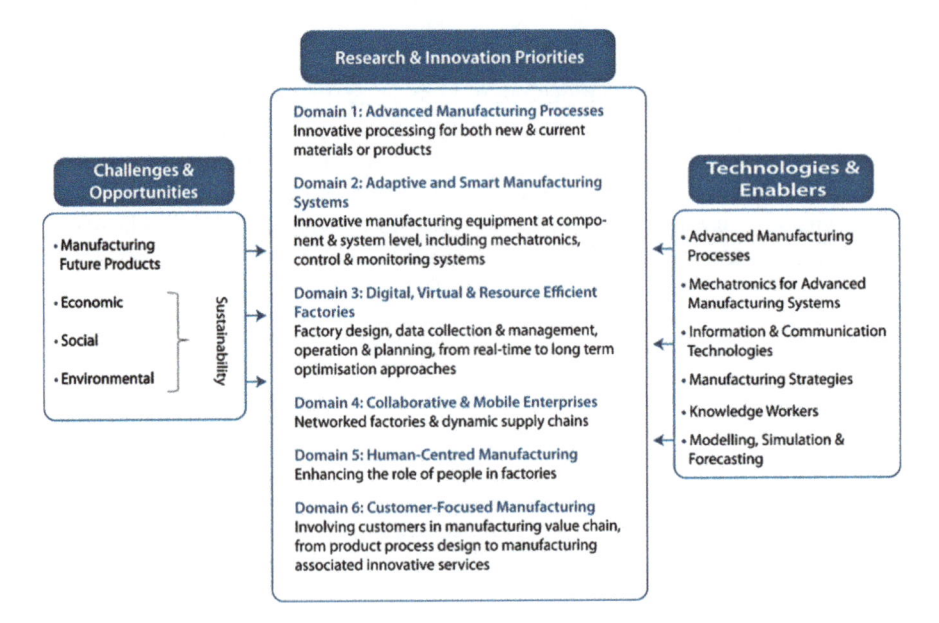

Challenges & Opportunities
- Manufacturing Future Products
- Economic
- Social
- Environmental

Sustainability

Domain 1: Advanced Manufacturing Processes
Innovative processing for both new & current materials or products

Domain 2: Adaptive and Smart Manufacturing Systems
Innovative manufacturing equipment at component & system level, including mechatronics, control & monitoring systems

Domain 3: Digital, Virtual & Resource Efficient Factories
Factory design, data collection & management, operation & planning, from real-time to long term optimisation approaches

Domain 4: Collaborative & Mobile Enterprises
Networked factories & dynamic supply chains

Domain 5: Human-Centred Manufacturing
Enhancing the role of people in factories

Domain 6: Customer-Focused Manufacturing
Involving customers in manufacturing value chain, from product process design to manufacturing associated innovative services

Technologies & Enablers
- Advanced Manufacturing Processes
- Mechatronics for Advanced Manufacturing Systems
- Information & Communication Technologies
- Manufacturing Strategies
- Knowledge Workers
- Modelling, Simulation & Forecasting

Fig. 13.1 Research and innovation priorities defined by EFFRA [3]

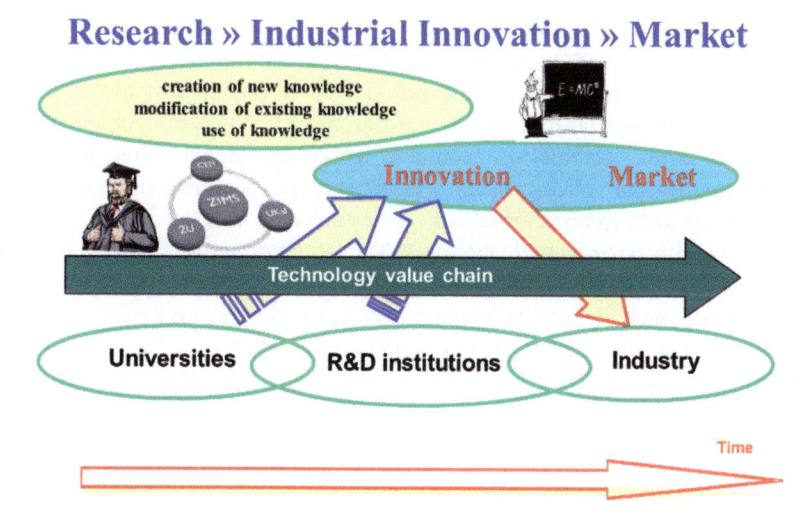

Fig. 13.2 Model of cooperation between universities, research institutes, and industry

To reach applicable and competitiveness increasing results, cooperation between all interested parties – universities, research institutes, and industrial companies – is necessary. Only this combination is supposition to realize research and development with potential to implement the results into industry.

Figure 13.2 illustrates the connection between industry, research institutes, and universities. Research institutes execute the role of "bridge" between industry with specific and realistic research tasks in one hand and universities producing new knowledge in other hand. Connection of research institutes and universities produces new knowledge or modifies existing knowledge which produces new innovation for market.

The University of Zilina systematically coordinates research activities with special focus on needs of European industry. Since 2000 the University of Zilina invested a big amount of money into research focused on digitalization, reverse engineering, rapid prototyping, and digital factory. One of the new development lines of research is the intelligent manufacturing system (IMS). New intelligent manufacturing systems will require new skills by workers. Based on this the University of Zilina focuses on innovative education of students with use of new education technologies and approaches.

Zilina Intelligent Manufacturing System

Current requirements for continuous reduction of products, processes, and systems life cycles increase the need of rapid design of "lean" and "flexible" production systems. This means that classical approaches of production systems design have to be extended by the application of advanced technologies and methods, such as digital factory, virtual and augmented reality, computer simulation, reverse engineering, etc. [9]. Coordination of research and development activities is a very important factor. Based on that the Central European Institute of Technology and University of Zilina established a common workplace named Zilina Intelligent Manufacturing System (ZIMS). ZIMS is established as open platform to cooperate with other interested organization. ZIMS was established in 2009, and in present it is represented with same-named laboratory. The laboratory covers an area of more than 1000 m^2 with disposition illustrated in Fig. 13.3.

ZIMS is an open collaborative environment supporting creativity, inventions of new solutions, and their practical implementation in the form of new innovative solutions. This environment fully supports experiments with new, unknown, and

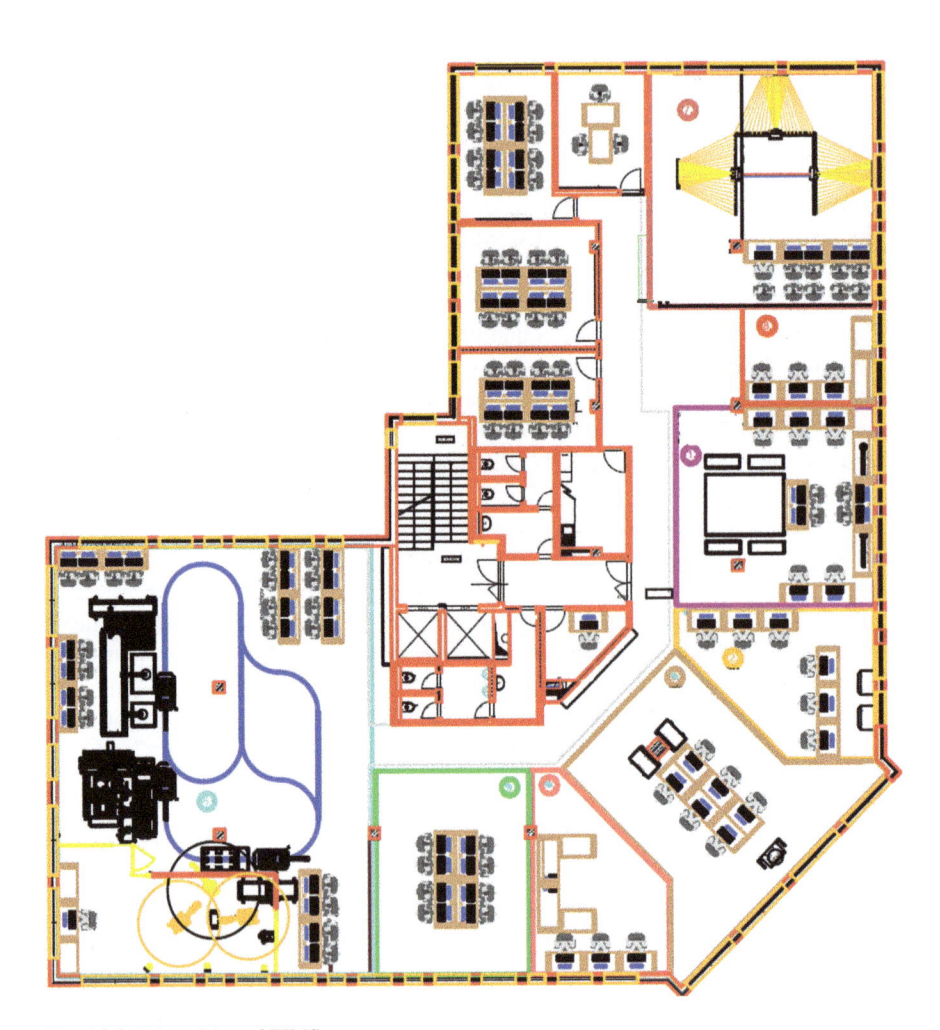

Fig. 13.3 Disposition of ZIMS

nontraditional approaches to solve industry-driven tasks. ZIMS has incubation role for potentially new technological and IT-related start-ups.

ZIMS laboratory is divided into three basic sections (Fig. 13.4) and other supporting sections based on product life cycle:

- Section of design, prototyping, and testing of product and technology. This section is based on immersive and reverse engineering technologies. Result of work in this section is related with product innovation.
- Section of design of manufacturing and logistic processes. This section is based on the use of methods and tools for implementation lean principles to manufacturing and logistic processes. Result of work in this section is related with process innovation.
- Section of production presented with implementation of innovation into manufacturing. This section is characterized as intelligent manufacturing systems.

The vision of ZIMS is to create integrated collaborative environment which is connecting three worlds – real, digital, and virtual. In this environment has the key

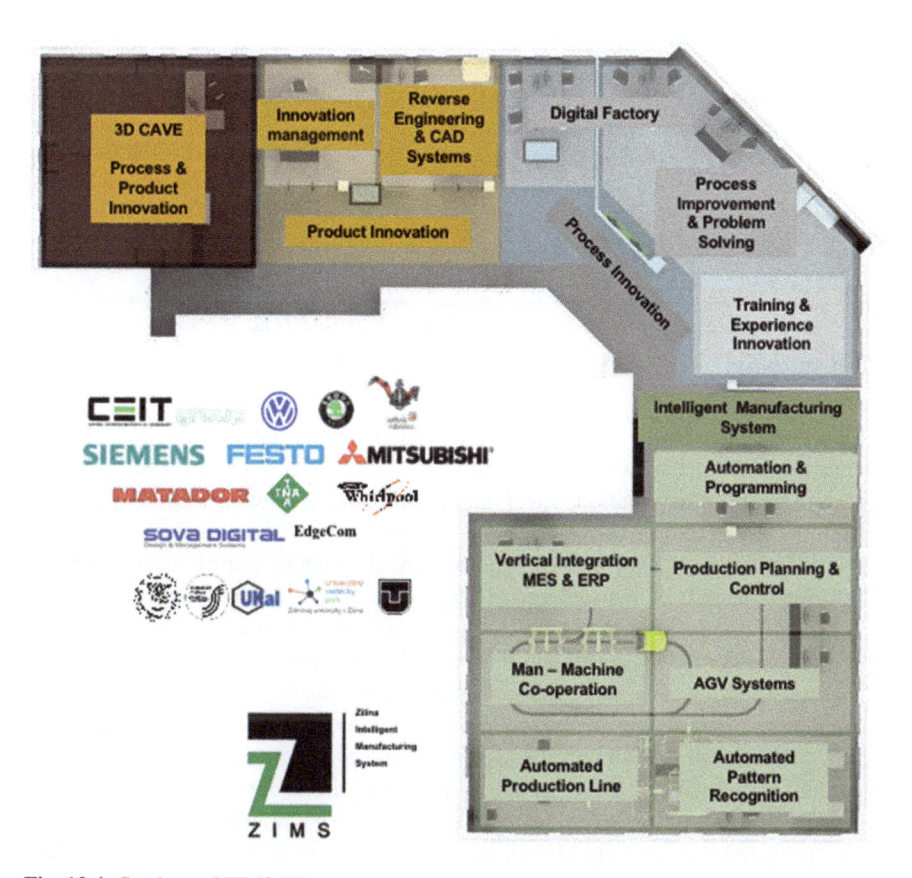

Fig. 13.4 Sections of ZIMS [5]

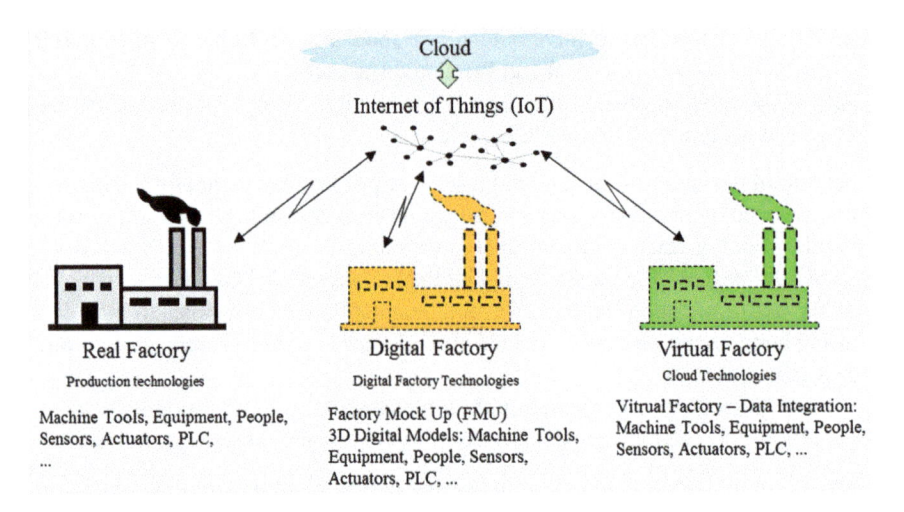

Fig. 13.5 Concept of CPS building at ZIMS [5]

role creation, management, and use of new knowledge. The target of researchers is to create a unique knowledge management system that includes automated management of knowledge. Connection of data from all three worlds creates new cyber-physical system (CPS) illustrated in Fig. 13.5.

ZIMS Cyber-Physical System

ZIMS, as a system, is built in three different worlds: the real, the virtual, and the digital, where their interface is created a cyber-physical system (CPS) ensuring a direct integration of virtual, digital, and real world:

- Smart factory – this is built as an agile system that is able to adapt rapidly to changing customer requirements. Intelligent features, automation, and robotics, reconfiguration, automatic control, simulation, and emulation technologies are used to create rapid change.
- Virtual factory – virtualization technology and data integration are used to represent the dynamics of real enterprise. Virtual factory represents cyber feature of real enterprise and virtual representation of all its elements. It uses data from sensors, actuators, video and audio information, biometric data, etc. In real time it creates a virtual image of functioning of enterprise.
- Digital factory – digitalization and digital technologies are used to the integration of all activities within product life cycle and production systems. Digitizing, modeling, simulation, and emulation are used to the understanding of comprehensive manufacturing processes and creation of new knowledge, which is used for optimization of real production systems. In contrast to virtual factory, digital factory does not use real data but use data, for example, from simulation. The concept of collaboration in digital factory is illustrated in Fig. 13.6.

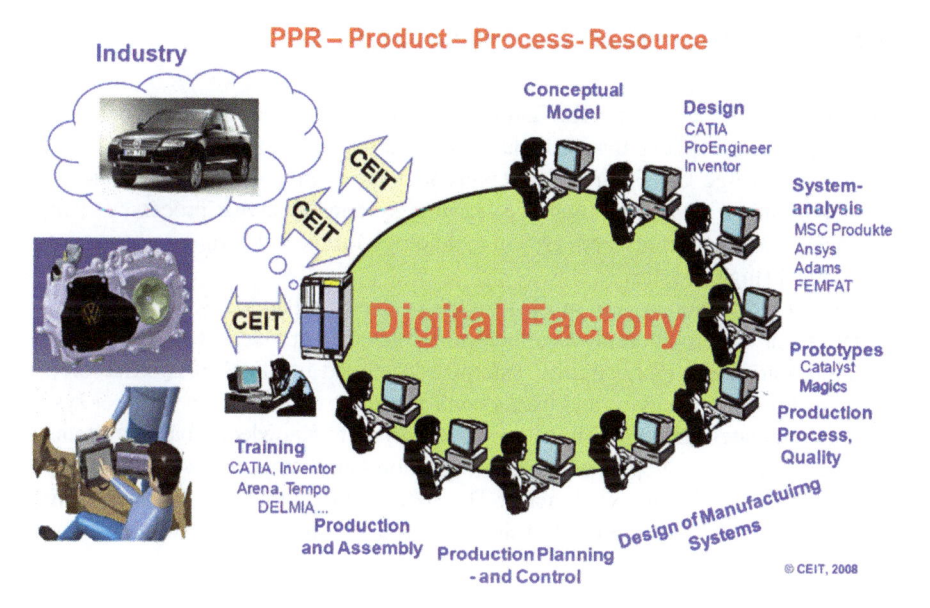

Fig. 13.6 Digital factory concept built at the University of Zilina [4]

ZIMS: Holonic Concept

ZIMS represents a pilot project of intelligent manufacturing systems, which is composed of workplaces that communicate with each other through holon. Holons form comprehensive holarchy. The strength of holonic organization, or holarchy, is that it enables the construction of very complex system that is nonetheless efficient in the use of resource, highly resilient to disturbance, and adaptable to changes in the environment in which they exist [1].

The proposed holonic concept of manufacturing system is used for control and monitoring of individual activities multi-agent system (MAS). The function of holonic systems is based on the use of autonomous ability of agent. The agents are considered to be autonomous entities of system. Their interactions can be either cooperative or selfish within the defined level of action. Agents receive tasks from higher level of holarchy, but their solutions are carried out autonomously. Intelligent agent is a natural or computing system that is able to perceive their environment, and on the basis of the monitoring carried out actions, it fulfills the global objectives of the system.

Multi-agent systems can be considered an elementary part of distributed artificial intelligence, which forms the conceptual framework for modeling of comprehensive systems. MAS is defined as a loosely bound network consisting of researchers of generated tasks. MAS platform represents distribution, autonomy, interaction (i.e., communication), coordination, and organization of individual agents.

Knowledge Environment: Learning from the Process

Simulations lend meaning to data and can be updated and adapted as further data come in. It often happens that provided historical data are not sufficient to derive relevant knowledge [2]. Figure 13.7 represents the principle of building a knowledge environment that will encourage learning systems of active processes. This approach is validated in research area of technology for the industrial production of large optical single crystals.

Physical production system in ZIMS is controlled using an MAS. When experimentation and development management for large-scale production of optical single crystals are used, logic control system is developed in ZIMS. It is based on a system of learning processes and uses a meta-modeling approach. The control system communicates directly with the knowledge system. Knowledge-based systems are used for decision support computer simulation (simulator). The simulator performs a set of simulation experiments. The statistical data that are obtained using multiple nonlinear regression analysis generated the desired meta-models. The

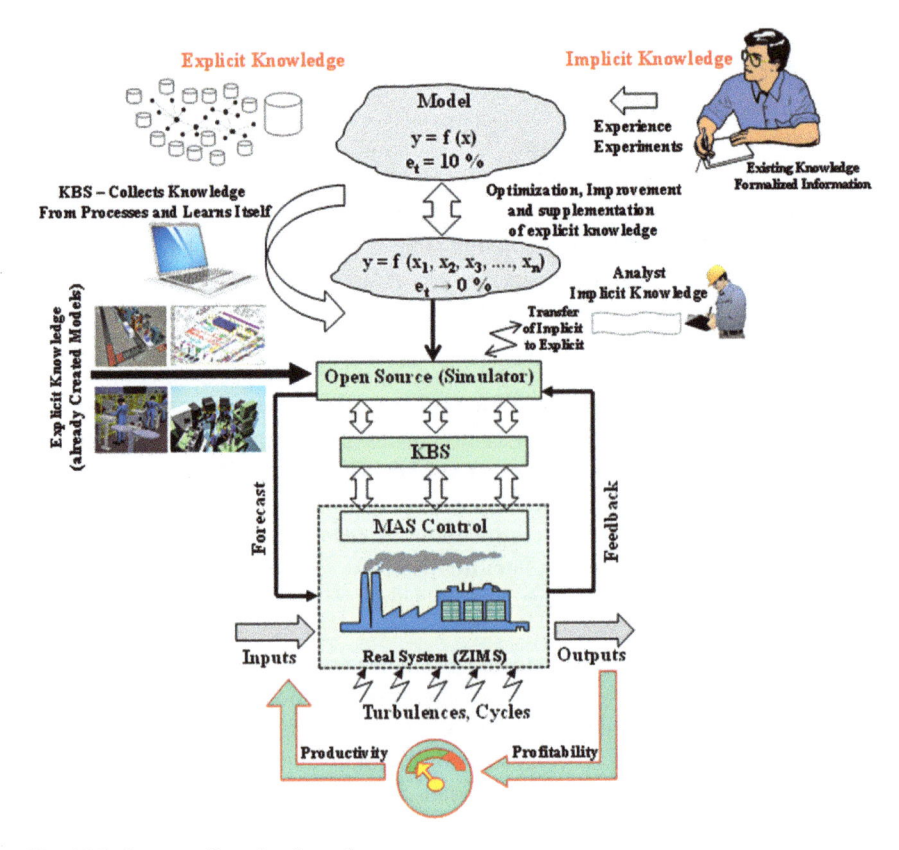

Fig. 13.7 System of learning from the process

resulting meta-models are used for making predictions about the future behavior of the controlled system (prediction) and these prediction amenities for approximative production management. Approximative (gross) production management, represented by a set of meta-models, is cyclically refined, using feedback (data) of real processes and new simulation. This creates a closed system of learning process, and its base is built and own knowledge system. It is integrating the explicit knowledge (e.g., models created in the past, the knowledge gained in the past, new explicit theoretical knowledge) and formalized implicit knowledge (conceptual system designers, analysts' existing system).

Technology Base of ZIMS

Technology base of ZIMS can be split into three groups of technologies – technologies for research and development of product, technologies for research and development of processes, and technologies for research and development of manufacturing system.

Technologies for research and development of product are technologies for virtual design, reverse engineering, simulation, virtual testing and validation, virtual prototyping, virtual and augmented reality, immersive technologies, etc.

Technologies for research and development of processes include technologies for virtual design of manufacturing systems, reverse engineering, computer modeling and simulation, ergonomic analysis, virtual and augmented reality, evaluation of performance, financial planning and analysis, etc.

Technologies for research and development of manufacturing system use real manufacturing technologies as adaptive assembly systems, industrial robots, systems for manufacturing planning, intelligent manufacturing systems, etc. (Fig. 13.8).

ZIMS uses technologies for research and development based on new trends in these areas:

- Advanced mechatronic systems
- Artificial intelligence
- Newest information and communication technologies
- Advance manufacturing processes
- Modeling, simulation, and prediction
- Manufacturing strategies
- Knowledge engineering

The concept of ZIMS was designed as a holistic system. Individual holon represents the main subsystems and the advanced corporate systems. Holon represents an individual, a unique part of the system (subsystem), which consists of a set of cooperating elements of the lower level. Practical examples of production holon are autonomous and cooperative subsystems, ensuring complete fulfillment of defined tasks (production, storage, transport, monitoring, logistic, etc.).

Fig. 13.8 Layout of ZIMS

Digital Factory Technologies

Digital factory technologies enable the development of the company's own decision-making approaches to analyze the potential for developing innovations (Fig. 13.9). The first step is to evaluate the innovation concept and technological feasibility – it finds out whether the innovation works. In the second step, the significance of innovation is tested, i.e., it is determined what the market potential and market interest are in innovation (e.g., does a customer buy this product?).

The users have the opportunity to try out the activities of their factory of the future production in the environment of virtual reality and augmented reality (CAVE – computer-aided virtual environment). It was labeled as Adaptive Haptic Virtual Collaborative Development Environment (HVACDE) (Fig. 13.10). There are also technologies for haptic, special virtual reality headset for 3D dynamic effects (Oculus Rift, HTC Vive) or the latest experimental technology for brain wave-reading device (EPOC).

Fig. 13.9 Decision-making on innovation

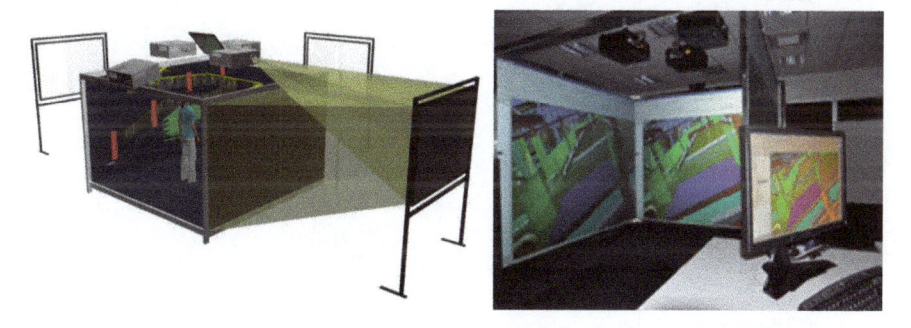

Fig. 13.10 CAVE architecture and CAVE internal perspective

The latest development is oriented to the wider use of digital factory technologies and advanced information and communication technologies. All devices in developing logistics system will automatically monitor through sensors. Their current status (operation, failure, downtime, etc.) will be available to any other element of the production system. In the development is the solution, in which each product will be carried (as one of the attributes) all the information, which will be required for processing in the base of the current status of the production system.

Result of Cooperation

Cooperation between the Central European Institute of Technology and University of Zilina is represented with many common grant projects and projects for industry. Most of them resulted in products which are commercialized. One of the most successful products is the concept of intelligent logistic management.

The concept named CEIT Intelligent Logistic Management is based on needs from industry and on artificial intelligence. Main contributors to this concept are several companies based in Slovakia and operate in automotive and electrotechnical

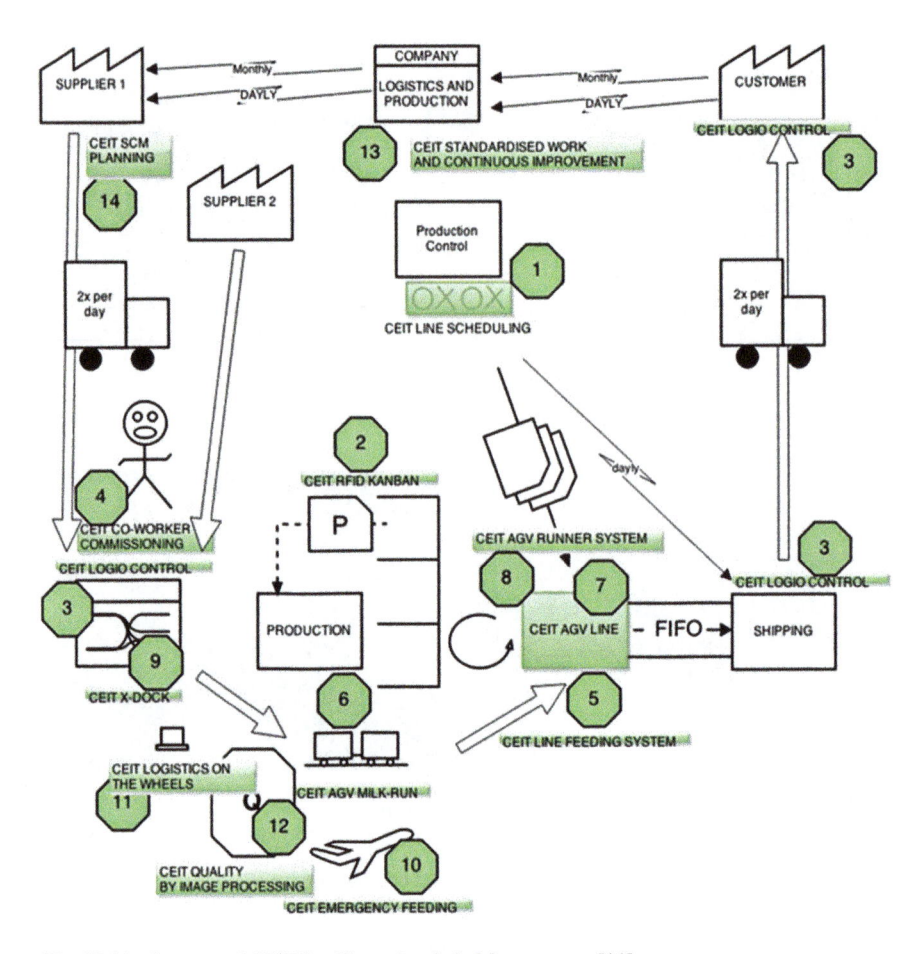

Fig. 13.11 Concept of CEIT Intelligent Logistic Management [11]

industry. These companies defined their needs, and CEIT with strong cooperation with the University of Zilina completed this concept. Artificial intelligence that ensures autonomous tasks directionally through agent communication should be integrated into an enterprise system [8, 9]. The principles and connections of this concept are described in Fig. 13.11.

An internal logistics system is basically a support service that provides material handling and transport for the superior, more significant, and value-creating manufacturing process. As such, it must undoubtedly be able to embody the changes that are taking place in the manufacturing environment. In Factories of the Future (FoF) or intelligent manufacturing systems (IMS), the production system is required to be flexible in adaptation to new products; therefore also logistics system must be able to swiftly and seamlessly adapt to those changes [7]. The concept of CEIT Intelligent Logistic Management is based on 14 solutions that are together connected and make common "logistics network." At this time, not all of these solutions are fully integrated and connected to IT infrastructure of companies. For the future is this concept facing big challenges like Internet of Things [6], Internet of Service, and smart factory.

Many of these solutions have been awarded directly to CEIT or to companies where solution has been implemented. For example, CEIT has been awarded by the Ministry of Economy as "Innovative Act of the Year" for practice game. This game has been used for practice new logistic concept to thousands of employees and suppliers of Volkswagen Slovakia. Next awards achieve Volkswagen Slovakia as "most lean company" in Volkswagen Group or "Lean & Green Efficiency Award" for they activities in field energy efficiency. The biggest award achieved by Volkswagen Slovakia with strong contribution of CEIT Intelligent Logistic Management is "mach18.FACTORY Oscar," which Volkswagen Slovakia achieved in 2015 as best plant of Volkswagen Group.

The key solution of CEIT Intelligent Logistic Management is automated guided vehicle (AGV). Currently there are two types of used AGV – underrun version and towing version (Fig. 13.12 shows an example). CEIT AGV systems have these advantages/features [11]:

- High speed – up to 2 m/s
- High towable weight – up to 3000 kg
- Brake energy regeneration
- Wireless monitoring and control system
- Automatic charging
- Safety scanners
- Towing and underrun version

The number of features during years of development and implementation in industry in Central Europe is growing. Since first generation with few basic features

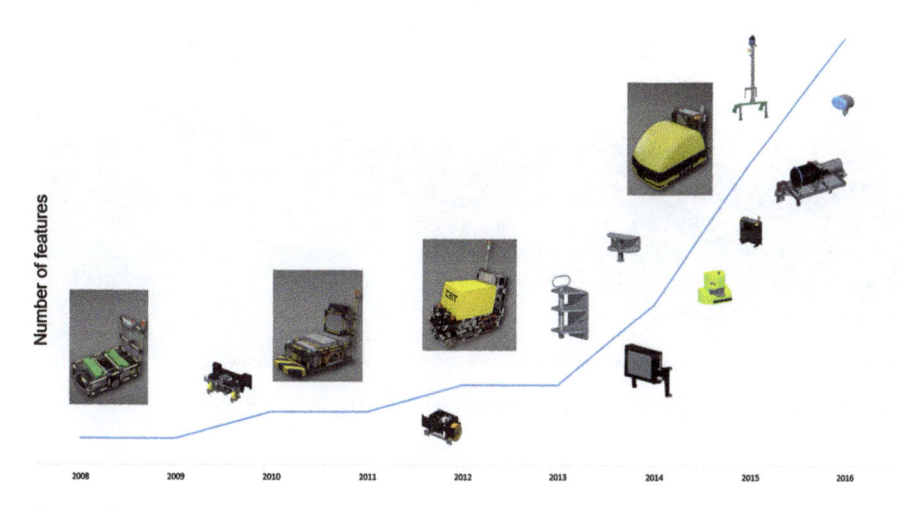

Fig. 13.12 Features of AGV

is current generation full of safety and operational features customized and developed based on industrial feedback.

Underrun AGV system can be used as a movable mounting table. AGV broke through device supporting frame, catches him and pulls in any place where he left it. This type of device is used in particular for the transport of parts which are sensitive to handling, respectively, during manipulation threaten to damage.

Towing AGV system (Fig. 13.13 shows an example) is able to take more wagons, and that means it would take much more material at once. This advantage is able to use with material, which the use in production is high [10]. This feature is supported with automatic system of connecting and disconnecting wagons.

Loading and unloading of pallets from/to AGV system are provided by peripheral devices. In case of big pallets, it used "c-frame." The principle of this pallet exchange is in exact stop of AGV in front of the frame, mechanical part from wagon takes the pallet on board, and AGV continues in route (Fig. 13.14 shows an example) [12].

One of the main benefits of the AGV is cost reduction by reducing work in process, as well as cost savings for employees who are required to operate the truck by manual logistics [11].

Fig. 13.13 CEIT AGV systems – towing version

Fig. 13.14 AGV systems automated pallet loading

Conclusion

Nowadays, active work in the area of cooperative management may help companies not only to keep their position on the market, but also it can even enhance its competitiveness and find new perspective partners [13]. Cooperation between research institutes and universities should bring high level of innovation for industry. This connection allows to transfer into practice the latest findings from research which have an impact on all stakeholders – universities have opportunities to use the acquired knowledge in the learning process and also in basic research; research institutions are able to transfer the latest knowledge directly to the industry, and the industry as end user can significantly increase their competitiveness and those to be successful in the global market.

Acknowledgments This paper was made about research work support VEGA 1/0936/16.

References

1. Botti, V., & Giret, A. (2008). *ANEMONA: A multi-agent methodology for Holonic manufacturing systems* (p. 214). London: Springer. ISBN 978-1-84800-309-5.
2. Bubeník, P., Horák, F., & Hančinský, V. (2015). Acquiring knowledge needed for pull production system design through data mining methods. *Communications: Scientific Letters of the University of Zilina, 17*(3), 78–82.
3. European Commission. (2013). *Factories of the future – Multi-annual roadmap for the contractual PPP under horizon 2020*. Brussels: European Commission.

4. Gregor, M., & Medvecky, S. (2010). Application of digital engineering and simulation in the design of products and production systems. *Management and Production Engineering Review, 1*(1), 71–84.
5. Gregor, M., Hercko, J., & Grznar, P. (2015). The factory of the future production system research. In *21st international conference on automation and computing: Automation, computing and manufacturing for new economic growth* (pp. 254–259). Glasgow: IEEE Press.
6. Gregor, M., Magvasi, V., & Gregor, T. (2015). Internet of things – IoT (in Slovak). *ProIN, 16*(2), 35–41.
7. Gregor, T., Krajčovič, M., & Więcek, D. (2017). Smart connected logistics. *Proceedings of Engineering, 192,* 265–270.
8. Micieta, B., Binasova, V., & Haluska, M. (2014). The approaches of advanced industrial engineering in next generation manufacturing systems. *Comm. vol., 16*(3A), 101–105.
9. Micieta, B., Gaso, M., & Krajcovic, M. (2014). Innovation performance of organization. *Communications: scientific letters of the University of Zilina vol., 16*(3A), 112–118.
10. Micieta, B., Zavodska, L., Rakyta, M., & Binasova, V. (2015). Sustainable concept for green logistics and energy efficiency in manufacturing. In B. Katalinic (Ed.), *DAAAM international scientific book 2015* (pp. 391–400). Vienna: DAAAM International.
11. Micieta, B., Hercko, J., Botka, M., & Zrnic, N. (2016). Concept of intelligent logistic for automotive industry. *Journal of Applied Engineering Science, 14*(2), 233–238.
12. Mleczko, J., Micieta, B., & Dulina, L. (2013). Identification of bottlenecks in the unit make to order production. *Applied Computer Science, 9*(2), 43–45.
13. Soviar, J. (2009). Cluster initiatives in Zilina region (Slovak republic). *Economics & management = Ekonomika ir vadyba, 14,* 528–534.
14. Soviar, J., & Zavodska, A. (2011). Knowledge and its creation – The case of introducing product to the market. *Business: Theory and Practice, 12*(4), 362–368.
15. Vodak, J., Soviar, J., Lendel, V., & Varmus, M. (2015). Proposal of model for effective management of cooperation activities in Slovak companies. *Communications, 17*(4), 53–59.

Chapter 14
Improvement of the Production System Based on the Kanban Principle

Ľuboslav Dulina, Miroslav Rakyta, Ivana Sulírová, and Michala Šeligová

Abstract The article deals with improvement of the production system based on the cooperation of enterprise and university. The improvement of production system in the selected enterprise included relocation of the pre-assembly workstation to the newly created supermarket – storehouse with narrow aisles – and implementation of new supplying system based on the principle of the Kanban method. An important part consists of financial investments and evaluation of their returns. The project has brought advantages to the enterprises described in conclusion of the article, where its benefits and return on investment are calculated as well.

Introduction

The present day is characterized by high degree of informatization of society and new trends in the field of business management and in the field of training of young people for the Factories of the Future (FoF). The key of success for industrial enterprises lies in cooperation with universities and R&D (research and development) centres. R&D does not relate only to universities but also companies which want to be innovative [1]. In present, cooperation as such represents for a company an important tool for increasing its competitiveness [2]. The result of such cooperation is a benefit for both sides. The enterprises will have new trends implemented, their processes will be managed more efficiently, which will bring them higher profit, and they will eliminate wastes. On the other hand, universities and development centres will acquire practical knowledge and a possibility to implement their development solutions into practice.

There are new tendencies in management of production processes showing in manufacturing enterprises. As purchasing market is pushing itself through and require-

Ľ. Dulina (✉) · M. Rakyta · I. Sulírová · M. Šeligová
University of Zilina, Faculty of Mechanical Engineering, Zilina, Slovakia
e-mail: luboslav.dulina@fstroj.uniza.sk; miroslav.rakyta@fstroj.uniza.sk; ivana.sulirova@fstroj.uniza.sk; michala.seligova@fstroj.uniza.sk

ments of customers are rising, together with the effort to shorten delivery periods and with plenty of variants of products' adjustments, it all causes the increase of the risk of keeping stock. New strategic concepts must fulfil the requirement of high reliability of supply, flexible production and the ability to adapt to new situations.

All enterprises create their own procedures of success, use similar methods, tools and techniques. The problem is the shortage of supplying systems supporting continuous flow, production in small batches and assembly workstations. To link continuous flow and zones of supplies, they lack pull signals. The result is insufficient supplying of processes with material, decrease of flow and high level of stock, which causes waste of effort and money [3].

Objective and Methodology

The result of cooperation of the enterprise and university was a complex project of implementation of several segments' production improvement. The highest savings were identified by translocation of pre-assembly workstation into another hall and by the transformation into the pull principle. The analysis of the current state of material flows in the enterprise was processed in the AutoCAD software using FactoryCAD and FactoryFLOW tools. In these modules of the Tecnomatix software, 2D models of production system were being created and also the optimization of the plant's layout was performed based on analyses of material flows. Based on analyses, also the ways of implementation of purchased parts' supermarket and of Kanban system were designed. The objective was to evaluate individual designs, benefits of implementation of supermarket concept, and to apply the best variant.

In enterprises, the most frequent tasks for improvement in production are [4]:

- Improvement of production processes
- Improvement of production batches
- Improvement of material composition
- Improvement of production procedures
- Improvement of production planning
- Improvement of supporting processes

To optimize material flows means to look for their ideal levels, for continuous motions of materials at minimal costs and at minimal consumption of time, energy and resources, with minimal number of workers or under criteria such as:

- Length of transport distance
- Transport volume per time unit, size, capacity and weight-bearing capacity of a transport unit
- Transport costs
- Utilization of cargo space
- Duration of transport operations

To implement a lean supplying system, it is necessary to understand the whole process of dealing with each and every part: How is the part being purchased? How is it being received? Where is it being stored? How is it being transported to the places of consumption? The right step at the beginning of process optimization is to collect all the relevant information into one entity – Plan for Every Part (PFEP), which makes the information visible for everyone.

The implemented project was focused on improvement of internal logistics, implementation of Kanban system, creation of supermarket and improvement of material flows.

The Analysis of the Current State of Selected Enterprise

The enterprise, where the project was implemented, produces components for automotive industry with historical experience and modern technologies. It produces wide range of products. For its spreading market operations, it needed to adapt its production system to rising amount of production. Within the project, improvement of four types of components was implemented. Each type of component is being produced in another hall. They have mutual input and output storehouse. These components were considered to be product representatives.

Based on the analyses, it was found out that technological procedures of production of representative components are composed of 18 technological operations on average and the production process lasts 502.5 min on average. The production of the smallest representative lasts 4.5 h, and production of the biggest one lasts 18 h. There are 6169 pieces of these representatives produced per year. In Table 14.1, technological procedure of one type of representative is described.

Figure 14.1 depicts the output of analysis of the current state of material flows in form of Sankey diagram.

Process diagram in the Fig. 14.2 shows production workstations and amounts of parts and unfinished products being transported amongst workstations, in that one type of components is being produced. Material flow starts in the input storehouse and ends in the output storehouse.

The input storehouse is located near production halls, and the material is being transported by fork-lift truck. The material is stored in metal gitterboxes, cardboard boxes with various sizes, rack boxes, on euro-pallets and in plastic packages or loosely. The inter-storage contains stock for workstations for 1 week (Fig. 14.3).

In the place of original storehouse (Fig. 14.4) with the total area of 391.0 m^2, there were these areas:

- Area of racks 145.1 m^2
- Area for transport 145.1 m^2
- Other areas 145.1 m^2

Table 14.1 Technological procedure

No. of operation	Name of operation	Duration of operation (min)
10	Soldering 1	22.14
20	Soldering 2	22.15
30	Soldering 3	49.98
35	Pre-assembly	4.74
40	Soldering 4	49.98
50	Pressure testing	4.02
60	Preparation of helium leak testing	4.96
70	Helium leak testing	12.00
80	Filling and vacuum forming	9.00
90	Isolating	28.02
100	Isolating of sides	13.98
105	Installation of electrical circuitry	25.02
120	Plugging in before testing	7.98
130	Complete testing	27.00
140	Plugging out after testing	10.02
150	Completion	19.98
160	Packing and dispatch	10.02

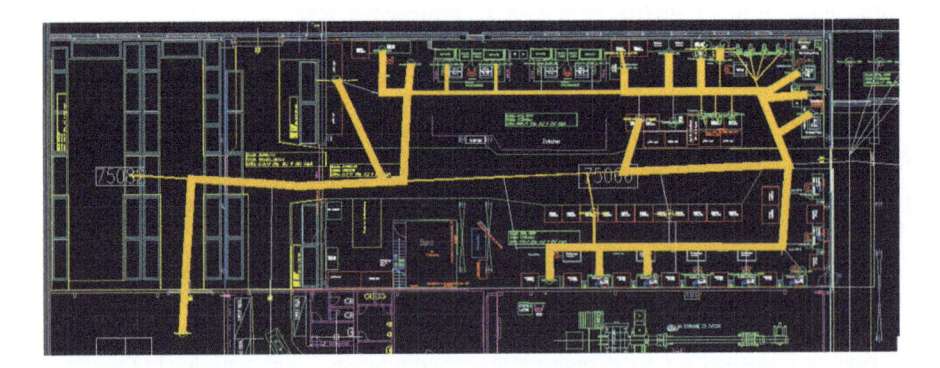

Fig. 14.1 Sankey diagram

To generate the calculations, it was necessary to gain the data on:

- Products – data on products and their individual parts
- Processes – data on activities in production, their linkages and order
- Resources – for the analysis of material flows (means of transport and handling)

The structure of data was created according to structured bills of material that started at the level of final product, and its assembly sets, subsets and components were defined according to production stages.

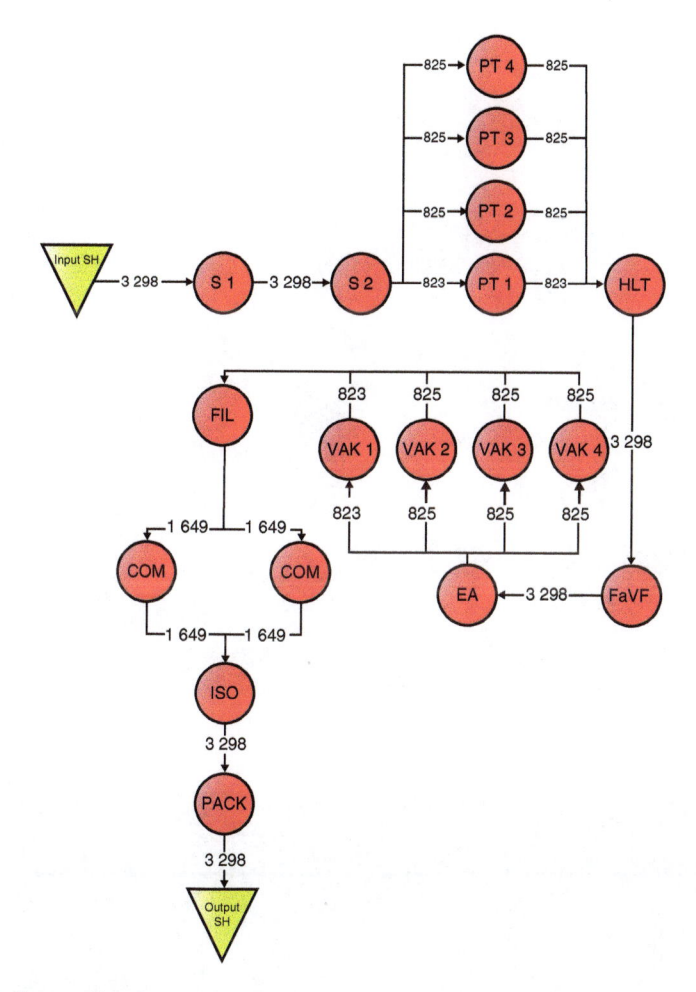

14 Improvement of the Production System Based on the Kanban Principle

Fig. 14.2 Process diagram

Subsequently, transport devices used in the production process were defined, because transport devices are assigned to every link of processes that follow after each other. Parameters of these devices then enable to perform the analysis of means of transport. The last part of the input data consists of the data on individual activities and their linkages. Before defining the relations, the schemes of transport patterns for individual products were processed. These schemes describe what items are being transported between what activities, in what volume, by what transport device, in what handling unit and in what number of handling units. An example of process diagram is displayed in Fig. 14.5.

Fig. 14.3 Rack storehouse

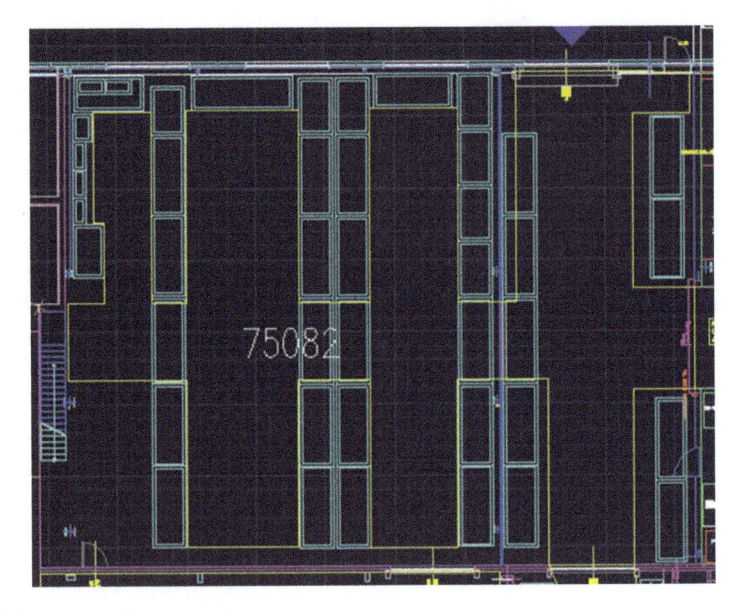

Fig. 14.4 Layout of input storehouse

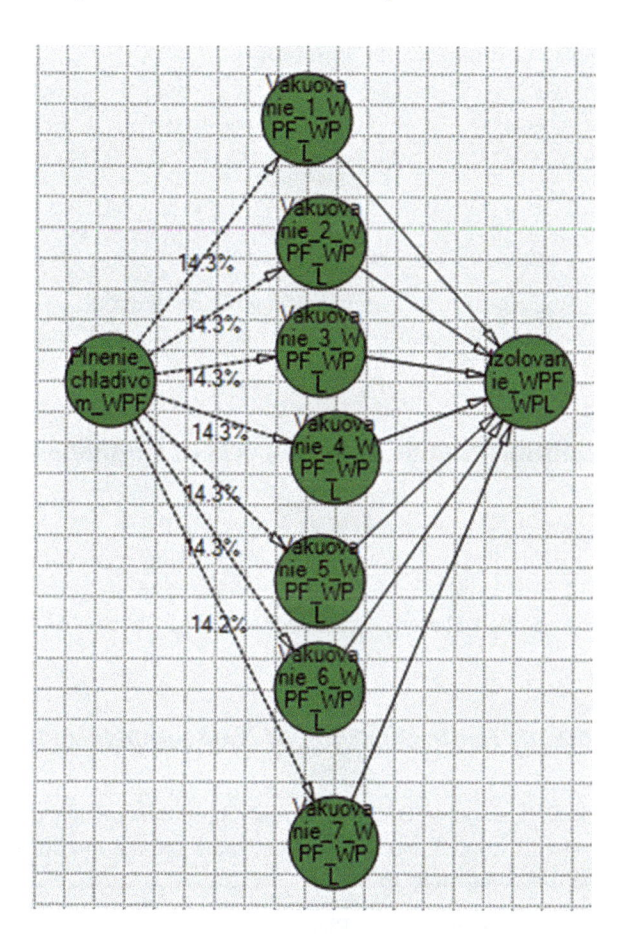

Fig. 14.5 Process diagram for the set of cooling aggregate

Analysis of Layout Design and Requirements of the Enterprise

Within the solution, the calculation of material flows was performed together with analysis of objects' layout, calculation of distances, costs of transport, intensity of routes' utilization, utilization of means of transport and of workstations and Sankey diagrams. The analysis showed the following results:

- The length of material flows was 293,924.41 m/year.
- Total annual transport costs were reaching 38,173.73 €/year.
- The number of movements of transport devices was 112,304.12 movement/year.
- Total time of manipulation and transport was 349,221.89 min/year.

The outcome of utilization of transport devices' analysis is shown in Table 14.2.

The enterprise's requirement was to implement supermarket of purchased parts in the current place of maintenance. The new storehouse would have the area of 393.0 m² at its disposal.

Table 14.2 Utilization of transport devices

Device	Quantity	Busy (min)	Avail. (min)	Utilization
Pallet truck	4	306,080.52	864,000.00	35.43
Fork-lift truck	2	43,141.37	432,000.00	9.99

Criteria for improvement were set as follows:

- Reduce costs of manipulation with the material.
- Increase the number of pallet places in the storehouse.

Identification of Main Aspects of Implementation

To make the new system work properly, it has to fulfil initial requirements. The Plan for Every Part will provide input information needed for designing, the supermarket will provide input material and the new supplying system must provide communication and flows between production and storehouse [5].

Plan of Implementation of New Supplying System

Based on experience and recommendations, sequence of steps of the new supplying system's implementation was designed for the enterprise. The new plan included the following steps [6]:

Step 1: Creation of the Plan for Every Part – PFEP – of the database containing material with numeric designation, specification of the material, supplier, supplier's location, with the place of storage and consumption, and other important information

Step 2: Creation of one storehouse for all parts entering the enterprise

Step 3: Creation of supply circuits

Step 4: Integration of the new supplying system's management with the management information system while using pull signals

The Plan for Every Part is a table with ordered data on materials, accessible in electronic form to every user. Its use has two main advantages:

- It enables to sort data based on categories (intensity of ordering, dimensions of transport units, hourly consumption).
- It enables to change and add categories with minimal effort.

The responsibility for accuracy and updating of the Plan for Every Part needed to be assigned to the PFEP manager. To establish the maintenance of this concept, a system of directions needed to be created. The Request for Change of PFEP Form was designed for the updating.

For the creation of supermarket, it was needed to set maximal amount of each component, which will be needed to provide normal operation of production system. The average daily consumption of parts was determined, together with transport batch and containers needed. By the calculation of transport units needed to store the material and then multiplying them by their physical sizes, the space needed for their storage was determined.

After implementing the PFEP, the enterprise gained the information needed concerning the parts, and it was able to continue in implementation of lean supplying system. The next phase was to create a supermarket with regulated level of each part being purchased. The supermarket is mainly used in implementation of the pull principle in the material flow. It represents the storehouse of finished products or inventory, in which the amount of material is precisely defined. It is used to establish a continuous material flow in cases where it would not be possible to create it otherwise. The material is taken from the storehouse only based on a Kanban card or other form of information supporting the pull principle. With the gradual continuing of implementation, the supermarket would spread, so it would include all the parts being purchased. The supermarket needed to be placed as close as possible to the receiving area, even though it required the relocation of production activities. The supermarket in the production represented a new form of storing, and so it replaced the conventional storehouse.

While preparing the supermarket, it was necessary to determine [6]:

- Storage system
- Positions labelling system
- Implementation of the process of storing and dispatching of parts
- Implementation of the process of reaction to excessive stock
- Minimal level of stock

The layout of new supermarket is shown in Fig. 14.6. Transport area has 217.42 m^2 and the area of racks has 115.5 m^2.

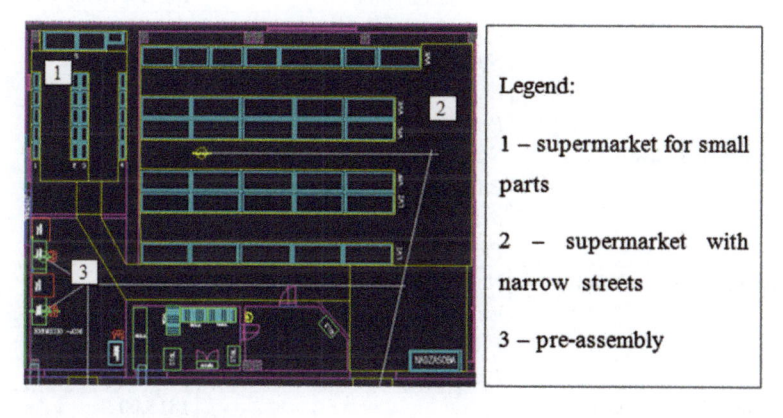

Fig. 14.6 Supermarket design

The objective of improvement of warehousing is to increase the number of pallet places. More efficient storing with maximal utilization of space and with efficient, fast and clear material flow will be provided by storehouse with narrow aisles. Carts for narrow aisles enable maximal utilization of storehouse's space. Because of their construction, they need an aisle only 1.52 m wide as their forks enable three-sided inserting [7].

In the new supermarket created for increase of the number of pallet places, the purchase of new racks was suggested. These would be used for storing the material needed, especially small parts of input material. The costs of new racks are summarized in Table 14.3.

The truck into the new supermarket was designed for inductive guidance. The inductive line is installed into the floor after installing the racks. Inductive line makes a closed loop powered by a frequency generator placed anywhere in the storehouse. It is powered from a common socket or power supply. The scheme of inductive line needed is in Fig. 14.7. The line of induction goes to the position in front of racks with the length of 3–4 metres for safe and fast finding of the machine for guidance. The inductive line will include:

- Four aisles with the length of 14.5 metres
- Three aisles with the length of 4 metres
- Four aisles of the line with the length of additional 4 metres as a ramp for easy getting on in front of racks
- About 10 metres of line to link the aisles

The calculation of the installation of inductive guidance and of the purchase of new system truck is listed in Table 14.4.

Table 14.3 Costs of new racks

Rack	Parameter	Value
Shelf rack with containers	Number of containers/pc.	64.00
	Number of shelves/pc.	12.00
	Dimensions/mm	1,850 × 1,000 × 424
	Price/€	*515.00*
Kanban rack	Number of shelves/pc.	10.00
	Effective depth of the rack/mm	400.00
	Price/€	*244.00*
Rack with wide shelves	Number of shelves/pc.	4
	Effective depth of the rack/mm	600.00
	Price of a basic panel/€	*199,00*
	Price of additional panel/€	*159,00*
Rack (straight and tilted shelves)	Number of shelves/pc.	4
	Effective depth of the rack/mm	1200.00
	Price of a basic panel/€	*389.00*
	Price of additional panel/€	*309.00*
Total price of new racks/€		**1815.00**

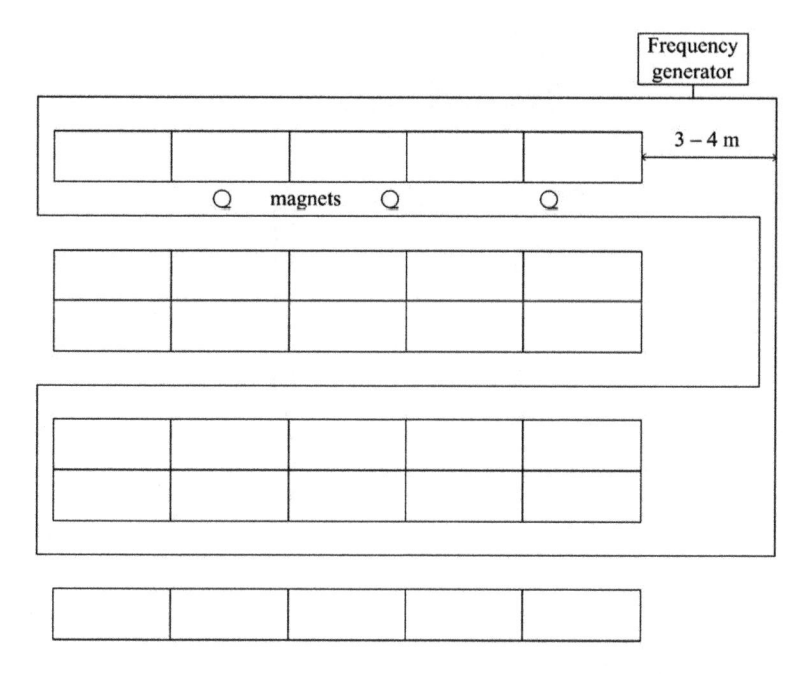

14 Improvement of the Production System Based on the Kanban Principle

Fig. 14.7 Scheme of inductive line

Table 14.4 Costs of inductive line

Item	Parameter	Value
Inductive line	Total length of the line/m	96.00
	Price of installation of 1 metre/€	14.02
Magnets	Number of magnets/pc.	9.00
	Price of 1 magnet/€	9.10
	Installation of 1 magnet/€	14.93
Generator	Price of generator and battery backup/€	1843.00
System truck	Price/€	33,000.00
Total costs of inductive line/€		**36,405.19**

Magnets for guidance were placed into three aisles, three magnets in each.

Designed performing of material's transport is provided by supply circuit with two-way aisles for supplying. In the design, existing aisles are being followed. Tugger with attachment is used. To save the time of supplying person and to serve more workstations, preliminary circuit was set, amounts of transported material were calculated and the circuit was tested. Necessary changes of truck's stops or places of supplies were made. The preliminary schematic circuit is depicted in Fig. 14.8.

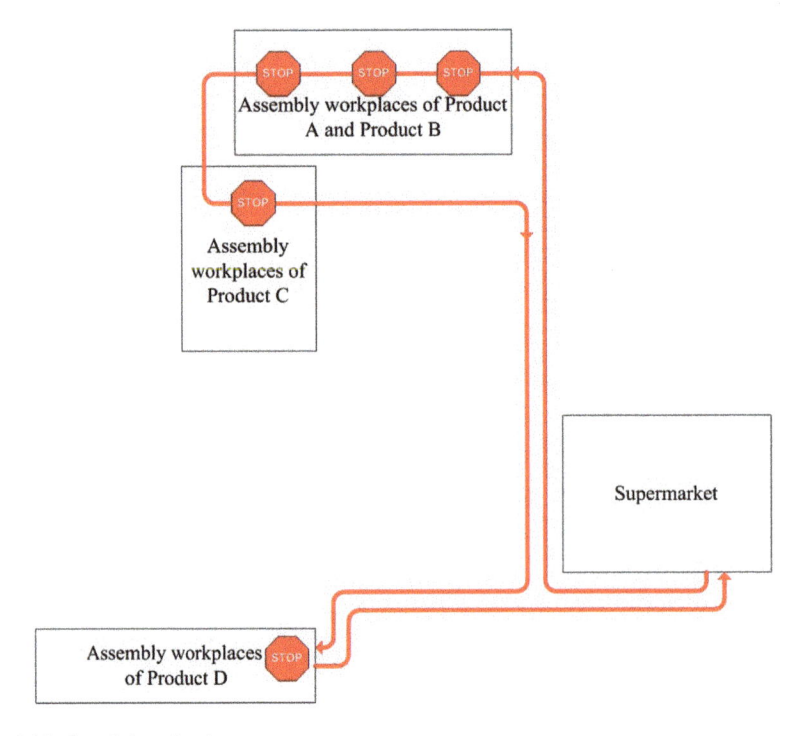

Fig. 14.8 Supplying circuit

Table 14.5 Costs of attachable device of the truck (taxi solution)

Item	Quantity/pc.	Price/€
Rack truck	1.00	490.00
Pallet truck	1.00	104.00
Towing trailer combination	1.00	500.00
Total price of taxi solution/€		**1094.00**

To transport the material from the supermarket to workstations in an efficient way, the tugger had to be implemented. It was suggested to buy an additional device to the existing truck so that greater amount of material can be transported at once. The costs of this extension are listed in Table 14.5.

To guide the material flow efficiently, it was needed to create position places and to label them. Making of the labelling of positions in the storehouse and at workstations is listed in Table 14.6. The cost of new over-Kanban production board is in Table 14.7.

In the enterprise, the pull system was implemented with the objective to create system that is able to flexibly react to changes in demand. The aim was to reduce costs related to consumption of goods and to material flow. The implemented Kanban system in the enterprise shall aim for fulfilment of [8]:

- Limited stock of material and components
- Delivering of 100% quality from supplier

Table 14.6 Costs of position labelling

Placement of labelling	Parameter	Value
Front labelling of racks	Number of labels/pc.	8.00
Labelling of horizontal rack sections	Number of labels/pc.	35.00
Labelling of vertical rack sections	Number of labels/pc.	30.00
	Price of 1 label/€	0.40
Labelling of each workstation	Number of workstations of type A/pc.	23.00
	Number of workstations of type B/pc.	26.00
	Number of workstations of type C/pc.	18.00
	Number of workstations of type D/pc.	2.00
	Price of 1 label/€	0.90
Total price of labels/€		**91.30**

Table 14.7 Costs of Over-Kanban Production Board

Board	Price
Board of over-supply – rack for boxes/€	50.00
Board of over-supply – pallet rack/€	40.00
Total price of boards/€	**90.00**

- Small and proper buffer stock between workstations
- Zero defects
- Delivery of final products into storehouse according to the need
- Small (in ideal case none) stock of finished components

To signalize the consumption of material and to move material from the supermarket, there are Kanban cards used containing information about [9]:

- Material's number
- Its position in the supermarket
- Place of its delivery
- Number of cards in the place of use

The implemented system with higher frequency of supplying of the circuit contains less stock, and it is more sensitive to changes. The work in supermarket is eliminated by smaller transport units. With lower frequency of supplying, resources are being used more efficiently, and costs of stock are being minimized [10].

In the designed disconnected circuit, there are dispatching person and supplying person needed. The supplying person, while supplying the circuit, collects Kanban cards, loads empty crates and pallets, passes collected cards in storehouse to the dispatching person, loads material according to previously collected cards and repeatedly supplies the circuit. The dispatching person in the supermarket dispatches (prepares) material according to the cards brought by supplying person.

Successful operation of a pull system requires the calculation of the number of Kanban cards for each material type and for each place of delivery, which requires information on:

- Frequency of supplying
- Type of supplying circuit
- Maximal amount of material transported in each supplying cycle
- Number of delivered transport units

Relocation of the Pre-assembly to the Supermarket

By the relocation of the detached workstation of pre-assembly of a component to the supermarket, one worker as a labour resource was saved because of the merger of two workstations. Original location of the pre-assembly workstation is shown in Fig. 14.9.

Costs related to the creation of new workstations near the supermarket are listed in Table 14.8.

Fig. 14.9 Workstation of pre-assembly

Table 14.8 Costs of creation of new workstations

Item	Parameter	Value
Workers	Workers needed/person	2.00
	Costs of operators/€	896.00
New work desk	Number/pc.	3.00
	Price of 1 table/€	300.00
Media feed	Price/€	4000.00
Total price of workstations/€		**5796.00**

Analysis of the Future State

After changing the input data, the new calculation of material flows was performed. The changes especially included:

- Decreasing the amount of material being manipulated, from the weekly stock to the daily need
- Changing the way of material's transport
- Reducing inter-storages in production halls
- Changing the plant's layout (creating the supermarket)
- Changing transport routes

Material flows after implementing the changes are displayed in Fig. 14.10. Material flow of type A products' production is depicted in red colour, material flow of type B products is blue, products of type C have the yellow one and material flow of type D products is bright blue.

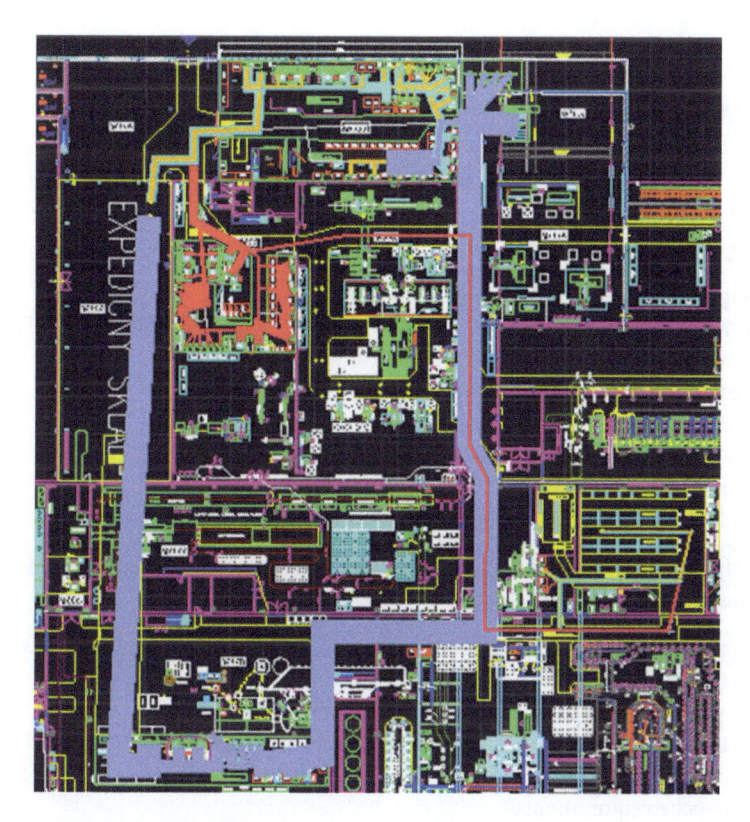

Fig. 14.10 Sankey diagram of the new layout

Table 14.9 Utilization of transport devices – future state

Device	Quantity	Busy (min)	Avail. (min)	Utilization
Pallet truck	4	229,963.21	864000.00	26.62
Fork-lift truck	1	49,800.93	216,000.00	23.06

The results of analysis show that:

- Total length of material flows is 2,841,074.88 m/year.
- Total transport costs are 28,896.68 €/year.
- Number of movements of transport devices is 91,120.15 movement/year.
- Total time of manipulation is 279,764.14 min (Table 14.9).

After generating the possible variants of solutions and based on their comparison, the most favourable variant was selected. After presenting the selected variant and analyses to the management of the enterprise, the decision was made that the enterprise will start the implementation of pull system and of the new supplying system.

Summary of Expended Costs and Benefits of Implementation of Designed Solution in the Enterprise

In Table 14.4, there are costs expended on the purchase of new devices and equipment into the supermarket listed. Its implementation will bring these advantages [6]:

- Managed amounts of reduced stock
- Increase of stock turnover rate
- Continuous service of multiple production workstations
- Decrease of the number of operators in the production
- Uninterrupted work of production operator
- Making some production areas free
- Decrease of storage areas
- Fast identification of the stock levels

By implementing the Kanban system, following advantages were created in the enterprise [11, 12]:

- Decrease of the number of production batches
- Decrease of the downtime
- Improvement of communication and improvement of quality
- Decrease of the amount of unfinished production
- Increase of labour productivity
- Lower space requirements
- Decrease of the costs of non-quality
- Making the flow in production clearer

14 Improvement of the Production System Based on the Kanban Principle

- Easier planning system
- Determination of minimal and maximal stock levels

Quantifiable benefits of the selected solution are listed in Table 14.10.

Besides the listed benefits, the number of fork-lift trucks needed dropped, inter-storages were eliminated, stock on production workstations was reduced by 80% and one production operator was made vacant by the creation of new assembly workstations by the supermarket, which represents saving of 18,000 €/year.

The return on invested funds according to the enterprise is listed in Table 14.11. One of the enterprise's criteria while selecting the projects is the necessity of the return on investment until 2 years. This was also the reason of choosing just this option by the enterprise.

Main benefits of the proposal:

- Shortening of the length of material flows by 98,149.44 metres
- Reducing the annual transport costs by 24.3%

Table 14.10 Comparison of the current and future state

Parameter	Unit	Current state	Future state
Total length of material flows	m/year	2,939,224.21	2,841,074.88
Total costs of transport	€/year	38,173.73	28,896.68
Number of movements of transport devices	movement/year	112,305.00	91,121.00
Total time of transport and manipulation	min/year	349,222.00	279,764.00
Busy time of pallet trucks	min/year	306,081.00	229,963.00
Percentage utilization of working time fund of pallet trucks	%	35.43	26.62
Busy time of fork-lift trucks (FT)	min/year	2 FT: 43,141.37	1 FT: 49,800.93
Percentage utilization of working time fund of fork-lift truck	%/year	9.99	23.06
Area allotted for storage racks in the storehouse	m²	145.10	217.42
Transport area in the storehouse	m²	221.80	115.50

Table 14.11 Return on investment

Cost of implementation	Price (€)
Costs of the supermarket	42,291.49
Costs of Kanban	580.00
Total costs	42,871.49
Annual saving of transport costs	9277.05
Annual saving of costs on production operator	18,000.00
Total savings	27,277.05
Calculation of return on investment	Total costs/total savings
Return on investment	**1.57 year**

- Reducing the transport time by 19.89%, which is 1157 h
- Increasing the utilization of fork-lift truck to 23.06%
- Increasing the storage area by 49.84%, which enables the elimination of interim storages
- Reducing the transport area in the storehouse by 52.07%
- Reducing the inventory level at production workstations by 80%

Acknowledgements This paper was made about research work support: VEGA 1/0936/16.

References

1. Šramová, V., Závodská, A., & Lendel, V. (2015). Reintegration of Slovak researchers returning to Slovak companies. In L. Uden et al. (Ed.), *KMO 2015, LNBIP 224* (pp. 353–364). https://doi.org/10.1007/978-3-319-21009-4_27
2. Vodák, J., Soviar, J., & Lendel, V. (2015). The proposal of model for building cooperation management in company. In *Verslas: Teorija ir praktika/Business: Theory and Practice*. (pp. 65–73). ISSN 1648-0627. https://doi.org/10.3846/btp.2015.535.
3. Krajčovič, M., Rakyta, M., Křížová, E., Bubeník, P., & Gregor, M. (2004). *Industrial logistics*. Žilina: EDIS. 2004, p. 375. ISBN 80-8070-226-8.
4. Harris, R., Harris, C., & Wilson, E. (2009). *We create material flows*. Žilina: SLCP. 2009, p. 92. ISBN 978-80-89333-11-0.
5. Weng, W., Song, Y., Yang, G., & Schnidt, R. (2010). PFEP – oriented in plant logistics planning method for assembly plants. In: *ICLEM 2010: Logistics for Sustained Economic Development – Infrastructure, Information, Integration*. ISBN 978-078441139-1. https://doi.org/10.1061/41139(387)192.
6. Smalley, A. (2009). *We make a balanced pull*. Žilina: SLCP. 2009, p. 115. ISBN 978-80-89333-10-3.
7. Sulírová, I., Rakyta, M., & Bjalončíková, J. (2016). Lean logistics trends. In: *InvEnt 2016: Industrial Engineering – Toward the Smart Industry: Proceedings of the International Conference* (pp. 148–151). Žilina: University of Žilina. ISBN 978-80-554-1223-8.
8. Nallusamy, S. (2015). Lean manufacturing implementation in a gear shaft manufacturing company using value stream mapping. In *International Journal of Engineering Research in Africa* (Vol. 21, pp. 231–237). ISSN 1663-4144. https://doi.org/10.4028/www.scientific.net/JERA.21.231.
9. Mičieta, B., Gregor, M., Quirenc, P., & Haluška, M. (2001). *Kanban – you are to pull!* (1st ed). Žilina: SLCP. ISBN 80-968324-2-5.
10. Mičieta, B., & Botka, M. (2001). KANBAN system and its implementation problems. In *Annals of DAAAM for 2002 & Proceedings of the 13th International DAAAM Symposium "Intelligent Manufacturing & Automation: Learning from Nature"* (pp. 353–354). Vienna: DAAAM International. ISBN 3-901509-29-1.
11. Krajčovič, M.(2010). Stocks in milk-run system. In *Produktivity and Innovation* (Vol. 11, No. 1, pp. 2–4). ISSN 1335-5961.
12. Gall, R., Minda, M., Baumgartner, M., & Rakyta, M. (2014). Process improvement of the production system by using lean manufacturing tools. In: *InvEnt 2014: Industrial Engineering – Navigating the Future: Proceedings of the International Conference* (pp. 178–181). Žilina: CEIT. ISBN 978-80-554-0879-8. – CD-ROM

Chapter 15
Quality Assurance in the Automotive Industry and Industry 4.0

Štefan Markulik, Juraj Sinay, and Hana Pačaiová

Abstract Industry 4.0 is more than just a phrase. A confluence of trends and technologies promises to reshape the way things are made. An illustrative example of new changes coming from the philosophy of Industry 4.0 is the automotive industry. Automotive industry is not only represented by final manufacturers of the car but the entire network of large and small suppliers who produce components for several automotive manufacturers. In order to implement this production according to customer requirements, they need to communicate with each other in a way and a language they can understand. This means special communication, its tools can be standards or standardized methodology to ensure quality throughout the production chain from the processing of raw materials to final assembly of the vehicle on the production line, its testing and the delivery to the customer. The basic precondition of Industry 4.0 strategy is communication in its broadest understanding. Communication or interconnection is conditioned by the application of generally accepted standards. Part of this connection is the focus on integrated management systems. The philosophy of Industry 4.0 is to minimize the risks as part of functional management systems. Timely and correct communication creates conditions for quality of a product at the end of the production and supply chain, i.e., a car according to customer requirements.

Š. Markulik (✉) · H. Pačaiová
Department of Safety and Quality Production, Technical University of Košice, Košice, Slovak Republic
e-mail: hana.pacaiova@tuke.sk

J. Sinay
Automotive Industry Association of the Slovak Republic, Bratislava, Slovak Republic
e-mail: juraj.sinay@tuke.sk

© Springer International Publishing AG, part of Springer Nature 2019
D. Cagáňová et al. (eds.), *Smart Technology Trends in Industrial and Business Management*, EAI/Springer Innovations in Communication and Computing,
https://doi.org/10.1007/978-3-319-76998-1_15

Introduction

The automotive industry is currently one of the industries that are ready to integrate principles of the Industry 4.0 within their structures. These are productive technologies which use automated systems, both in the production and in the logistics chain. Meeting the requirements of customers during their purchase is part of a comprehensive information system that starts with receiving orders and continues by requesting the production in different parts of the world through Internet communication so that the final recipient gets the product at the agreed time and in the agreed quality.

One kind of lost value that is sure to interest manufacturers is process effectiveness. Industry 4.0 offers new tools for smarter energy consumption, greater information storage in products and pallets (called intelligent batches), and real-time yield optimization (Fig. 15.1). Manufacturing technologies applied in automo-

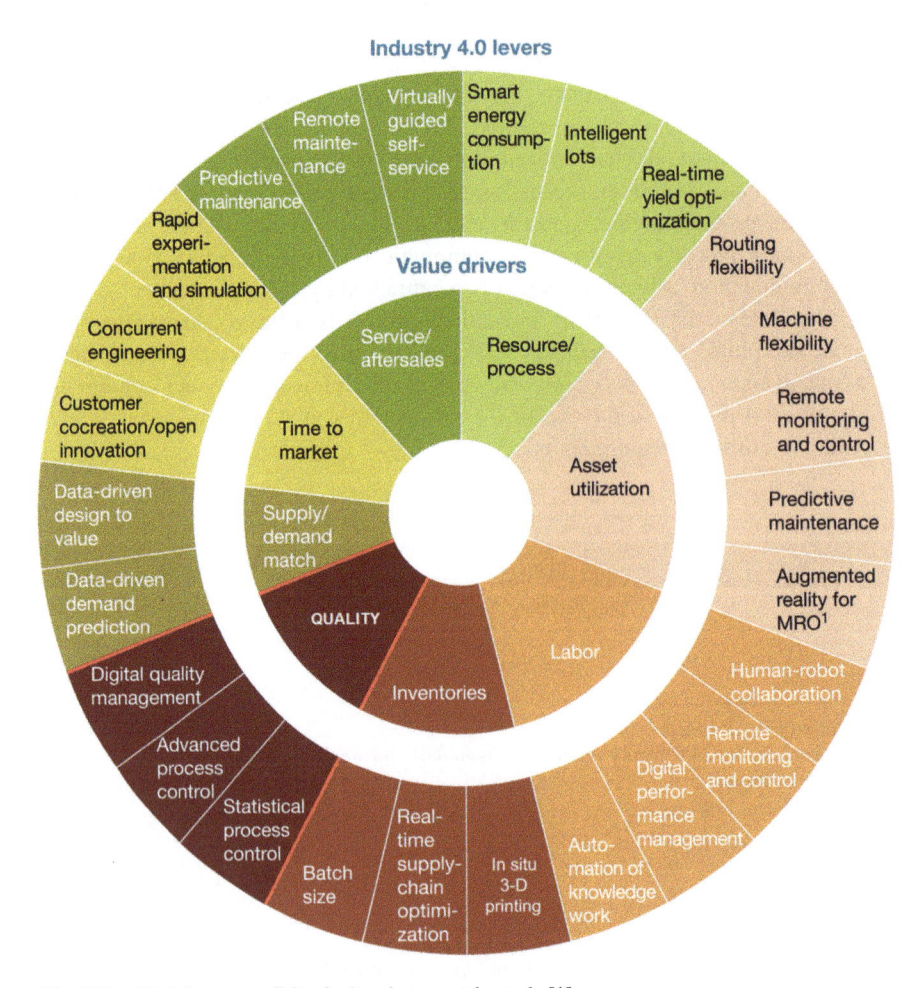

Fig. 15.1 Digital compass" for find tools to match needs [1]

tive production, intensively and effectively apply quality management principles deriving not only from the standards, but also supporting effective tools for quality assurance in production.

Structure of Supply Chain in Automotive Industry

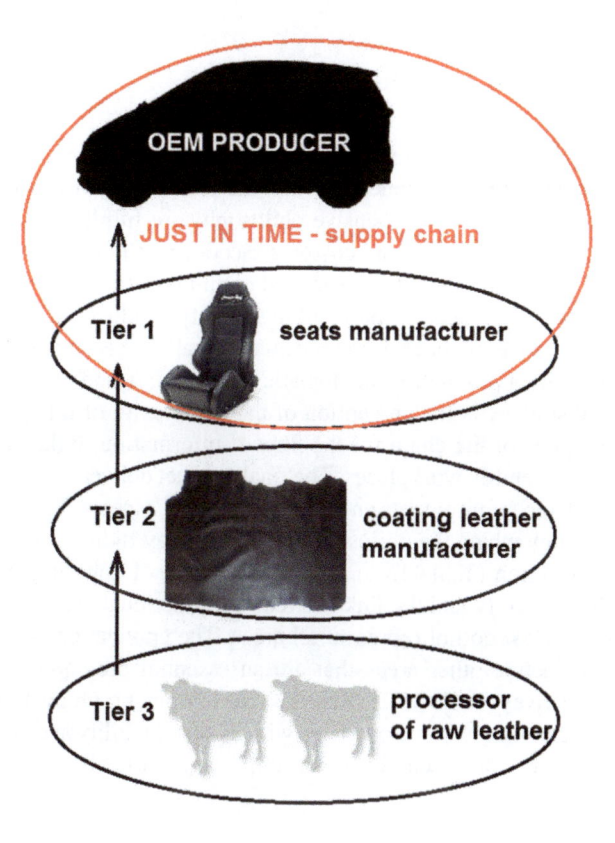

There are two principles of quality assurance currently applied within the structure of the automotive industry. It is American and German approach. American approach represents the methodology issued by the American Society for Quality. German approach represents the methodology issued by the German Automotive Industry Association [2]. Despite the attempt to unify the requirements of the automotive industry, which led to the adoption of technical specification IATF 16949, some car manufacturers use the approach which they prefer.

All of these standards and methodologies create mutual communication system throughout the supplier-customer system, from the processors of raw materials to final manufacturers (i.e., OEM—original equipment manufacturer). In the automotive industry, the classification of suppliers (more precisely the degree of supplier-customer relations) uses well established manner of marking (Fig. 15.2).

Fig. 15.2 Structure of supply chain in automotive industry

As nowadays there is no final car producer able to produce the whole car completely by themselves, they use external suppliers for the manufacture and supply of various components, which are divided as follows (Fig. 15.2):

- Tier 1—a group of external suppliers who supply their products directly to final car manufacturers. Essential requirement for this group (suppliers) is to provide their products by means of just-in-time (or just-in-sequence) system, i.e., always at the required time and at the appropriate place on the assembly line. In this regard, the geographic location (distance) to the car producer is the key factor (red line area on Fig. 15.2) (e.g., seat manufacturer/supplier of seats for OEM).
- Tier 2—a group of external suppliers who supply their products to the external suppliers in Tier 1 (i.e., subsuppliers). Therefore, their customer is not OEM but Tier 1 (e.g., manufacturer of leather and textile lining/supplier for Tier 1).
- Tier 3—a group of suppliers who deliver essential components, for example, supplier of leather and textile/supplier for Tier 2.
- Tier 4—a group of suppliers who supply raw materials (e.g., steel, glass, etc.).

Suppliers from Tier 2 to Tier 3 may not be located close to the OEMs, but they may be located throughout the territory (region, state). One external service may be categorized as Tier 1 for one (OEM) customer but also may be categorized as Tier 2 for another (Tier 1) customer with another product.

Philosophy of Industry 4.0

Based on the structure of production technologies in the context of automotive production, it can be concluded that the conditions for the application of Industry 4.0 strategy as a comprehensive philosophy are fulfilled, e.g., delivering final product (automobile) to the end customer according to their requirements. Implementation of this requirement has expanded communication level to all those working within the logistics chain. The resulting quality product can be achieved only when the same tools of quality management at all levels are applied in horizontal cross-linking of production and logistic structures. Based on the basic idea of the Industry 4.0 strategy is the perception of the complexity of this philosophy, which involves all parts of the chain for the flow of information, linked to specific manufacturing and assembly workplaces. The final product delivered to the customer in accordance with the requirements encompasses several aspects. These are the fragments (areas) through which the philosophy of the strategy Industry 4.0 can be applied in the real production (Table 15.1). The precondition of applying strategy Industry 4.0 is its functionality and the link between the individual areas. The areas shown in the figure below do not represent all areas. The application of the philosophy in practice will define other areas that are an essential prerequisite for the fulfilment of its objectives (Table 15.1). Research (development, innovation) is an area essential for the company to develop in line with its sustainability and competitiveness. Therefore, the area of research is an important part of the strategy. Products in car

Table 15.1 Fragments of philosophy of industry 4.0

Philosophy	Fragments
Industry 4.0	Research, development, innovation
	Automatization
	Quality Control
	Maintenance
	Logistics
	Security/safety
	Education

manufacturing have a defined life, whether in terms of functionality or design. Car innovation time is about 6–8 years. OEMs must therefore continually develop their products, and this imposes severe conditions for all participants in the production and logistics chain. To meet this market demand, applied research in new production technologies must be developed in order to offer innovative products and maintain long-term customer favor and increase production efficiency.

Quality Control focuses on the delivery of conforming product. This area involves the application of various tools and methods which are well established in the automotive industry and represents a kind of standardized basis—a precondition for the conformity of the product during its output from the process by means of their application.

Logistics focuses on ensuring the appropriate supply of the process by input materials, in the right quality, quantity and time, as well as the supply of various production processes. Finally, it includes the distribution of the finished product to the (taking into account requirements) customer in the given amount and within given time.

Industry 4.0 strategy includes applications of information and communications technology methods that ensure effective communication—the exchange of information in real time—based on which the well-defined entry requirements (parameters) are applied to finished products that are provided to the customer in the right quantity and at the right quality at the right price, at the right time, and at the right place. The precondition of reliable communication by means of adequate information and communications technology is a major challenge for effective and reliable communication within supplier-customer chain in order to streamline processes to meet the requirement on time of delivery and cost of the product for the customer with regard to their, maybe operationally changing, requirements.

Security/safety is an area that can be seen within "safety-security." In the context of "safety," i.e., occupational safety, the focus will be placed on flawless work of operators, where it is not possible to exclude the human factor from the process [3]. Any accident can lengthen the time of delivery of the product or suspend its delivery completely. In the context of "security," it is the protection of property, persons, and information [4]. Production machinery and equipment are in many cases complicated technological complexes, where any damage (e.g., sabotage) can have a liquidation effect for the manufacturer. The Industry 4.0 strategies are activities in relation to the Internet of Things, big data processing technologies, and the use of cloud technology. In this regard, a strategic role is played by information security, failure of which may cause information noise or it may be subject to industrial espionage.

The Internet of Things (IoT) is a system of interrelated computing devices, mechanical and digital machines, and objects or people that are provided with unique identifiers and the ability to transfer data over a network without requiring human-to-human or human-to-computer interaction [5].

Big data is a phrase used to mean a massive volume of both structured and unstructured data that is so large it is difficult to process using traditional database and software techniques.

Cloud computing is a type of Internet-based computing that provides shared computer processing resources and data to computers and other devices on demand. It is a model for enabling ubiquitous, on-demand access to a shared pool of configurable computing resources (e.g., computer networks, servers, storage, applications, and services) which can be rapidly provisioned and released with minimal management effort.

A new approach to quality management is based on risk. Each organization (manufacturer) will have to perform a risk analysis on their processes, to be able to define preventive measures to effectively prevent and thereby reduce the negative effects and also identify areas for possible opportunities. Risk analysis can become a strategic weapon in the competitive battle, because it can provide competition with vulnerabilities in processes or information about where the organization (manufacturer) perceives their opportunities.

Education is an area that includes the preparation of competent staff for carrying out activities as part of the Industry 4.0 strategy. Most social systems face challenges to ensure sufficient capacity of qualified labor. In the Slovak Republic, it is primarily due to the fact that the once successful model of dual education (professional/vocational) has gradually disappeared from the educational system. Therefore, the area of education as an integral part of the philosophy of Industry 4.0 is one of the priorities in the education process of a new, young, qualified, and competent generation, which should meet the requirements of the labor market within the terms of Smart industry. Education is not a short-term activity but a comprehensive strategic model of education beginning with preschool and ending at universities. Universities as educational and scientific institutions must work together so that the research and development in the automotive industry is an effective tool for its market competitiveness.

On the basis of Table 15.1, it is clear that the above-described areas are not the only areas covered by the philosophy of Industry 4.0 strategy. There are far more areas that can be developed and linked with others with a focus on streamlining the entire supply chain. This can include such issues as, for example, automation (robotics), human resources management, maintenance, and environment, and so on.

Support Tools for Quality Assurance in Terms of the Automotive Industry

One of the basic preconditions of Industry 4.0 strategy functionality is the already mentioned Internet of Things (IoT) and the mobility of data within it, preferably in real time (Fig. 15.3). The scope and volume of data defining the communication between departments within the complex production and logistics systems as well as individual machines themselves, including robots with operators at various levels, is in such volume and such quality level of information that it's processing in digital form can only be carried out by Big Data technologies. These data are efficiently processed by cloud technology. This involves one of the distinctive aspects of modern production processes (based on widespread deployment of robots and automation systems within logistics operations—storage in modern warehouses or warehouses on wheels). The quality management systems must be applied to be compatible with the procedures previously applied within conventional manufacturing technology and logistics (handling flows). These processes stem from use of digital information. To obtain them, it is necessary to select locations for their collection and apply modern scanning equipment allowing digitization of signals and subsequently technologies for their subsequent mobility to customers via the Internet (Fig. 15.3).

This fact conditions the application similarly as in the context of risk management systems, quality management systems (a new concept that has not been used yet, but is based on the principles of safety integration within the complex management systems) for all operations throughout the production chain of a car. System activities implementation within an integrated quality management system requires the application of both traditional and new tools for quality assurance.

These are difficult and complex production systems, as well as logistical approaches within the technical life of a final product (e.g., from planning, purchase of materials, processing, assembly, testing to the customer delivery) that are linked horizontally. An important role is played by the just-in-time requirement (or

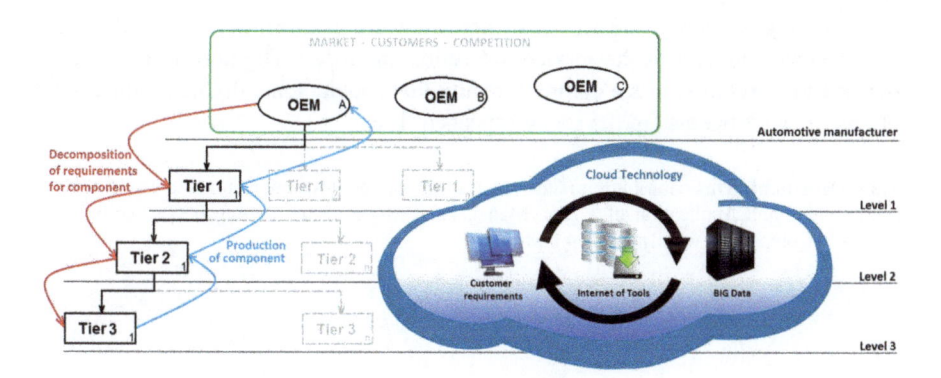

Fig. 15.3 Principle of Industry 4.0 [6]

alternatively just-in-sequence) for supplies, which places high demands on the quality of communication flows in a broad sense (collection of relevant information via machine interfaces, systems, people). In this context, various tools must be applied (methods and methodologies for their application) as part of comprehensive quality management systems or a development of procedures for the development of new tools and methodologies.

Conclusions

Closer look at what's behind Industry 4.0 reveals some powerful emerging currents with a strong potential to change the way factories work. Industry 4.0 is gathering force, and executives should carefully monitor the coming changes and develop strategies to take advantage of the new opportunities.

Digitalization, communication, information, and its sharing are the main pillars of Industry 4.0 strategy as well as the key factors in the effort to deliver the finished product to the customer/product requirements at the required time. Due to the efficiency of production and logistics processes in the automotive industry, the management attention needs to focus on shortening production times, taking into account the minimization of losses in terms of production (e.g., the lack of inputs in the production process, industrial accidents, downtime due to failure within production technology, or a product at the output that does not conform with the product specification). Car manufacturers within the application Industry 4.0 strategy do not only focus on their production and the improvement of efficiency but also on their customers and meeting their requirements for comfort when using the car. An example is the application of mobile phones, the application of which increases with the number of different applications, for example, Daimler AG has decided to develop a mobile application "Mercedes me" for owners of Mercedes-Benz vehicles [7]. It offers not only communication between the owner and the vehicle (which is typically the lock control, tire pressure, fluid levels, closing windows, etc.), but it can also share information about technical condition with service centers and vice versa. By monitoring of vehicle, the service center can warn the driver of a hidden defects in the vehicle and advises the service inspection and repair. Digitalization is not a phenomenon anymore; it is a common reality that should generally help humanity not only in work but also in the social aspects of life.

Acknowledgment This contribution is the result of the project implementation APVV-15-0351 "Development and Application of a Risk Management Model in the Setting of Technological Systems in Compliance with Industry 4.0 Strategy."

References

1. McKinsey & Company. (2017). [Online]. http://www.mckinsey.com/business-functions / operations/our-insights/manufacturings-next-act
2. Automotive Industry Association of the Slovak Republic. (2017). [Online] http://www.zapsr.sk/
3. Sinay, J., & Pačaiová, H. (2017). Sicherheit in der industrie 4.0-strategie – eine einführung / 2017. *Sichere Arbeit, 1*, 40–46. ISSN 0037-4512.
4. Kliment, J., & Šolc, M. (2016). The process of identification security risks in the automotive in-dustry / 2016. In *SGEM 2016* (pp. 475–482). Sofia: STEF92 Technology Ltd.. ISBN 978-619-7105-58-2.
5. Sova Digital. (2017). [Online]. http://www.sova.sk/sk/riesenia/industry-40
6. Sinay, J., Markulik, Š., & Pačaiová, H. (2017). Quality as a part of modern technology in the automotive industry / 2017. In *Smart City 360 2016* (pp. 1–8). Gent: EAI. ISBN 978-1-63190-149-2.
7. Daimler, AG. (2017). [Online]. https://www.mercedes-benz.com/en/mercedes-me/

Part III
Smart Technology Trends Business Management

Chapter 16
Potential of Human Resources as Key Factor of Success of Innovation in Organisations

Stacho Zdenko, Stachová Katarína, and Cagáňová Dagmar

Abstract Effective exploitation of the human resources potential in innovation efforts of organisations in an environment characterised by a high degree of turbulence is becoming fundamental in terms of maintaining competitiveness. The key to success is not information as such anymore, but correct knowledge, associated with a particular bearer, i.e. the employee, who has to constantly develop it in a highly competitive environment and in the context of organisation's orientation. Work with talented employees is becoming a priority as a result of market unpredictability and flexibility, unfavourable demographic development, as well as realisation of the fact that technological progress can be achieved especially by investment in people and their development. Different levels of employee engagement in innovation are characterised in the paper. The present situation in organisations operating in Slovakia is subsequently analysed, and a simplified analytical instrument for the needs of organisations is proposed, offering a possibility to identify bottlenecks preventing them from developing their innovation potential.

Introduction

Competitive advantage is an actual capability which is not only reflected in an ability of a company to design and develop a product but also to produce and sell it for more advantageous prices and (or) in greater amounts or quality than competitors.

S. Zdenko (✉) · S. Katarína
Department of Management, Institute of Economics and Management, School of Economics and Management in Public Administration in Bratislava, Bratislava, Slovakia
e-mail: zdenko.stacho@vsemvs.sk

C. Dagmar
Institute of Industrial Engineering and Management, Faculty of Materials Science and Technology in Trnava, Slovak University of Technology in Bratislava, Bratislava, Slovakia

© Springer International Publishing AG, part of Springer Nature 2019
D. Cagáňová et al. (eds.), *Smart Technology Trends in Industrial and Business Management*, EAI/Springer Innovations in Communication and Computing,
https://doi.org/10.1007/978-3-319-76998-1_16

It is not provided spontaneously by external environment but is also dependent on an ability of internal environment of a company to identify and response flexibly to changes of external conditions [36]. Tidd et al. [44] and Franková [16] deal with factors affecting the quality of internal environment with regard to exploitation of human potential of an organisation in innovation, highlighting permanent education and development of employees, i.e. their preparation for changes, which are significant in terms of a speed of implementing changes, while pointing to a necessity to create organisational culture encouraging creativity and creation of innovation. Employees in such an environment are positively encouraged to creative behaviour on the one hand [22] and they feel safe in terms of their willingness to undergo a certain risk without worrying about sanctions in an event of failure on the other [15, 35].

This statement was also supported by Berger and Berger [2], Covey [13], and Leary-Joyce [26], who included creativity, with reference to creative human potential, among nine representative key competences of talented employees. Creativity has been more and more comprehended as a social phenomenon in recent years [49]. As E. Franková [15] states, it primarily results from the fact that the source of all creative ideas is the human mind. A number of empirical studies have confirmed the fact that creative potential is integral to every human and that every person is, or can be, creative [14, 16, 19, 28, 30, 31, 51]. The level of creativity or the level of the development and application of a creative potential achieved by an individual primarily depends on themselves and the influence of their environment. Organisations have a direct impact on both, as they should focus on the potential of a hired employee, i.e. their level of abilities, commitment and aspiration, already during recruitment [12, 34], as well as on what environment and conditions, whether spatial or social, are created in the organisation for this purpose. If the organisation ensures suitable conditions, employees are motivated and apply a creative approach in a considerably higher extent; they are able to come up with new and creative ideas and subsequently introduce innovations in daily practice [24]. It is the lack of creativity or innovative behaviour that is frequently understood as one of the key reasons of a failure of the organisation [50]. It is therefore necessary to create and maintain an organisational culture supporting innovation and creativity [16]. The same characteristics should be also required from managing employees, as not only their creative potential but also the ability to inspire their subordinates and co-workers and the ability to create adequate conditions are important in their position [48].

Organisations go through several stages with different engagement levels upon engaging their employees in innovation. Tidd et al. [44] defined *five stages of building an innovation model with a high engagement level*, the *first stage* being so-called unconscious engagement, in which people occasionally participate in innovation (e.g. in cases of defective new products), however without a formal process encouraging such behaviour. Managers frequently apply authoritative methods of management [21, 33], which results in the fact that when employees submit innovative proposals they are reviewed by management before even starting considering their implementation [41]. If employees participate in innovation, it is typically only because they are at the right time on the right place, however frequently without even realising that they are participating in a form of innovation. In the *second stage*, processes focused on troubleshooting are developed. Proposed ideas are implemented in the greatest possible extent, and employees are subsequently remunerated or otherwise motivated (e.g. creation of creative environment [11], encouragement of open communication with effective share of information [16], application of instruments confirming the competence of employees [11]), thus *creating a habit of engagement of employees* in innovation. The *third stage aims* at *interconnecting the habit with corporate strategic objectives*, which practically means that corporate strategy is communicated within the whole company from management downwards and is divided into particular goals in the context of engagement in innovation-related activities. Strengthening the powers of individuals and groups upon experimenting and their own initiatives, so-called empowerment, are the priorities of the fourth stage - maximum encouragement of effective freedom of individuals and teams in decision-making [41]. In the *fifth stage, everybody is fully engaged* in experimenting and improvements, in sharing information and creating an actively learning organisation. In this stage, values connecting people and enabling their participation in development of their organisation are created and shared, i.e. organisational culture encouraging creativity and innovation is created. As long as corporate management succeeds in creating pro-innovative environment, it is capable of taking innovative decisions quickly [29] which is considered a significant competitiveness factor in the present-day turbulent environment [7, 10].

Effective exploitation of the human resources potential in innovation is based on the fact that although individual employees might be capable of generating only restricted innovations, the summary of such activities might have far-reaching consequences [3]. It is therefore obvious that the priority task of management is to create an environment encouraging creativity of employees and to create an atmosphere where dialogue among managers and other team members is of importance and where interests of employees in the issues related to decision-making and management are respected. Such a motivating approach creates room for innovative proposals of employees [8], for work of innovative teams [25], or suggestions for improvement [10].

Development of Most Prospective Employees

Talent management is considered to be one of the key tools of achieving and enhancing of organisational performance. *Talent* (as a person) can be defined as an employee with a *high potential* that they can exploit for the fulfilment of objectives and an improvement of the performance of their company [1, 17, 23, 48]. The main idea of current talent management is that organisations do not choose the best but look for a hidden potential possible to complement according to their particular needs.

Organisations focus on recruitment of the right person from internal as well as external sources, they further work with them, and subsequently look for appropri-

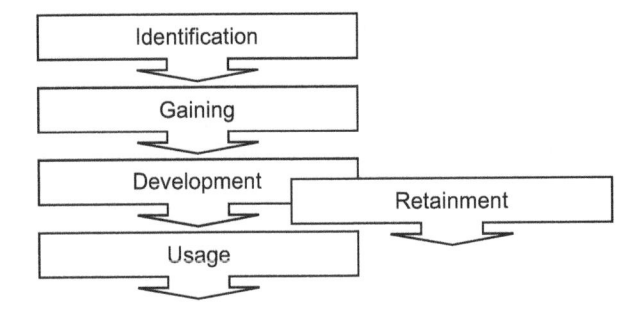

Fig. 16.1 Talent management process. (Source: Hroník [20])

ate ways how to involve the gained talent in organisational objectives so that this process brings expected results [42].

Last but not least, it is necessary that organisations are able to ensure that the talent is not encouraged to be an advantage for competition but is motivated to stay and perform in the organisation which provided them development possibilities. In order for an organisation to be able to implement talent management successfully, it is necessary to have set an appropriate talent management strategy, based on which environment is created, and processes, activities, and objectives are defined.

To attract, educate, and especially to retain quality talented employees makes the company distinguished and enables it building a position well ahead of other organisations [48]. There is a limited number of talented individuals in the whole population, and, therefore, companies should seek to gain and retain them by all available means [39].

Talent management process can generally be described as five steps shown in Figure 16.1. The process begins by the identification of talents in internal or external sources; it is then necessary to win the selected person for the programme (project). After entering the programme, their intense development follows, leading to talent usage, its transformation to results, and it is already important to focus on talent retainment in the organisation [1] concurrently with the third and fourth steps.

The process begins by identifying talents in internal or external sources, the selected people subsequently need to be enthused about the programme (project), which is followed by their intense development, aimed at exploiting their talent, by the transformation of talent into results, and, concurrently with the third and fourth steps, it is necessary to focus on the retaining of talents in the company.

The identification of talents in the *internal sources* of the company is made by employee appraisal. It is important to realise that not only the *performance* but also the potential of employees in the context of their *identification with the organisational culture should be appraised.*

Employee Performance When long-term performance is concerned, it is necessary to consider that the present performance correlates only slightly with the future potential of the employee. In addition, performance needs to be appraised with regard to the fact *that high performance does not guarantee high potential; however, high potential typically guarantees high performance.* The tendency of managers to identify future talents based on their present performance is frequently seen in practice. When appraising potential, they tend to overestimate the potential of highly productive employees, to underestimate their own motivation and aspiration, and promote them only on the basis of performance, loyalty, and reliability. However, such employees may have achieved their best performance, they might not be able to advance, especially not at positions requesting completely different skills compared to those they currently excel at, i.e. not every acknowledged expert can become an excellent manager or leader. Completely different requirements for skills, knowledge, and experience are concerned. When such a remarkably productive employee is promoted to a managerial position, they may fail. Practical consequences of such a situation include the loss of a qualified and motivated expert, and gaining a demotivated manager, or even their loss, if they leave the company [34].

Researches have shown that only 30 % of highly productive employees have the potential—abilities, commitment, aspiration—to advance at a more senior position in their career in the company [45].

Employee potential is a function of a number of variables, typically including the following: expertise, long-term above-standard performance, above-standard level of relevant competencies, willingness and ability to learn fast, personal ambition, aspirations, motivation to advance in the career in the company, and commitment and loyalty to the company [34]. Employee potential can be evaluated on the basis of three criteria [12]:

- *Abilities*: Abilities are the combination of innate and unlearned personal characteristics, which are used in doing everyday work—mental and cognitive abilities, emotional intelligence, technical and functional know-how, interpersonal abilities, abilities to process and solve complex situations, willingness to learn new things [12], as well as the ability to think beyond one's responsibilities or beyond a problem the employee is dealing with, and openness to feedback and to changes [46].
- *Commitment*: Commitment is the extent to what the employee is committed to someone or something in the company, the deployment and enthusiasm they work with, and how long they remain in the company as a result of such a commitment. This commitment can be emotional or rational. Emotional commitment reflects the level to which the employee appreciates their company and has trust in it. Rational commitment represents the level to which employees are persuaded that staying in the given company corresponds to their own interests and ensures achieving their personal objectives and interests [12].
- *Aspirations*: Aspirations express the need and desire of a person to gain prestige and raise their profile in the company, to be promoted and have influence, to be very well financially rewarded, to "like" their job and have a work-life balance [12].

Therefore, it is necessary for the future success of employees to succeed in all the aforementioned criteria—abilities, commitment, and aspirations. If the candidate has weaknesses in any of the areas, it substantially reduces their prospects of succeeding at a senior position, and company development resources are thus used for wrong people [34].

Identification with organisational culture is important especially because such employees receive both internal attention (other employees compare to them, or try to act like they do in order to ensure the same possibilities of their career or professional growth), and external attention (the employees included in talent management participate in competitions, conferences, or different discussion fora more frequently). They are often the leaders, ideals, or even heroes of their company, and the achievement of their agreement with the content of organisational culture is thus desirable [40].

There are many external sources where organisations can *find talents* necessary for their development and competitiveness retainment, e.g. university students, sportspeople, employees of non-profit organisations, customers, or suppliers. Possibilities and reasons are as follows:

Universities Participation in career days (e.g. a fair called National Career Days is held in Slovakia) and waiting for students to respond and show interest in writing their theses for the organisation, or offering of an internship abroad and involvement in a development programme are insufficient at present. Strong organisations try to do something more, they try to prepare students for practice, and they influence the content of their studies. School sponsorship is conditioned by setting up a field of study or subject with lecturing experts from the given organisation, who monitor talented students and encourage them to cooperate with their organisation.

Sportspeople There are many sportspeople doing mainly team sports, training from a very young age, who know already in their teenage years that they will not become the best however they have at least ten-year experience in creating habits, which can be beneficial for their activities in the organisation. It is desirable that consultancy organisations create special managerial trainings for top sportspeople.

Non-Profit Organisations Many qualified and talented people work in the non-profit sector, which can represent an interesting resource of talents for organisations.

Business Partners Can also be a great resource of talents for organisations; however, this talent gaining process should in this case be agreed in advance so that mutual cooperation is not disrupted.

Subsequently, after talent identification, there is a time to *create an offer* by the organisation for the given talent, contract negotiation defining duties of both parties. An agreement confirming that participation in the given programme (project) neither means any relief at work itself nor is a career growth guarantee is a part of such contract; the guarantee can only be results achieved in working on the programme. It is also necessary to clearly define in the contract what is required from the chosen talent.

Hroník [20] defines two types of *development programmes*, one focusing on the development of potential employees, so-called *trainee programme*, the other is for people with a certain company history and credit, so-called *talent development*. It implies that the number of employees involved in different career management programmes grows with the length of their employment duration.

Trainee programme is an orientation programme implemented by organisations to support social and working adaptation of university graduates. It represents a departmental and company preparation of potential employees. The objective is to ensure potential managerial recruits and experts in individual professional spheres of the given organisation [32].

Talent development normally concerns a yearly development programme for talented employees with a great potential, whom their organisation considers to be the key specialists or future managing employees. Employees are nominated in this programme on the grounds of specific criteria approved by organisation's executive board. Programme participants develop by means of various educational forms, e.g. mentoring and coaching, and they participate in seminars with external experts [20].

Departure of talented employees can have a significantly negative impact on further operation and competitiveness of organisations. When an employee with key knowledge, skills, and abilities necessary for the organisation now as well as in the future is leaving the company, it is losing a part of its intellectual capital. Therefore, if the organisation wants to retain these employees they need to be able to agree with them and cooperate with them; however, it is obvious that certain boundaries need to be respected. It is essential that the organisation encourages talents in the greatest possible participation in organisational objectives and rewards them on the basis of their performance. Besides, the organisation should also focus on the fulfilment of objectives of talents (e.g. career growth, satisfaction at work or the feeling of safety) and strengthening their loyalty to the organisation.

Snipes [37] created a successful *"high potential" programme* and declares that companies should create so-called *leadership package* within their talent development programme, which should include the following:

- *Specialised leadership programme,* i.e. a modular programme focused on the development of managerial skills (change and crisis management, building and development of teams, strategic management, coaching, etc.).
- *Multidisciplinary programme,* rotations across disciplines, divisions, regions.
- *Unlimited learning possibilities.* Logically, most of the companies limit the number of courses and trainings available to their employees, which is mainly related to costs. The results of the research show that highly productive companies ensure their high-potential employees' unlimited access to learning and trainings in order to support the speed of their growth as well as to verify the ability of "self-management" in the learning process.
- *On the job learning* (Action learning) is learning by doing. It represents a shift from classical learning methods (simulation case studies, etc.) to particular problems of the real world and particular projects. The group of "high potentials" has to elaborate a particular development project relating to a new product, service,

or business analysis that helps a particular department, which is followed by its implementation, and by presentation of the results to company management.
- *Mentoring*. Internal senior mentors are involved in the programme in order to exchange their long-standing knowledge and experience with talents.

In the case companies have not elaborated a particular development programme for talented employees, however have understood the strategic significance of the need to focus on such employees, it is appropriate to apply some of the methods of employee education for this purpose. The most effective educational methods focused on talent management include coaching, mentoring, counselling, and manager shadowing.

Talent exploitation represents the exploitation of the potential of the most capable employees not only in common work process but especially in corporate development. The more effectively the company is able to exploit talents, the better competitive position it gains.

If employees feel that they are fairly financially rewarded, also in comparison to employees of competitive organisations [38], if they perceive their work as interesting and meaningful, supported by good managerial methods, if they have good relationships with direct superiors and feel support by top management, if they are provided sufficient room for development and growth, they are not tempted by an interesting offer from other organisation, where they cannot be certain about so good working conditions [4].

It can be stated on the basis of the facts provided in this sub-chapter that it is important upon building and maintaining the innovative organisation to focus on talent management especially because the search and development of talents cannot start at the moment when it is necessary to find a solution or a way how to do things differently and more effectively; but it is necessary to deal with the search and development of talents continuously and in the long term so that the organisation disposes of such people before a problem occurs, respectively that these "talents" are able to foresee the given problem and prevent its occurrence [41]. Organisation's properly working talent management should not only lead talents to prevention and solution of problems, but it should also provide them room and means to create innovations.

Materials and Methods

The research necessary for elaboration of this paper was conducted from 2010 to 2014 at School of Economics and Management in Public Administration in Bratislava in collaboration with Faculty of Materials Science and Technology in Trnava STU in Bratislava. Organisations participating in the research were questioned in a form of questionnaires personally delivered to a person responsible for human resource management in the given organisation. The questionnaire survey contained approximately 90 questions, and the results obtained from answers to the questions focused on analysing the present focus of organisations on engagement of employees in innovation were used for the needs of this paper.

Two stratification criteria were determined to identify a suitable research sample; the first criterion was a minimum number of employees of an organisation, which was determined to 50 employees. The second stratification criterion was a region of operation of a company, while the structure of the research sample was based on the data provided by the Statistical Office of the Slovak Republic.

According to the Statistical Office of the Slovak Republic, the number of organisations with 50 and more employees was between 3261 and 3359 over 2010–2014. The regional structure of organisations with over 50 employees in the given years is shown in Table 16.1.

Identifying an optimum research sample from the given base set of organisations, confidence level was determined at 95%, and confidence interval of the results was determined at H = +/− 0.10. On the basis of the aforementioned criteria, a relevant research sample for individual regions of Slovakia was determined in the analysed years, which are provided in Table 16.2.

Approximately 500 organisations participated in the research every year; however, only approximately 65% of the questionnaires were received comprehensively completed due to a great extent and form of data collection. Subsequently, 259 organisations were selected as the optimum research sample determined on the basis of the stratification criteria.

Table 16.1 Regional structure of organisations with more than 50 employees

Region	Whole Slovakia	Western Slovakia	Central Slovakia	Eastern Slovakia
Districts	All districts	BA, TT, TN, NR	BB, ZA	KE, P
Number of organisations 2010	3308	2031	655	622
Number of organisations 2011	3359	2061	666	632
Number of organisations 2012	3295	2025	652	618
Number of organisations 2013	3268	2017	645	606
Number of organisations 2014	3261	2005	644	612

Source: Elaborated based on data of the Statistical Office of the Slovak Republic

Table 16.2 Size of research sample for individual regions of Slovakia

Region	Western Slovakia	Central Slovakia	Eastern Slovakia
Districts	BA, TT, TN, NR	BB, ZA	KE, PO
Number of organisations 2010–2014	2005–2061	644–666	606–632
Size of research sample	92	84	83

Source: Own elaboration

Results

Our research was focused on analysing the attributes of motivation and level of engagement of employees in innovation processes. With regard to the fact that one of the underlying assumptions of an innovative organisation is an ability of their managers to enthuse their subordinates about changes, one of the questions was dealing with whether their organisations create room for their engagement in innovation, primarily from the viewpoint of a purposeful regulation of their innovative behaviour towards an organisational strategy by both remuneration and encouragement of knowledge share.

In relation to the level of employee engagement in innovation, the questionnaire was also dealing with whether employees are perceived as a preferred source of innovation initiative or whether all employee categories are engaged in innovation (Table 16.3).

The research indicated that the level of employee coordination, i.e. their engagement in innovation activities aimed at achieving corporate strategic objectives in the monitored period, was gradually growing from 19% in 2010 to 36% in 2014; however, most of the organisations only engage their employees in innovation processes occasionally, upon occurrence of failures in new processes, or employees are engaged in innovation regularly however without a common strategy-based coordination. In this relation, remuneration for knowledge of employees, reaching approximately 40%, or share of all information encouraged by management, was among the monitored attributes. The research showed that employees are motivated to share information in only a minimum number of organisations without any significant improvements recorded in the monitored period.

Positive findings of the research included the fact that organisations most frequently obtain innovation initiatives from their employees (in 61%–70%); however, the result was not as positive upon finding out whether all employee categories are used for this purpose.

On the grounds of the uncovered theoretical information, talentent is considered to be one of the key tools of achieving and increasing organisation's performance. In the pro-innovative atmosphere in the company, the ability to enthuse oneself about an idea

Table 16.3 Motivation and engagement of employees in innovation

Focus of organisations on motivation and engagement of employees in innovation	Share of organisations in %				
	2010	2011	2012	2013	2014
Employees are regularly engaged in coordination with a focus on strategic objectives	19	21	25	29	36
Knowledge of employees is monitored, evaluated and remunerated	42	33	33	37	36
Employees are encouraged to fully share their knowledge	13	6	11	7	14
Preferred sources of innovation initiatives are employees	61	61	62	63	70
All employee categories are engaged in innovation	32	33	31	48	41

Source: Own research

and to influence other co-workers by one's behaviour is becoming one of the factors of prospectiveness of employees for the company. The most prospective employees are thus not only able to fulfil their work assignments with maximum effectiveness, but they also frequently help the effective implementation of changes. With regard to the aforementioned, our research sought to uncover whether companies realised the need to focus on talents and their development, and whether they actually dealt with talent management in practice (Table 16.7).

The answers indicated that approximately half of the companies considered talent management to be important, and even more organisations actually dealt with the management of the career of their employees; however, actual activities focused on the identification, gaining or development of the most talented employees were only executed in a small number of the interviewed organisations. The number of organisations with a set talent management strategy increased in the monitored period from 5 % in 2010 to 20 % in 2014. This value is not favourable; however, we hope that this result can indicate a positive change of the approach of management.

With reference to the relation between talent management and innovative approach of the company, we sought to uncover whether the organisations gained talents from the external environment as potential bearers of new ideas applicable for the benefit of the organisation. However, the research showed that talent management was not perceived in this relation in almost 90 % of the interviewed organisations, as they preferred the internal environment as the source of talents.

Discussion

The results of our research can be considered as positively developing over time; however, they are insufficient in the context of their level. The need to improve the present state also results from the findings concerning organisations such as U.S. Steel Košice and *Continental Matador* Rubber, s.r.o., as well as from the results of researches conducted by authors such as Blašková [4], Cagáňová, et al. [5], Chidambaram [9], Hansen, Winther and Hansen [18] and Urbancová [47].

Company U.S. Steel Košice declares that the exploitation of the improvement potential of regular employees from operation is of special importance for organisations, as their proposals for improvement resulted in significant savings for their company. "It represented a saving in the amount of almost 20 dollars for a ton of produced steel", said the speaker of the company Ján Bača [27]. Between 2004–2005, implemented improvements proposed by the employees of company *Continental Matador* Rubber, s.r.o., which is based in Púchov, resulted in the economic benefit amounting to more than EUR 700,000, while only EUR 50,000 were paid as bonuses [27]. Researches conducted by authors such as Chidambaram [9] and Hansen, Winther and Hansen [18], which dealt with the position of organisations in knowledge economy, indicated that human capital had a positive impact on economic growth.

In the results of her research, Urbancová clearly declared that only 20.5% of the analysed organisations tried to maintain the knowledge of every employee, although they were aware of the competitive advantage they had achieved due to knowledge [47].

With regard to the fact that talented employees are highly productive and they put by 20 % more efforts in their work compared to others [34], organisations need to deal with not only their identification but also their retaining. Campbell and Smith [6] examined talent management in organisations from the perspective of managers who were identified as having high potential. The research involved 199 leaders who had participated in a development programme within the talent management programme; 77% of the interviewed declared that the "formal identification" of high potentials was very important. The study proved a notable difference between the managers clearly identified as having high potential and the managers who were persuaded that they had high potential; however, they had not been identified as such by their company. Fourteen per cent of the identified managers and 33 % of the unidentified managers were seeking a new job.

As the aforementioned indicates, the key to success is not information as such anymore, but correct knowledge, associated with a particular bearer—the employee, who has to constantly develop it in a highly competitive environment. Active use of knowledge is undoubtedly a competitive instrument for individual organisations.

Recommendation

Based on findings, the participating organisations are recommended to apply a simplified analytical instrument created by us, helping them identify their bottlenecks in motivating employees to creativity and levels of their engagement in innovation. Revealed bottlenecks can be an incentive for organisations to adopt remedial measures helping them boost their innovation potential and competitiveness.

The following table provides the proposed instrument, dealing with the motivation of employees to creativity and levels of their engagement in innovations (Tables 16.4 and 16.5).

Within the sphere "high commitment in innovations", companies were divided into the following three groups:

A. All company employees are supported by management in their commitment in innovations and different improving processes or procedures. All information in the company is fully shared by all employees, i.e. individuals and teams work on innovations in coordination, focusing on company strategic objective. Atmosphere supporting dialogue between managers and other employees is created. Powers of employees related to decision-making and management are respected. Managers perceive all employees involved in the innovation process as equal colleagues, and they appreciate and support their ideas. That is a reason why employees feel the companionship with their company and they

Table 16.4 Questions analysing motivation of employees to creativity and levels of their engagement in innovation with scores

Questions and response options	Score
1. *Does the management show support of engagement of employees in innovative proposals?*	
(a) Yes, constantly	10
(b) Yes, occasionally	5
(c) No	0
2. *Does the company use its employees in search for innovations?*	
(a) Yes, it uses all employees by inviting them to participate in different competitions or establishing a bonus remuneration in relation to successful innovations	10
(b) Yes, but only some employees are used (specialised departments, project teams)	5
(c) Yes, there is a possibility to submit a proposal to a superior or through an anonymous box, however mostly without implementing any of such suggestions	0
3. *How and when are employees engaged in innovation?*	
(a) At every occasion (creation of ideas, planning of implementation, implementation itself) with common coordination based on corporate innovation strategy	10
(b) Regularly, however without common coordination based on corporate innovation strategy	5
(c) They are only engaged unconsciously (upon occurrence of failures in implementing new processes)	0
4. *Are innovative proposals of employees checked and approved by the management of the organisation?*	
(a) No, employees have full confidence of the management	10
(b) Yes, they are approved; however, employees have full confidence of the management upon their implementation	5
(c) Yes, they are thoroughly checked and approved by the management	0
5. *How does the management proceed in an event of a failure upon implementing innovations?*	
(a) The organisation tries to remove such failures and prevent their repetition, and employees revealing such failures are remunerated in some cases	10
(b) The organisation tries to remove such failures and prevent their repetition, and employees responsible for such failures are not sanctioned	5
(c) The organisation tries to remove such failures and prevent their repetition, and employees responsible for such failures are sanctioned	0

Source: Author

Table 16.5 The level of employee commitment in innovations on the grounds of a sum of the scores of individual questions

Feature of an innovative industrial enterprise	Your result	Your level
High commitment in innovations	50–40	A
	39–20	B
	19–0	C

Source: Author

willingly accept and implement innovations. Encouragement of employees to innovate is related to rewarding or other motivation of employees for their innovative proposals.

B. The company occasionally supports individuals to make innovative proposals; however, communication of strategic objectives within the whole company is minimal, which results in the fact that individuals and teams work on innovations without a common strategically oriented coordination. All innovation activities carried out in the company are managed and approved by top management. Managers perceive all employees involved in the innovation process only as subordinates, i.e. they listen to their ideas; however, they decide themselves whether to support the given proposal or not. Management trusts employees, which enables them to undergo risk at creating innovations without worrying about unfair sanctions from their superiors in case mistakes or failures, which frequently accompany innovations, occur.

C. There is practically no formal support of the commitment of employees in innovation processes in the company. Employees occasionally commit in the innovation process, however unintentionally, e.g. when mistakes occur in relation to a new product. Each innovative proposal submitted by an employee is checked and assessed by management before considerations of its implementation are initiated. Management's encouragement of employees to behave innovatively is minimal; mistakes are often sanctioned, even when they are unforeseeable, which causes the atmosphere of uncertainty in the company.

Table 16.6 aims at revealing bottlenecks, as it enables to recognise a particular part of the area of engagement in innovation which needs to be focus on in order to achieve a higher level.

The proposed method results from an analysis and enables, in a short time interval, to carry out a self-assessment of an actual level of an organisation as well as identify bottlenecks preventing them from developing their innovation potential.

The bjective of the first step is to analyse the present level of talent management in organisations on the basis of five closed questions. The questions with scoring evaluation are provided in Table 16.8.

The objective of the second step is to specify talent management level in organisations. For this purpose, answers in Table 16.8 were individually assigned points from 15 to 0 points, while the higher score organisation gets, the higher is the level

Table 16.6 Revealing bottlenecks in employee engagement in innovation

Question number/answer	1	2	3	4	5
Excellent	a	a	a	a	a
Average	b	b	b	b	b
Insufficient	c	c	c	c	c

Source: Author

Table 16.7 Level of focus of companies on talent management

Focus of companies on talent management	Share of organisations in %				
	2010	2011	2012	2013	2014
Company considers talent management to be important	52	50	31	46	49
Company focuses on the career planning of all employee categories	55	59	65	57	54
Talent management is executed on the basis of a determined strategy	5	14	15	16	21
Company gains new talents primarily from the external environment	10	5	9	11	10
Company uses an internship programme to gain new employees	3	5	4	5	6

Source: Own research

of its focus on talent management. On the basis of the total point score, it can be individually stated at what level organisation is. Individual point intervals providing the basis of particular level specification attributed to the analysed organisation are shown in Table 16.9.

From the viewpoint of the development of the most prospective employees, companies can be divided into the following three levels:

(a) The company carries out all activities related to talent management in full extent, while for this purpose, it has elaborated particular talent management strategy. The company focuses on talent identification from internal as well as external sources. It subsequently focuses its effort on gathering these talented people. After entering the programme, intense talent development follows; it is further worked on and a suitable way is being searched to incorporate the talent in strategic objectives in order to achieve expected results. The company has a functional process of career planning, succession planning, and knowledge continuity management, which implies that quality performance of talented employees is an impulse for company management regarding their career growth. With regard to the fact that the company has identified talents who are going to grow in terms of their career, it can purposefully ensure knowledge transfer continuity at the key positions in the company. The company is continuously trying to ensure that talents are motivated to stay and bring further benefits for the company which provided them room for personal development.

(b) The company identifies talents only randomly on the basis of intuition, predominantly from internal sources. It subsequently focuses its efforts on intense development of talent with whom it further cooperates and searches for a suitable way how to incorporate the talent in strategic objectives in order to achieve expected results. However, talent management is not interconnected with strategy of employee career growth, which implies that even quality performance of talented employees is not going to ensure their career growth. With regard to the fact that company is not able to identify talents to grow in terms of their career, it cannot ensure knowledge transfer continuity at the key

Table 16.8 Questions analysing the sphere of talent management implementation and their scoring evaluation

Questions and answer variants	Points
Does your company have an elaborated talent management strategy?	
(a) Yes, talent development is strategy-based	10
(b) No, talent is developed on the basis of intuition	5
(c) We do not deal with talent management	0
Is talent management interconnected with career growth?	
(a) Yes	10
(b) No	5
(c) Company has not implemented talent management	0
Which sources does your company use to find talents?	
(a) Internal as well as external sources	10
(b) Almost exclusively internal sources	5
(c) Company does not search for talents	0
Are the development and support of the creativity of talents part of their management?	
(a) Yes, the identification of talents includes discovering the ability to search for new solutions to problems, and creativity development is subsequently part of development programmes	15
(b) Yes, creativity development is part of development programmes, although discovering the ability to search for new solutions to problems is not included in the identification of talents	10
(c) Yes, the identification of talents includes discovering the ability to search for new solutions to problems	5
(d) No, creativity is neither discovered nor developed for managed talents	0
How does your company avoid talent outflow?	
(a) Regular financial or non-financial motivation	10
(b) Avoiding through different contracts or threats	5
(c) Company does not solve talent outflow	0

Source: Author

Table 16.9 Specification of the level of talent management implementation on the basis of point score resulting from evaluating individual questions

Feature of innovative organisation	Your result	Your level
Level of talent management	55–40	A
	39–25	B
	24–0	C

Source: Author

positions in the company. The company does not pay sufficient attention to motivation of talented employees, which results in the fact that it responses too late and the talent ends up with competition.

(c) The company does not identify talents due to either the fact that it does not consider talent management as important, and very incorrectly so, or it is financially very demanding for it. Even quality performance of talented employees is not a guarantee of career growth for them. Knowledge transfer continuity is

Table 16.10 Table to reveal bottlenecks in the sphere of talent management implementation level

No. of question/answer	1	2	3	4	5
Very well	a	a	a	a	a
Standard	b	b	b	b	b
Below average				c	
Bad	c	c	c	d	c

Source: Author

not ensured at the key positions in the company, and it is therefore impossible to eliminate negative impact of knowledge loss. Quality of decision-making in management process as well as performance of whole organisation is lowered after withdrawal of employee with critical knowledge. The company does not solve motivation of talented employees, which often results in their switching over to competing organisations.

Table 16.10 was designed to reveal bottlenecks in the management and development of the most prospective employees. It can help respondents specify which part within talent management implementation is necessary to focus on in order to achieve a higher level in this sphere. Colour scale ranges from yellow, representing the best condition to red, representing the worst condition.

Conclusion

The submitted paper underlines a need of organisations to comprehensively focus on motivation of their employees to creativity and on the level of their engagement in innovation. It analyses the focus of organisations operating in Slovakia on the given issue and provides an instrument which is recommended in this relation. Theoretical contributions of the paper include a set of questions aimed at analysing the focus of organisations on motivation of their employees to creativity and levels of their engagement in innovation. The main practical contribution of the paper is a proposed method evaluating the focus of organisations on motivation of their employees to creativity and levels of their engagement in innovation, and a possibility of its immediate application in organisations in a form of identification of bottlenecks preventing them from developing their innovation potential, and the levels of both their focus on motivation of their employees to creativity and their engagement in innovation.

Acknowledgements The article is written within the projects: Grant Agencies of VSEMvs project No 2/2010 Human Potential Development in Central and Eastern EU States and Grant Agencies of VSEMvs project No 2/2016 The selected attributes of managerial work of creating the internal environment conducive to business competitiveness and project KEGA *056 STU-4/2016* Public Portal for the Support of the Connection between the Education Process at the High Schools and Requirements of the Industrial Practice—institutional project.

References

1. Armstrong, M. (2007). *Řízení lidských zdrojů: nejnovější trendy a postupy. 10. vyd.* Praha: Grada Publishing.
2. Berger, L., & Berger, D. R. (2003). *Talent Management Handbook: creating organizational excellence by identifying, developing, and promoting your best people.* New York: McGraw-Hill.
3. Bessant, J. (2003). *High Involvement Innovation.* Chichester: Wiley.
4. Blašková, M. (2011). *Rozvoj ľudského potenciálu Motivovanie, komunikovanie, harmonizovanie a rozhodovanie.* Žilina: Edis.
5. Cagáňová, D., Bawa, M., Šujanová, J., & Saniuk, A. (2015). Innovation in industrial enterprises and intercultural management, *Scientific Monograph* Publisher : IIZP, ISBN: 978-83-933843-4-1
6. Campbell, M., & Smith, R. (2010). *High – potential talent.* A view from inside the leadership pipeline. Centre for Creative Leadership.
7. Cantwell, J. (2009). *Innovation and competitiveness (Book chapter) The oxford handbook of innovation Publisher.* Oxford: Oxford University Press.
8. Çekmecelioglu, H. G., & Günsel, A. (2011). Promoting creativity among employees of mature industries: The effects of autonomy and role stress on creative behaviors and job performance. *Procedia - Social and Behavioral Sciences, 24,* 889–895.
9. Chidambaram, R. (2014). To become a knowledge economy. *Current Science, 106*(7), 936–941.
10. Clark, J., & Guy, K. (1998). Innovation and competitiveness: a review. *Technology Analysis and Strategic Management, 10*(3), 363–395.
11. Collins, M. A., & Amabile, T. M. (2008). Motivation and creativity. In *Handbook of creativity.* Cambridge: Cambridge University press.
12. Collins, S. K., & Collins, K. S. (2007). Succession planning and leadership development: Critical business strategies for healthcare organizations. *Radiology Management, 29*(1), 16–21.
13. Covey, S. R. (2011). *7 návyků skutečně efektivních lidí.* Praha: Management Press.
14. Florida, R. (2004). *The rise of the creative class and how it´s transforming work, leisure, community and everiday life.* New York: Basic Books.
15. Franková, E. (2003). Creativity and innovation supportive organizational culture. In *Business development and European community.* Brno: Brno University of Technology.
16. Franková, E. (2011). *Kreativita a inovace v organizaci.* Praha: Grada Publishing.
17. Gelens, J., Hofmans, J., Dries, N., & Pepermans, R. (2014). Talent management and organisational justice: Employee reactions to high potential identification. *Human Resource Management Journal, 24*(2), 159–175.
18. Hansen, T., Winther, L., & Hansen, R. F. (2014). Human capital in low-tech manufacturing: The geography of the knowledge economy in Denmark. *European Planning Studies, 22*(8), 1693–1710.
19. Hitka, M., & Sirotiaková, M. (2011). Impact of economic crisis on change of motivation of Ekoltech s.r.o. Fiľakovo employees. *Drewno, 185*(54), 119–126.
20. Hroník, F. (2007). *Rozvoj a vzdelávaní pracovníků.* Praha: Grada Publishing.
21. Kampf, R., Hitka, M., & Potkány, M. (2014). Medziročné diferencie motivácie zamestnancov výrobných podnikov Slovenska. *Communication, 4,* 98–102.
22. Kim, J.-G., & Lee, S.-Y. (2011). Effects of transformational and transactional leadership on employees' creative behaviour: Mediating effects of work motivation and job satisfaction. *Asian Journal of Technology Innovation, 19*(2), 233–247.
23. Kotter, J. P. (2008). *Vedení procesu změny: Osm kroků úspěšné transformace podniku v turbulentní ekonomice.* Praha: Management Press.
24. Králíková, A. (2010). Chcete motivované zaměstnance? Jděte příkladem! *HR Forum* [cit. 2015-12-17]. zpravodaj.feminismus.cz/cz/clanek/chcete-motivovane-zamestnance-jdete-prikladem.

25. LaRoche, L. S., Whitesell, M. V., Gilbert, F. E., & Smock, A. P. (2005). Creative action teams – Innovative opportunities for team work. *Proceedings ACM Siguccs User Services Conference*, California, pp. 160–163
26. Leary-Joyce, J. (2011). *Psychologie úspěchu*. Brno: Computer Press.
27. Marčan, P., & Slovák, K. (2005). *Aj jednoduché zlepšováky môžu priniesť milióny*. News and Media Holding [cit. 2016-02-07]. https://www.etrend.sk/podnikanie/aj-jednoduche-zlepsovaky-mozu-priniest-miliony.html.
28. Morongová, A., & Urbancová, H. (2014). Talent management as a part of employee development – case study. Efficiency and Responsibility in Education 2014 Czech University of Life Sciences Prague, pp. 471–477.
29. Petrakis, P. E., Kostis, P. C., & Valsamis, D. G. (2015). Innovation and competitiveness: Culture as a long-term strategic instrument during the European Great Recession. *Journal of Business Research, 68*(7), 1436–1438.
30. Petrowski, M. J. (2000). Creativity research: implications for teaching, learning and thinking. *Reference Services Review, 28*(4), 304–312.
31. Philipps, A. (2010). Drawing Breath: Creative elements and their exile from higher education. *Arts and Humanites in Higher Education, 9*(1), 42–53.
32. Poláková, I. (2007). Trainee program. *Moderní řízení*, Praha: Ekonomia 2006/07. pp. 69–71
33. Potkány, M., & Benková, E. (2008). *Company processes - the basic condition for outsourcing use*. International Conference on Wood Processing and Furniture Production in South East and Central Europe – Innovation and Competitiveness Belgrade, Serbia, Jun 25–27, 2008, pp. 85–90.
34. Rošková, E. (2012). Ako pracovať s talentovanými zamestnancami v organizácii? *Sociálno-ekonomická revue, 10*(3), 112–117.
35. Sanner, B., & Bunderson, J. S. (2015) When feeling safe isn't enough: Contextualizing models of safety and learning in teams *Organizational Psychology Review 5*(3), A002, pp. 224–243
36. Slávik, Š. (2013). *Strategický manažment*. Bratislava: Sprint 2.
37. Snipes, J. (2005). Identifying and cultivating high-potential employees. *Chief Learning Officer Magazine*. http://network.clomedia.com/
38. Stacho, Z. (2012). Proper setting of performance evaluation decreases overall labour costs. *Visnyk of Volyn Institute for Economicsand Management, 3*(1), 173–183.
39. Stacho, Z., & Stasiak-Betlejewska, R. (2014). Approach of organisations operating in Slovakia to employee's performance evaluation. *Ekonomičnij časopis – XXI, 5–6*(1), 82–85.
40. Stachová, K. (2015). *Návrh súboru funkcií riadenia ľudských zdrojov ako nástroja pre tvorbu organizačnej kultúry v kontexte udržateľného rozvoja* [Habilitačná práca] Univerzita Komenského v Bratislave. Fakulta managementu; Bratislava: UK,
41. Stachová, K., & Stacho, Z. (2015) The extent of education of employees in organisations operating in Slovakia. 12th International Conference on Efficiency and Responsibility in Education (ERiE) Location: Prague, Czech Republic. Jun 04-05, 2015, p. 548–555.
42. Stachová, K., & Stacho, Z. (2014). *The role of innovations in the development of organisation Kvaliteta, rast i razvoj* (pp. 17–27). Zagreb: Hrvatsko društvo menadžera kvalitete.
43. Štatistický úrad Slovenskej republiky. www.statistics.sk.
44. Tidd, J., Bessant, J., & Pavitt, K. (2007). *Řízení inovací*. Brno: ComputerPress.
45. Timmerman, M., & Sabbe, C. (2007). *High potential. The competetive edge within your company* (p. 216 s). Mechelen: Wolters Kluwer ISBN 987-90-465-0889-3.
46. Ulrich, D., Smallwood, N., & Sweetman, K. (2010). *Kód lídrov - Päť pravidiel úspešného riadenia*. Bratislava: Eastone Books.
47. Urbancová, H. (2016). Knowledge transfer in a knowledge-based economy. *E+M. Ekonomie a Management, 19*(2), 73–86.
48. Urbancová, H., Vnoučková, L., & Smolová, H. (2016). *Talant management v organizacích v České republice*. Praha: Vysoká škola ekonomie a manažmentu.
49. Watson, E. (2007). Who or What Creates? A Conceptual Framework for Social Creativity *Human resource. Development Review, 6*(1), 419–441.
50. Yapp, M. (2009). Measuring the ROI of talent management. *Strategic HR Review, 8*(4), 5–10.
51. Zelina, M., & Zelinová, M. (1990). *Rozvoj tvorivosti detí a mládeže*. Bratislava: SPN.

Chapter 17
Environmental Policy as a Competitive Advantage in the Global Environment

Zuzana Tekulova, Zuzana Chodasova, and Marian Kralik

Abstract In a constantly changing market environment, competitiveness is a pillar requirement and fundamental characteristic of a company. In difficult economic conditions, only the strongest enterprises will survive. Thus, companies should be encouraged to control costs in order to maintain a competitive advantage. For manufacturers and service providers, environmental management is one of the most effective tools to achieve this objective because it minimizes the negative impacts of production activity on the environment. Environmentally-oriented business management can overcome discrepancies in the market, society, and environment. Environmental management programs increase the economic efficiency of a business entity to increase its profit potential and improve its environmental profile. Furthermore, a company that intends to operate in a foreign market will be faced with more intense environmental certification requirements regarding the quality of products, manufacturing, and services.

Introduction

A company's profitability is based on its competitiveness in key indicators, such as concept; however, competitiveness remains difficult to measure because of its complexity and specificity. The research on competitiveness has focused primarily on the identification and description of decisive factors. At the microlevel, competitiveness is most often measured by the indicators of productivity, profitability, export performance, or (more precisely) market share. A common feature of these

---

Z. Tekulova (✉) · M. Kralik
Slovak University of Technology, Institute of Manufacturing Systems, Environmental Technology and Quality Management, Bratislava, Slovakia
e-mail: marian.kralik@stuba.sk

Z. Chodasova
Institute of Management, Slovak University of Technology in Bratislava, Bratislava, Slovakia
e-mail: zuzana.chodasova@stuba.sk

indicators is significant inertia; thus, the current value of these indicators largely depends on their past value. From a macroeconomic perspective, statistically significant indicators can be identified by regression analysis, including the essence of competitive advantage and the related technological level of production processes, innovation capability, research and development costs, the credibility of the police, and the willingness to delegate authority, among others.

The business sector, which is the main source of the country's competitiveness, can perceive many of these factors as irrelevant. It follows that competitiveness is actually "competing with other suppliers in a given branch." Kissova defines competitiveness of the company as the ability to produce and sell a specific product on condition of profitability preservation. A competitive company must be prepared, if necessary, to reduce the final cost of the product and offer a higher quality than its competitor [4]. An important factor in enhancing business competitiveness is to ensure the improvement of current production, increasing utility value, functionality and simplifying manageability while reducing production costs. Therefore, in the process of transition to production with higher added value and more complicated processes, amount of investment in research and development grows as another of the main factors. Competitiveness in a constantly changing market environment at the present time is one of the pillar requirements and therefore is regarded as one of the fundamental characteristics of a company. An essential feature of the market economy is the freedom of the customer to decide not only what product to buy, but mainly, from whom. This leads to a potential supplier's competition for a customer. It can be said that the supplier chosen by the customer has higher competitive ability than other competitor participants. Every business should strive to create such competitive advantages that enhance the competitiveness of offered products [2]. Marinom says that the basis for the definition of competitiveness is a high quality item, real interest rates, and marginal costs. The basis, however, remains that insufficient quality of goods reduces the overall quality of a company and thus the competitiveness of company.

If companies want to take a strong position in the market in a dynamic economy, their behavior must be disciplined in providing quality, whether for products or services. First, a competitive marketplace is one in which someone can properly communicate and present his/her products. All businesses are exposed to competition [6].

Indicators of Profitability Versus the Company's Environmental Policy

An indicator of profitability is achieving the same or better results as compared to the competitors or, more precisely, not lagging from the competition. Being profitable means achieving higher revenues, having improved cost management, and holding better position in rankings than potential competitors [5]. Competitive advantage arises from the value that the company is able to create for its customers. This may take the form of lower prices for identical products than its competitors or providing specific advantages, which compensate higher price. Competitive advantage enables enterprises in specified fields of business to maintain average results over a long period. Successful operation of the market environment must be created by the competitive ability of a particular enterprise. The competitiveness of an enterprise is the result of level and operational ability of its business activities. It increases with innovation activity, a productive and knowledgeable workforce, payment ability and readiness, and worker qualification.

Efforts for a high level of competitiveness begin with strategic goals in research, development and marketing, proceed with relationships with suppliers and supply strategy, is reflected in the use of cutting-edge techniques and technology in the transformation process, continues in the innovative production variety and high quality, and culminates in customer satisfaction, good business reputation and positive assessment of the public [2]. The foundation of competitiveness is a competitive advantage. The origins of competitive advantages are:

- Differentiation through the instruments of marketing mix on the basis of new ecological attributes
- Cost advantages, whether through design or technological innovations, which are the result of reduction in the use of materials, energy, pollution or waste associated with a product or service
- Entering new markets, as markets for sustainable products currently present the fastest-growing areas
- Use of market gaps
- Innovation and creativity in product development

According to Karpissová the competitiveness of business indicators can be divided into two groups: external and internal. External indicators are those which the enterprise cannot affect directly and thus the enterprise fails only marginally or indirectly. These indicators include bargaining powers of suppliers and customers, competitive contests, market products, corruption in the environment, interest in employment in the enterprise, and support of local and national authorities [2]. Internal competitiveness indicators, as opposed to external, can be influenced by the firm. These include factors of scientific and technological development, marketing and distribution of production and its management, working resources, as well as financial and budgetary aspects of the business. On the basis of the research studies of the Department of Research of the National Bank of Slovakia, the most important factors in the competitiveness of Slovak enterprises were identified in 2008 and divided into three areas at the internal, sectoral and macro levels. In Table 17.1, the factors of competitiveness are listed according to their importance, from the most important factors to the least significant [5].

Clearly the most important factor, not only within the group of internal, but also within the entire identified set, is professionalism of management of the enterprise. The second highest rated is the quality of management in the company.

In the top five most influential factors was also the efficiency of management. Other important internal components focus on reducing the cost and the scope of using communication technologies. A strong or very strong factor is reducing costs in three quarters of enterprises as per a SWOT analysis of competitiveness of Slovak enterprises.

Table 17.1 Most important factors in the competitiveness of Slovak companies [5]

Internal factors	Branch factors	Macrolevel factors
Professionalism of management Quality of enterprise management Orientation on reducing costs (production costs) Efficiency of enterprise management Scope for utilisation of communications technologies	Customer demands Availability of skilled and experienced managers The essence of competitive advantage An adequately trained workforce The existence of developed customer industries	Membership in the European Union Energy costs The adoption of the euro Exchange rate stability The quality of transport infrastructure

Table 17.2 Comparison of the average indicator industry in manufacture in other machine and manufacturing

Strong aspects	Weak aspects
The use of IS and communications technology Professionalism and effectiveness of leadership Customer orientation Proportion of export and sale Emphasis on modernization of production	Control of international distribution Horizontal and vertical integration Use of marketing Strategy for differentiating themselves from the competition Human resource management and controlling costs
Opportunities	Threats
The difficulty of customers Strength of supplier and customer branches Membership in the EU and EUR payment system Accessibility and quality of telecommunication infrastructure Access to loans and accessibility of office space	Availability of workforce with international experience and adequate levels of education Quality and availability of specialized educational and research services High cost of energy Lack of functionality of legal system, allotting of investment incentives Quality of transport infrastructure

These results more or less agree with the generally observed facts. Slovakia is still the country with the attribute "assembly shop". It also documents the structure of foreign trade. In addition to the high proportion of exports to GDP it is characterized by a high share of imports to GDP, while the bulk of imports are in industrial production. The weakest influence on current competitiveness is associated with property connections with other enterprises in the industry, customers and suppliers. Exports outside the EU, control over international distribution and expenditure in research and development also have relatively low impact. Based on the SWOT analysis of the competitiveness of Slovak enterprises [6], strengths, weaknesses, opportunities and threats for Slovak companies were identified, which are summarized in Table 17.2. Strengths in almost all the most important areas can be identified in Slovakian enterprises. Slovak companies have professional management using effective leadership.

Indicator	Average indicator in the industry of manufacturing of other machines	Average indicator in the industry of manufacturing [6]
Return on equity	6.28	1.69
Return on assets	3.12	0.55
Operating return on sales	3.20	1.57
Share of value added in sales	26.99	20.46
EBITDA share in sales	9.56	4.58

Source: Own calculations [6]

Indicator 'Return on equity' concerns the return on own resources invested into the business in the conversion of net profit. On average the result is 6.28% which is 4.59% higher than the profitability throughout industrial production. In comparison with other possible alternative businesses, e.g. use of funds for investment, term deposits or purchase of securities, we can characterize the result as satisfactory given the current interest rate yield fixed at five years to range from 2.5% to 3%. This implies the result is in favour of the business where the percentage recovery is on average 3% higher than the non-business activities. The average value of return on equity for the business is 0.16%, which is among the most profitable belonging to the rated industries.

The 'Return on assets' indicator evaluates the general assets contribution to the company, regardless of origin or source of coverage. From this perspective it represents recovery of funds invested in the business, as well as evaluating the overall economic activity of the company. It is 2.57% higher than the average for the sector. In this sense we can say that the sector of mechanical engineering 'Manufacturing of other machines' is an attractive business perspective.

The 'Operating return on sales' indicator shows the profitability of the main business of the company; therefore, how much of an effect the company can produce with 1 € sales. The result is an average indicator of industry 'Manufacture of other machinery' operating profit of € 0.32 per euro of revenue, compared with an average indicator in manufacturing which is 50% higher. A 32% profit per one euro in sales compared with the previous analysis supports the argument of attraction to business in that sector.

The indicator 'Share of value added in sales' is the ability of a company to establish a value on purchased inputs; precisely this figure is a significant indicator of GDP in developing and determining the significance of countries in the creation of value. This figure thus indicates how much added effect is created by the euro from sales and is calculated as 26.99%. This figure is among the highest in all production areas and suggests an attractive environment. The aim is to promote the interest of the state GDP growth and thus the sector where the added value is significant. Support of this indicator is in the favour of the industry's future.

Indicator 'Share of EBITDA in sales' is a measure of profit before tax, interest and depreciation cost of sales in euro, and the effectiveness of profit, but also the ability to cover payments of the company and the costs resulting from depreciation. This indicator is 50% higher compared to the average indicator in manufacturing and also concerns payment ability for interest and amortization of fixed costs. This calculated indicator considers the average values of individual sectors as satisfactory and argues in favour of the company in the reporting sector. Resource efficiency is a strategic priority of the Europe 2020 Strategy, a policy response to address a wide spectrum of important economic and environmental concerns [10]. In 2010, a flagship initiative for a resource-efficient Europe was adopted [14] and the resulting 2011 Roadmap to a resource-efficient Europe identified milestones for specific areas and almost 100 individual actions to be taken by the European Commission and Member States [4]. One of the priority objectives of the 7th Environmental Action Programme, which will guide European environmental policy until 2020, is to 'turn the Union into a resource-efficient, green, and competitive low-carbon economy' [9]. However, no targets have yet been adopted for resource use or resource efficiency at a European level. In the recent communication, "Towards a circular economy: a zero-waste programme for Europe" [10], the European Commission proposed the adoption of a resource-productivity target, and it is hoped that this will provide an impetus for countries to also adopt targets. At present, only a few individual countries (e.g. Germany) have concrete and measurable targets accompanied by a deadline [11]. Many European countries have developed their own national programmes or strategies for resource efficiency. These initially tended to address individual topics such as energy consumption or waste recycling. However, they have gradually expanded to cover wasteful production and consumption patterns; the increasing cost of energy and raw materials; the rising global demand for raw materials; concerns over depletion of resources and the security of supply; environmental pollution; and global impacts of greenhouse-gas emissions. A review of national initiatives shows that there is a great variety of regulatory settings and organisational arrangements in place in relation to resource-efficiency policies [15].

The Environmental Priorities and Economic Concerns

National policy priorities and responses are guided by EU regulations but vary widely and are driven by a combination of local economic and geographic conditions, environmental priorities, and economic concerns. The total use of material resources is strongly correlated with the population of a country and the size and structure of its economy. In 2012, the three countries with the largest total DMC were Germany, France, and Poland, while those with the lowest were Malta, Luxembourg and Cyprus.

The economic crisis that started in 2008 has been a major factor in shaping trends in resource use. In individual countries and at the European level, the most significant changes in resource use took place during 2007–2011 (Fig. 17.1). In the EU-27, DMC grew from 15.6 tonnes/capita in 2000, peaked at 16.7 in 2007, before declining by 19% to 13.7 in 2012 (Fig. 17.1). In 2012, the countries with the highest per

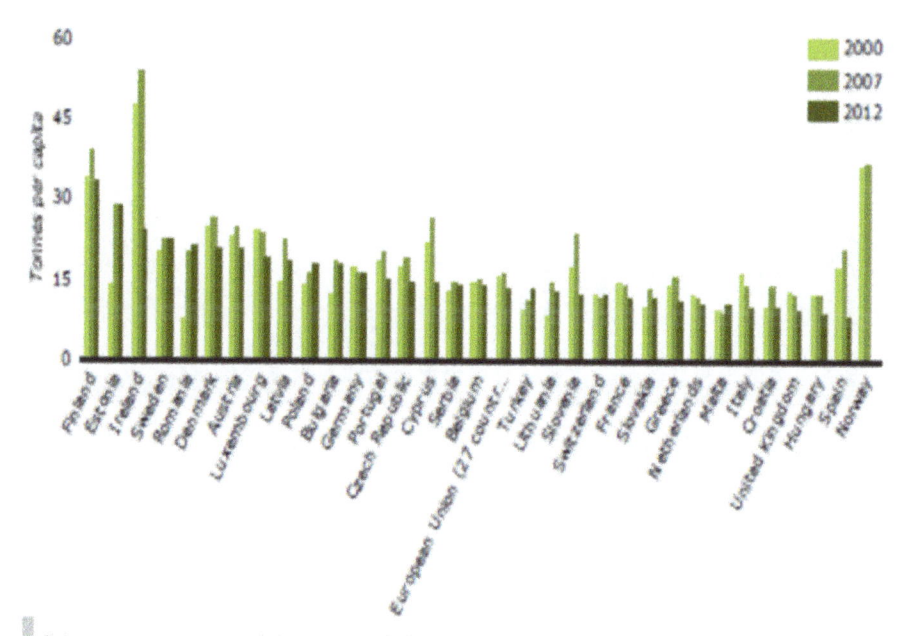

Fig. 17.1 Material resource use (DMC) per capita in 32 European countries (2000, 2007 and 2012)

capita DMC were Finland, Estonia and Ireland, while the lowest were Spain, Hungary and the United Kingdom.

There was a reduction in per capita DMC in the majority of countries over the period 2000 to 2012. The largest declines were recorded in Ireland (50%) (Box 1) and Spain (49%) — mainly caused by a collapse in construction activities — followed by Italy (38%) and Cyprus (32%). Per capita DMC increased in 13 countries, and the largest per-capita increases over this period — primarily due to large-scale infrastructure investments — were recorded in Romania (178%), Estonia (104%), Lithuania (54%), Bulgaria (46%) and Turkey (44%).

Resource productivity, expressed as a ratio of GDP to DMC, links overall resource use to economic activity (Fig. 17.2). Between 2000 and 2012, it increased markedly in the European Union (+29% for the EU-27 and +39% for the EU-15), a sign that European economies are creating more wealth out of the material resources that they use, although it also reflects changes in material use and the structure of economies. There are large differences amongst countries, with little evidence of convergence of resource-productivity rates between 2000 and 2012. Resource productivity is lower in the new member states and in non-EU members. This is partly due to construction sector activity, which dominates material use in many countries [16].

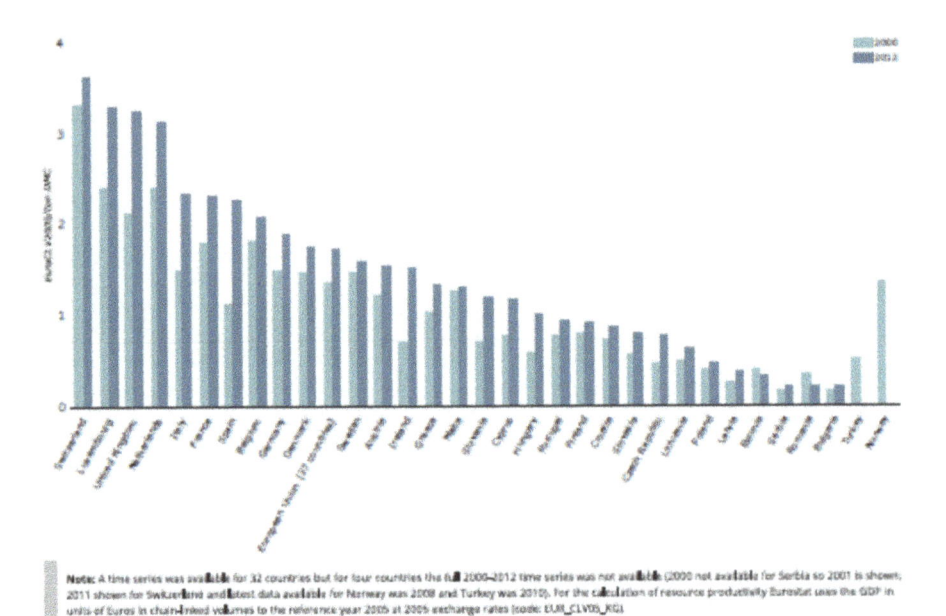

Fig. 17.2 Resource productivity (GDP/DMC) in 32 European countries (2000 and 2012)

The management process is therefore complicated as it is interdisciplinary and dynamic in character. It must account for various aspects of the creation of the area of focus – economic, political, technical, sociological, legal and others, all which carry significant risk. This risk may be suitably moderated by the employment of methods such as controlling, which represents a high-quality advisory capacity for the boards of management. Controlling is a methodological tool of management as its employment may result in quality information necessary for efficient management. High-quality controlling is aimed at the implementation of a cooperative style of management. This doesn't mean the use of only routine management activities, but the use of information obtained to point out the so-called 'weak spots' due to which the company fails to achieve its objectives. Therefore, controllers must be acquainted with the company as a whole and be able to reveal specific features of individual departments. Controlling prepares the information base for planning solutions, implementation and control tasks, which is why controlling is regarded as a subsystem of enterprise management.

Business in the Market Economy

Business in a market economy is under great pressure from competition. It is forced to continually improve internal processes and management systems and respond to new situations and new functions of management methods that allow:

- Assessment of how the company sub-serves projected goals
- Identification of risks, and highlighting the threat and actual deviations from the desired development
- Analyzing and evaluating the effects of business activities and decisions
- Planning and program development of the business in the aggregate and analytic indicators
- Inspiring corporate governance to detect new business activities that have economic effects

These tasks also help management properly utilize their available instruments and other modern methods in the management of business entities. Decisions that are implemented by the management board must be expeditious and efficient. Correct decision-making requires the supply of appropriate, precise and timely information, which may be affected by controlling as a progressive management tool. Flexibility is one of the basic desirable qualities that an organization should have because it enables a prompt reaction to the market demands. New requirements mean new approaches which allow a company to enter the market competition successfully [3].

A permanently changing environment requires the company management to be able to make decisions on short notice and ground them on reliable information. The fast changing conditions of the company environment make the work of company management difficult. Controlling may be of assistance in performing the following tasks in the management of organizations:

- Comparing plans with reality and budget control
- Employment of information from cost/management accountancy
- Financial planning
- Reporting
- Analysis of deviations
- Information interpretation
- Budgeting
- Investment planning
- Balancing accounts
- Advisory services in decision-making processes, etc.

It is important to determine the relationship between the functions of management and controlling in a business entity. Management of the organization makes decisions and assumes responsibility, whereas controlling inspires, evaluates, analyses, controls and gives recommendations. The main instruments of controlling include budgeting, costing, accounting and financial planning, which are important sources of information in control work. They provide important information about the costs and benefits over productivity and form the basic elements of a business. These tools must be interconnected so that their information sheets are productive in controlling activities [8].

Financial management is essential in terms of internal business management. It should be planned and managed first to prevent sub-sequential detection of variations. We therefore define the fundamental objectives that have to support internal management tools:

- Control, i.e. recording variations and their analysis
- Providing information for decision making

Executors of individual instruments are those whose position is identical with the inclusion of these instruments. The role of controller is to attune and coordinate their work, give them motivation and provide guidance to use the tools of economic management that created such links, which would allow capturing the real financial and capital flows within management organizations. These instruments provide important information on the costs and benefits over their real flow, and are therefore essential elements of an in-house information system [3].

The environment in which companies conduct business undergoes constant changes, triggering necessary changes inside a business unit, its management, and evaluation of achieved results. The task of controlling is 'to help' the company management to reach and maintain enterprise processes within set limits in order to provide for performance of the defined goals. Thus, controlling as a management tool should also help in complying with decisions of the management. The necessity of continuous maintenance of balance between the internal and external world of the company, especially with regard to framework conditions such as 'complexity, dynamism, and differentiation', primarily results in the following requirements on the current company management:

- The speed of creation of the unified enterprise and in particular its promotion is the critical factor of economic success. This requires quick decision making and also fast resoluteness of management.
- The ability of anticipation and adaptation has an initiate character, which should guarantee timely adaptation to changes of surroundings before the occurrence of imbalances endangering the company existence. The company must be sensitive to weak signals from the enterprise surroundings. This can be reached in particular by higher awareness of individual problems and with higher vigilance of employees. It is also suitable to implement a certain system of early warning.
- Reactive capability and structural flexibility for adaptation to unforeseeable events.
- Reduction and control over the company in complexity, where the company's objective must be acquiring knowledge on how to deal with diversity of relations by targeted creation or influence on factors of the production system and the entire company surroundings, through inclusion of self-regulating mechanisms without managerial interventions.

A vital company manages to create a balance between development transformation and stability, reaching it by means of concurrent maintenance, creation and elimination. Maintenance of the basic level is aimed at healthy basic development of the company. Its task is the continuous improvement and rationalisation of products and processes through that what was successful in the past, should be maintained, and strengthened in the future. The opposite is the release of committed resources and reduction in size of the company in order to remove parts not necessary to survival, or parts endangering the company's survival. In this manner the company heads for new potential opportunities for its future [9]. Companies capable of adapting to progress manage the art of concurrent maintenance, creation and elimination by means of differentiated management cultures, structures, procedures, systems, planning and management tools.

Conclusion

How is it then that even with efficient engineering production, output may be contingent on analyzed indicators which characterize the attractiveness of the environment? Defining the attractiveness of the environment must come from other factors, mainly from the growth potential of the sector, industry prospects, stability and variability of competitive forces, as well as uncertainty or risk of future development of the sector. The given data contain strong explanatory power of earnings and profitability which, when considering entering into a business, play an important role. When assessing the attractiveness of the environment with routines and methods, however, the emphasis should be placed on the use of modern approaches to the management of the company across all management structures as a condition for a well-functioning company and assets that support the future of continuous ongoing development and improvement of all management and executive activities of the company. Among the most effective methods (although in the current business practice in Slovakia a few are implemented) seems to be the method of benchmarking, which provides models towards excellence. Its focus is to set goals so that the organization can implement a realistic picture of improvement and understand the changes that are necessary for improving not only internal evaluations, but also in the context of societal conditions in which it carries out business. Many factors determine resource use and productivity, including climate, population density, infrastructure needs, domestic availability of raw materials versus reliance on imports, prevailing fuel in the power generation sector, the rate of economic growth, technological development, and the structure of the economy [12].

There is also the long-term tendency for absolute amounts of resources used to increase in tandem with economic growth despite technological progress (the 'rebound' effect). The long-term objective of current European environmental policies is that the overall environmental impact of all major sectors of the economy should be significantly reduced, and resource efficiency increased [13]. This policy goal — a double decoupling of resource use from both economic growth and environmental impacts — provides a framework and direction for national policies [11]. The large differences in resource-efficiency performance amongst countries — and the fact that the same half dozen countries have remained at the bottom of resource efficiency rankings since 2000 — indicates opportunities for improvements and actions. Efforts to support the exchange of good practice in policy design could be one tool to facilitate faster uptake of the most effective solutions.

In addition, the use of indicators such as RMC will give a broader perspective on resource productivity, incorporating upstream material use. However, the link to the overall environmental impact of resource use is still not easily captured within available indicators.

In the current global environment, it is increasingly difficult to ensure long-term success of a company. The globalization of the world economy is causing significant changes in the conditions of businesses not only in Slovakia but also abroad. Globalization is still a significant trade liberalization, with integration of subjects into larger units in a significant international environment, internal culture or agreement establishing strong economic areas. Today, enterprises cannot trust their momentary performance, but must seek and find ways to permanently increase it. Performance measurement is the process of supporting the development of the company using an evaluation of the performance indicators systematically trying to adapt to change in order to maintain long-term competitiveness.

In today's dynamic business world companies are in a very difficult position. Market calls for maximum performance, optimal adaptation, as well as prospective opportunities. Company performance is becoming a very hot topic today. If companies want to achieve top positions and maintain a competitive advantage, they need to set such control systems that can ensure controlled use of their resources towards achieving their vision.

Acknowledgement This contribution is the result of the project *VEGA 1/0652/16*, "The influence of territorial location and sectoral focus on the performance of business subjects and their competitiveness in the global market".

References

1. Ďurišová M., & Kucharčíková A. (2014, October 23–24). The quantitative expression of factors which affect the cost of transport enterprise. In: Transport means 2014: Proceedings of the 18th international conference. Kaunas University of Technology, Lithuania, Kaunas, pp 190–193. ISSN 1822-296X.
2. Nenadál, J., Noskievičová, D., Petříková, R., Plura, J., & Tošenovský, J. (2002). Moderní systémy řížení jakosti. Praha: Management Press. 282 p. ISBN 80-7261-071-6.
3. Chodasová, Z.,Tekulová, Z., & Králik, M. (2016, September 8–9). Importance of management accounting knowledge in decision making. Medzinárodná vedecká konferencia: Znalosti pro tržní praxi 2016, Olomouc, pp 162–167. ISBN 978-80-87533-14-7.
4. Mateides, A., & Kol, A. (2006). Manažérstvo kvality. Bratislava, Ing. Mračko, 2006, 751 p. ISBN 80-8057-656-4.
5. Sedláčková, H. (2000). *Strategická analýza* (1st ed.). Praha: C. H. Beck.
6. Statistical Office of the Slovak Republic. (2013). Regional statistical yearbook of Slovakia 2013, 522 p. ISBN 978-80-8121-301-4.
7. Scarborough, H., Swan, J., & Preston, J. (1999). *Knowledge management: A literature review*. London: Institute of Personnel and Development.
8. Tekulová, Z., Kralik, M., & Chodasová, Z. (2016). Analysis of productivity in enterprises automotive production. In *Production management and engineering sciences* (pp. 545–549). London: Taylor & Francis Group. ISBN: 978-1-138-02856-2.

9. Tokarčíková, E., Poniščiaková, O., & Litvaj, I. (2014, October 23–24). Key performance indicators and their exploitation in decision-making process. Transport means – Proceedings of the 18th international conference. Kaunas Univiversity of Technology, Kaunas, Lithuania, pp 372–375. ISSN: 1822-296X.
10. Slovak Credit Bureau. (2012). *Stredné hodnoty finančných ukazovateľov ekonomických činnosti v Slovenskej republike za rok 2011*. Bratislava.: ISBN 978-80-971109-0-1.
11. EC. (2010, March 3). European Commission, Europe 2020: A strategy for smart, sustainable and inclusive growth, communication from the commission, COM (2010) 2020, Brussels.
12. EC. (2011, January 26). European Commission. A resource-efficient Europe– Flagship initiative under the Europe 2020 Strategy, COM(2011) 21, Brussels.
13. EC. (2013, December 28). Decision no 1386/2013/EU of the European Parliament and of the Council of 20 November 2013 on a General Union Environment Action Programme to 2020 'Living well, within the limits of our planet'. Official Journal of the European Union, L354/171.
14. EC. (2014). Communication from the Commission to the European Parliament, the Council, the European Economic and Social Committee and the Committee of the Regions "Towards a circular economy: A zero waste programme for Europe", COM/2014/0398 final.
15. EEA. (2011). *Resource efficiency in Europe- policies and approaches in 31 EEA member and cooperating countries* (EEA report 5/2011). Copenhagen: European Environment Agency.
16. EEA. (2012). *Material resources and waste: 2012 update of SOER2010 thematic assessment*. Copenhagen: European Environment Agency.

Chapter 18
Sustainable Organization of Cooperation Activities in a Company: Slovak Republic Research Perspective

Jakub Soviar, Viliam Lendel, Josef Vodák, and Jana Kundríková

Abstract The goal of the article is to use detailed literature analysis and findings of an empirical research and to propose efficient organization of cooperation activities in a company. The proposed efficient organization enabled authors to describe the process of creating a cooperating company and the individual recommended types of organizational structures. The article thus provides a tool for company managers for managing their cooperation projects and activities. The use of this tool is meant to help minimize the occurrence of conflict situations and to support smooth progress of cooperation activities from the organizational perspective. This tool means also a strong sustainable aspect concerning the cooperation's overall stability.

Introduction

The topic of managing cooperation activities is currently highly up-to-date. In present, cooperation as such represents for a company an important tool for increasing its competitiveness. Companies no longer develop their cooperation activities based on "impressions" or "gut feelings" but rather based on knowledge derived from the opinions of their customers, employees, and partners. They collect the necessary information, support creation of knowledge, explore market opportunities, and make decisions about the need to cooperate. Companies aim to fully utilize their cooperation potential. In order to be successful, it is needed to effectively manage these activities and to dynamically react to the ongoing market development. Here we can use the proposed matrix of cooperation organizational structures, which will ensure efficient organization of the emerging market opportunities using the created cooperation.

J. Soviar (✉) · V. Lendel · J. Vodák · J. Kundríková
University of Žilina, Faculty of Management Science and Informatics,
Žilina, Slovak Republic
e-mail: jakub.soviar@fri.uniza.sk; viliam.lendel@fri.uniza.sk;
josef.vodak@fri.uniza.sk; jana.kundrikova@fri.uniza.sk

© Springer International Publishing AG, part of Springer Nature 2019
D. Cagáňová et al. (eds.), *Smart Technology Trends in Industrial and Business Management*, EAI/Springer Innovations in Communication and Computing,
https://doi.org/10.1007/978-3-319-76998-1_18

The article aims to offer, in an understandable form, a coherent perspective on the management of cooperation activities in a company as well as a methodology of organizing its cooperation projects. Both of these would be based on a comprehensive mapping of theoretical and practical findings in the area of cooperation management as well as the performed research about its utilization in Slovak enterprises.

The main goal of the article is to produce new insights in the area of cooperation management, with particular focus on its definition within management and its potential use for managing cooperation projects and activities of a company. Identification of suitable organizational structures may significantly contribute to minimizing the occurrence of conflict situations in the process of managing cooperation projects and company activities.

We use several *methods* to perform our research:

- Method of document analysis: for analyzing current as well as historical data about the topic.
- Questionnaire survey method and a method of semi-structured interview: for gathering data in an empirical research.
- Method of observation: used during visits of selected companies.
- Method of quantitative evaluation: for processing the data – statistical methods and tools were applied.
- Method of comparison: for comparing data gathered by empirical research and data from the analysis of secondary information sources.

The performed research focused on medium and large enterprises active in the Slovak Republic. The actual respondents were company managers on the mid to top management level within the managerial hierarchy of companies. In total, 273 respondents took part in the research focused on diagnostics of the level of use of cooperation management. Research included companies active in multiple sectors of the Slovak economy. Included companies were categorized by the Statistical Office of the Slovak Republic as medium or large enterprises. The actual respondents were company managers on the mid to top management level within the managerial hierarchy of companies. The size of the sample was 345 respondents, with the required 95% interval of reliability and the maximum allowable error of 5%. Since 273 respondents actually took part in the research, the maximum allowable error reached 5.72%. Data was gathered exclusively via personal interview. Partially results of this research were already published in several papers, e.g., Lendel, 2015 [20]; Vodak, Soviar, Lendel, 2013 [45]; and Vodak, Soviar, Lendel, 2015 [46]. We consider this paper as a continuation of the abovementioned works.

Current State of Dealing with the Issue

Cooperation is one of the key tools for achieving strategic competitiveness of companies. It is a complex system whose elements are stakeholders striving to achieve certain benefits which would be individually hardly achieved [16, 36]. The benefits of cooperation are less likely to emerge in a short time; they show up after a certain lapse of time, so from a long-term perspective, it is more profitable to choose cooperation rather than a selfish strategy [26].

Cooperation needs to be properly managed in order to be successful. However, there is still an ongoing scientific discussion regarding the term cooperation management, its definition, and the scope of use. Several definitions of cooperation management can be found in the scientific literature, such as the following. It is a philosophy of management that can be applied irrespective of ownership structure [4]. Cooperation management offers effective management of cooperative processes between independent organizations with the aim of continuously improving interorganizational activities and providing flexibility for companies which are facing the challenges of today, so that opportunities for cooperative development don't remain unused [3]. However, these definitions of cooperation management typically address only a subset of the whole task of cooperation management. We built this chapter based on our previous published works, mainly [20, 45, 46]. High variability in interpretation of the term *cooperation management* can be supported by the following examples:

- It is a way of managing and developing collaboration in a competitive environment [19].
- It represents a term for integrated management of company networks [31].
- It is a cooperative decision-making process within heterogeneous preferences [38, 44].
- It provides conditions for creating a system of cooperation based on the effective use of resources and technologies [49].

The most important *characteristics* of cooperation management are following [44]:

- It is a complex decision-making process, and the decisions are made on all managerial levels.
- Primary goal of cooperation management is to satisfy the needs of the members of cooperation.
- All activities need to occur according to the agreed principles of management and cooperation.
- Suitable balance needs to be established between the efforts for commercial success and maintaining the goals of the cooperating parties.
- Management focused on reaching a goal via the effective use of resources.

Based on the performed detailed analysis, as described above, we can define a more precise definition of this term [21, 37, 47, 48]: *Cooperation management is an effective and efficient management of relationships in a cooperation between separate and relatively independent organizations or individuals, with the goal of improving their competitiveness.*

Building of relationships is based on cooperation and having the following *attributes* [2, 7, 9, 13, 14, 21, 22, 31, 42]: cooperation and partnership; seriousness; non-disturbance of mutual competitive relationships; focusing particularly on long-term time horizons, long-term cooperation; continuous learning and knowledge transfer; and effective and efficient combination of resources, the ability to integrate external resources through networking.

To build a cooperation management in a company is a real challenge that company managers need to deal with. Among a number of *factors* influencing the process of establishing cooperation management in a company belong the following:

- *The role of innovation:* effective cooperation management processes have significant impact considering creating of successful innovations (product, market, services, processes, etc.) [11, 18, 24, 32, 34].
- *Mutual trust – trust between the partner organizations:* trust is mainly based on previous positive experience, and it is also strongly connected with reference power of an organization [10, 47, 48].
- *Information background (as a cooperation processes support) and knowledge creation:* effective management of cooperation processes is strongly dependent on ensuring quality information within the company and to enable its sharing for the decision-making needs of the managers [1, 6, 23].
- *Impact of wider, regional environment:* a sum of companies in a region, which could be used as an integral part of cooperation management activities; already existing cooperation networks (alliances, clusters); state of the regional economy, etc. [17, 25, 27, 30, 40].
- *Organizational factors:* all necessary changes in the organizational structure to support cooperation; new dynamic organizational structures (e.g., clusters) [15, 27, 28, 35, 39].
- *Geographical proximity of the partners:* it is not a necessity.
- *Sharing a common purpose*, values, and objectives; knowledge in the area; and reaching a consensus [33].

Assuming that the abovementioned identified factors are taken care of within a company, then its cooperation management will bring expected results, such as better product quality, shorter delivery times, and higher customer satisfaction [41]. This will contribute to the overall competitiveness of the company.

Situation in Slovak Enterprises: Results of Empirical Research

We conducted our research between September 2012 and February 2013 (further information about this research could be found in our previous published papers [20, 45, 46]). Our main goal was to gather and interpret information about the level of using cooperations in the environment of Slovak enterprises. In order to reach that goal, our research had to identify the key aspects of efficient management and functioning of cooperations, related issues, degree of satisfaction of companies within cooperation, and the opportunities for improvement of already functioning cooperations. All gathered data provided complete the picture about readiness of Slovak enterprises to use (implement) cooperation management. In total, 273 managers of small, medium, and large enterprises took part in the research, from companies active in Slovak republic. Data from the respondents was gathered via personal interviews. Table 18.1 at the end of the chapter provides an overview of the main results for the individual researched areas.

Main goal of the research was to identify the:

(a) Key aspects of efficient management and functioning of cooperation
(b) Degree of satisfaction of companies within cooperation
(c) Opportunities for improvement of already functioning cooperation

We identified areas in which cooperation is most developed (Fig. 18.1). Not surprisingly it concerns mainly supply relationships. Relationships with suppliers are critical for every business. They are often in a form of long-term partnerships with a positive experience. Respondents expressed their opinion of relationships with suppliers as an effort to keep them stable. They are trying to create functional and mutually beneficial partnerships.

Regarding other categories, technical cooperation is important. Basically, it represents external experts – partners, who supply technology, software, or ICT solutions. Respondents consider it extremely important. The category included in the term education is interesting. It consists of cooperation in order to perform diverse trainings and achieve required certificates. Organizations often need them by law (e.g., safety, hygiene, etc.); on the other hand, they invest in employee training through external experts.

Respondents were also asked to choose their most preferred areas of interest for a more intense cooperation in the near future (Fig. 18.2). Responses shown in the following chart demonstrate that there is still an attractive capacity for cooperation in the monitored system. Respondents have shown interest in new cooperative links, even in areas where they consider them to be in a sufficient condition. This is mainly due to the natural interest in new, more attractive opportunities, as well as the interest in working with more serious or more significant partners.

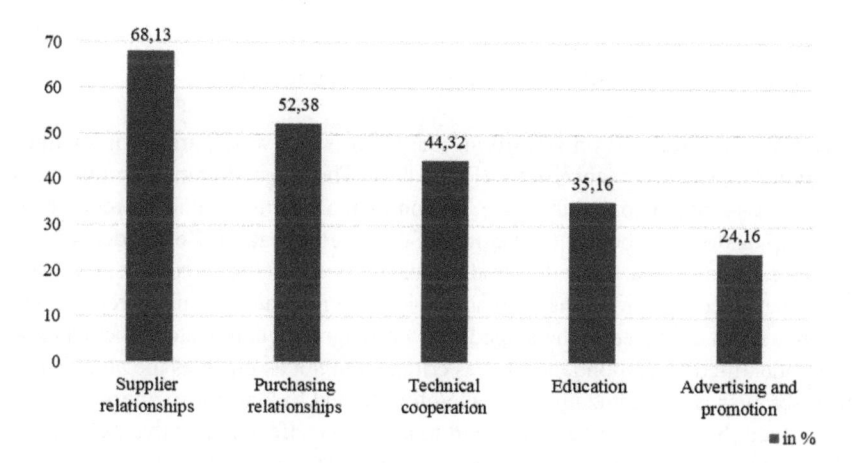

Fig. 18.1 Areas of the most developed cooperation

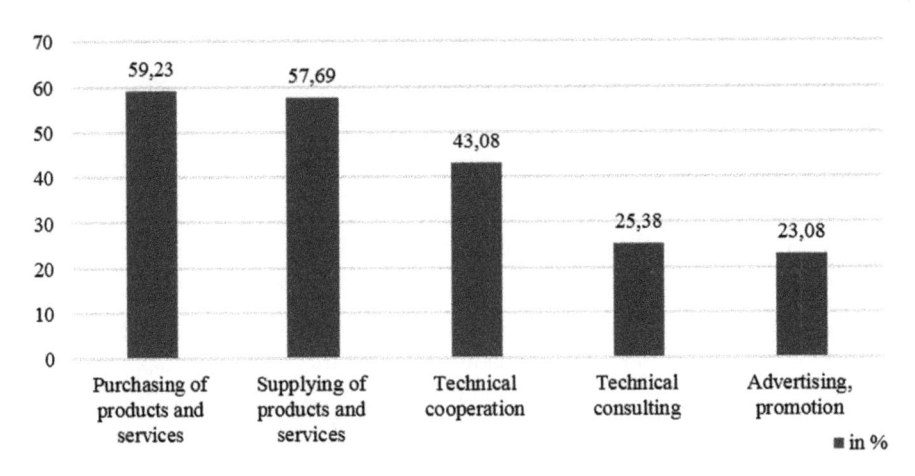

Fig. 18.2 The most preferred areas of interest for a more intense cooperation in the near future

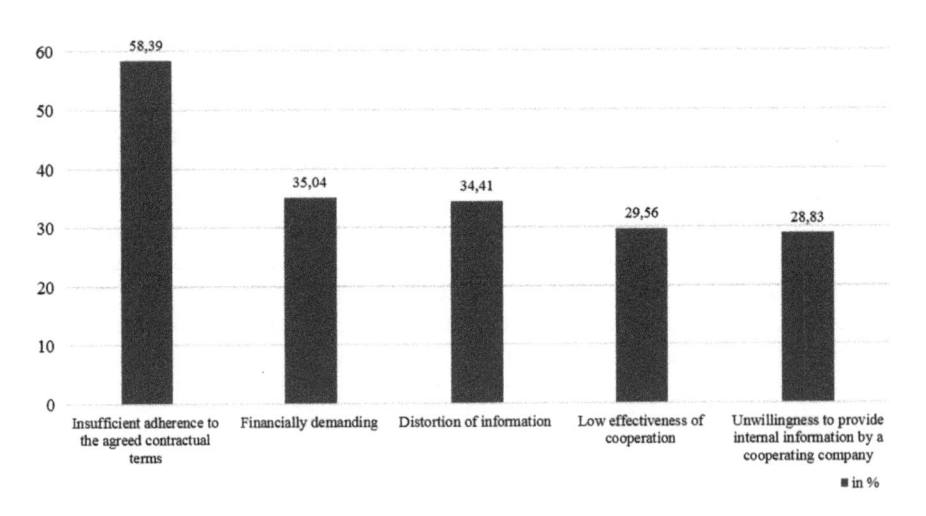

Fig. 18.3 Main issues arising when cooperating with other organizations

Next question focused on identifying the main issues which are arising within cooperation between organizations (Fig. 18.3). The most dominant problem is seriousness – insufficient adherence of contractual terms, which were agreed, written up, and signed before the beginning of the cooperation. Respondents were very critical in this area. They declared this category as a systemic problem. In particular, they specified problems such as invoices are not paid in time, agreed goods or service is not delivered at the agreed time or in agreed quality, etc. Other categories are identified by respondents also as critical, but not as much as the abovementioned seriousness. Organizations are therefore forced to look for partners who are serious enough, which can be considered as a positive effect. A positive experience is the most highlighted and at the same time desirable criterion in this area.

Major benefits resulting from cooperation and partnership were identified by respondents and are referring to previous findings (Fig. 18.4). This is mainly about good mutual relations. There is a direct relation to the main problem described above. Good relationships might seem to be a too general term at first sight, but they are not. They can be defined as progressively forming relationships (medium-term to long-term horizon) and are based in particular on positive experience, willingness to overcome and solve problems (as some problems occur almost always), the validity of mutual agreements, and their adherence. Therefore, it can be briefly said that it is primarily about seriousness.

We consider as a very positive finding that respondents identified better mutual communication as the most important factor in improving their cooperative relationships (Fig. 18.5). It suggests they think critically and rationally. They know that with a better, which means more open, concrete, and clear – accurate and serious – communication, they will achieve common far more effective results.

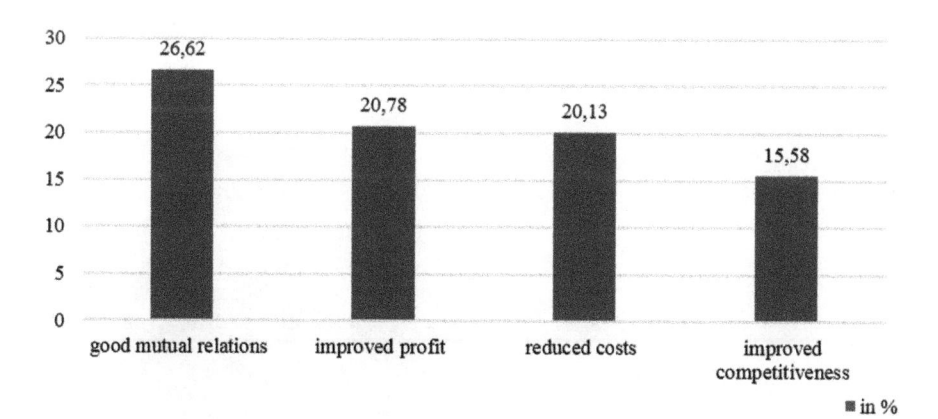

Fig. 18.4 Main benefits resulting from cooperation

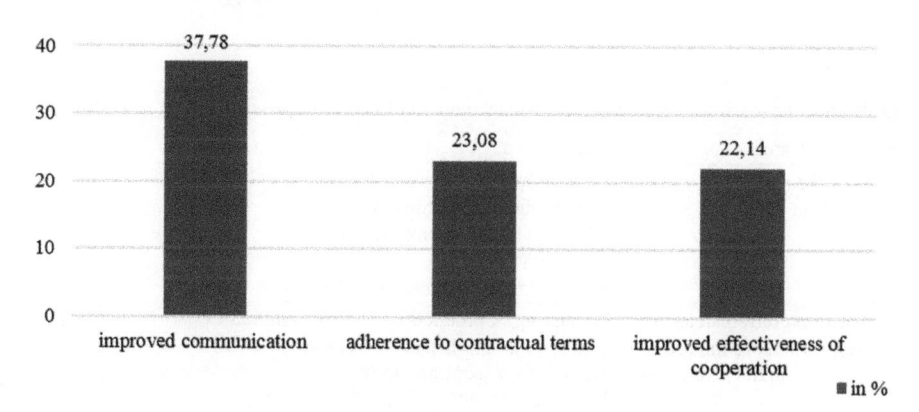

Fig. 18.5 Areas for improvement in cooperation

The second selected category has already been captured and described in the previous questions. Mutual seriousness appears also here to be extremely critical.

Improving effectiveness of cooperation is mainly about its organizational and technical assurance, e.g., better trained staff, secure and effective software support solutions, etc.

The following chart illustrates the interest of respondents to build up a close partnership in 1-year horizon (Fig. 18.6). Up to 48% of respondents expressed themselves that they were interested. We consider this to be a fact that confirms the existence of an even more attractive capacity for building up cooperative management structures.

Table 18.1 provides an overview of the main results for the individual researched areas, which consist of areas described on Figs. 18.1, 18.2, 18.3, 18.4, and 18.5.

Let's recall some important findings. It could be considered positive that almost half of the respondents (47.62%) plans in the near future (within 1 year) to establish a more intense cooperation with a company or an organization. When selecting partners for cooperation, companies make decisions based on the following factors: costs (8.12), insolvency (8.03), market position (7.25), profitability (7.18), and certificates (7.05). In contrast, the lowest importance was assigned to the factors such as the legal form (4.16) and company seat.

Fig. 18.6 Interest for building up close cooperation (partnership) within 1 year

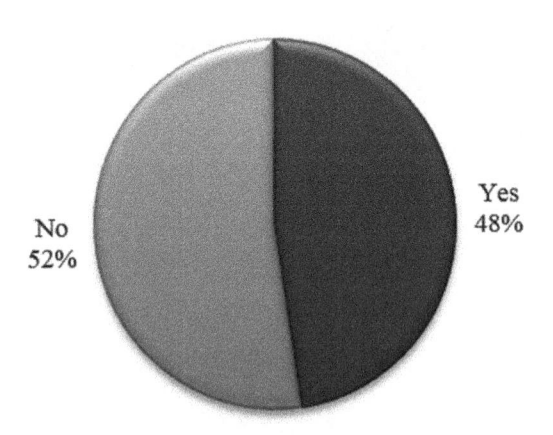

Table 18.1 Level of use of cooperation management in Slovak enterprises

Researched area	Main results
Area of the most developed cooperation	Supplier relationships (68.13%) Purchasing relationships (52.38%) Technical cooperation (44.32%) Education (35.16%) Advertising and promotion (24.18%)
The most preferred areas of interest for a more intense cooperation in the near future	Purchasing of products and services (59.23%) Supplying of products and services (57.69%) Technical cooperation (43.08%) Technical consulting (25.38%) Advertising, promotion (23.08%)

(continued)

Table 18.1 (continued)

Researched area	Main results
Main issues arising when cooperating with other organizations	Insufficient adherence to the agreed contractual terms (58.39%) Financially demanding (35.04%) Distortion of information (34.41%) Low effectiveness of cooperation (29.56%) Unwillingness to provide internal information by a cooperating company, i.e., concerns about providing internal information to a company (28.83%)
Main benefits resulting from cooperation	Good mutual relations (26.62%) Improved profit (20.78%) Reduced costs (20.13%) Improved competitiveness (15.58%)
Areas for improvement in cooperation	Improved communication (31.78%) Adherence to contractual terms (23.08%) Improved effectiveness of cooperation (22.14%)

Proposal for Effective Organization of Cooperation Activities of a Company

During the process of managing cooperation processes in a company, company strategy is revised and modified so that it reflects the plans of top management regarding the management of cooperation activities. However, such a change can end up influencing the roles of multiple employees. Depending on the character and number of realized cooperation projects, it is necessary to revise the currently used company organizational structure and to adapt it to the current situation.

Given a great variability of cooperation projects, it is possible to use multiple types of organizational structures for their organization. The general rule is that the organizational structure adapts to the cooperation project (content, complexity, extent, time needs) and not vice versa.

Organization remains of key importance in the process of managing cooperation activities, especially in today's turbulent environment. Cooperation management aims to ensure competitiveness of the company in such environment. However, for this to happen, it is needed that the company is capable to dynamically react to the arising changes.

Here is a room for using *dynamic cooperation organizational structures* that offer immediate reaction and consequent change in configuration of employees and processes, as necessary. Therefore, such organizational structures enable cooperation with partners in the area of research and development, marketing, etc., as well as work on multiple projects at the same time. Dynamic cooperation organizational structures (champions, purpose teams, project teams, project centers, etc.) are characterized by the following *properties*:

- *Ability to rapidly react to changes*: Changes (opportunities or threats) occur on the market in the organization's external environment. Cooperation organizational structures allow dynamic responses. This means purposefully creating a structure to meet current needs – respond to changes. There can be observed three options for responding to changes in practice: (a) the organizational structure adapts (a new partner joins the existing structure, the personal constitution changes, etc.), (b) a new organizational structure is created (new partners suitable for taking advantage of the given opportunity, e.g., constructional consortia for big public contracts such as highways or research teams requiring multidisciplinary approach, etc.), (c) the existing structure disintegrates – terminates (the objectives are achieved and there is no need to continue, market conditions have changed, the structure is inconvenient, etc.).

- *Decentralized management*: Management of a dynamic cooperation organizational structure is resulting from a partnership agreement. Each of the partners also contributes to the management of the structure. It is necessary for each of them to participate to a certain extent. Of course, the main control unit exists, but it is rather a coordinator of activities. Project management often takes place by individual cooperating partners. In principle, the decentralization means that to the management of the cooperation structure are involved all interested parties to a certain extent.
- *Use of the creative approach*: Research is often present in cooperation organizational structures. It requires a creative approach and solutions. An agreement of several partners wanting to achieve common objectives also often requires creative solutions in organizing activities as well as in their performance. This creativity is crucial to formation of innovative solutions.
- *Flexibility in content and activities of the groups and individuals*: Ability to respond quickly to changes (described above) must be supported by the flexibility of an existing organizational structure. This means that people in the structure are mostly involved in multiple projects – multiple activities. It presents an opportunity to better use their experience. Activities within cooperation structures are also very specific and unique, which means that they do not have to be repeated in each case. Each project can be, for example, in the field of engineering research, but it will always concern something else – another part and another unique research in the end.
- *Acceptance of higher degree of uncertainty and risk in management*: Because of cooperation structures being dynamic, they are also mainly project type. Uncertainty is typical here. It concerns particularly subvention project types – projects funded from EU financial resources. In this case frequent correction of budgets (money must be refunded) is based on the results of an audit. Uncertainty also relates to partnership itself: individual partners may choose otherwise, they may stop behaving seriously, etc. That is why previous experience and references are so important.
- *Direct evaluation and testing of new ideas*: In majority of the cases are cooperation organizational structures working together to achieve specific objectives (e.g., product development, joint research, subcontracting alliance in construction field, etc.). Although the environments are specific, results can be relatively accurately evaluated in each of them.

- *Focus on results*: The cooperation organizational structure is objective-oriented. When the objective is attained, it is necessary to reevaluate its existence. Its termination does not have to be considered a failure. On the contrary, it may mean that the structure has fulfilled its objective and it is no longer necessary to cooperate. It is important to remember that the whole organizational structure is adapted according to main set objectives and serves mainly for their achieving. It also does not resist against adaptation when changing conditions (external or internal).
- *Adequate number of management levels*: In order for the structure to be effective (in terms of helping achieving common objectives), it needs to be effectively managed. Management levels adapt to the main cooperation objectives. They serve as a means for achieving an effective functioning of the entire structure. Representatives of concerned parties should be adequately involved in the management. They may delegate performance of management to set joint organizational units (including external management).
- *Administratively undemanding methods of management*: This is related to the effectiveness mentioned in the previous case. If the structure should be dynamic and effectively reflecting on the cooperative goals of the partners, it is also logically necessary to make it administratively as undemanding as possible. Of course, for example, EU subvention projects are often administratively demanding. It depends on the decision of the partners whether it is acceptable for them (e.g., as a risk). In practice, this is solved by hiring people or companies that have sufficient experience with this type of project – administration. Therefore it is a cooperative and dynamic solution.
- *High added value*: By cooperation of several independent entities, often direct competitors, the potential for achieving objectives increases. Added value is above all represented by the ability to achieve objectives that could individual organizations achieve alone only very hardly, if at all.
- *Informal team work*: In addition to the formal structure and rules, there is, of course, also an informal one. In case of a cooperation organizational structure, informal relationships between cooperating units are important – their mutual getting to know each other. It affects the effectiveness of cooperation, and it is also important for future experience. Simply said, whether we want and will work together with a given partner in the future or whether the experience is positive or negative is affected not only by qualification and professional performance. It is a complex question (behavior by solving problems, accommodation in communication, etc.).
- *Lower number of organizational elements and connections*: Each organization participating in a particular cooperation has its own organizational structure and only puts into cooperation what is necessary for it: people, devices, parts of departments, and so on. The cooperation organizational structure should be dynamically formed so that it contains only what is necessary – essential elements.
- *Lower requirements on the management system*: As in the previous case, the entire management system is target-oriented. Therefore, it focuses only on managing the given cooperation. According to the difficulty and busyness, the man-

Fig. 18.7 Matrix of cooperation organizational structures

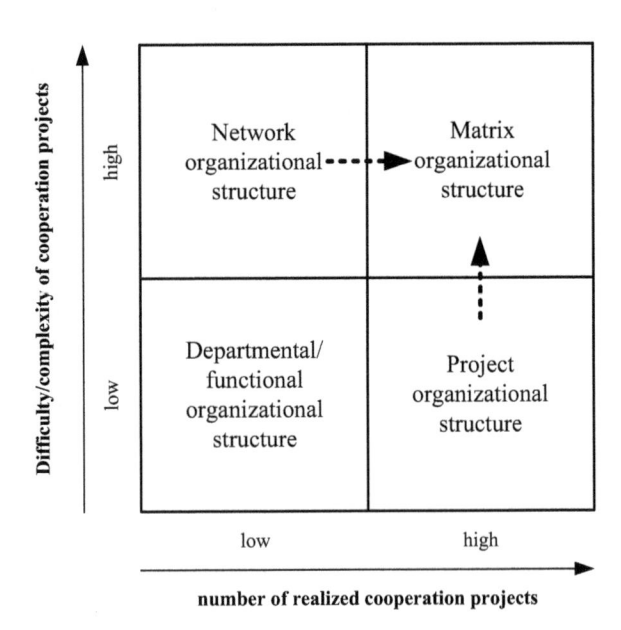

	low	high
high	Network organizational structure ----▶	Matrix organizational structure ▲
low	Departmental/ functional organizational structure	Project organizational structure

Difficulty/complexity of cooperation projects (vertical axis)

number of realized cooperation projects (horizontal axis)

agement can at the same time devote itself to managing other cooperation or executing other tasks.

For the purpose of fulfilling the organizational needs related to managing cooperation activities, *a matrix of cooperation organizational structures* was created (Fig. 18.7). Cooperation organizational structures are located in the matrix based on two main parameters – number of realized cooperation projects and how demanding/complex are the cooperation projects.

In the first quadrant, we can find the *functional organizational structure*. This is a classic organizational structure, suitable for situations with low number of realized cooperation projects with relatively low complexity. Company employees are managed by their superior within a department to which they are assigned. Their work position does not change, i.e., they stay on their linear positions. Communication in this organizational structure takes a form of coordination work meetings of cooperation teams. The role of line managers is to ensure the process of planning, realization, and control of cooperation activities.

In the second quadrant, we can find the *project organizational structure*. It is used mainly in situations when company realizes multiple projects with relatively low complexity. If necessary and if existentially important for the company, it is possible to use this organizational structure to deal with demanding and complex cooperation projects (represented by the arrow in Fig. 18.7). In this organizational structure, members of project teams are freed from their permanent work position.

In the third quadrant, we can find *network organizational structure*. It enables to deal with complex and demanding cooperation projects and if necessary also multiple projects at the same time (represented by the arrow in Fig. 18.7). This organizational structure is characterized by high degree of flexibility and dynamics. Cooperation projects are managed in required time and quality, while a relationship is being established with the main organization.

In the last quadrant, we can find *matrix organizational structure*. Due to its structure, it enables dealing with multiple cooperation projects with high degree of complexity. It also enables efficient use of company resources. Employees are managed by a project leader, while they also remain on their functional positions.

Discussion

Organization, whether commercial or not, is a social group. Its goal is to fulfill the set goals. Cooperating organizations have certain categories in common. Most often these are common goals that can be reached more effectively via cooperation [45]. Organizations assume culture of the society from which they stem, and at the same time, they create their own (organizational or company culture). Success represents an important aspect – this is represented in a way by company survival, market success, profit, etc. If organization is not in the long-term successful in fulfilling its goals or it is not competitive, one of the solutions is to connect with other organization or organizations.

Figure 18.8 represents basic steps necessary for creating cooperating organization. Organizations exist in a state of mutual competition. In case a certain problem turns out to be significant enough, it may represent a potential stimulus for establishing mutual cooperation. Mutual discussion and agreement leads to cooperation. Organizations exist in a dynamic environment that creates further changes that in

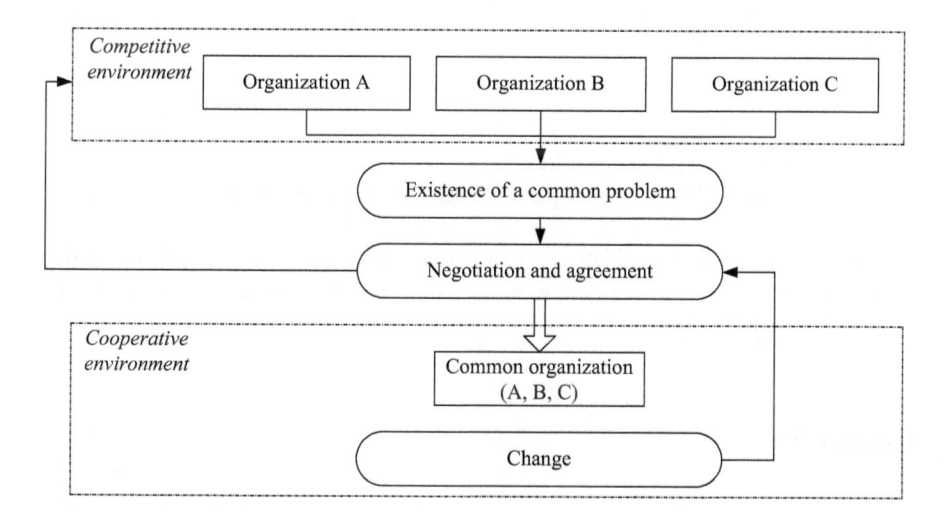

Fig. 18.8 Process of creating cooperation organization

turn create the need for another discussion (planning and decision-making). This may result in a decision to continue the cooperation, to modify it, or to terminate it and to return to mutual competition.

The described aspects form *dynamic cooperation organizational structures*, which are created, modified, and terminated, depending on the current goals and tasks. One organization could participate in multiple dynamic organizational structures. It could also be the case that only a part of organization participates. This arrangement enables individual structures and employees to work on tasks from multiple projects, depending on the current needs. The cooperation organization itself takes on standard organizational structures.

Frequently we encounter *matrix organizational structure*, which suitably addresses the needs created by the environment dynamics. This type of organizational structure is also partially defined by the management literature: "Virtual organization or organization with virtual organizational structure is a special type of organization. It differs significantly from the hierarchical organizations. It is a temporary connection of companies, based on information technologies. Its purpose is to rapidly and efficiently use available entrepreneurial opportunities. Subjects connected within the virtual organization are not connected via ownership, and do not form formal organizational structures. Rather, they are independent and each of them contribute to taking advantages of the opportunity by its specific skill and obtains that what could not be obtained in being isolated." [43]

Although the sustainability concept is mainly oriented on ecological or environmental issues, it is widely used also in socioeconomic context [5, 8, 12]; for example, Delai and Takahashi refer to "consensus around three main dimensions of sustainability - economic, social, and environmental" [5]. They define the social and economic dimensions as follows: "The **economic dimension** assesses short and long term value generation by a company and its relationship with shareholders. It is related with the long-term sustainability of an organization." "In the organization point-of-view, the **social dimension** of sustainability concerns impacts on the social systems within which it operates or its stakeholders" [5]. Effective cooperation management between independent organizations has also strong *sustainable aspect*:

- It is oriented on long-lasting solutions.
- It is mutually beneficial for all participants (common sources or value creation, etc.).
- Cooperation-based organizational structures (e.g., clusters) have high probability for the creation of positive externalities and have significant potential for innovations [28, 29].

Conclusion

New cooperating organization creates new quality of *culture*, which will be based on the cultures of the cooperating subjects. Cultural similarity plays a certain role here, as cooperation strategy is often used by small and medium organizations that are regionally concentrated. On the other hand, e.g., joint venture is often established by transnational corporations. Culture also determines values that are attractive for the subjects to such a degree that they decide to cooperate (competitiveness, effectiveness). *Inequality* will manifest particularly in the organizational structure of the new organization. It can also manifest depending on the division of decision-making influence between the cooperating parties. *Conflict* should be seen here more broadly, as it is mainly a negotiation. Cooperating subjects may have different opinions about the future direction of the cooperating organization. If agreement or compromise is not found, change will take place. *Change* is here understood as a modification of the organization (new goals, change of partners, etc.) or as its termination, in case it loses its relevance for the cooperating subjects (or at least for the critical number of involved subjects).

Proposed matrix of cooperating organizational structures is meant to serve as a tool for managers of cooperating companies for managing their cooperating projects and activities. Its use is meant to help minimize the occurrence of conflict situations and to support smooth progress of cooperation activities from the organizational perspective.

Acknowledgments This paper is an output of the science projects – Slovak Republic scientific grants VEGA 1/0617/16 and APVV-15-0511.

References

1. Biggiero, L. (2006). Industrial and knowledge relocation strategies under the challenges of globalization and digitalization: the move of small and medium enterprises among territorial systems. *Entrepreneurship and Regional Development, 18*(6), 443–471.
2. Chesbrough, H. W., Vanhaverbeke, W., & Wes, J. (2006). *Open innovation: Researching a new paradigm*. Oxford: Oxford University Press.
3. Davis, P. (1999). *Managing the cooperative difference: A survey of the application of modern management practices in the cooperative context*. Geneva: International Labour Office.
4. davis, P., & Donaldson, J. (1998). Co-operative management: A philosophy for business. In *Cheltenham*. Cheltenham: New Harmony Press.
5. Delai, I., & Takahashi, S. (2011). Sustainability measurement system: A reference model proposal. *Social Responsibility Journal, 7*(3), 438–471.
6. Díaz, Piraquive, F. N., Medina Garcia, V. H., Crespo, R. G., & Liberona, D. (2014). Chapter 29. Knowledge management, innovation and efficiency of service enterprises through ICTs appropriation and usage. In L. Uden, D. F. Oshee, I. Ting, & D. Liberona (Eds.), *Knowledge management in organizations, Lecture notes in business information processing* (Vol. 185, pp. 300–311). 9th International Conference, KMO 2014 Santiago, Chile, September 2–5, 2014 https://link.springer.com/content/pdf/10.1007%2F978-3-319-08618-7.pdf.
7. Doz, Y. L. (1996). The evolution of cooperation in strategic alliances: Initial conditions or learning process? *Strategic Management Journal, 17*(S1), 55–83.
8. Dyllick, T., & Hockerts, K. (2002). Beyond the business case for corporate sustainability. *Business strategy and the Environment, 11*(2), 130–141.
9. Eisenhardt, K. M., Furr, N. R., & Bingham, C. B. (2010). CROSSROADS—Microfoundations of performance: Balancing efficiency and flexibility in dynamic environments. *Organization Science, 21*(6), 1263–1273.

10. Fawcett, S. E., Jones, S. L., & Fawcett, A. M. (2012). Supply chain trust: The catalyst for collaborative innovation. *Business Horizons, 55*(2), 163–178.
11. Felzensztein, C. H., Gimmon, E., & Aqueveque, C. (2012). Cluster or un-clustered industries? Where inter-firm marketing cooperation matters. *Journal of Business & Industrial Marketing, 27*(5), 392–402.
12. Galpin, T., & Whittington, J. L. (2012). Sustainability leadership: from strategy to results. *Journal of Business Strategy, 33*(4), 40–48.
13. Gumilar, V., Zarnić, R., & Selih, J. (2011). Increasing competitiveness of the construction sector by adopting innovative clustering. *Inzinerine Ekonomika-Engineering Economics, 22*(1), 41–49.
14. Gurrieri, A. R. (2013). Networking entrepreneurs. *Journal of Socio-Economics, 47(C*, 193–204.
15. Jassawalla, A. R., & Sashittal, H. C. (1998). An examination of collaboration in high-technology new product development processes. *Journal of Product Innovation Management, 15*(3), 237–254.
16. Ketels, C. H., & Sölvell, Ö. (2007). *Innovation clusters in the 10 new member states of the European Union*. Luxembourg: European Union's publisher.
17. Kowalski, A. M., & Marcinkowski, A. (2014). Clusters versus cluster initiatives, with focus on the ICT sector in Poland. *European Planning Studies, 22*(1), 20–45.
18. Kultti, K. (2011). Sellers like clusters. *Journal of Theoretical Economics, 11*(1), 1–28.
19. Lafleur, M. (2009). *A model for cooperative challenges [online]*. Cooperative Grocer Network 116 [ref. 17 November 2014]. Available at: http://www.cooperativegrocer.coop/articles/2009-01-21/model-cooperative-challenges
20. Lendel, V. (2015). Chapter 5. Application of cooperative management in enterprises: Management approach, problems and recommendations. In W. Sroka & Š. Hittmár (Eds.), *Management of network organizations – Theoretical problems and the dilemmas in practice*. Springer International Publishing. http://www.springer.com/gp/book/9783319173467.
21. Lydeka, Z., & Adomavičius, B. (2007). Cooperation among the competitors in International Cargo Transportation Sector: Key factors to success. *Inzinerine Ekonomika-Engineering Economics, 51*(1), 80–90.
22. Malakauskaite, A., & Navickas, V. (2010). Relation between the level of clusterization and tourism sector competitiveness. *Inzinerine Ekonomika-Engineering Economics, 21*(1), 60–67.
23. Monczka, R. M., Petersen, K. J., Handfield, R. B., & Ragatz, G. L. (1998). Success factors in strategic supplier alliances: The buying company perspective. *Decision Sciences, 29*(3), 553–577.
24. Mustak, M. (2014). Service innovation in networks: a systematic review and implications for business-to-business service innovation research. *Journal of Business & Industrial Marketing, 29*(2), 151–163.
25. Nemcova, E. (2004). The function of clusters in the development of region. *Ekonomicky casopis, 52*(6), 739–754.
26. Perru, O. (2006). Cooperation strategies, signals and symbiosis. *Comptes Rendus Biologies, 329*(12), 928–937.
27. Perry, M. (2007). Business environments and cluster attractiveness to managers. *Entrepreneurship and Regional Development, 19*(1), 1–24.
28. Porter, E. M. (1998). *Clusters and the new economics of competition [online]*. Harward Business Review Nov-Dec: 25–26. [ref. 26 September 2016]. Available at: https://hbr.org/1998/11/clusters-and-the-new-economics-of-competition
29. Porter, M. E. (1998). *On Competition*. Boston: Harvard Business School.
30. Ramanauskiené, J., & Ramanauskas, J. (2006). Economic management aspects of cooperatives. *Economics of Engineering Decisions, 49*(4), 15–21.
31. Ray, P. K. (2002). *Cooperative Management of Enterprise Networks*. New York: Kluwer Academic Publishers.
32. Ritala, P., & Sainio, L. M. (2014). Coopetition for radical innovation: technology, market and business-model perspectives. *Technology Analysis & Strategic Management, 26*(2), 155–169.

33. Robson, M., & Kant, S. (2006). The development of government agency and stakeholder cooperation: A comparative study of two Local Citizens Committees' (LCC) participation in forest management in Ontario, Canada. *Forest Policy and Economics, 9*(8), 1113–1133.
34. Sahut, J. M., & Peris-Ortiz, M. (2014). Small business, innovation, and entrepreneurship. *Small Business Economics, 42*(4), 663–668.
35. Schmoltzi, C., & Wallenburg, C. M. (2012). Operational governance in horizontal cooperations of logistics service providers: Performance effects and the moderating role of cooperation complexity. *Journal of Supply Chain Management, 48*(2), 53–74.
36. Solvell, O., Lindqvist, G., & Ketels, C. H. (2003). *The Cluster Initiative Greenbook*. Stockholm: Bromma tryck AB.
37. Soviar, J. (2012). *From Cooperation to Management – Cooperative management* (in Slovak). [Habilitation thesis]. University of Zilina, Faculty of Management Science and Informatics.
38. Staatz, J. M. (1983). The Cooperative as a Coalition: A Game-Theoretic Approach. *American Journal of Agricultural Economics, 65*(5), 1084–1089.
39. Staber, U. (2010). Imitation without interaction: How firms identify with clusters. *Organization Studies, 31*(2), 153–174.
40. Szekely, V. (2008). Regional industrial clusters and problems (not only) with their identification. *Ekonomicky casopis, 56*(3), 223–238.
41. Valenzuela, J. L. D., & Villacorta, F. S. (1999). The relationship between the companies and their suppliers. *Journal of Business Ethics, 22*(3), 273–280.
42. Varmus, M. (2009). Comparison of selected concepts strategies. In Š. Hittmár & W. Sroka (Eds.), *Theory of management 1* (pp. 169–173). Žilina: University of Zilina.
43. Veber, J., et al. (2006). *Management. Foundations, prosperity, globalization (in Slovak)*. Praha: Management Press.
44. Veerakumaran, G. (2006). *COCM 511 – Management of cooperatives and legal systems*, Faculty of Dryland Agriculture and Natural Resources, Mekelle University.
45. Vodák, J., Soviar, J., & Lendel, V. (2013). Identification of the main problems in using cooperative management in Slovak enterprises and the proposal of convenient recommendations. *Communications – Scientific letters of the University of Žilina, 15*(4), 63–67.
46. Vodák, J., Soviar, J., & Lendel, V. (2015). Proposal for effective planning of cooperation activities in a company. In *Proceedings of electronic government and the information systems perspective, 4th International Conference, EGOVIS 2015, Valencia, Spain,* (Vol. 9265, pp. 351–363). https://www.springer.com/it/book/9783319223889.
47. Weck, M., & Ivanova, M. (2013). The importance of cultural adaptation for the trust development within business relationships. *Journal of Business & Industrial Marketing, 28*(3), 210–220.
48. Wicks, A. C., Berman, S. L., & Jones, T. M. (1999). The structure of optimal trust: Moral and strategic implications. *Academy of Management Review, 24*(1), 99–116.
49. Zhang, W. (2011). Cooperation system constructing and model of its operation mechanism. *In Proceedings of the International Conference on Business Management and Electronic Information (BMEI), Vol., 3,* 784–787.

Chapter 19
Green Markets and Their Role in the Sustainable Marketing Management

Katarína Gubíniová, Gabriela Pajtinková Bartáková, and Jarmila Brtková

Abstract The global market environment is marked by hyper-competitive market players' uncompromising efforts and activities. In recent years declaring social responsibility principles has become a tool to differentiate their business plans in terms of sustainable development of the future society. One of the significant cultural trends on the threshold of the third millennium is a shift in customers' attitudes towards environmental issues. This trend is documented by numerous studies from around the world. In Slovakia public attitudes mapping towards solving issues around the impact of their consumer behaviour and organizational behaviours have gradually become a more crucial problem. The aim of the paper is an indirect comparative synthesis and mapping of the emerging green market representatives' manifestations and attitudes to consumption.

Introduction

Environmental awareness and environmentally friendly consumer behaviour is influenced not only by economic impulses themselves but also by a number of incentives of different types and intensities [16]. These include, for example, nature conservation, health protection, ensuring life chances for the future generations and stable economic development, etc. Implementation of such goals in real life is often made difficult by unclear current and future daily impact on the environment, often under the influence of expected short-term benefits [10].

If an organization decides to factor the principles of sustainability into its activities and create and improve attributes of sustainable marketing management, then it must firstly identify those customers who will be approachable and favourably disposed to sustainable products and in particular those who will be willing to pay a

---

K. Gubíniová (✉) · G. P. Bartáková · J. Brtková
Comenius University in Bratislava, Bratislava, Slovakia
e-mail: katarina.gubiniova@fm.uniba.sk

premium price for such products. The term green market is used to distinguish sustainably oriented customers and organizations which are able to attend to them.

Nationwide representative surveys on consumer attitudes towards the environment and studies on environmental awareness clearly indicate attentive perception of ecological problems, as well as accepting environmental protection as an important part of social responsibility. A great majority of respondents is persuaded that the growth limits have been reached, and precautions to prevent natural catastrophes (e.g. global warming, air, water and soil pollution) are vitally important [11].

Materials and Methods

The results presented in the paper are based on primary, representative, quantitative and qualitative research, while the main role in the qualitative research is played by motivating factors, which customers find currently applied in sustainable consumption. The research was conducted from September to November 2014 on a sample of 1820 respondents. The reliability of the results of the conducted research was at the level of 95% with precision of 3%, while the sample size represented 1820 respondents. The sample comprised an adult population of Slovakia based on gender, age, education, nationality, regional representation and size of seat. Evaluation of the questions in the section findings and discussion has been extracted from the research results.

Problem Identification

From a marketing point of view, the concept of consumption is very closely related mainly to two elements of the marketing concept – needs and wishes. Almost all definitions of marketing are based on crucial customer focus and efforts to meet his/her needs. Customer focus should remain at the forefront and also on the threshold of the third millennium, but it is important to reconsider satisfaction of customer needs taking into account many negative attitudes of social critics of marketing. They emphasize that marketing, in addition to identification, creates and predicts customer needs and thereby makes a wide offer leading to meeting the needs (often artificially evoked).

Driving Forces of Dynamization in Sustainability of Consumption

Current customer behaviour models need to be reassessed in accordance with the criteria for sustainable consumption [7]. This objective appears at the level of the professional public, governments, supranational groupings and non-governmental organizations.In the scientific literature, there are many marketing activities defined within the so-called sustainability audit that contribute to a more rational, sustainable consumption. These include [1]:

- *General aspects of sustainability principles implemented in marketing management:* integrating sustainability into strategic plans of an organization, creating a sustainable marketing strategy, market segmentation that takes into account the environmental awareness of customers, creating strategic alliances for more effective implementation of sustainable marketing management and meeting the environmental certification criteria
- *Concrete sustainable strategies of the marketing mix elements:* creating product packaging and labelling, creating a sustainable product, genuine product innovations, creating distribution channels that take into account the phase after the end of the product life cycle, identifying environmental responsibility within the distribution channel elements, marketing communication tools that should emphasize positive aspects of a sustainable product and activities of an organization towards sustainable marketing and customer education

At the national level or at the level of supranational groupings of states, governments and parliaments, they adopt a range of legislation that seeks either directly or indirectly to influence consumption towards sustainable consumption. These include, for example:

- *Directive 2005/32/EC of the European parliament and of the council establishing a framework for the setting of ecodesign requirements for energy-using products:* the directive deals with the fact that "energy-using products account for a large proportion of the consumption of natural resources and energy", "Energy efficiency improvement is regarded as contributing substantially to the achievement of greenhouse gas emission targets."

In Article 8 there is the directive goal formulated which is "to achieve a high level of protection for the environment by reducing the potential environmental impact of energy-using products, which will ultimately be beneficial to consumers and other end-users. Sustainable development also requires proper consideration of the health, social and economic impact of the measures envisaged. Improving the energy efficiency of products contributes to the security of the energy supply, which is a precondition of sound economic activity and therefore of sustainable development."

- *Regulation (EC)No 1980/2000 of the European parliament and of the council on a revised community eco-label award scheme:* these European Union institutions are aware of the importance of customer education in this regulation – "It is necessary to explain to consumers that the eco-label represents those products which have the potential to reduce certain negative environmental impacts, as compared with other products in the same product group, without prejudice to regulatory requirements applicable to products at a community or a national level."
- In the document that is not legally binding, the Commission of the European Communities has notified the European Parliament, the Council, the European

Economic and Social Committee and the Committee of the Regions of *the Sustainable Consumption and Production and Sustainable Industrial Policy Action Plan*. In the introduction the Commission states the following: "Sustainable development aims at the continuous improvement of the quality of life and well-being for present and future generations. It is a key objective of the European Union. Yet, increasingly rapid global changes, from the melting of the icecaps to growing energy and resource demand, are challenging this objective. (...)The way we produce and consume contributes to global warming, pollution, material use, and natural resource depletion. The impacts of consumption in the EU are felt globally, as the EU is dependent on the imports of energy and natural resources. Furthermore, an increasing proportion of products consumed in Europe are produced in other parts of the world. The need to move towards more sustainable patterns of consumption and production is more pressing than ever."
- Legal regulations adopted by the National Council of the Slovak Republic contain provisions which aim to directly or indirectly influence consumption in a positive way. The indirect influence is embodied, e.g. in the Act No. *147/2001 Col. on Advertising and Amendments of Some Laws*. The lawmaker in § 3 of this act states the general requirements for advertising. In § 3 Article 4, it is stated that advertising *may not promote products harmful to the environment or products harmful to life or health of people, animals or plants unless attention is specifically drawn to such fact.*
- In the *Code of Ethics for Advertising Practice*, which is issued by the Council for Advertising, in Article 13 a requirement for social responsibility in advertising is defined as the basic requirement for advertising; in Article 24 there is a specific requirement for advertising to take into account – the environment. "An advertisement shall not intentionally promote unjustified waste or irrational consumption of raw materials or energy, particularly of the energy derived from non-renewable resources."

The level of non-governmental organizations deals with the issue of the impact of consumer behaviour patterns on consumption particularly intensively. For example:

- *European Environment Agency* that understands household consumption in a number of contexts, such as external links and political connections, and monitors changes in consumer behaviour by dividing it in few areas (food and drinking habits, tourism, energy and waste)
- *Slovak Environment Agency* that understands sustainable consumption in the context of an integrated product policy aimed at reducing environmental impacts through environmental measures focused on the whole life cycle of the product
- *World Wildlife Fund* that points out unsustainable consumption at present ("People only have one planet, but consume products in the same way as they would have three") and necessity of its reassessment in symbiosis with sustainable production.

The previous list of scientific literature, legislation and NGOs is not comprehensive but only highlights the most important driving forces that should be involved in building sustainable consumption.

Responsibility for the Development of Sustainable Consumption

The model, which focuses on subjects responsible for unsustainable consumption nowadays and at the same time subjects whose responsible actions can lead towards sustainable consumption, highlights coordination of activities and collective approach towards sustainable consumption.

Barriers found on the side of people or customers and causing the development of unsustainable consumption include, e.g.:

- People do not like to embrace changes in their routine models of behaviour (not only buying behaviour) [6].
- They are greatly influenced by social models (unwritten norms) of behaviour.
- Sustainability is associated with a very high price.
- When making buying decisions, they only take into account the short-term household budget.
- They often lack trust to organizations and the government, who should directly or indirectly promote sustainable consumption.

When deciding what to buy, customers can be well aware of sustainable products and sustainable consumption, but they are still unable to change their organizations' existing activities. The change towards sustainable consumption can only be reached if all the participating parties will be interested in it – customers, organizations and governments.

The key role in the shift towards sustainable consumption remains with organizations. It is the organizations which are the first element of the distribution channel deciding which products and what amount and variations of them will be available to customers. The barrier which causes slowing of sustainable consumption is competition. In the environment of current markets, it is very difficult for organizations to be pioneers in inspiring towards more intelligent and far-sighted consumption. There are several reasons which should make organizations take into account aspects of sustainable consumption in their activities:

- The value of organization grows for different stakeholders.
- Sustainability in the distribution channel means savings of energy, water, resources and materials and thus eventually savings of expenditures.
- Building the brand through meeting or even surpassing customers' expectations: it is incorrect to assume that customers do not want sustainable products because they do not ask for them – they can only choose from the portfolio of the offered range of products.

- Efforts to fulfil moral duties towards the society, e.g. not to harm future generations of customers.

Governments of countries or supranational groupings should play the role of the so-called invisible hand of market while building models of sustainable consumption in the markets.

The 2007 and 2008 studies of the *World Business Council for Sustainable Development* state that the green market size is questionable and depends on different understanding of such a market. However, one fact is indisputable – the size of this market shows a growing tendency [17].

In the 2008 report [2], the research organization *Hartman Group* following the trends in the field of customer perception on sustainability states that more than 90% of customers want to contribute to sustainability in some way. However, only half of them understand the concept of sustainability, and they expect leadership and responsibility from organizations. The report shows a very important fact on the customer side, namely, that sustainability issues are important to customers if they affect them personally and directly, e. g. for essential consumption of products.

Another research organization *GfK Roper Consulting* publishes an annual *Green Gauge Report* for organizations that are interested in customer attitudes and behaviour in the green markets. The 2008 report [1] also confirmed a growing trend in the population of the United States, who is interested in sustainability. Seventy-two percent of Americans (a year-on-year increase of 10%) declared fairly clear knowledge of environmental issues. Twenty-eight percent of Americans (a year-on-year increase of 20%) seek out environmental information, and they make so-called green purchases that are savings inspired, but there is also an increasing number of those who buy such products at a premium price. Almost a third of Americans believe that they should be more concerned about the environment.

The organization *Natural Marketing Institute* that leads *LOHAS Consumer Trends Database* explores the global trends in consumption/sustainable consumption as follows. In all countries surveyed (the European countries represented in the survey: France, Germany, Spain, Belgium, the Netherlands, Portugal), price or consumption expenditure is the most frequently reported barrier to purchase decisions for sustainable products and services on the customer side. Approximately two-thirds of customers in all countries deal with the impact on the environment. However, price still remains the critical element of the marketing mix in their purchase decision-making.

Relatively few customers in all countries surveyed are not price sensitive, while customers in the United States are the least willing to pay a premium price.

In the 2009 report, the research organization *Hartman Group* presented another psychographic segmentation scheme for green markets which consists of four segments. These segments [12] differ based on several criteria: *concern for social and environmental issues and the frequency and intensity of pro-sustainability behaviour.*

The customers belonging to the outside-sustainability segment have the following characteristics:

- They believe that benefits of recycling and reusing products are overstated.
- They rarely, if ever, factor ecological and social factors into their buying behaviours.
- They are not familiar with the term *sustainability*; they do not understand the concept.

The customers belonging to the inside-sustainability segment have the following characteristics:

- They are twice more than others convinced that it is important to buy and consume environmentally friendly or otherwise sustainable products and services.
- They are four times more than others willing to pay a final price increased by 10% for sustainable products.

This segment is subdivided into four subsegments as follows.

The *core customer* segment (13%) shows the highest level of involvement in a sustainability lifestyle, including purchasing environmentally friendly products and adhering to pro-sustainability attitudes. The *mid-level* segments (35% + 31%) demonstrate a lower level of commitment to sustainability principles and less active buying behaviour. *The periphery customers* show only minimal concern for environmental and social sustainability.

GfK Roper Consulting is another research organization that conducts segmentation research in the area of consumer behaviour and environmental sustainability.

Results

Question 1: To what extent are sustainability principles and environmental characteristics of products important to you in the process of purchase decision-making?

As it follows from the research results, 16% of the surveyed respondents consider environmental sustainability principles to be important in the process of purchase decision-making (of which 5% very important). These principles are not important to 84% of the respondents.

Question 2: Are you willing to pay a premium price if it is a sustainable product (environmentally friendly product and the like)?

The research results point out that Slovak consumers are not willing to pay a premium price for sustainable products, despite their benefits for the society and environment. Up to 87% of respondents gave a negative opinion in their answers (i.e. rather no or no). On the contrary, only 13% of respondents said that they are willing to pay a premium price.

Question 3: To what extent would company activities beneficial to the society and environment affect you (e.g. financial contribution to charities or reducing the environmental burden caused by company activities)?

More than two-thirds of respondents said that company activities which are beneficial to the society and environment do not affect their purchase decisions.

Seventeen percent of respondents would prefer products of such a company to competition. For 5% of respondents, it is definitely the reason to buy a product of such a company. Two percent of respondents consider such activities to be a PR stunt, and they definitely would not buy a product.

Question 4: In your opinion are consumers (and general public) well informed about sustainable consumption principles?

Up to 88% of respondents think that they are not well informed about sustainable consumption principles (of which 54% rather no, 34% no). Slovak consumers have lack of awareness of how they can reduce their environmental and social impacts for the sake of sustainable consumption. Only 12% of respondents replied that there is sufficient awareness in this area.

Question 5: On a scale of 1 to 5 with 1 being the best evaluation and 5 being the worst evaluation, how would you rate availability of sustainable products in the Slovak market?

Forty-six percent of respondents rate availability of sustainable products negatively (as 4 and 5). Forty percent of respondents consider availability to be average. Only 14% of respondents rated availability as 1 and 2. It follows from the above that respondents spend a lot of time and energy searching for truly sustainable products.

Concerning the position of a customer in the concept of sustainable marketing management, it is necessary to identify discrepancies between attitudes and proactive consumer behaviour. Based on these discrepancies, there are five main barriers to sustainable consumer behaviour.

1. *Lack of information and awareness*: many customers declare that they do not know how to reduce their environmental and social impacts [15].
2. *Negative perception*: many customers are convinced that sustainable products are inferior with regard to various attributes such as design, style and performance.
3. *Distrust*: many customers have no confidence in statements of marketing managers (e.g. in the form of marketing communication, CSR and sustainability reports). They are often right, because many statements and claims are false and misleading (Hesková, [3, 5]).
4. *High prices*: many customers (and they are often right) consider sustainable products to be expensive and overpriced, and they are not able/willing to pay such a price for them.
5. *Low availability*: customers spend a lot of time and energy searching for truly sustainable products.

Conclusion

Building a market position in green consumers' minds should have a distinctive product/brand identity compared to competition [14]. In order to create the market position of products correctly and strongly enough in consumers' minds (in green markets), an organization must analyse benefits of offered products and services and determine the extent to which they correspond to sustainability values of the target segment.

The way that enables the market position creation through sustainability attributes is the development of a product portfolio which will demonstrably meet the sustainability criteria, in other words to create such value attributes complying with the values which the target segment shares. Another strategy is product positioning by emphasizing efficiency and financial savings.

It is important to emphasize that *for implementation of sustainability principles into positioning strategies, it is not necessary to position products as green or sustainable.* It is extremely vital to highlight the fact that the core of the market position is a good reputation of an organization in the market. This reputation is built not only on using traditional and new marketing communication tools but nowadays mostly on what "others say about the organization" [13]. In the age when there are so-called channels on the Internet controlled by customers, *it is important for organizations to act in a sustainable way not only passively declare such acting.*

Solutions corresponding to the above discrepancies can be derived from sustainable marketing management:

- *Customer education* [4]
- *Production and extensive distribution of better, sustainable products*
- *Honest marketing communication* [8]
- *Clearer declaration of elements building value*

Consumer education should start with specially adapted teaching tools because of the growing need for information – critical awareness, social and environmental responsibility in terms of sustainable consumption, rights and obligations, well-thought-out decisions, the will to act and all these for realizing and asserting their interests, as well as general interests of consumers/individuals or whole groups [9, 11].

Acknowledgements This paper was elaborated with the support of the research project VEGA 1/0205/14 – The Prospect of the Existence of Dynamic Service Industries in the Slovak Republic in the Context of the Application of the Principles of Innovation Union.

References

1. GfK Roper Consulting (2008). Green gets real ... current economic environment subduing green enthusiasm but driving practical action.
2. Hartman Report (2008). Presentation made during the Northwest sustainability discovery tour in Portland, Oregon.
3. Hesková, M., & Štarchoň, P. (2009). *Marketingová komunikace a moderní trendy v marketingu* (180 p). Oeconomica: Praha.
4. Hitka, M. – Balážová, Ž. (2015). The impact of age, education and seniority on motivation of employees, Journal Business: Theory and Practice. 15, 1. Litva: Vilnius
5. Kubičková, V. (2009). *Inovačné aktivity podnikov služieb* (162 p). Bratislava: Ekonóm.

6. Martíšková, P. – Švec, R. (2017). E-shops and customer feedback: Experience by Czech B2C customers. Littera Scripta, Vol. 10, No. 1, 2017.
7. Nagyová, Ľ. – Košičiarová, I. – Holienčinová, M. (2016). Sustainable consumption of food: A case study of Slovak consumers. Proceedings of the 2016 International Conference Economic Science for Rural Development, no. 43, 2016.
8. Olšavský, F. (2013). Generation approach in operating of the target market – Opportunities and risks. In *Theory and practice in management* (pp. 122–131). Bratislava: Univerzita Komenského v Bratislave.
9. Solarová, P. (2014). Angažovanost spotřebitelů v maloobchodě – získání námětů. *Marketing Science & Inspirations,* 9(1), 16–25.
10. Stachová, K. – Stacho, Z. (2013). Employee allocation in Slovak companies, Business: Theory and Practice, Vol. 14, No. 4, pp. 332–336. Litva: Vilnius, 2015.
11. Steffens, H. (2006). Správanie spotrebiteľov a spotrebiteľská politika. *Bratislava : Ekonóm, p., 36.*
12. The Hartman Group (2009). Sustainability: The rise of consumer responsibility, executive summary.
13. Treľová, S. (2014). Manažér a jeho právne postavenie. In *MMK 2014* (pp. 58–64). Magnanimitas: Hradec Králové.
14. Vilčeková, L. (2014). Etnocetrizmus slovenských spotrebiteľov. *Marketing Science & Inspirations, IX*(3), 53–59.
15. Vilčeková, L. – Štarchoň, P. – Sabo, M. (2013). Segmentation process in determining Slovak consumers' attitudes toward brands Mathematics and computers in contemporary science. Athens: WSEAS, pp. 206–211.
16. Volná, J. – Papula, J. (2013) Analysis of the behavior of Slovak Enterprises in the context of low innovation performance. Procedia – Social and behavioral sciences, Vol. 99. Amsterdam: Elsevier, pp. 600–608.
17. World Business Council for Sustainable Development (2008). Sustainable consumption facts and trends from a business perspective.

Chapter 20
A New Approach to Sustainable Reporting: Responsible Communication Between Company and Stakeholders in Conditions of Slovak Food Industry

Mária Holienčinová and Ľudmila Nagyová

Abstract Companies are tasked with monitoring the progress of initiatives and disclosing information about company efforts by preparing corporate social responsibility (CSR) reports and promoting dialogue with stakeholders. Actively informing the public has become an indispensable part of responsible behavior of companies that lead their business activities in the sense of CSR. This issue is particularly important in the food sector as these businesses are involved in the production and sale of food products to final customers. Our idea connected to the title of this paper is based on the authors' opinion that claims – idea that CSR can afford only strong multinational companies is already overcome because CSR is perceived in a wider context (environmental, social, and economic area), which ultimately mean for businesses more savings than just the additional costs. We focused mainly on the area of reporting and stakeholder involvement. The aim of the presented paper is to evaluate the level of CSR reporting toward stakeholders in the conditions of Slovak food industry. To the research are involved different sized Slovak food manufacturing companies with the aim to point out to the most significant differences between them. Part of the paper is to determine the key stakeholders, to whom the communication should be targeted. In order to achieve the formulated aim, marketing research was conducted (total number of food companies was 138; companies were from all regions of the Slovak Republic. The research outcomes confirmed that there are significant differences between different sizes of enterprises (micro, small- and medium-sized, large) in the area of sustainable reporting toward stakeholders.

M. Holienčinová (✉) · Ľ. Nagyová
Slovak University of Agriculture in Nitra, Nitra, Slovak Republic

© Springer International Publishing AG, part of Springer Nature 2019
D. Cagáňová et al. (eds.), *Smart Technology Trends in Industrial and Business Management*, EAI/Springer Innovations in Communication and Computing,
https://doi.org/10.1007/978-3-319-76998-1_20

Introduction

Stakeholder engagements, collaboration, shared learning or fact-finding have become buzzwords and hardly any environmental assessment or modelling effort today can be presented without some kind of reference to stakeholders and their involvement in the process [1]. The issue of stakeholders is a key part of the concept of corporate social responsibility (CSR). The strategy of CSR concept is basically building of relations and trust between the company and the individual groups. They are either directly or vicariously involved in the development of the organization [2]. Business organization today, particularly the modern corporation is currently made up of many people with many interests, expectations, and demands [3]. Members of the surrounding society expect quality and sustainable lifestyle. Companies must respond to individuals and groups that are entitled to create pressure and demands [4]. The challenge to the sustainability and the new role of business in society and increased expectations and new rules and tactics, management is required to come into contact with the main stakeholders in the area of corporate responsibility, global versus regional and local needs, and different national cultures [5, 6]. Inclusion of stakeholders into its strategies is the key to the success of organization in the twenty-first century [7].

Because the stakeholders are an important part of corporate responsibility, the company will not only focus on its values and principles but will try first of all to understand the expectations of those who have an impact on business and are influenced by it [8]. Involvement of stakeholders brings certain advantages:

- *Setting objectives and monitoring performance* – setting realistic goals and evaluating the actual performance
- *Innovative environment* – anticipating new trends and allowing to adapt to the difficult business environment and in advance to set new strategic objectives with respect to market development
- *Risk management* – communication with external subjects can early identify potential risks, especially if the company will build relationships with individuals/groups who perceive the company negatively
- *Information value* – good and strong relationships with stakeholders mean for organization the source of valuable information
- *Mutually beneficial relationship* – personal encounters and the ability to develop individual relationships are the best way to build mutual trust between the company and key stakeholders [9].

Dialogue with Stakeholders

A disconnect in communication between CSR initiatives and public awareness will impede any potential benefits to a company; so, it is important to intelligently and strategically communicate this to the public and to all stakeholders [10, 11].

Company creates a certain system through which can obtain a list of potential stakeholders and based on individual activities of business can identify their sphere of interest. The relationship between the company and the stakeholder becomes more concrete. These may be the following activities and areas, respectively [12, 13].

- Environmental impacts on society
- The future development of the company
- Media campaigns
- Social impacts
- Risks arising from anticipated behavior of existing or potential partners / competitors

Each organization has a different circle of stakeholders. The primary step in the implementation of CSR should be the identification, analysis, and continuous dialogue with all stakeholders. Special focus, in the issue of CSR, should be devoted to key groups. Figure 20.1 provides a way how to identify the key stakeholder groups.

One of the possible positive results of the dialogue with stakeholders is, for example, dissemination of good practices, mutual inspiration, and motivation. Very noticeably, this aspect can be seen in the area of supplier-customer relations. If the customer is an enterprise with high profile of corporate social responsibility and the same standards will also require from its suppliers, it is a direct and very effective way to spread the principles of CSR [15]. Leadership of effective dialogues of company with stakeholders is a very complex process that requires a high degree of readiness and precision. It signals the openness and interest of company of its surroundings and helps to bridge the mutual distrust. Properly conducted dialogues lead to mutual partnerships between the private, public, and civil sector, so-called cross-sector relationship [16].

Sustainability and Reporting

Sustainability reporting is a broad term considered synonymous with others used to describe reporting on economic, environmental, and social impacts (e.g., triple bottom line, corporate responsibility reporting, etc.) [17].

The level of expectations			
	High	Average informing	**Have a dialogue**
	Low	Answer the questions	Ensure the satisfaction
		Low	High

The level of influence on company

Fig. 20.1 Stakeholder matrix [14]

Corporate sustainability reporting has evolved rapidly since the first environmental reports appeared in the late 1980s. Many of the early efforts were health and safety reports modified to account for impacts on the environment. In the early days, there were no accepted standards for corporate reports, so there were wide variations in the content and format of the reports produced. Nowadays, environmental issues have been joined on the agenda by social considerations, and the reporting process has been broadened into an audit of what is loosely termed "corporate responsibility" [18].

The Global Reporting Initiative (GRI) has pioneered corporate sustainability reporting since 1997, transforming it from a niche practice to one now adopted by a growing majority of organizations. It is an international, independent organization that helps businesses, governments, and other organizations understand and communicate the impact of business on critical sustainability issues. It produces the world's most trusted and widely used standards for sustainability reporting, the GRI Standards, which enable organizations to measure and understand their most critical impacts on the environment, society, and the economy. Thousands of reporters in over 90 countries use the GRI Standards – a free public good – for their reporting [19].

Clear and open communication about sustainability requires a globally shared framework of concepts, consistent language, and scheme. The mission of this independent institution of the Global Reporting Initiative (GRI) is to meet this need by providing credible and trusted reporting framework [20]. GRI reporting framework is intended to serve as a generally accepted framework for reporting on organization's economic, environmental, and social performance. It is designed for use by organizations of any size, sectors, or location. It takes into account the practical considerations faced by a diverse range of organizations – from small enterprises to those with extensive and geographically dispersed operations [17].

Sustainability reports based on the GRI reporting framework disclose outcomes and results that occurred within the reporting period in the context of the organization's commitments, strategy, and management approach. Reports can be used for the following purposes, among others:

- *Benchmarking* and assessing sustainability performance with respect to laws, norms, codes, performance standards, and voluntary initiatives
- *Demonstrating* how the organization influences and is influenced by expectations about sustainable development
- *Comparing* performance within an organization and between different organizations over time

Material and Methodology

The aim of the presented paper is to evaluate the level of CSR reporting toward stakeholders as an important component of the CSR strategy in the conditions of Slovak food industry. The paper deals with key stakeholders to whom the communication should be targeted from the point of view of food companies and also refers

Table 20.1 Classification of companies [21]

Company category	Staff head count	Turnover	Balance sheet total
Micro	< 10	≤ € 2 mil.	≤ € 2 mil.
Small-sized	< 50	≤ € 10 mil.	≤ € 10 mil.
Medium-sized	< 250	≤ € 50 mil.	≤ € 43 mil.
Large	> 250	> € 50 mil.	> € 43 mil.

on incorporation of annual/special reports about responsible marketing activities in company guidelines. Furthermore, in our research, we want to highlight main differences between different size groups of companies connected with the area of stakeholders and reporting as a positive contribution between the society and company itself.

In order to achieve the formulated aim of the paper were collected and used primary and secondary sources of information. Underlying secondary data were obtained from available literature sources, i.e., from professional book publications from domestic and foreign authors and organizations. When processing of individual underlying data and formulating conclusions of the paper were used methods of analysis, synthesis, induction, deduction, and the comparative method.

In order to meet the objectives of the paper, marketing research was conducted. Marketing research was conducting in the period from November 2014 to March 2017 by method of interview using a structured questionnaire. The research was focused directly on food manufacturing companies in Slovakia. Totally 350 food companies operating in the Slovak food industry were asked. The questionnaires were sent to company email address. In some companies we visited personally, and managers were asked to fill in the questionnaire in controlled interview. Finally, in the research outcomes, 138 food companies (totally 32 micro companies, 93 small- and medium-sized companies/SMCs, 13 large companies) from all regions of the Slovak Republic were involved.

Individual companies were classified by size according to excerpt from Article 2 of the Annex of Recommendation 361/2003 / EC (Table 20.1).

The questionnaire was evaluated with the use of contingency tables, which were prepared by Excel, under which they were subsequently developed graphic representations. For a deeper analysis of the obtained results, there were set out two assumptions about the dependence resp. independence between tested variables. Subsequently, Friedman test, nonparametric test of contrasts that uses Nemenyi method, and Fisher exact test were used for testing the formulated dependencies.

Results and Discussion

Primary data for meeting the aim of the paper were obtained through marketing research, in which 138 food manufacturing companies operating in Slovakia were included. Through the following question, companies were asked to assign to their stakeholders the importance of influence (power), with regard to their marketing

Table 20.2 Results of Friedman test

Friedman's test:	
Q (Observed value)	895,1132
Q (Critical value)	26,99,632
DF	15
p-value (Two-tailed)	< 0,0001
Alpha	0,05

Source: Own processing, XLStat

Table 20.3 Results of Nemenyi test

Categories	Groups							
Surrounding community	A							
NGOs	A	B						
Creditors	A	B						
State, governmental institutions		B	C					
Investors			C	D				
Employees				D	E			
Media				D	E	F		
Suppliers					E	F		
Competition						F	G	
Management							G	H
Customers								H
Company owners								H

Source: Own research and processing, XLStat

activities. We set out an assumption: *There are differences in the importance of influence of different stakeholder groups from the perspective of food companies.* The importance of stakeholders influence was analyzed by using Friedman test; the results are shown in Table 20.2. Based on the theoretical level of significance, which was compared with a significance level of alpha = 0.05, the H_0 hypothesis of the absence of differences in the importance of stakeholder influence was rejected. Based on these facts, we can conclude, there are statistically significant differences in the importance of stakeholder influence from company view in consideration of marketing activities.

More detailed information which shows the results of nonparametric test of contrasts that uses Nemenyi method is available in Table 20.3. The various stakeholders were sorted into eight groups of at least important influence (A) to the strongest influence (H), which was marked by individual food companies with respect to their marketing activities.

For the best graphical representation of mentioned facts, we decided to put in a graph (Fig. 20.2) a modal value – mode, the value of quantitative trait of statistical series, which has the highest relative frequency, and what means that occurs most frequently. Companies were asked to indicate with 5-point scale (5 max. – 1 min. importance) the most important stakeholders. The results show that the highest

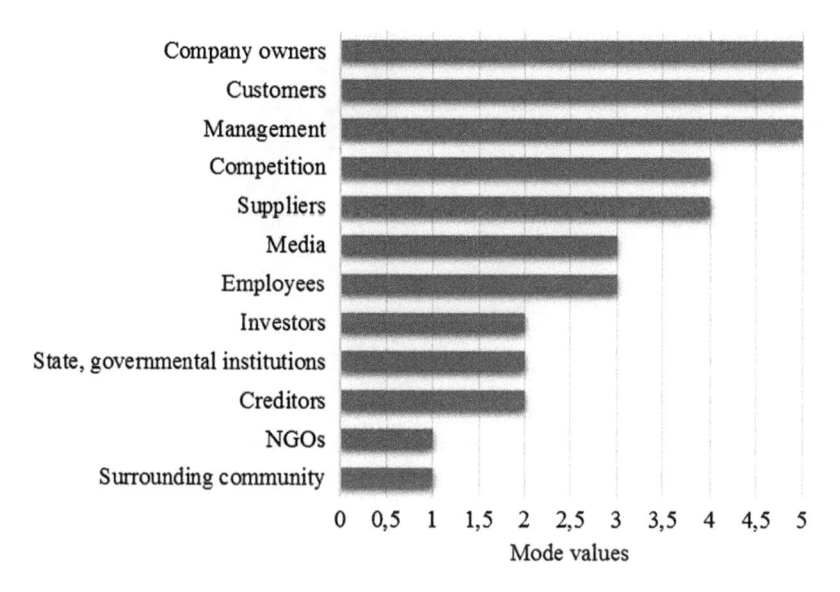

Fig. 20.2 The importance of stakeholder influence. (Source: Own research and processing)

value (5 – max. importance) was achieved by company owners followed by customers, management, competition, suppliers, media, and employees. The lower values of importance (closer to 1) were attributed to stakeholders as investors, state/governmental institutions, creditors, NGOs, and surrounding community.

The CSR reports provide to a company an opportunity to communicate its CSR efforts to the company's stakeholders and to discuss certain company successes and challenges on a wide array of CSR issues. The CSR report is also a medium for transparency (which often improves a company's reputation with certain stakeholders, particularly shareholders, employees, suppliers, and communities within which the company operates) and may be used as an effective outreach tool as part of an ongoing shareholder relations. In addition, the CSR report provides existing and potential investors with CSR information to assist in analyzing investment decisions. From the above-mentioned follows the importance of informing stakeholders about their responsible business and also choosing the right or any access to company CSR records.

The purpose of further question was to find out if companies provide annual or special reports in which they inform about their responsible and sustainable marketing activities or whether companies inform the public at all.

Of the total sample of companies (Fig. 20.3), totally 71% companies reported that they don't provide any CSR reports. Fifteen percent of respondents issue special CSR reports, and annual CSR reports were only cited by 9% of companies, the majority of which were large companies. Three percent of companies reported that stakeholders could obtain information through a formal request.

We intended to find out if there is dependence between sustainable reporting and the size of the company. We assume that the activities about which companies would

Fig. 20.3 Access to company environmental records. (Source: Own research and processing)

- Annual CSR reports 11%
- Special CSR reports 15%
- Formal Request 3%
- No CSR records 71%

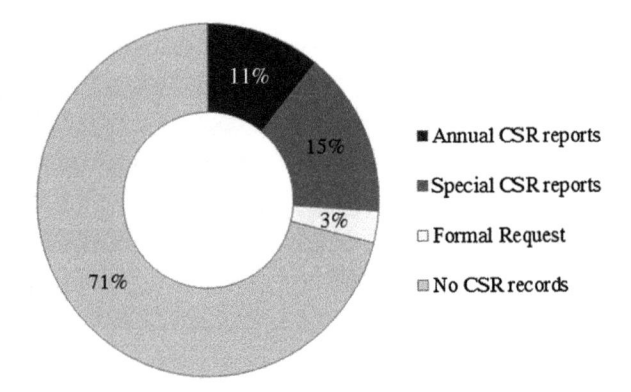

like to inform the general public are activities precisely those companies that dispose with sufficient financial capital and have own department for organization of this activity. We expect that smaller companies do not pay too much attention to the issue of such reports and all their resources devote to just the core business performance. Morsing and Perrini argue that the original idea that CSR can afford only strong and rich multinational companies is already overcome because CSR is perceived in a wider context (pillars CSR), which ultimately mean for businesses more savings than just the additional costs. The aim of our contention is not that small companies are not oriented toward socially responsible and sustainable activities but insufficiently inform about these activities and should focus their attention on the area of reporting (sustainable reporting).

For a deeper analysis of the obtained results, there was an assumption set out: "Annual respectively special reports, in which companies inform about sustainable and responsible marketing activities, are issued to a greater extent by large companies."

This question was tested with the use of Fisher exact test (Table 20.4). Degree of dependence is reflected in Table 20.4 by phi coefficient, contingency coefficient, and Cramer's coefficient. The values of these coefficients indicate moderate up to strong dependence.

According to results of Fisher exact test, H_0 hypothesis must be on the level of significance 5% rejected and adopted must be alternative H_1 hypothesis talking about the dependence between tested variables.

To better illustrate, we present graphic processing of the results from questionnaire research (Fig. 20.4). As the chart shows, the answer "a" is marked mainly a group of large companies (77%), which means that they provide a report, containing information about all responsible marketing activities. Fifteen percent of large companies say they inform partially and only in requested reports, and the smallest percentage of large companies (8%) argue they do not provide reports about socially responsible and sustainable marketing activities.

An interesting result is apparent when comparing the responses of large companies to micro companies. The percentage structure of responses is almost exactly

Table 20.4 Results of Fisher exact test

Statistic	DF	Value	Prob
Chi-square	4	69.411	<0.0001
Likelihood ratio chi-square	4	45.2651	<0.0001
Mantel-Haenszel chi-square	1	25.2563	<0.0001
Phi coefficient		0.781	
Contingency coefficient		0.6122	
Cramer's V		0.5997	
Fisher exact test			
Table probability (P)		3.15E-12	
$Pr <= P$		8.58E-09	

Source: Own processing, XLStat

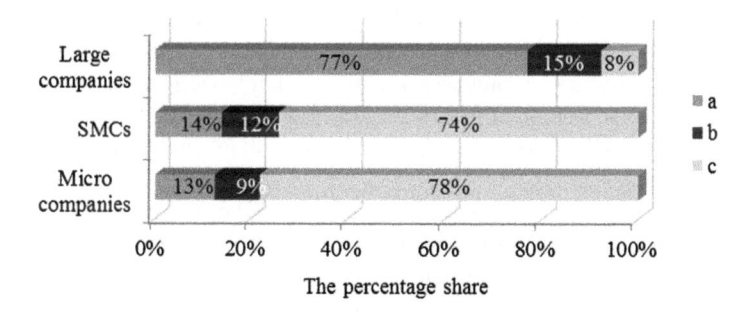

Fig. 20.4 Issuing of reports – by company size. *Legend*: (a) yes, we provide a report containing information about all marketing activities; (b) partially, we inform about marketing activities in requested reports; and (c) no, we do not provide reports about our marketing activities. (Source: Own research and processing)

opposite to responses of large companies. SMCs also do not appear like businesses that regularly issue this kind of reports. In summary we state that 26% (answer "a" and "b") of SMCs, either regularly or at least partially, apply some form of reporting related to sustainable responsible performance. Based on these facts, we can conclude that our scientific assumption was confirmed.

Conclusion

Transparency about the sustainability of organizational activities is of interest to a diverse range of stakeholders, including business, labor, nongovernmental organizations, investors, accountancy, and others. This is why GRI has relied on the collaboration of a large network of experts from all of these stakeholder groups in consensus-seeking consultations. These consultations, together with practical experience, have continuously improved the reporting framework since GRI's founding in 1997. This multi-stakeholder approach to learning has given the reporting framework the widespread credibility it enjoys with a range of stakeholder groups.

It is obvious that if the potential of social responsibility is fully exploited, it cannot become a "prerogative" of large companies but must become a matter of the whole business sector.

This issue is particularly important in the area of food sector, as these companies are directly or vicariously involved in the production and sale of food products to final customers and provide nutrition of the population. The concept of corporate social responsibility together with application of sustainable marketing opens new business opportunities. A dialogue with diverse stakeholder groups represents a rich source of ideas for new products, processes or markets, thus contribute to the creation of long-term competitive advantage.

From marketing research it can be stated that the most important stakeholders for food companies are company owners followed by customers, management, competition, suppliers, media, and employees. The lower values of importance were attributed to stakeholders as employees, investors, state/governmental institutions, creditors, and NGOs. Food companies should lead effective dialog in great extent with these groups, as they marked them as subjects with the highest degree of influence on their business.

Many authors consider that the original idea that CSR can afford only strong and rich multinational corporations is already overcome. The aim of our contention is not that small businesses are not oriented toward socially responsible and sustainable activities but insufficiently inform about these activities and should focus their attention on the area of reporting (sustainable reporting). Companies tended to be a little more guarded about providing any kind of information like environmental performance information in light of possible penalties and a bad image they may acquire. This trend is likely to change for companies that are interested in moving toward a sustainability report, where all aspects of CSR, including environment, are highlighted in one comprehensive document.

The research confirmed that there are big differences in issuing of reports between different size groups of companies. Seventy-seven percent of all large companies claim that they provide a report containing information about all responsible marketing activities. Fifteen percent of large companies say they inform partially and only in requested reports, and the smallest percentage of large companies (8%) argue they do not provide reports about socially responsible and sustainable marketing activities. We can watch an interesting result when comparing the responses of large companies to micro companies. The percentage structure of responses is almost exactly opposite to responses of micro companies.

In formulating the second assumption, we started from a global vision for issuing the reports on the level of sustainable development. Its essence lies in the fact that the administration of comprehensive reports on economic, environmental, and social performance of organizations, or even countries, will become so routine matter such as financial reporting and so on. However, the reporting in real practice of industry – particularly in small- and medium-sized businesses – still faces to several obstacles.

References

1. Voinov, A., & Bousquet, F. (2010). Modelling with stakeholders. *Environmental Modelling & Software, 25*(11), 1268–1281.
2. Svoboda, J. (2010). *Význam konceptu společenské odpovědnosti organizací (CSR) a jeho využívání v České republice*. Zlín: Univerzita Tomáše Bati ve Zlíně.
3. Kubicová, Ľ., & Kádeková, Z. (2013). Impact of consumer prices and cash income on consumption of dairy products. *Acta scientiarum polonorum, 12*(3), 61–71.
4. Carroll, A. B., & Buchholtz, A. (2014). *Business and society: Ethics, sustainability, and stakeholder management*. USA: Cengage Learning.
5. Blowfield, M., & Googins, B. K. (2006). *Step up: A call for business leadership in society: CEOs examine role of business in the 21st century*. Chestnut Hill: Boston College Center for Corporate Citizenship.
6. Ubrežiová, I., Kapsdorferová, Z., & Sedliaková, I. (2013). Competitiveness of Slovak agrifood commodities in third country markets. *Acta Universitatis Agriculturae et Silviculturae Mendelianae Brunensis, 60*(4), 379–384.
7. Weiss, J. W. (2009). *Business ethics: A stakeholder and issues management approach*. Mason: Cengage Learning.
8. Sedliaková, I., & Lenčéšová, L. (2012). Corporate culture as a competitive tool. In *Drive your knowledge be a scientist*. Zlín: Tomas Bata University in Zlín.
9. Sakál, P., et al. (2013). *Udržateľné spoločensky zodpovedné podnikanie*. Trnava: Alumni Press.
10. Maign, I., & Ferrell, O. C. (2004). Corporate social responsibility and marketing: An integrative framework. *Journal of the Academy of Marketing Science, 32*(1), 3–19.
11. Morsing, M., & Schultz, M. (2006). Corporate social responsibility communication: Stakeholder information, response and involvement strategies. *Business Ethics: A European Review, 15*(4), 323–338.
12. Zelený, J. (2008). *Environmentálne manažérstvo a spoločenská zodpovednosť (organizácií)*. Banská Bystica: Univerzita Mateja Bela.
13. Jankajová, E., Kotus, M., Holota, T., & Zach, M. (2016). Risk assessment of handling loads in production process. *Acta Universitatis Agriculturae et Silviculturae Mendelianae Brunensis, 64*(2), 449–453.
14. Steinerová, M. (2008). Koncept CSR v praxi: průvodce odpovědným podnikáním. Aspra, Praha, http://www.spolecenskaodpovednostfirem.cz/wp-content/uploads/2013/09/Konc ept %20C SR%20v%20praxi.pdf.
15. Trnková, J. (2004). Společenská odpovědnost firem: kompletní průvodce tématem & závěry z průzkumu v ČR , http://www.csr-online.cz/co-je-csr/kam-pro-vice-informaci/
16. Franc, P., Nezhyba, J., & Heydenreich, C. (2006). Když se bere společenská odpovědnost vážně. Ekologický právní servis, Brno, http://www.eps.cz/sites /default/files/publikace/kdyz_ se_bere_csr_vazne.pdf.
17. Sustainability Reporting Guidelines, https://www.globalreporting.org/resourcelibrary/ G3.1-Guidelines-Incl-Technical-Protocol.pdf.
18. IISD'S Business and Sustainable Development: A Global Guide, https://www.iisd.org/business/issues/reporting.aspx
19. Global Reporting Initiative, https://www.globalreporting.org/information/news-and-press-center/press-resources/Pages/default.aspx
20. Smernice reportovania trvalo udržateľného rozvoja, https://www.globalreporting.org/ resourcelibrary/Slovakian-G3-Reporting-Guidelines.pdf.
21. European Commission, http://ec.europa.eu/growth/smes/business-friendly-environment/ sme-definition/index_en.htm.

Chapter 21
Cooperative Relations and Activities in a Cluster in the Slovak and Czech Automotive Industry

J. Vodák, M. Varmus, P. Ferenc, and D. Zraková

Abstract Nowadays, it is difficult for an individual to achieve success, mainly because of a lot of competition. One of the solutions is to see cooperation with other individuals or groups. Since collaboration can be beneficial not only for people but also for companies, it is important for organizations too to search for key partners for their business. Collaboration with other organizations can bring many benefits – for example, small companies can improve their competitiveness on the market. Within multi-company cooperation, it is now often for companies to group into homogeneous or heterogeneous groups. One of the possibilities of grouping companies is cluster. In this group of businesses, different relationships arise between companies. These relationships, as well as business activities, are based on the focus of the cluster. In this paper, based on the analysis of two clusters, we pointed out how the cluster group should function properly. This solution is shown in a general, comprehensive model.

Introduction

Nowadays, full of fierce competition, we can move forward only with a sufficient dose of a sense of something different and something new and prove to apply it in practice. Basically, the basis for the competitiveness of the company is continuous development and innovation. This competitive advantage can also be achieved through well-designed collaboration with stakeholders or by cooperation with similar organizations and other stakeholders in cluster.

"Clusters are geographic concentrations of interconnected companies and institutions in a particular field. Clusters encompass an array of linked industries and other entities important to competition" [1].

J. Vodák (✉) · M. Varmus · P. Ferenc · D. Zraková
Faculty of Management Science and Informatics, University of Zilina, Zilina, Slovakia
e-mail: josef.vodak@fri.uniza.sk; michal.varmus@fri.uniza.sk; patrik.ferenc@fri.uniza.sk; diana.zrakova@fri.uniza.sk

Based on an analysis of selected clusters in the automotive sector, the sense of the existence of clusters and activities which the clusters perform has been specified. Subsequently, the general model of managing cluster was suggested.

In view of the foregoing, the main aim of this article is to identify new opportunities for strategic direction and improving cooperation structures which have been processed into a uniform general model. New options have been proposed based on the analysis of two real existing and relatively well-functioning associations, Automotive Cluster Slovakia (ACS) and Moravian-Silesian Automotive Cluster (MAC). Mentioned clusters were selected based on the similarity of the cluster itself but also the states on whose territory they operate.

The Reasons for the Existence of Clusters

According to authors of Business Dictionary, a cluster and a network of interconnected companies, vendors, and collaborators in a particular field are all located in the same geographical area [2]. The main idea of clusters is to provide increased efficiency and productivity to make companies more competitive at national and global level [3–5]. Thus, cluster represents a group of independent entities that are linked through cooperation on a common purpose. The goal is to gain competitiveness and increase profits for all stakeholders. Companies that work together in a cluster can create a more tricky and complex product than any of them can do separately. Clusters are recently seen as a means of increasing the innovation performance of which show the increase of the competitiveness of associated enterprises and the whole region [6, 7]. Through clusters, small- and medium-sized companies can group their activities into one. By grouping, companies create smart specialization and unrivalled prices as well as products and services. Feser also points out that it is advisable to create clusters even in a highly competitive environment, not just for small businesses [8].

Clustering is a phenomenon where companies are in close proximity (in the same locality) from the same industry [9, 10]. Because of this proximity, they are better able to create synergy [11]. Apparently, the most frequent occurrences of clustering are industries such as banking, engineering or tourism. Clusters consist of companies, suppliers and service providers but also government agencies and other institutions that, for example, provide specialized training and education, information, research and technical support. Regional economies are the cornerstone of competitiveness [12]. For the regional and national economy to be competitive, it is necessary to create valuable products and services. For this reason, it is sometimes better to connect strong companies with good ideas for a common goal, i.e. products or services that will be able to compete with products in European or global dimensions. Through clustering, not only is it possible to achieve competitiveness and create valuable products and services but also to strengthen and accelerate innovation activity. The innovation potential stems from a combination of technologies, information, academic institutions and, finally, talent and creative ideas that can be pinpointed through the joining of companies.

In general, it can be say that, the sense of cluster is development of new technologies and innovation, development of new products and services, observation and exploration of new trends but also acquisition of new members from different fields which will bring new knowledge and also the increased education of employees, which is related to attracting new business opportunities. Clusters are also beneficial in achieving cost savings of scale [13].

The sense of the creation of the association was the promotion and development of its members – improving market conditions for members and increasing their competitiveness. Specific meaning of individual association describes their mission (the reason for the existence of the cluster). The mission of ACS is to promote the development of subcontractors in the automotive industry and help to ensure their continued competitiveness at home and abroad through peer grouping of industrial enterprises, universities, scientific research institutions and other entities of the private and public sectors [14].

MAC's mission is to create conditions and support the competitiveness of the members for the region's sustainable development.

The cluster fulfilled this mission by the realization of a vision to become the integrator companies, educational and research institutions and other stakeholders whose activities support the development of the automotive industry in the region [15].

Activities of Clusters

Activities in clusters as well as in other clusters are directly influenced by the current situation in the industry. The current situation in the automotive industry in the Czech Republic and Slovakia is positive, because the automotive industry has a strong tradition in both countries and has become the most important sector and driving force of both economies [16, 17].

In both countries, it is beneficial for the business to invest in the automotive industry. Both countries have a strategic location in Central Europe (300 million clients in the radius of 1000 km and 600 million clients in the radius of 2000 km). Also, both countries have a stable business and political environment, which contributes to business development. In both countries exist well-developed transport and telecommunication infrastructure, wide supplier base, a highly educated workforce and last but not least also strong innovation potential for R&D projects [16–18].

Together, the automotive industry employs more than 350,000 people in both countries. In both countries, more than 100 cars per 1000 inhabitants (178 cars per 1000 inhabitants in Slovakia, 118 cars per 1000 inhabitants of the Czech Republic) have been produced. On this basis, it can be said that both countries are among the world's leading automotive manufacturers [16, 18].

Overall, it can be said that the business environment in the automotive industry is very attractive in these countries. For better negotiating position and competitiveness, companies are linked to various groups such as the cluster.

According to Chudoba and Svač, the exchange and transfer of information as well as their mutual dissemination and sharing are essential activities in the cluster. The authors also reported some cluster activities that have been identified in the ACS.

These activities are [7]:

- Training and classification (transfer of know-how and dedicated actions)
- Internationalization (collaboration with global markets)
- Cooperation with enterprises and educational institutions
- Marketing activities (improvement of image, identity creation and branding, public relations, etc.)
- Activity centres of excellence (research, development, tests, exams, projects, etc.)
- Information and communication with businesses in finding solutions

Activities that members of ACS and MAK expect from the association are identical and include eliminating the disadvantages arising from their size (costs of research and development, marketing, training of staff, differentiation of production), removal of the difficulties in obtaining loans and subsidies, expansion of production for border region, the support of members in the research, development and innovation, provision of the know-how, synergies in the implementation of common activities, sharing of services and human resources development [19, 20].

Activities of clusters processed into a model can be seen in Figs. 21.1 and 21.2. Figure 21.1 shows the activities carried out within the Automotive Cluster Slovakia.

Automotive Cluster Slovakia is regionally oriented. Its scope is mainly in the Trnava Region. ACS tries to create a favourable business environment for enterprises. Various cooperation relationships should be formed among the organizations in the cluster. Manufacturing companies cooperate with recruitment agencies that give them employees. These agencies gain employees from educational institutions. Educational institutions have a direct relationship with the innovation centres, which in turn cooperate with manufacturing companies. These enterprises have a relationship with the Chamber of Commerce.

The ACS with their activities seeks to fulfil the following objectives: linking production requirements with the academic and scientific research environment and internationalization in cooperation with foreign companies.

Figure 21.2 shows the activities carried out within the Moravian-Silesian Automotive Cluster.

Moravian-Silesian Automotive Cluster also creates for businesses a favourable business environment. However, in this cluster, the relationships between individual entities are different than in the ACS. Educational institutions are directly related with research organizations and the whole industry that includes automobile, mechanical and metallurgical industries. Among these components' sectors are mutual supplier- customer relations. Development and research organizations

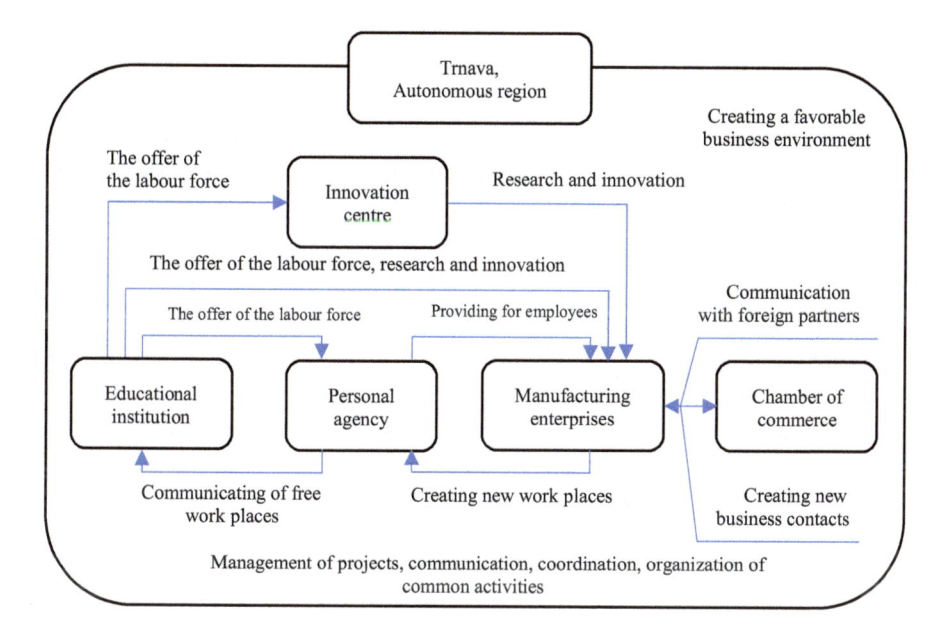

Fig. 21.1 Automotive Cluster Slovakia

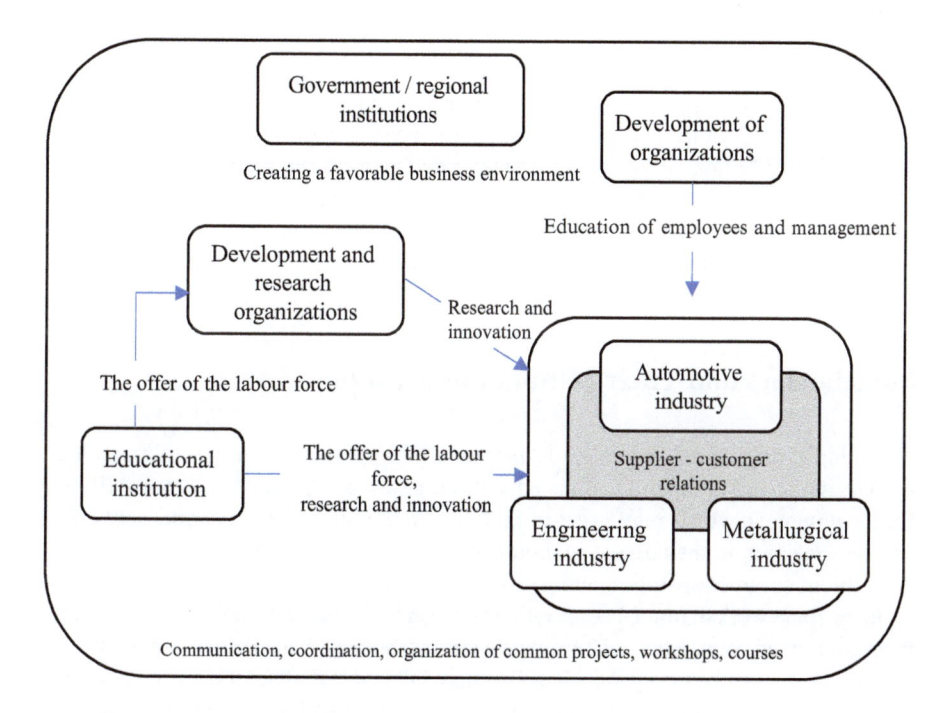

Fig. 21.2 Moravian-Silesian Automotive Cluster

impact with their activities on the industry and also the possibilities for development of organizations in the form of employee training and management of enterprises.

The MAC with their activities seeks to fulfil the following objectives: development of human resources of cluster members and the support of cluster members in the area of research and innovation.

Management and Marketing Activities Within the Analysed Clusters

Marketing and management activities in each cluster are important, like in each company. One of the most important management activities, respectively, a collection of activities, is cooperation. Cluster would not work without cooperation. Thanks to the good cooperation cluster uses all capacities of all associated enterprises. Cluster merge the innovation capacity for all own members, thanks to which it implements science and research and improve competitiveness. Creation projects and laboratory research belong to the second group of very important management activities in cluster. The projects represent common activity and the involvement of associations of undertakings in the cluster, so this is a common product that creates a cluster. Projects are often linked to the need of analysing and proposing solutions which implies the needs for research. Marketing and marketing activities in the cluster are aimed at promoting the entire cluster and its individual members not only of regional scope but especially abroad. Among the most used marketing activities include public relations, participation in conferences, exhibitions and fairs and creation and distribution of promotional materials such as catalogues or brochures. Marketing activities do not have to perform only companies in the cluster but also the other stakeholder, for example, the booklet that released the Ministry of Economy of the Slovak Republic for support Slovak clusters (for more resources [21]).

Stakeholders and Their Mutual Cooperation

"Building clusters and networks is particularly important in terms of potential growth interested companies, their education, new opportunities and possibilities for development which may arise in building cooperation. Clusters and networks can be a reaction to the current economic crisis and strengthen the company which act without broad cooperation and partners yet" [7].

In the framework of mutual cooperation between individual enterprises of the cluster, there are formed cooperative relationships with stakeholders. Stakeholders are individuals, groups of persons or organizations that may influence or affect the business or are affected by the activities of the enterprise [22, 23]. Therefore, it can be said that individual companies in the cluster are not only partners but also stakeholders. To achieve success, you must venture the key stakeholders who can then use to achieve success [24].

The collaboration with the various stakeholders is one of the core activities for enterprises in the cluster. Thanks to the good control of cooperative relations, the company may achieve the desired success or elimination of one of their weaknesses (e.g. low competitiveness).

The above-mentioned relationship management in the cluster can be difficult for businesses; therefore, to simplify the management of cooperative relations with companies within the cluster and individual stakeholder, it is appropriate to use the DSS (decision support system).

To ensure the effective functioning of the competitiveness of joint ventures in the cluster, or entire industries, is needed to support decision-making by means of IS/ICT. Under conditions of globalization, it is not basically possible to obtain information without the use of IT. Through integrated IT applications, it is possible in the area of management, for example, to interconnect IS stakeholders, to ensure the smooth operation of the JIT (just-in-time), to link the bank data and knowledge of cooperation partners, to ensure product and service innovation, and to be able to streamline and speed up a slow human labour; thanks to integrated IT applications, it is possible to better coordinate geographically decentralized operations, which, in the case of clusters, is very important [25].

According to Škorecová, in a modern economy, late recommendation to minimize competitive clashes and neutrality of small- and medium-sized enterprises is secured by strategic alliances. These strategic alliances are based on information and communication processes, which show the imperative need to IS/ICT. From the above, it can be concluded that nowadays, there should be an emphasis on cooperation and the creation of competitive associations of small- and medium-sized enterprises that support its operation through the IS/ICT. Because alliance is an association of undertakings like the cluster, it also implies for the cluster to use support their decisions by IS/ICT, respectively, directly with decision support systems (DSS) [25].

According to Shim et al., DSS is an interactive information system that supports the decisions by solution of well-structured and unstructured problems.

The outputs from the DSS are various variants of solutions that may be in the form of model, reports, graphs, etc. [26].

In spite of the fact that DSS is made up of models, a variety of structured and unstructured data, databases where this information is located but also the people and the various processes, it can be concluded that DSS can support decision-making in the cluster as a whole. So if DSS associates models, people, databases, etc., it can be able to associate also associated enterprises with the cluster and support their common decision [27].

Models of Cooperative Connection in the Cluster

Based on previous findings, the model of collaborative links within the cluster was proposed. This model was named as model six components because the identified information was six main components affecting the cluster as a whole. The main idea of this model is connection factors Porter's diamond and main components of the cluster. Porter's diamond reflects the effects of four components (firm strategy, structure and rivalry, demand conditions, factor conditions, related and supporting industries) with respect to the geographical coverage of the cluster. However, the Porter Diamond Framework cannot be regarded as a new theory, because it is rather a framework that improves the understanding of international competitiveness of companies [28]. A model including the influence of Porter's diamond can be seen in Fig. 21.3.

Cluster, as well as businesses that are grouped, carries on business. In carrying out business activities, the clusters proceed mostly in traditional way (planning, organization, implementation, evaluation and monitoring activities). According to the proposed model, the clusters should be conducting its business activities, first to define the objectives, then carry out the activities to achieve

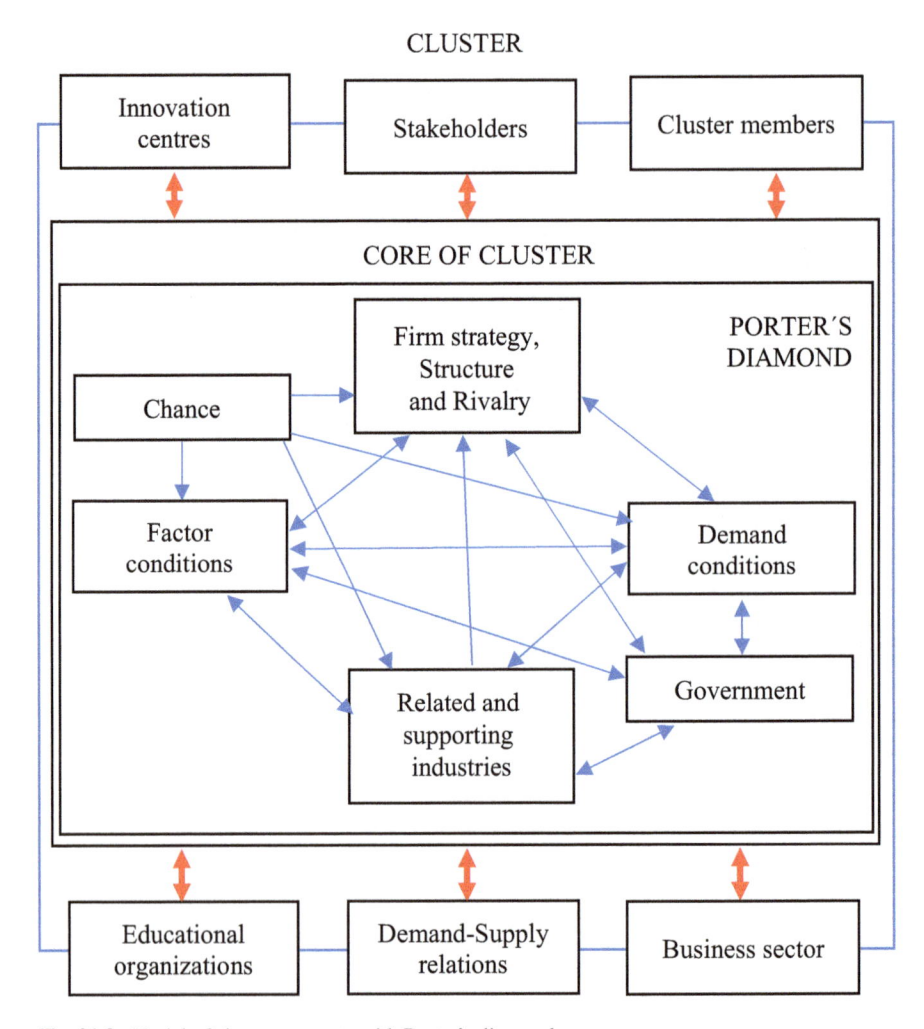

Fig. 21.3 Model of six components with Porter's diamond

the set objectives and evaluate the attainment of goals. Consequently, it is necessary that the findings are compared with market requirements. Finally, it needs to be adapted to the requirements of cluster activities. To the external world, cluster should perform various marketing and management activities. The main marketing activities, which should perform automotive clusters include participation in competitions, public relations and organizations of conferences, fairs or exhibitions. Management activities are creating projects and research in the laboratory (Fig. 21.3).

However, the suitability of both of these models can be questioned because the core of cluster is only counted with the influence of Porter's diamond. Porter's diamond has a very narrow view and does not take into account the foreign activity [29], which is very important in the clusters and the automotive industry. It is therefore better to consider the impact on the core of cluster by the generalized double diamond.

The generalized double diamond includes domestic and global view too. The size of the global diamond is determined by prediction, and the size of the domestic diamond is determined depending on the size of the country and its competitiveness in the sector (which is changing dynamically).

The diamond dotted line between these two diamonds is an international diamond. This diamond represents national competitiveness determined by both domestic and international parameters.

The generalized double diamond calculates competitiveness as the ability of companies to engage in industry-specific activities in a given country to maintain added value over long periods of time despite international competition [29].

As can be seen, new model solutions differ only in perceptions of the impact on the cluster core. All other cluster functions remain the same as for the model solutions shown in Figs. 21.3 and 21.4.

Looking at the automotive industry in Slovakia, it is possible to see its great dynamics, global competitiveness, continual growth and the creation of added value. It is therefore advisable to think more about the impact of the generalized double diamond on the cluster core than on the impact of Porter's diamond. This modified model can be seen in Fig. 21.5.

The Benefits, Anticipated Benefits and Planned Effects

Clusters bring many benefits to associated enterprises. The very small businesses by themselves could not be as successful as where they form a single unit. By the concentration of core competencies, they form clusters of specialized suppliers and stakeholders who can collectively create a unique value and they be able to be more flexible than individual enterprises. Enterprises grouped into clusters can promptly respond to emerging changes in global markets, and they can also share costs and risks that are spread to cluster members, so they are reduced. An activity that constitutes a further advantage is also non-public dissemination of knowledge among

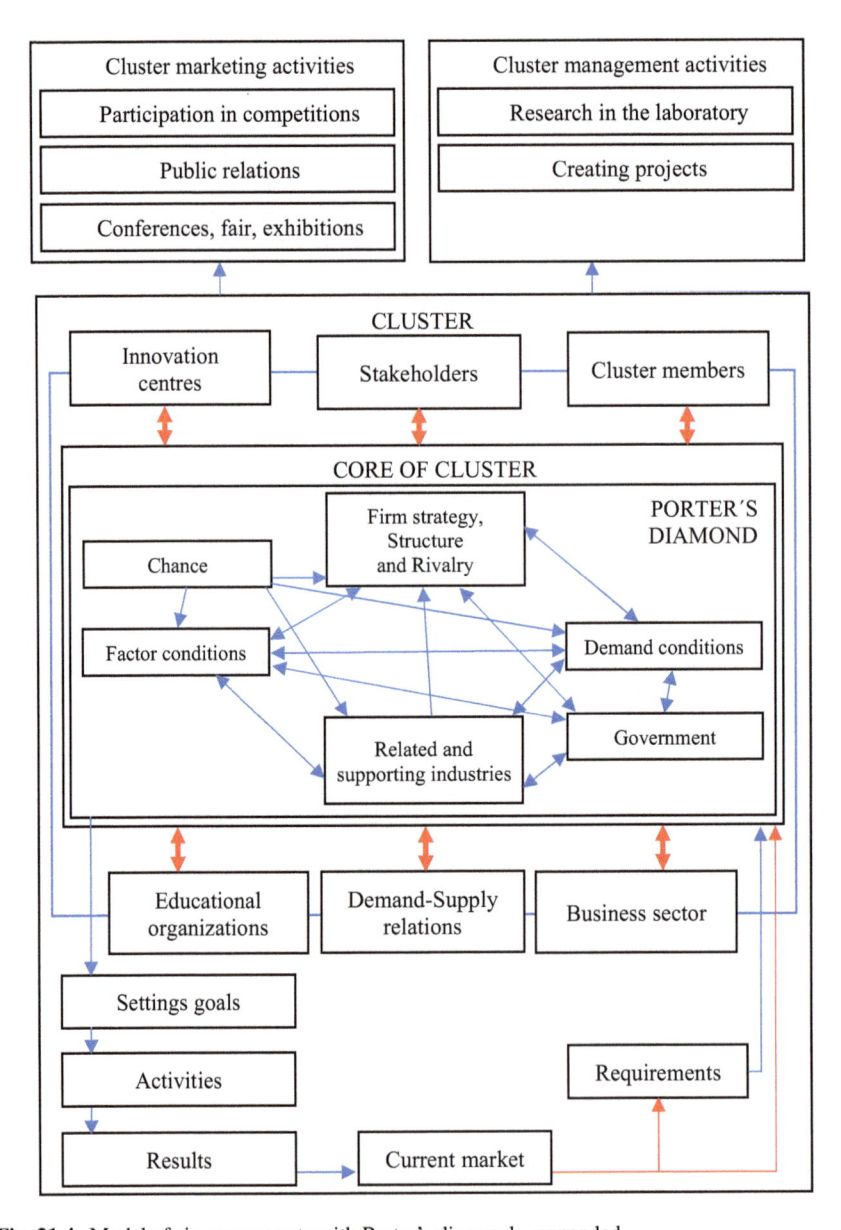

Fig. 21.4 Model of six components with Porter's diamond – expanded

cluster members. This activity brings the better mutual education, improving the knowledge of channels and also increasing the overall level of cluster members. The cooperation between firms in the association, so as in a cluster, creates a competitive advantage for the whole and for individual companies, reducing risks and costs [7].

The introduction of these models should bring a new international contacts, expand the membership base, improve counselling for members, improve learning

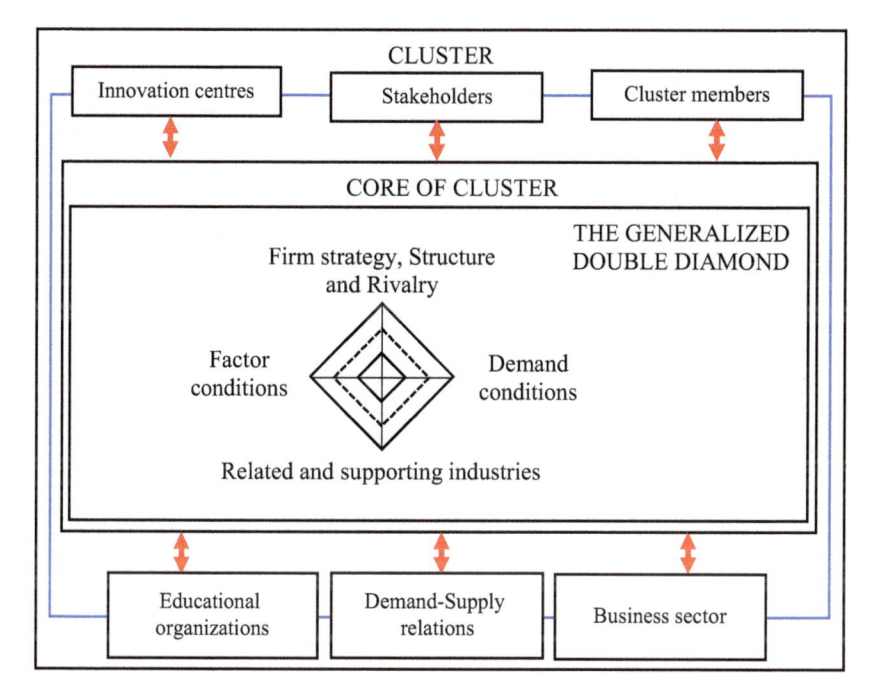

Fig. 21.5 Model of six components with the generalized double diamond (Modified extended model of six components (with the generalized double diamond) can be seen on Fig. 21.6)

opportunities for members of cluster and improve the scientific and research activities of the cluster for cluster.

Implementation Risks

The implementation of each model brings with it certain risks. In the case of implementing the extended model, four main potential risks were identified. The first risk is the negative change in the external environment. This environment can be affected by cluster positive as well as negative, such as with different laws and regulations, either in the region or abroad. The cluster can be also negatively affected by economic crisis, which may impact on the operation of the whole cluster and to its individual members. Competition, which belongs to the external environment, affects the cluster at all times but may occur as a potential threat to unfair fights. The second major risk is ineffective marketing which result in a lack of membership, if it cannot attract businesses and institutions from all main components (educational institutions, R&D institutions and production companies). Inefficient marketing could also affect the enforcement of the cluster in foreign markets, while the companies do not achieve added value in the form of competitive advantage. The third risk is the failure to fulfil the objective/purpose of

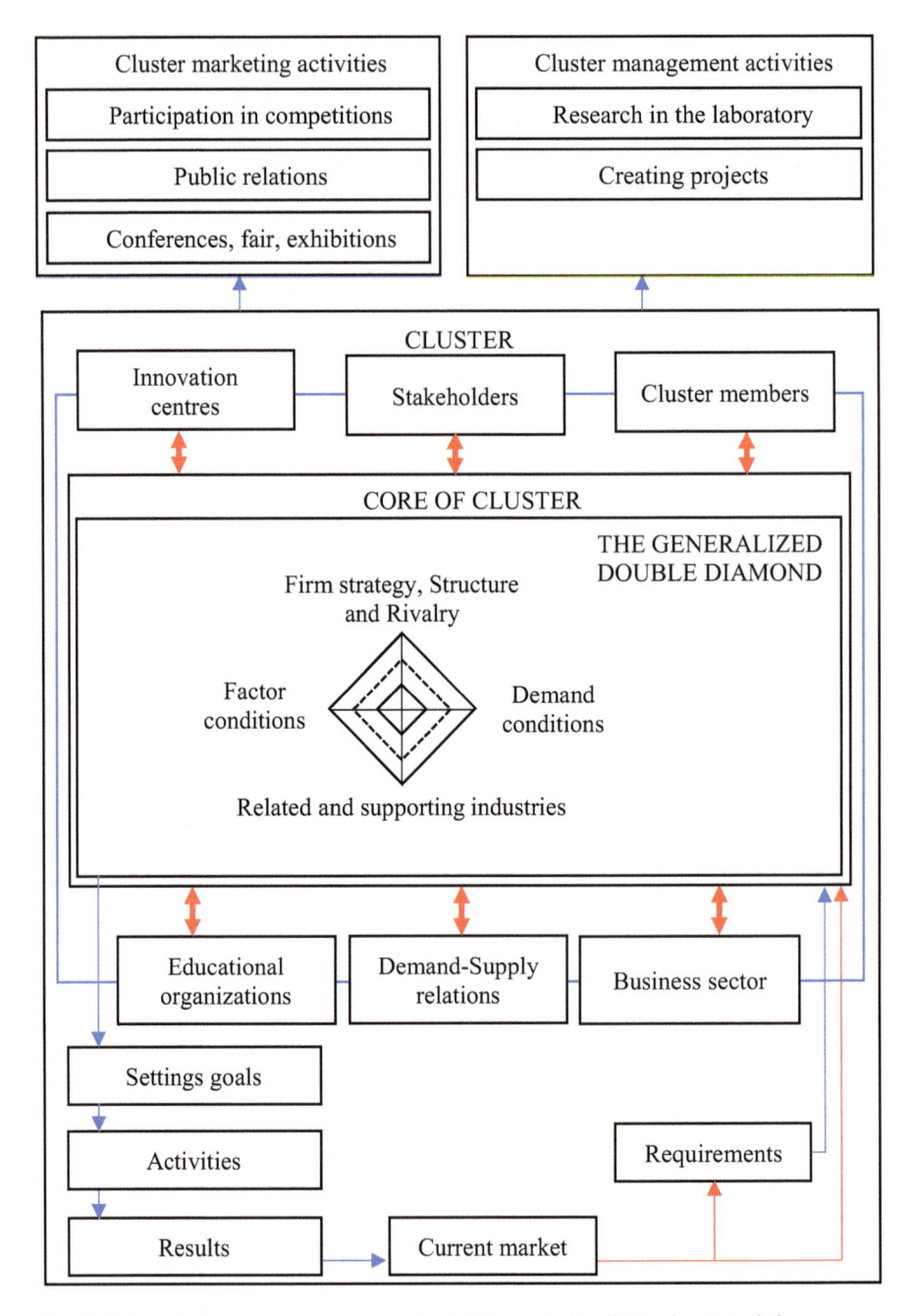

Fig. 21.6 Model of six components with the generalized double diamond – expanded

formation of the cluster, namely, large businesses in the cluster are at risk of stagnation, if they do not engage to the whole membership. By failing to comply with the objective/purpose, we can expect the risk that the cluster will not be able to create your own project or to engage in other solutions; the consequence may also be an unsuccessful research and development of innovation. The last risk is the reluctance of cluster membership. The reluctance may arise in several areas – the unwillingness of communication and cooperation with other members of the cluster, also reluctance to actively participate in the project or unwillingness to share information, know-how, or producing and innovating capacities.

Conclusion

Clustering is a modern way of increasing the competitiveness of regions, companies and other institutions in its territory. Cooperation of companies associated in the cluster given the rapidly evolving markets in the future is almost inevitable. There were designed two model solutions arising from the analysis of two clusters in the automotive sector. Designed models point to the need to improvement of relations between the individual components of the cluster. Model solutions have been proposed universally, and therefore they are suitable for usage in various business sectors. Expansiveness of solved issues opens the way for further investigation. As part of a possible future investigation, it would be interesting to think about questions such as: How to increase competitiveness of cluster? Which specific marketing and management activities will be carried out in a cluster? How could be possible to achieve synergies in cooperation in a cluster? What effect have stakeholders on the performance of the cluster? What types of system to support decision (decision support systems) should be used by the cluster to gain greater competitive advantage?

Acknowledgements The paper was supported by the Slovak Research and Development Agency under the contract no. APVV-15-0511 and the project VEGA 1/0617/16 – Diagnosis of Specifics and Determinants in Strategic Management of Sporting Organizations.

References

1. Harvard Business Review (1998). *Clusters and the new economics of competition.* November–December 1998 Issue. Available at: Https://hbr.org/1998/11/clusters-and-the-new- economics-of-competition
2. Business cluster. Available at: http://www.businessdictionary.com/definition/business-cluster.html
3. Porter, M. (2002, April). *Regional foundations of competitiveness and implications for government policy.* Paper presented to Department of Trade and Industry Workshop.
4. Rocha, H. (2004). Entrepreneurship and development: The role of clusters. *Small Business Economics, 23*, 363–400.

5. Bergamn, E., & Feser, E. (1999). *Industrial and regional clusters: Concepts and comparative applications*. University of West Virginia. Retrieved from http://www.rri.wvu.edu/WebBook/Bergman-Feser/contents.htm. Accessed 24 May 2017.
6. Šmíd, J. *Connecting companies*. Available at: http://www.nanosvet.sk/_paper/zdruzovanie_podnikov_odvetvova_a_geograficka_blizkost.pdf
7. Transfer of Innovations. (2009). *Projekt Automotive Cluster Centrope*. Transfer of Innovations volume 13/2009: 160–161. Available at: https://www.sjf.tuke.sk/transferinovacii/pages/archiv/transfer/13-2009/pdf/160-161.pdf
8. Feser, E. (1998). Old and new theories of industry clusters. In M. Steiner (Ed.), *Cluster and regional socialisation: On geography, technology and networks* (pp. 18–40). Londres: Pion.
9. Perry, M. (2005). *Business clusters: An international perspective* (256 p). New York: Routledge. ISBN: 0-415-33962-6
10. Swann, G., & Prevezer, M. (1996). A comparison of dynamics of industrial clustering in computing and biotechnology. *Research Policy, 25*, 1139–1157.
11. Rosenfeld, S. (1997). Bringing business clusters into the mainstream of economic development. *European Planning Studies, 5*(1), 3–23.
12. Cluster mapping. Available at: http://www.clustermapping.us/content/clusters-101
13. Cluster initiative as a tool for increasing the efficiency and prosperity of the Liptov region enterprises. Available at: http://www.cutn.sk/Library/proceedings/mch_2014/editovane_prispevky/15.%20Littvov%C3%A1.pdf
14. Vision and goals ACS. Available at: http://www.autoklaster.sk/sk/o-autoklastri/vizie-a-ciele-autoklastra
15. Strategy of MAC. Available at: http://autoklastr.cz/strategie
16. Czech Automotive Industry. Available at: http://www.czechinvest.org/en/1automotive-industry
17. Slovak Automotive Industry. Available at: http://www.sario.sk/en/invest/sectorial-analyses/automotive-industry
18. Summary of Slovak Automotive Industry. Available at: http://www.sario.sk/sites/default/files/content/files/sario-automotive-sector-in-slovakia-2017-03-02.pdf
19. Membership in ACS. Available at: http://www.autoklaster.sk/sk/clenstvo
20. Website of MAC. Available at: http://autoklastr.cz
21. Clustering. Available at: https://www.siea.sk/materials/files/inovacie/slovenske_klastre/SIEA-brozura- Klastrovanie.pdf
22. Trnava Industrial and Technological Park. Available at: http://www.trnava.sk/sk/clanok/mestsky-priemyselny-a-technologicky-park-trnava-1#sthash.VA3E8XSG.dpuf
23. Participation at the fair ACS. Available at: http://www.autoklaster.sk/sk/news/487-veltrh-hannover-messe
24. Scientific laboratories MAC. Available at: http://www.rozhlas.cz/ostrava/aktualne/_zprava/menza-vsb-skryva-velmi-neobvykle-vedecke-laboratore--1106244
25. Škorecová, E. *Information technology as a factor of competitiveness*. Available at: http://www.slpk.sk/eldo/2005/011_05/sekcia4/skorecova.pdf
26. Elsevier Science. (2002). Past, present, and future of decision support technology. *Decision Support Systems, 33*, 111–126.
27. Tripaty, K. P. (2011). Decision Support System is a tool for making better decisions in the organization. *Indian Journal of Computer Science and Engineering, 2*(1), 112–117.
28. Porter's Diamond Framework. Available at: https://www.ajol.info/index.php/sabr/article/view/76358
29. Porter, M. E., & Armstrong, J. (1992). Canada at the crossroads: Dialogue (Response by Porter). *Business Quarterly, 56*(4), 6–10.

Part IV
Smart Technology Trends in Materials

Chapter 22
Composites Manufacturing: A New Approaches to Simulation

Lucia Knapčíková, Michal Balog, Alessandro Ruggiero, and Jozef Husár

Abstract The paper deals with new approaches of the composite manufacturing used in simulation software. First and perhaps the most important step is a right definition of targets simulation. After model verification, we can make experiments. The advantage of presented aper is the proposal of simulation model design used by composite manufacturing. Advantage of the simulation program is in the field of possibility to choose the right technology of the composites manufacturing, using the material flow and working times of course. Simulation process saves financial resources in the enterprises and especially decreasing the work time for preproduction stage in the composites industry.

State of the Art

Technological processes are important part of the production system. The behavior and functioning of those systems cannot predict with certainty, because it belongs to a group of determined probability systems [1]. If we say about using and increasing the efficiency, then it is generally to minimize costs and maximize benefits. If we want to know the exact behavior of these systems, we would need to know how to mathematically describe them or observe the behavior of the real object [2]. Simulation software environment consists of four basic parts. The basic menu includes a panel necessary for working with files and functions associated with modeling activities. The simulation panel shows the progress of the work with the model through a tree structure [2, 3]. The section entitled modeling window has a squared base, which facilitates positioning of the imagination department. At the

---

L. Knapčíková (✉) · M. Balog · J. Husár
Technical University of Košice, Department of Industrial Engineering and Informatics, Prešov, Slovak Republic
e-mail: lucia.knapcikova@tuke.sk; michal.balog@tuke.sk; jozef.husar@tuke.sk

A. Ruggiero
University of Salerno, Department of Industrial Engineering, Fisciano, Italy
e-mail: ruggiero@unisa.it

bottom of the working environment is a panel element. The panel element is used to form the model [4]. The elements are arranged according to type as basic, transportation, data, delivery equipments, graphs, and statistics [2].

Experimental studies of manufacturing process can be now implemented by using available simulation software. In general, simulation software is used for optimizing of manufacturing process. Using simulation software, it is possible to experimentally simulate many variants of model situations and to prioritize interventions in the course of the process.

Composite Manufacturing Process

By simulation program it was simulated the preparation process of composites manufacturing (Figs. 22.1 and 22.2) [5]. The important part was to use cleaning process of fabrics from used tires. After cleaning, a composite material from waste tires was prepared. The composite consists of two components. We used thermoplastics material, the polyvinyl butyral as a matrix and the filler were fabrics from used tires [6, 7].

Table 22.1 describes main parameters by homogenization (Fig. 22.3) of recycled polyvinyl butyral. Table 22.2 describes pressing parameters of material pressing (Fig. 22.4) [7].

Simulation Program, Results, and Discussion

The productive system should meet "3E condition."

- Economy
- Efficiency
- Efficiency (outputs and customer demand)

Fig. 22.1 Fabrics from used tires [5]

Fig. 22.2 Recycled polyvinyl butyral [5]

Table 22.1 Homogenization of recycled polyvinyl butyral by Brabender lab station

Parameter	Characteristic
Preheating in [min]	10
Work temperature in [°C]	150
Homogenization PVB in [min]	25
Homogenization temperature PVB in [°C]	150
Homogenization PVB and fabrics in [min]	30
Homogenization temperature of mixture in [°C]	180

Making productivity increase means producing more products or reducing the cost of the product over time. The company's controlling strategy is to set the production in the company to "leanness," which means eliminating waste from, for example, overproduction, unnecessary handling, material waiting, machine and equipment dwellings, tool searching, complex material transport, machine tracking by the operator, and others. Used simulation software by material flows was able to set up the system so that it has the least developed workplaces to pass through the product, or the number and associated workloads are optimal [5, 7, 10]. Table 22.3 shows input of materials used by simulation process of composite manufacturing.

After homogenization it has achieved a mixture material. After mixture pressing it obtained a final product. Average working value on the machine by each process for a particular component (fabrics and PVB) is 81%. The following graphic representation (Fig. 22.5) describes values for each component entering the process, especially for fabrics, PVB, mixture of PVB and fabrics, and finally a composite [8].

Average time for recycled polyvinyl butyral and fabrics from used tires is 100% for each component [9]. The other key factors (work in process, entered, shipped, etc.) in this technological process are insignificant [11, 12]. In Fig. 22.6 we can see the layout of manufacturing process used simulation software.

Fig. 22.3 Homogenized polyvinyl butyral reinforced by fabrics from used tires

Table 22.2 Pressing of homogenized polyvinyl butyral by Brabender W 350 E Laborpresse

Parameter	Characteristic
Work temperature in [°C]	150
Preheating in [min] and temperature in [°C]	25
Pressing temperature in [°C]	150
Cooling in [min]	30

Fig. 22.4 Pressing of composite material used Brabender W 350 E Laborpresse

22 Composites Manufacturing: A New Approaches to Simulation

Table 22.3 Materials input used by simulation process

Name	Fabric	PVB	PVB and fabric	Composite
Input materials	1	1	1	1
No. of materials	1	1	1	0
Work in process(WIP)	0	0	0	1
Avg. WIP in [%]	0.81	0.81	0.16	0.02
Avg. time in [min]	100.00	100.00	20.00	3.00

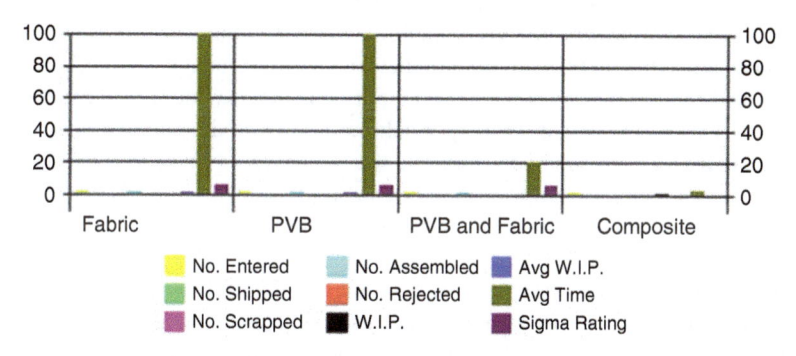

Fig. 22.5 Simulation of material input

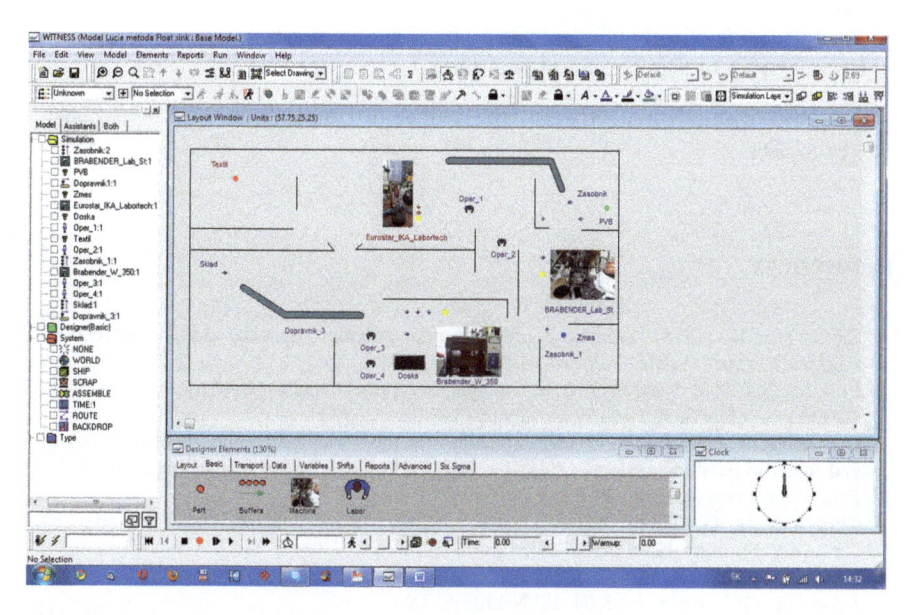

Fig. 22.6 Simulation of material flow by composite manufacturing

The working time [13] as an input parameter for individual operations was needed too. It was also necessary to set an amount in the working process [14, 15]. The simulation process [16, 17] is realized for composite product. Working times used by simulation process are defined as operating time, the time needed for each operation [18], and setting up time is the time for heating and cooling.

Conclusions

Used of variety simulation software is available in the portfolio of software analytic tools and consulting services to enable experts and managers to design forecast and evaluate the planned production before it is implemented. With a validated and verified model, it is possible to carry out considerably more experiments, as well as enrich the model for tracking machine failures in the manufacture or servicing of machinery by technical staff and the associated waiting for the operator.

The focus of simulation tools is mainly on enterprises with a significant share of foreign capital, especially in the new business start-up phase and in the start-up phase of new production.

The aim of this paper was to show how the simulation software can help the composite manufacturing by planning and used and manufactured of composite materials, for example, composite from recycling materials.

The other important reasons are:

- Reorganizing company in order to achieve a more efficient organizational structure
- Reengineering of technological and all business processes
- Restructuring of the production program

The need of simulation software can significantly improve production and business processes.

References

1. Zgodavová, K. (1998). Simulačné projektovanie systémov riadenia kvality, QPROJEKT PLUS, s. r. o., p. 2. ISBN 80-967144-4-9.
2. Lešková, A.(2003) Nástroje počítačovej podpory projektovania pracovísk na báze AL – stavebnicových prvkov, Transfer inovácií, 6, 3.
3. Devise, O., & Pierreval, H. (2000). Indicators for measuring performances of morphology and material handling systems in flexible manufacturing systems. International Journal of Production Economics, 64, 209–218.
4. Chryssolouris, G. (2006). Manufacturing systems: Theory and practice (2nd ed.p. 9). New York: Springer. ISBN 0-387-28431-1.
5. Václav, Š., & Benovič, M. (2011). Simulation assembly in teaching. Journal of Technology and Information Education, 3, 17–21. ISSN 1803-537X.

6. Kara, S., & Kayis, B. (2004). Manufacturing flexibility and variability: An overview. *Journal of Manufacturing Technology Management, 15*(6), 466–478. ISSN 1741-038X.
7. Knapčíková, L. (2011). *Optimalization of technological processes by recovery of plastics materials*, Diss. Thessis, FVT TUKE with a seat in Prešov.
8. Mihalíková, J. (2007). Problém výberu simulačného nástroja pre simulačný projekt, *Novus Scientia*, TU Košice, TU, pp. 393–394. ISBN 978-80-8073-922-5.
9. Knapčíková, L., Lazár, I., & Husár, J. (2011). Using of simulation by fabrics separation (Využitie počítačovej simulácie pri výbere optimálnej metódy dočistenia). *ATP Journal, 8*(2011), 38–40. (in slovak).
10. Knapčíková, L. (2012). Využitie simulačného programu v oblasti reverznej logistiky, Trendy v podnikání 2012: mezinárodní vědecká konference: 15. – 16.11.2012, Plzeň. - Plzeň: Západočeská univerzita, 2012, pp. 1–5. (in slovak).
11. Koste, L. L., Malhotra, M. K., & Sharma, S. (2004). Measuring dimensions of manufacturing flexibility. *Journal of Operations Management, 22*(2), 171–196. ISSN 0272-6963.
12. Paholok, I. (2008). Simulácia ako vedecká metóda. E-Logos, *Electronic Journal For Philosophy*. Vol.1, ISSN 1211-0442, p. 1–19.
13. Law, A. M. (2007). *Simulation modeling and analysis* (p. 768). McGraw-Hill Companies, New York, ISBN-13 978-0-07-298843-7
14. Malindžák, D. (1991). *The process simulation (Simulácia procesov)* (p. 298). Košice: TU. (in slovak).
15. Panda, A., Jurko, J., & Pandová, I. (2016). *Monitoring and evaluation of production processes an analysis of the automotive industry* (p. 117). Springer International Publishing Switzerland, ISBN 978-3-319-29441-4
16. Straka, M., Trebuňa, P., Rosová, A., Malindžáková, M., & Makyšová, H. (2016). Simulation of the process for production of plastics films as a way to increase the competitiveness of the company. *Przemysl chemiczny., 95*(1), 37–41. ISSN 0033-2496.
17. Semančo, P., & Fedák, M. (2013). Assessment of material flow in foundry production by applying simulation analysis. *Applied Mechanics and Materials, 308*, 185–189. ISSN1662-7482.
18. Wang, W., & Koren, Y. (2012). Scalability planning for reconfigurable manufacturing systems. *Journal of Manufacturing Systems, 31*, 83–91.

Chapter 23
Study of the Cutting Zone of Wood-Plastic Composite Materials After Different Types of Cutting

Dusan Mital, M. Hatala, J. Zajac, P. Michalik, J. Duplak, J. Vybostek, L. Mroskova, and D. Knezo

Abstract This chapter deals with composite materials, particularly wood-plastic composites. The introductory section describes composite materials and their distribution, as well as wood-plastic composites and their history, usage, composition and production technology. The practical part deals with monitoring the cut surface obtained using water jet cutting and band saws. The samples were measured using a Mitutoyo SJ 400 roughness tester. Measurements revealed that the percentage of the wood in wood-plastic composites has a strong influence on surface roughness.

Introduction

Composite materials are materials that are composed of two or more materials. More specifically, a wood-plastic composite (WPC) is any composite material containing two basic components: cellulose-based natural fibers and thermoplastics/thermosets. Natural fibers are used as reinforcements for polymers because of their relatively high strength, rigidity, low cost, and biodegradability (reducibility). Production technologies for WPCs include extrusion for linear profiles, injection molding to make three-dimensional parts with regular and irregular shapes, and a calendering process for producing flooring. The key component of their manufacture is to establish a sufficiently effective link to the interface of components. One method to improve this interaction is the inclusion of tie substances (as additives) in the process of mixing (before the extrusion), thereby increasing the compatibility of

---

D. Mital (✉) · M. Hatala · J. Zajac · P. Michalik · J. Duplak
J. Vybostek · L. Mroskova · D. Knezo
Department of Automobile and Manufacturing Technologies, Technical University of Kosice Presov, Slovakia
e-mail: dusan.mital@tuke.sk; michal.hatala@tuke.sk; jozef.zajac@tuke.sk; peter.michalik@tuke.sk; jan.duplak@tuke.sk; jaroslav.vybostek@tuke.sk; lenka.mroskova@tuke.sk; dusan.knezo@tuke.sk

© Springer International Publishing AG, part of Springer Nature 2019
D. Cagáňová et al. (eds.), *Smart Technology Trends in Industrial and Business Management*, EAI/Springer Innovations in Communication and Computing, https://doi.org/10.1007/978-3-319-76998-1_23

hydrophobic and hydrophilic components (creating a single-phase composite). Unlike plastic products, WPCs are considered to be environmentally friendly or "green" materials because recycled plastics (or waste from the wood-processing industry) can be used for their production.

From the point of view of machinability, WPCs are easier materials to work with than are metals and plastics, as shown in many studies [1–4]. However, in machining, the heterogeneity of material plays an important role (resulting from the combination of wood particles and a polymer). Any surface defects arising as a result of "incorrect" machining lead to additional costs for repair. During machining, the set parameters of the machining process technology, the mechanical properties of the material, and the compatibility of the material are all important, as shown in the study by Hutyrova and Zajac [5]. In the experimental research described in this chapter, the mechanical properties of a composite were assessed from images acquired during non-destructive testing. A radiographic method was used, which can evaluate air bubbles in the profile volume and cracks on the surface after machining.

The term *wood-plastic composite* is used to describe composite materials consisting of a wood filler and polymer matrix. Wood filler is mainly used in wood flour from various wood types to reinforce the thermoplastic matrix. The main advantages are its cost and low density. WPCs are biodegradable and recyclable, which places them in the group of green materials. The most common thermoplastic polymers used in WPC materials are polyethylene, polypropylene and polyvinylchloride. Specific combinations of wood fillers and thermoplastics can be custom designed and produced for selected applications. Properties of composite materials can be set by various concentrations of the fillers, matrix and process variables such as temperature, pressure and stir. Manufacturing of the composite materials involves high costs related to technology and equipment (form, machine, maintenance and so on). The amount of wood in the composite is usually in the range of 30–70%. WPC is characterized by less water absorption and higher resistance to molds and fungus. Wood-plastic composites do not require additional surface treatment, and with wood ratio over 40% the materials are characterized by higher temperature resistance [3, 7].

Today, WPC are mainly used in the construction industry as a substitute for impregnated wood for terraces, railings, wood frames and so on. Currently, WPC materials also replace components in the automotive industry, aerospace industry, and the electrical industry. Appropriate selection of raw materials is the basic requirement to achieve the desired properties [1].

Wood flour is obtained from wood particles from waste products in the wood processing industry. Wood mass comes from sawmills, furniture and joinery factories, as well as forestry; it comes in dissimilar sizes and moisture content and is blended with other substances [5, 8].

Wood flour (Table 23.1) density depends on several factors including moisture content, particles size and wood type. Wood flour is unusual for its compressibility, because during the fabrication process, temperature and high pressure causes material compression. The specific weight of wood flour is around 1300 kg/m^3 [6].

23 Study of the Cutting Zone of Wood-Plastic Composite Materials After Different...

Table 23.1 Parameters of wood flour

Density [kg/m^3]	Dimensions classes [μm]	Ratio length/toughness	Humidity [%]
190–220	50–150; 100–200; 200–450; 250–700	2:1–5:1	4.VIII

Table 23.2 Parameters wood fibers

Fiber type	Density [kg/m^3]	Length [mm]	Diameter [μm]
Linen	1440	25–50	15–18
Cotton	1520	15–55	10–17
Cannabis	1480	15–20	15–50
Soft wood	1440–1550	0.7–11	2–12
Hard wood	1440–1550	0.1–7	2–7

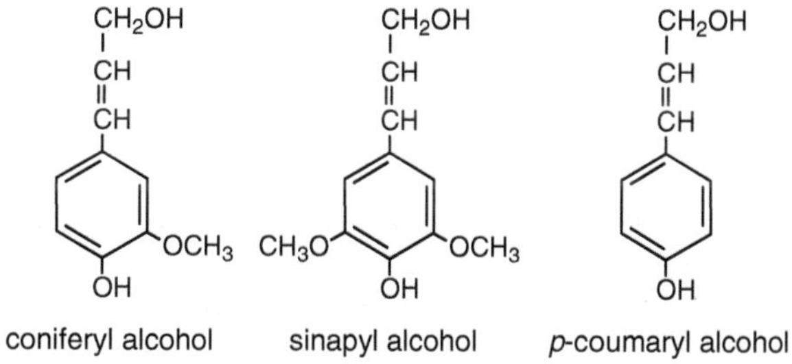

Fig. 23.1 Structure of the lignin derivates

coniferyl alcohol   sinapyl alcohol   p-coumaryl alcohol

Wood fibres are cellulose fibers which are obtained from wood, straw, bamboo, cotton seed, hemp, sugar cane and other natural materials. Due to the better mechanical properties, wood fibres began to be used for the production of WPC. In general, a higher slenderness ratio leads to better mechanical properties. The physical and mechanical properties of the fibers are dependent on the type of wood, and the choice is largely influenced by the properties of the final product. The dimensions and the length of wood fibers are quite variable (Table 23.2) [7].

Lignin is a polymer of three-dimensional phenyl propanol which is aromatic and amorphous. The basic building blocks are phenyl propanol derivatives (e.g. P-coumaryl alcohol, coniferyl alcohol, sinapyl alcohol) (Fig. 23.1). Lignin acts as a type of glue, which is used to join the material in the cell wall, and supplies the appropriate stiffness and strength. Soft woods are comprised of about 22–33% lignin, but in hardwood, this value is about 16–25%. If we want to produce a quality material, the lignin needs to be removed from the wood [1, 5].

Thermoplastic matrices used in WPC production have to be characterized by a melting temperature under 200 °C because of degradation of the wood particles at higher temperatures. Polyethylene with high density (HDPE), polypropylene (PP) and polyvinylchloride (PVC) are the currently used thermoplastic matrices for manufacturing WPC [3].

Polyethylene (PE) is the most produced plastic in the world. It has a relatively low melting point varying between 106 and 130 °C depending on the density. PE is produced by the polymerization of ethylene achieved using different methods (e.g. radical polymerization or anionic, cationic additional polymerization and the ionic coordination polymerization), and each of these methods provides a polyethylene with different physical-mechanical properties [4].

Polypropylene or polypropene (abbreviation PP or POP) is a thermoplastic polymer from the polyolefin and is used in many sectors including the food industry, textile industry and laboratory equipment. It is sold under the trade names Tipplen and Tatren. Its has a low density of about 0.90–0.92 g/cm^3 [3, 6].

Polyvinyl chloride, abbreviated PVC, is a plastic polymer made artificially. PVC is a plastic material that is widely used in construction, transportation, electrical-electronic and medical devices. PVC is extremely durable and a long-lasting construction material, which can be used in various applications, either rigid or flexible. Due to the properties of PVC it is used in many sectors to fabricate a number of popular and useful products.

Additives used in the manufacturing process of WPC materials eliminate deficiency and problems connected with joining two different materials. Adhesion between hydrophobic plastic materials and hydrophilics of the wood mass is quite low [5, 7].

Reasons for use of additives:

- Improve conditions during processing
- Connection of raw materials; higher level of resistance against biotic and abiotic factors
- Improve the mechanical and physical properties

Stabilizers:

- UV stabilizers – absorb UV radiation, which depredate polymers
- Thermo oxidation stabilizes – increase temperature level and viable duration of use

Biocides:

- Significantly increase degradation of biotic factors, and eliminate growth of fungus and molds

Polymer modifications:

- Substances that enhance impact resistance and extensibility of the composite

Plasticizers:

- Volatile organic substances, which affect ductility, flexibility and a reduction in viscosity, increase the thermo-plasticity.

Experimental Study

As an experimental material, wood filled plastic was used with a volume ratio of plastic matrix (HDPE) to wooden filling of 30/70% (the material was supplied by a commercial company). The size of the wooden particles of WPC was from 420 μm (35 meshes).

The experimental study was based on analyzing the cutting zone after cold cutting technologies (band saw and water jet). Cutting wood composite samples were made with the Ergonomic 275.230 DG band saw, which has a sufficiently dimensioned drive motor, a gearbox with helical gears and oil fill, and a large range of angular cuts from −45° to +60°. Cutting is ensured by a weighted arm with hydraulic speed control. Prerequisites for excellent cutting performance include precise hard metal rolling drives on the band saw, a joint saw head mounted on tapered roller bearings, a 27-mm saw band and a synchronized-running brush to remove chips. Adjustable clamping is designed to prevent the workpiece from moving during cutting.

Technical and cutting parameters of the band saw were:

- Minimal cutting diameter 5 mm; Shortest length of the rest: 20 mm; Material laying height: 760 mm; Blade dimensions (length × height × thickness): 2720 mm × 27 mm × 09 mm; Drive power band: 1.1/1.5 kW; Saw band speed: 40/80 m/min; Total installed power: 2.7 kVA; Dimensions (width × length × height): 640 mm × 1400 mm × 1270 mm
- Cutting condition: Saw band speed: 80 m/min 1; Feed: 0.180 m/min

Cutting samples made of WPC by water jet were done on machine WJ 3020b – 1 Z (Fig. 23.2).

Cutting conditions for the experiment were:

- Cutting speed 100 mm/min-1; Type of abrasive: Australian grenade; Granularity abrasives: MESH 80; Diameter of the nozzle: 0.25 mm; Guidance tube diameter: 1.12 mm; Guidance tube length: 76 mm

Samples used for the experiments were selected by various grades of wood manufactured by extruded technology. Individual samples are shown in the Table 23.3 below (Fig. 23.3).

The Mitutoyo SJ 400 was used to measure the surface roughness of the pre-cut wood composite samples.

Fig. 23.2 Water jet 3020b – 1Z

Table 23.3 Measured values of the surface roughness for sample 1

Method	Roughness (µm)	Place of Measuring	Number of measuring									Mean
			1	2	3	4	5	6	7	8	9	
Water Jet	Ra	Beginning	2.4	2.4	2.4	2.4	2.4	2.4	2.4	2.4	2.4	2.38
	Rz	Beginning	14	14	14	14	14	14	14	14	14	14.06
	Ra	End	3.1	3.2	3.2	3.1	3.2	3.2	3.2	3.2	3	3.125
	Rz	End	18	17	17	18	18	17	18	17	17	17.65
Band saw	Ra	Beginning	2.8	3.3	3.3	2.9	3	3.2	2.9	3.3	3.1	3.06
	Rz	Beginning	18	17	17	IS	IS	17	IS	17	17	17.25
	Ra	End	2.9	2.8	2.8	2.8	2.8	2.9	2.8	2.8	2.8	2.81
	Rz	End	17	16	17	16	17	17	17	17	17	16.65

Sample number	Sample	Composition	Sample number	Sample	Composition
1		Wood 60% Plastic 30% Additivies 10%	4		Wood 50% Plastic 45% Additivies 5%
2		Wood 65% Plastic 35% Additivies 5%	5		Wood 55% Plastic 35% Additivies 8%
3		Wood 70% Plastic 25% Additivies 5%	6		Wood 63% Plastic 30% Additivies 7%

Fig. 23.3 Experimental samples

The arithmetic mean deviation of the profile Ra is a height parameter calculated as the arithmetic mean of the heights of the absolute values Z (x) within the regular length lr.

Technical parameters during the measuring process were set on the following values: Measuring rate: 0.05; 0.1; 0.5; 1.0 mm/s; Rate of return of 0.5; 1.0; 2.0 mm/s; Measuring direction: backward; Positioning: ±1.5° (slope), 10 mm (up/down); Range/measurement resolution: 800/0.01 microns; 80/0,001 microns; Supply type: via AC adapter; Evaluated parameters: P (primary), R (roughness), W (filtered waviness); Digital Filter: 2CR, PC75, Gauss; Cutoff length: 0.08; 0.25; 1.8; 2.5; 8 mm. Each sample was measure nine times with mean values calculated for top, middle and end of the sample (Table 23.3). Basic statistical indicators were also determined for each type of cutting with dependence on wood percentage with correlation index over 95% (1) (2) (Table 23.4) **Calculated curves were overlaped on minimal 95% with measured**.

Figure 23.4a shows a comparison of the surface roughness after technological operation of the water jet and band saw cutting. The graphs have similar curves

Table 23.4 Statistical indicators

Measuring place	Dispersion	Standard deviation
Beginning WJ	2.23	1.57
End WJ	1.23	1.11
Beginning BS	1.44	1.2
End BS	2.27	1.5

$W\%$ of the wood, $P\%$ of the plastic

$$R_{awj} = +0.4241 \cdot W^2 - 24{,}6272 \cdot P + 473{,}5791 \quad (1)$$
(Water jet)

$$R_{abs} = 0.4946 \cdot W^2 - 29{,}1777 \cdot P + 570{,}5326 \quad (2)$$
(Band saw)

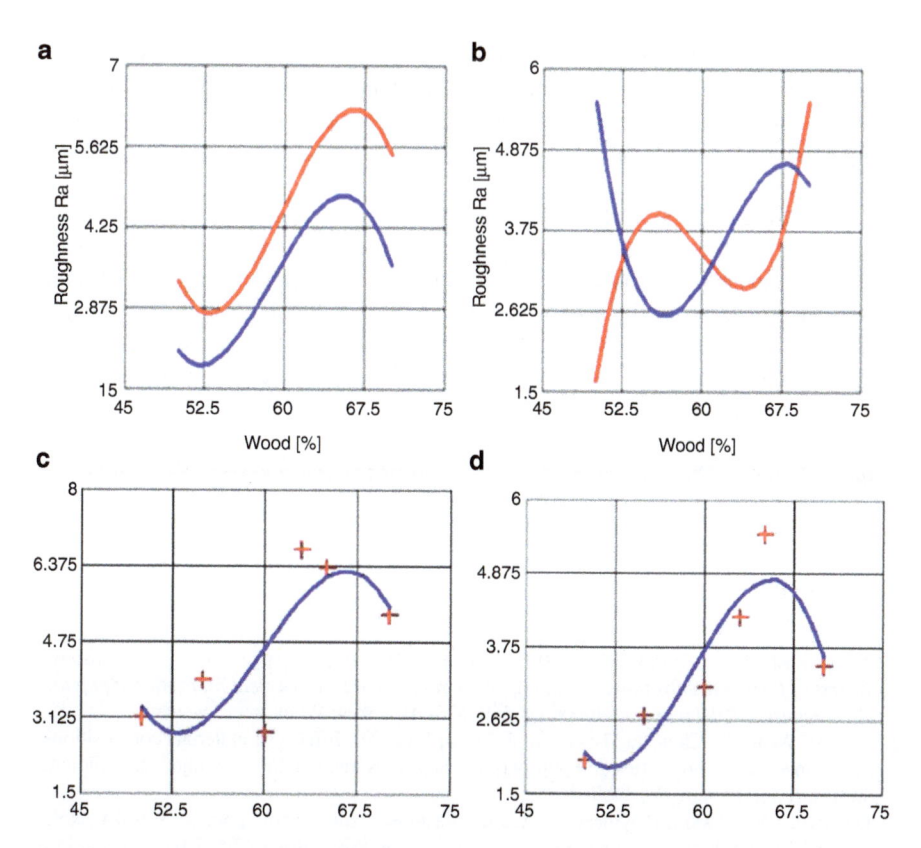

Fig. 23.4 Results of the experiment. (**a**) Roughness Ra measured at the end of the experiment depends on wood ratio. (**b**) Roughness Ra at the beginning of the experiment (*red line* bend saw; *blue line* water jet). (**c**) Measured and evaluated values of roughness depend on wood ratio for water jet cutting. (**d**) Measured and evaluated values of roughness depend on wood ratio for band saw cutting

except that the surface roughness of the cutting band saw is greater than the water jet cutting. Figure 23.4b compares surface roughness after cold cutting in the zone of the first contact of the saw or water with the workpiece. From the results it can be concluded that different technologies affects variable surfaces differently and the boundary is a wood ratio of 55%. Varying surface curves also change the character (e.g. convex to concave). Thus, the correct technology should be chosen to obtain the best quality surface, in other words, one should use the band saw for materials with wood ratio under 55% and over that value water jet technology should be used.

Conclusion

The height of roughness profile Rz was measured using the Mitutoyo SJ 400 roughness tester to determine the amount of wood in a sample. The graphs show the wood percentage dependence on the arithmetic mean deviation of the profile Ra. Comparison of these graphs reveals that the percentage of wood in the composite influences the surface quality.

Surface roughness depends on the type of cutting materials used. For sample no. 4, with 50% wood, it is preferable to use a cutting band saw which can be used with a less rough surface compared to water jet cutting. For samples no. 5, with 55%, and the sample timber no. 1 with 60% wood, it is better to cut with water flow; although the surface arising from the cutting band saw has only slightly higher values. For samples no. 6 with 63% wood, sample 2 with 65% wood and in a sample of 70% wood, it is better to cut with the water jet.

Acknowledgement This research was financially supported by the project APVV–15–0700.

References

1. Nourbakhsh, N., & Ashori, A. (2008). *Fundamental studies on wood-plastic composites: Effects of fiber concentration and mixing temperature on the mechanical properties of poplar/ PP composite, Polymer Composites* (pp. 569–573). Hoboken: Wiley InterScience.
2. Kuo, P., Wang, S., Chen, J., Hsueh, H., & Tsai, M. (2009). Effects of materials compositions on the mechanical properties of wood-plastic composites manufactured by injection molding. *Materials and Design, 30*, 3489–3496.
3. Ducháček, V. (2006). Polymery - Výroba ,vlastnosti, zpracování, použití, Vysoká škola chemicko-technologická v Praze, Praha, vyd. 2, str. 279, ISBN: 80-7080-617-6.
4. Werk,J. (2016). A new material is conquering the world: wood plastic composite, Online[09.09.2016] Available on http://www.jelu-werk.com/technical-industry/applications/wood-plastic-composite/.
5. Hutyrova, Z., & Zajac, J. (2013). Turning of composite material with organic reinforcement (wood plastic composite). *Advanced Science Letters, 19*(3), 877–880.
6. Valíček, J., Harničárová, M., Hlavatý, I., Grznárik, R., Kušnerová, M., Hutyrová, Z., & Panda, A. (2016). A new approach for the determination of technological parameters for hydroabrasive cutting of materials. *Materialwissenschaft und Werkstofftechnik, 47*(5–6), 462–471.

7. Kravec, M. (2016) Evaluation of cutting area after machining of wood plastic composite, Diploma thesis. Technical University of Kosice, Presov.
8. Dobránsky, J., Baron, P., Kočiško, M., & Vojnová, E. (2014). Monitoring of the influence of moisture content in thermoplastic granulate on rheological properties of material. *Applied Mechanics and Materials, 616*, 207–215. ISSN 1660-9336.

Chapter 24
Risk Analysis Causing Downtimes in Production Process of Hot Rolling Mill

Marcela Malindzakova, Dagmar Cagáňová, Andrea Rosova, and Dusan Malindzak

Abstract This article is focused on the application of quality management tools in the production process of hot-rolled steel products, namely, metal sheets and strips. The actual production of hot-rolled products involves a set of main, support, and management processes and network of interconnected relationships. The production process of hot-rolled sheets and strips is characterized by strictly one-way direction of the production process from the beginning to end. It is a relatively simple, linear process based on precise continuity, time sequence, and capacity settings. On the other hand, considering the technology involved, it is a system heavily dependent on stability of the important process parameters, such as temperature, etc.

Introduction

In this process, it is not possible to repeatedly perform any operation. On top of that, for each sub-process, there is only a limited time window available. For the production and non-production companies, the quality is an integral tool used to maintain and improve competitiveness, either by maintaining the product quality or by providing outstanding quality of service. Suppliers should respect the requirements of customers and of the market. Their primary aim should be a continual effort not to lose customers and provide the highest quality products and services. Especially during the preproduction phases of a new product, there is a significant pressure put on the management that involves plans for marketing, logistics and the

---

M. Malindzakova (✉) · A. Rosova
Technical university of Košice, Institute of Logistics, Park Komenskeho 14, Kosice, Slovakia
e-mail: marcela.malindzakova@tuke.sk

D. Cagáňová
STU Bratislava, J. Bottu 25, Trnava, Slovakia

D. Malindzak
U. S. Steel Kosice, Vstupny areal U. S. Steel Kosice, Kosice, Slovakia

© Springer International Publishing AG, part of Springer Nature 2019
D. Cagáňová et al. (eds.), *Smart Technology Trends in Industrial and Business Management*, EAI/Springer Innovations in Communication and Computing,
https://doi.org/10.1007/978-3-319-76998-1_24

actual drafting of a new product. In the market environment for a successful operation of any new or a well-established company, it is important to rely on an analysis in combination with the support of quality management tools (security, evaluation, continuous improvement required parameters).

The main production flow of hot-rolled sheets and strips consists of the following sub-processes:

- Handling and heating of slabs
- Hot strip rolling
- Modification of coils (hot strip cutting into a sheet of metal tapes and discs) [1]

Experimental Tests

The input materials for this manufacturing process are the continuously cast slabs from the steelworks. To transport these slabs, railway wagons are used to move them into the warehouse. The slab warehouse is logistically divided into two areas defined by the width of the slabs. The main machinery facilities of this operation are four pusher furnaces, where the slab heating takes place, with the slabs coming from the warehouse. The slabs are heated to the rolling temperature of about 1250 °C. The input temperature of slabs coming from the warehouse varies from 20 °C to 400 °C, depending on the time they were imported by wagons. Reheated slabs from the pusher furnace move next by the conveyor and are arranged and prepared into a correct rolling sequence, for the hot rolling mill [2]. During the first part of this process, the slab changes into a so-called coble (intermediate rolling slab – defective work). The slabs then continue into the main rolling segment, where by continuous rolling, it changes into a long rollout hot strip.

After the main sequence follows the segment of laminar strip cooling. The task here is to rapidly reduce the temperature down from 900 °C to about 400–600 °C, depending on customer requirements. By this controlled cooling, the required mechanical properties are achieved as requested by the customer. The final step is to roll the hot strip in one of the three coilers in to create the so-called coils. An integral part of the production process is the regular control checks of selected process parameters [3]. Divided by the position within the production process, where the controls are carried out, these checks are divided into an input control, interoperational control, and checkout. Based on the character, the controls can be divided into control checks performed by a man and control checks performed by a machine.

The input control – the pusher furnaces – checks the identity of the given slab; checks all the dimensions, namely, the length, width, and thickness of the slabs; and looks for possible surface defects.

Interoperational control – the pusher furnaces – temperature, duration time, and excess air checks. Furthermore, checks are carried out on the rolling pin, and the next control checks the thickness, width, flatness, profile, and strip temperature.

Checkout – after winding the coil right after the coilers, the HRM checks the thickness, width, and the surface quality of the strip. After cutting the strip into sheets or tapes by the slitting lines, there is a check for width, thickness, and surface quality of rolls of tapes. The employee that performs a visual inspection is annually tested in regard to his abilities to perform visual inspection (kappa methodology) [4]. Right after the laminar cooling, there is a checkpoint performing control by a camera inspection system (CIS).

Due to basic physical principles, when at the beginning of the production process, the heated slab from the pusher furnace during the rolling process gradually loses the temperature until it reaches the critical temperature for the hot strip winding into a coil. Any subsequent delivery of heat into the slab during the manufacturing process is possible. If the intermediate temperature drops below a critical value for given process, inevitably occurs a nonconforming product. Such a product is immediately removed from the production process, and a scrap is returned back into production process as part of the initial charge. Any nonconforming product therefore worsens the yield rate for hot rolling process.

In an effort to reduce the occurrences of nonconforming products during the main process of hot-rolled sheets and strips, a wide range of statistical analyses of tools and quality management are used. Management uses the most basic methods such as the position analysis (median), or analysis of variance (dispersion), but also more sophisticated tools, such as the Shewhart control charts.

Management solves operational problems of production and logistics, as well as problems of a qualitative nature by brainstorming and subsequently developed Ishikawa diagram, to help determine the causes of this problem. Subsequently, using the Pareto principle, they identified the cause of the most serious problems, for which it is necessary to take a corrective action. A so-called risk number is calculated. C-E diagnosis is the basis to create a so-called FMEA (failure mode and effect analysis) [5].

FMEA process generally comprises the following steps:

1. The determination of individual processes
2. Definition of possible errors and their causes and consequences (C-E diagnostics)
3. Assessment of the current state (occurrence, detection, meaning of errors)
4. The calculation of risk number
5. Determination of activities requiring action – determination of critical sites (Pareto diagram of the most important issues)
6. Proposals for corrective action [6]

The higher the risk number, the more urgent are the corrective actions. Also the selection of design, manufacturing, and testing measures are the most appropriate for given situation is best determined by a qualified evaluation of factors, where the occurrence frequency, importance, probability, and possibility of exposing the defect is tested.

- The probability of defects occurrence, which represents the probability that a particular cause will lead to the occurrence of the next defects
- The importance of defects which can be reduced by changing the design or concept
- The likelihood of detecting defects that can be increased by design improvements

The order and the necessity of improvement are given by the value of the risk number indicator – RNI/P – priority. As an illustrative example, the following case study scenario wishes to present the solution to taperness – wedging (difference of the strip thickness at the right and left side) of hot-rolled sheets and strips. The C-E diagram is divided into seven main groups, which are organized by Ishikawa principle of succession and continuity of the process: materials, methods, machines, measurements, management, men and environment.

After processing of a comprehensive FMEA table within the corresponding software, it is necessary to monitor the risk number of individual causes. However, the creation of just one Pareto diagram followed by a direct evaluation and selection of a corrective action carries a risk of incorrect logic [7].

Therefore, it is preferable to create as many Pareto diagrams as there are basic groups (7) and begin the corrective actions from the material – successively followed by technology A, B, and C. This "step-by-step" procedure for the implementation of corrective measures reduces the risks involved in the follow-up groups. The question remains in regard to the repetition frequency of creating the FMEA table with options such as:

- After any corrective action
- After a certain period of 3/6/12

FMEA results show that 27 identified processes may cause 50 different defects with risk numbers values ranging from 6 to 336. The sheer number of different defects indication in each category and the average risk in these categories shows that it is preferable to create a Pareto distribution for each category, rather than a single total for all 50 defects indications. The Pareto charts explain the "most problematic" and riskiest possible problems where the company must focus their efforts to maintain a defect-free production process.

Results and Discussion

The Pareto analysis identified the causes of various categories which are the roughing mill, finishing mill, and roll mill inspection [8, 9].

For the roughing mill, the following processes were specified:

1. Rolls cooling
2. Pressure removal of scaling before the roughing mill stands
3. Guiding the coble between stands of the roughing mill
4. Vertical rollers – rolling of the coble in roughing mill (vertical rollers)
5. Horizontal rollers – rolling of coble in roughing mill

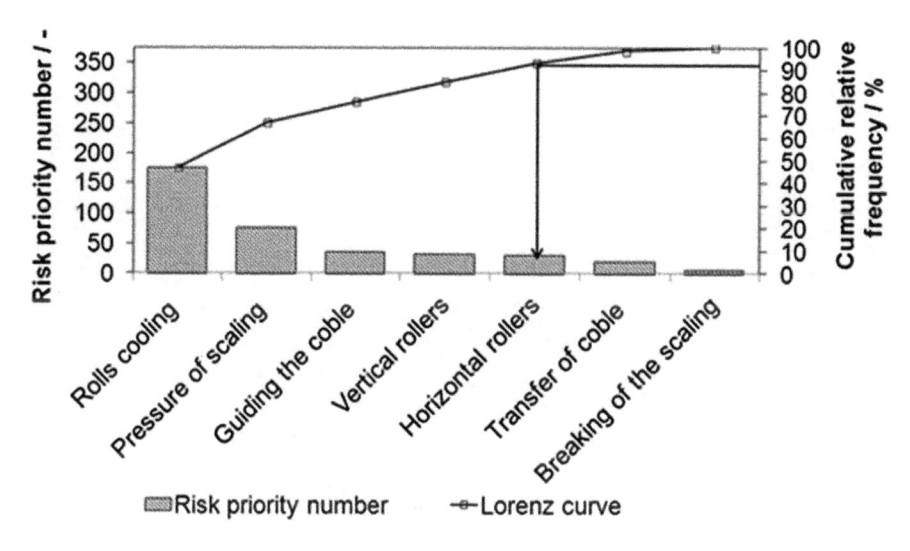

Fig. 24.1 Pareto analysis for roughing mill

The results are presented in Fig. 24.1.

The results of the Pareto analysis show that the following processes influence the finishing mill:

1. Keeping the strip between the stands of the finishing mill
2. Pressure removal of scaling for the finishing mill
3. Cooling of finishing mill rolls
4. Front trimmer
5. Establishment and transfer of coble
6. Cooling of the strip between stands
7. Greasing of rolls

The results are presented in Fig. 24.2.

The results of Pareto analysis show the importance of these processes for the roll mill inspection:

1. Transfer of the coil
2. Winding the coil on a coiler
3. Centering the strip before coilers
4. The handling of the coil
5. Inspection of the strip surface
6. Coil check

The results are presented in Fig. 24.3.

The problem of the excessive taperness, at the broad-line hot rolling mill is caused by the occurrence of slabs exposed to drafts as shown by the Pareto analysis results. This type of slabs occurs if the downtime lasts longer than 30 min dashboard based on web technologies.

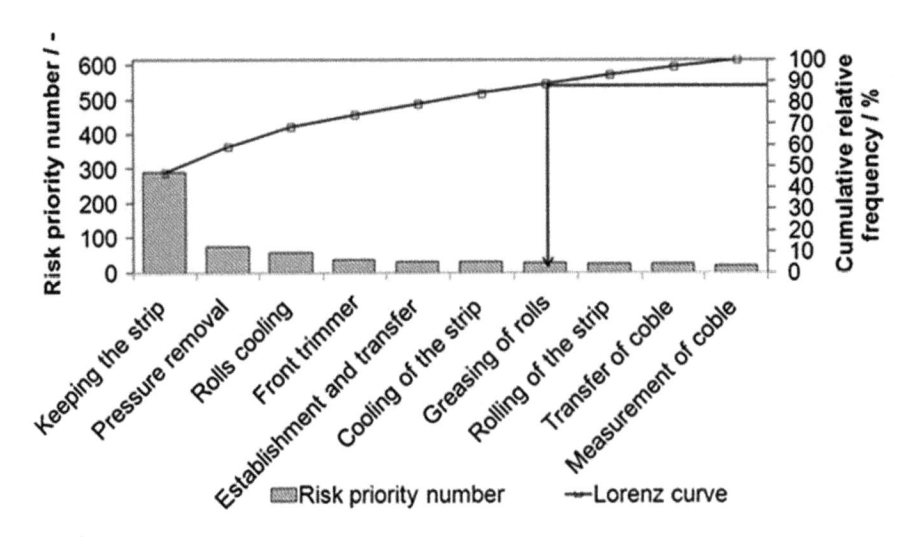

Fig. 24.2 Pareto analysis for the finishing mill

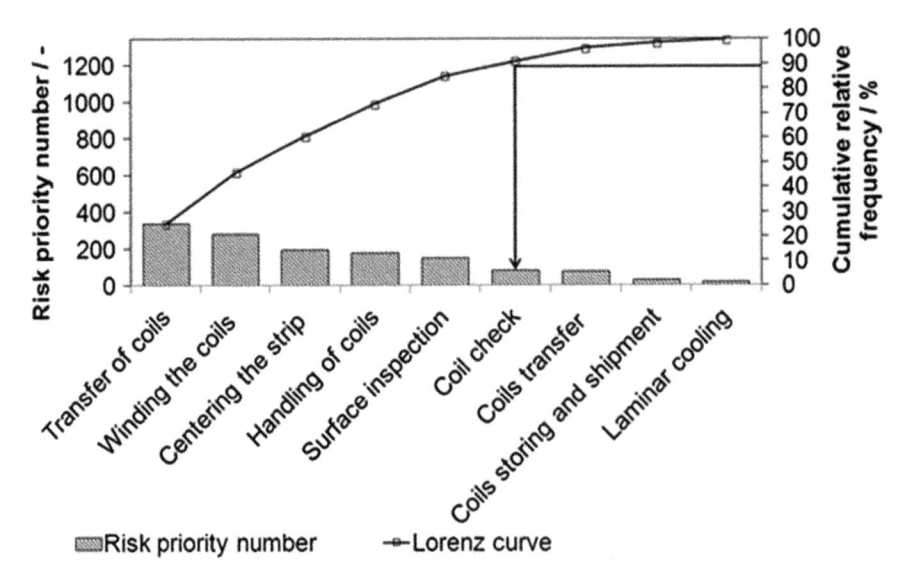

Fig. 24.3 Pareto analysis for the roll mill inspection

Pareto analysis results for the finishing mill category identifies as the main cause for the excessive taperness, the formation of wedge-shaped coble [10]. A wedge-shaped coble occurs by an uneven left/right wedge correction during the rolling in roughing mill HRM. This type of coble must be marked as a nonconforming roll and must be removed into the warehouse of nonconforming production. The problem in the process of coil transfer into the HOLD warehouse is omission of a physical inspection of the entire length of the strip. The impacts of human errors, or a wrong adjustment settings, damaged roller surfaces, power or bearing failures have an negative effect also on the second process, which is the winding of the coil on a coiler.

The main cause for the third process called centering the strip before coilers is a jammed setting mechanism of rulers, communication failure in the control system, or a lack of segments renovation. Difficulties in coil handling are caused by the malfunction of binding machine, a human error in the description of coils, defected weighing machine, improper weights calibration, and failure of data transmission. Failure in the process of inspection of the strip surface is caused by insufficient lighting and unsuitable conditions for inspection (steam, water on the strip surface, temperature maps).

Conclusions

For the process analysis with the focus on the automotive industry, there is a requirement to use the norm "ISO/TS 16949:2009 Quality management systems – Particular requirements for the application of ISO 9001:2008 for automotive production and relevant service part organizations". The article is focused on assessing the degree of risk of downtime. The FMEA method was used to assess such risk on a particular node in the manufacturing process of rolled steel on hot rolling mill. To highlight the most limiting factors hindering the production process optimization and the logistics planning, the Pareto analysis was used. The result of Pareto analysis presents the most urging problems for each assessed areas of influence.

This work was supported by the Slovak Research and Development Agency under the grants VEGA No. 1/0216/13.

References

1. Laciak, M., Durdán, M., & Kačur, J. (2012). Utilization of indirect measurement in the annealing process of the steel coils. *Acta Metallurgica Slovaca, 18*(1), 40–49.
2. Baricová, D., Pribulová, A., & Demeter, P. (2012). Utilizing of the metallurgical slag for production of cementless concrete mixtures. *Meta, 51*(4), 465–468.
3. Balog, M., & Husár, J. (2016). Methodical framework of flexibility production evaluation in terms of manufacturing plant. *Key Engineering Materials: Operation and Diagnostics of Machines and Production Systems Operational States 3, 669*(2016), 568–577. ISBN 9783038356295.
4. Balog, M., Knapčíková, L., & Husár, J. (2016). *Plánovanie v strojárskej výrobe, 1. vyd* (p. 86). Brno: Tribun EU. ISBN 9788026310785.
5. Futáš, P., & Pribulová, A. (2014). Quality criteria as implement for advisement of metallurgical quality of grey iron. In *Evaluation of people and products features* (pp. 99–118). Celje: University of Maribor.

6. Malindzakova, M. (2014). *Process approach – a synergy of influences to address issues of quality and environmental management*. Kosice: Technicka univerzita, F BERG, Habilitation thesis.
7. Gašpar, M. (2015). *Analysis tools of quality management in the logistics of the selected company*. Kosice: Technicka univerzita, F BERG, Bachelor thesis.
8. Company information U.S. Steel Košice, c. l. (2011). b, [online]. Available on the Internet: www.usske.sk/corpinfo/corpi-s.htm.
9. Saniuk, S., & Saniuk, A. (2008). Rapid prototyping of constraint-based production flows in outsourcing. *Advanced Materials Research, 44-46*, 355–360.
10. Husár, J., & Dupláková, D. (2016). Material flow planning for bearing production in digital factory. *Key Engineering Materials : Operation and Diagnostics of Machines and Production Systems Operational States 3, 669*(2016), 541–550. ISBN 9783038356295.

Chapter 25
Evaluation of Roughness Parameters of Machined Surface of Selected Wood Plastic Composite

J. Zajac, F. Botko, S. Radchenko, P. Radič, A. Bernat, J. Roman, and B. Zajac

Abstract Nowadays new materials are gaining ground, which, compared to traditional materials, can be made by combining materials with different mechanical and chemical properties. Example of such material is wood plastic composite (WPC). WPC represents a combination of wood in the form of wood flour or wood fibers and polymer with different additive substances. Wood plastic composite brings some improved properties compared to pure wood. An example of this feature is high resistance to rot, which is caused by a combination of polymer plastic material and organic wood material in the form of flour or fibers. Experimental part of presented article is focused on turning of wood plastic composite material using custom tool and subsequent evaluation of surface roughness parameters depending on technological conditions.

Introduction

Presented experimental research is focused on turning of wood plastic composite and evaluation of surface roughness parameters. Nowadays composite materials are used to replace traditional materials. These new materials are made by combining several materials with different attributes [1].

Development of wood plastic composites (WPC) started at the beginning of the 1980s in the twentieth century located in the USA. Wood-filled plastics are mixed and shaped using extruders. Many plastic materials, such as polyethylene, polypropylene, and polyvinylchloride, are used in production of WPC [3, 5, 7].

Wood plastic composites are materials, which combine properties of wood and polymer matrix. In general WPC materials contain a variable ratio of wood, plastics,

---

J. Zajac · F. Botko (✉) · S. Radchenko · P. Radič · A. Bernat · J. Roman · B. Zajac
Faculty of Manufacturing Technologies with a Seat in Prešov, Prešov, Slovak Republic
e-mail: jozef.zajac@tuke.sk; frantisek.botko@tuke.sk; svetlana.radchenko@tuke.sk; andrej.bernat@tuke.sk; jan.roman@tuke.sk; branislav.zajac@tuke.sk

© Springer International Publishing AG, part of Springer Nature 2019
D. Cagáňová et al. (eds.), *Smart Technology Trends in Industrial and Business Management*, EAI/Springer Innovations in Communication and Computing,
https://doi.org/10.1007/978-3-319-76998-1_25

and additives. Since 1990 the WPC market share has been continuously rising for different applications. Nowadays WPC is mainly used to replace wood in building industry for outdoor applications. Next field of WPC usage is in automotive industry for interior panels. When machining WPC, it is crucial not to exceed temperature 200 °C due to thermal decomposition of wood.

Nowadays, machining all kinds of composite materials is the trend. The focus of presented article is machining and evaluation of surface roughness parameters for material Megawood. This knowledge could be used in future research for determination of surface quality in composite machining process. Quality control in machining process is very important [2, 4].

Preparation and Machining of Experimental Samples

Experimental procedure consists of several steps from cutting on band saw to final turning and subsequent measurement on optical profilometer.

Description of experimental material is listed in the table below (Table 25.1). Preparation of test samples consists of three steps, which are listed below:

Step 1: Cutting (band saw BOMAR ERGONOMIC 275.230 DG) – cutting to length 150 mm (Fig. 25.1)

Table 25.1 Description of experimental material

Commercial name	Megawood
Material	Wood plastic composite
Dimensions	40 × 60 × 3600 mm
Color of material	Brown pigment
Matrix	HDPE
Composition of wood filling	75% of wood + HDPE
Size of particles	From 420 µm (mesh 35)
Stability	Decomposition at 180 °C

Fig. 25.1 Cutting of experimental samples on band saw

Fig. 25.2 Turning to baseline diameter

Fig. 25.3 Final turning ap = 2.5 mm

Step 2: Turning (lathe SUI 40) – turning of square profile to diameter 36 mm; equipment, three-jaw chuck and tail stock (Fig. 25.2)

Step 3: Final turning – turning to final diameter with depth of cut ap = 2.5 mm (Fig. 25.3)

Experimental samples were turned using custom tool with geometrical parameters as listed below:

- Rake angle on the back plane $\gamma_p = 30°$
- Side range $\alpha_p = 10°$
- End cutting edge angle $\kappa_r = 45°$
- Nose radius $r_\varepsilon = 5$ mm
- Tip angle $\varepsilon_r = 90°$
- Side relief angle $\gamma_f = 0°$

Presence of cooling fluid in machining process was undesirable due to high content of wood in experimental material.

Machining was performed with constant depth of cut ap = 2.5 mm, and variable parameters were set: feed speed per revolution (*f*), revolutions per minute (RPM), and cutting speed [m.min^{-1}] (Table 25.2).

Table 25.2 Cutting conditions

Sample marking	f [mm]	RPM [min^{-1}]	Vc [m.min^{-1}]
B1	0.1	450	50.89
B3	0.3	450	50.89
B5	0.61	450	50.89
B6	0.1	900	101.79
B8	0.3	900	101.79
B10	0.61	900	101.79
B11	0.1	1400	158.34
B13	0.3	1400	158.34
B15	0.61	1400	158.34

Fig. 25.4 Optical profilometer MicroProf FRT

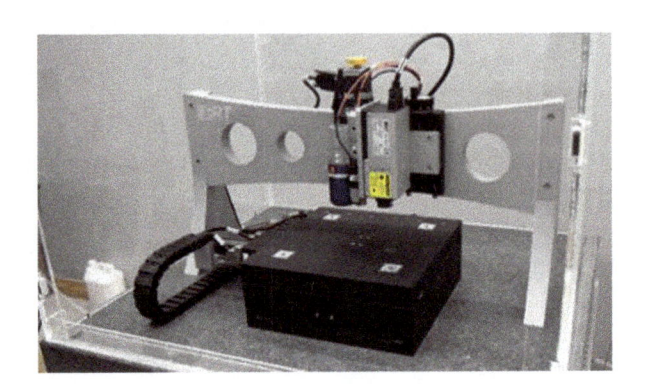

Evaluation of Surface Roughness

Surface roughness parameters of each sample were scanned using optical profilometer MicroProf® FRT (Fig. 25.4). Measuring with optical spectrometer belongs to noncontact and nondestructive methods. Experimental sample is located on machines' worktable, which is moving rapidly and accurately and is controlled by microprocessor. Topography data can be obtained from surface areas. Optical profilometer works with fixed focus (do not use autofocus technology), and data are obtained from each single point independently, which leads to a high-speed measuring process [6, 8].

Maximum height of profile Rz was measured with nine repeats, and subsequently arithmetical average was calculated for each sample. Graphical dependencies of Rz on feed speed rate f and cutting speed vc were created. During machining process, several flat cracks occurred on machined surface. Cracks were excluded from measuring process because they were causing distortion of final values.

Figures 25.5, 25.6, and 25.7 show examples of scanned surface. Area of scanning was selected to 2 × 2 mm. Individual samples from B1 to B15 show recognizable differences. Graphical dependence (Fig. 25.9) illustrates change of surface roughness parameter Rz with increasing cutting speed.

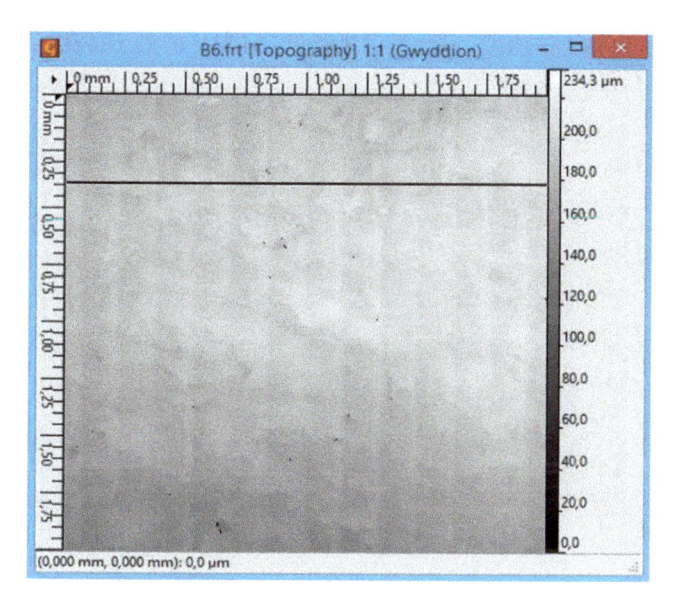

Fig. 25.5 Experimental sample B6 ($f = 0.1$ mm)

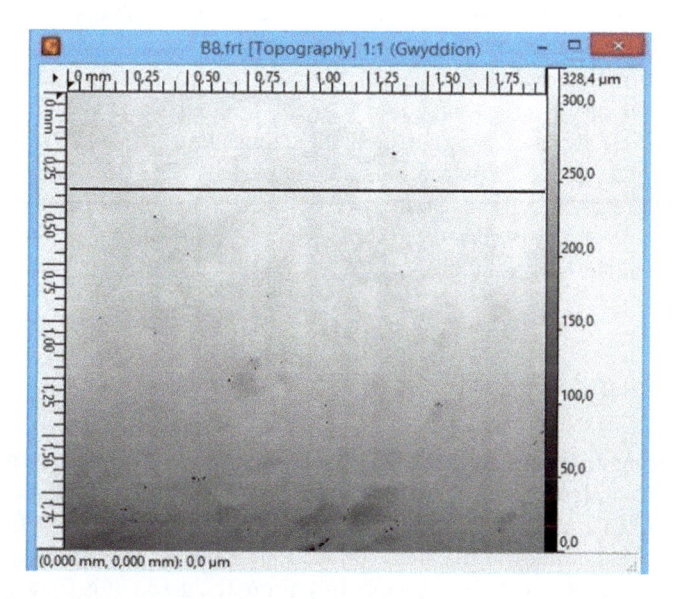

Fig. 25.6 Experimental sample B8 ($f = 0.3$ mm)

Values of roughness parameters were obtained using software Gwyddion. Subsequently arithmetical averages were calculated and graphical dependencies were created.

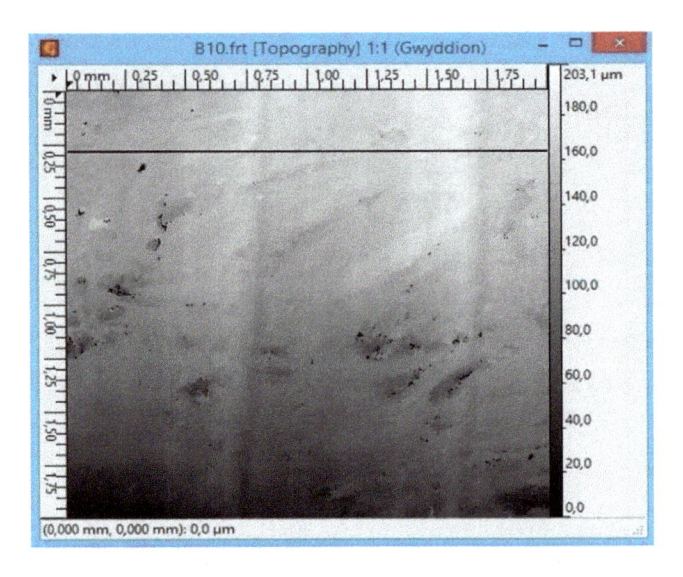

Fig. 25.7 Experimental sample B10 ($f = 0.61$ mm)

In the following picture (Fig. 25.5), scan of experimental sample B6 machined with parameters $f = 0.1$ mm and vc = 101.79 m.min^{-1} can be observed. Dense grid is visible which is caused by low values of feed speed.

The following figure (Fig. 25.6) is an example of experimental sample manufactured with feed speed per revolution $f = 0.3$ mm. In comparison with the previous figure (Fig. 25.5), there are approximately three times less visible tool marks.

Experimental sample displayed on following figure (Fig. 25.7) was produced with highest preset of feed speed rate ($f = 0.61$) used in this experiment. Three tool markings can be clearly observed on whole scanned area (feed speed rate was 0.61 and size of scanned area 2 × 2 mm).

Conclusion and Discussion

Graphical dependencies of surface roughness on machining conditions were created based on measured roughness characteristics of machined surface. Graphical dependence (Fig. 25.8) illustrates change of surface roughness parameter Rz with increasing cutting speed. Values of maximal surface roughness Rz are in the range from 11.31 μm (sample B1) to 13 μm (sample B6). From the graph, it is evident that for minimal cutting speed vc = 50.89 m.min^{-1}, values of Rz exhibit a decreasing trend for the whole range of feed speed rates (from 0.1 to 0.61 mm) (Table 25.2).

Similar trend can be observed for cutting speed vc = 101.79 m.min^{-1}. For cutting speed at maximal value (vc = 158.34 m.min^{-1}), values of maximal surface roughness Rz significantly increase for feed speed rate $f = 0.3$ mm and afterward decrease to value 14.85 μm (difference 4.02 μm).

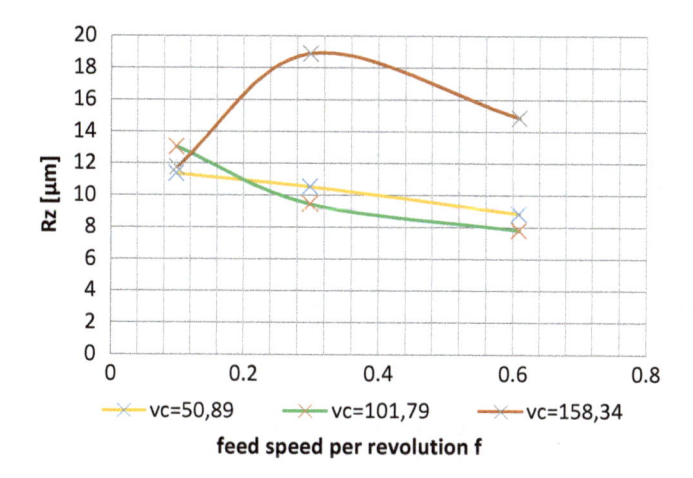

Fig. 25.8 Graphical dependence Rz on f

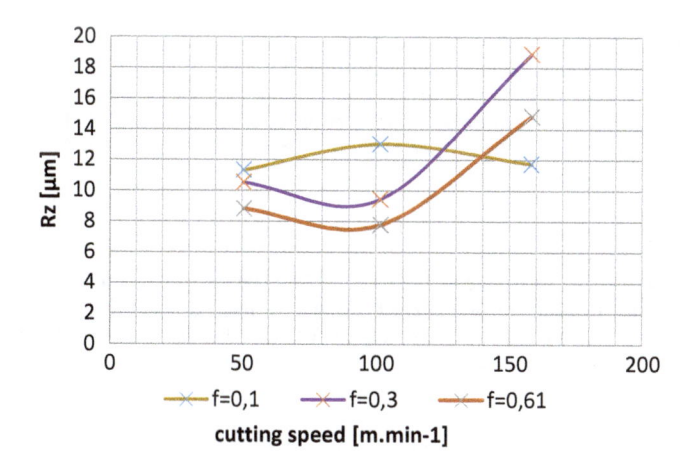

Fig. 25.9 Graphical dependence Rz on vc

Graphical dependence (Fig. 25.9) represents influence of decreasing cutting speed vc on machined surface quality. For middle and maximal settings of feed speed rate, curves of very similar course are obtained. Values of maximal surface roughness Rz slightly decrease for cutting speed 101.79 m.min^{-1} and afterward rapidly increase for cutting speed 158.34 m.min^{-1}. For minimal feed speed rate, curve convex characteristics with local maximum are obtained on cutting speed 101.79 m.min^{-1}.

From a theoretical point of view, increasing of feed speed rate results to a decreasing trend in surface quality. Vice versa increasing cutting speed should improve the quality of machined surface. Machining of wood plastic composite paradoxically does not confirm this theoretical axiom. Courses of graphical dependencies indicate that variable surface qualities for nonhomogeneous materials are obtained. This paradoxical situation is caused by the fact that composite materials have random distribution of individual components in every cross section.

Acknowledgment This work was supported by the Slovak Research and Development Agency under the contract No. APVV-15-0700.

References

1. Angelovič, M. (2016). *Machining of composite material with natural reinforcement*, Diploma Thesis, 61pages.
2. Hutyrová, Z., Zajac, J., Mital, D., Harničárová, M., Valíček, J. (2015). Evaluation of share material after turning of wood plastic composite, *Proceedings – International Conference on Solar Energy and Building*, ICSoEB 2015, art. no. 7244946, Chapter: SURNAME, A. (2010) Title of chapter. In SURNAME, C. [ed.], *Title of Book* (Publisher location: Publisher), Chapter 3.
3. Zajac, J., Mital, D., Radchenko, S., et al. (2014). Short-term testing of cutting materials using the method of interrupted cut. In S. Fabian & T. Krenicky (Eds.), *Operation and diagnostics of machines and production systems operational states II, Applied mechanics and materials* (pp. 236–243). Pfaffikon: Trans Tech Publications Ltd.
4. Zivcak, J., Petrik, M., Hudak, R., et al. (2009). Embedded tensile Strenght test machine FM1000-an upgrade of measurement and control. In Z. Gosiewski & Z. Kulesza (Eds.), *Mechatronic systems and materials, Solid state phenomena* (Vol. 3, pp. 657–662). Switzerland: Trans Tech Publications.
5. Cep, R., Janasek, A., Petru, J., et al. (2014). Surface roughness after machining and influence of feed rate on process. In J. Kundrat & Z. Maros (Eds.), *Precision machining VII, Key engineering materials* (pp. 341–347).
6. Valicek, J., Drzik, M., & Hryniewicz, T. (2012). Non-contact method for surface roughness measurement after machining. *Measurement Science Review, 12*, 184–188.
7. Nourbakhsh, N., & Ashori, A. (2008). Fundamental studies on wood-plastic composites: Effects of fiber concentration and mixing temperature on the mechanical properties of poplar/PP composite. *Polymer Composites, 29*, 569–573. Wiley InterScience.
8. Kuo, P., Wang, S., Chen, J., Hsueh, H., & Tsai, M. (2009). Effects of materials compositions on the mechanical properties of wood-plastic composites manufactured by injection molding. *Materials and Design, 30*, 3489–3496.

Chapter 26
Evaluation of the Transverse Roughness of the Outer and Inner Surfaces of the Thin-Walled Components Produced by Milling

Peter Michalik, Jozef Zajac, Michal Hatala, Dusan Mital, and Łukasz Nowakovski

Abstract The article deals with the measurement and evaluation of transverse roughness of the outer and inner surfaces of thin-walled components made of aluminium alloy ENAW2007. The components were manufactured using CAM applications. On the CNC machining centre, VMC 650 Pinnacle S. Roughness was measured by a measuring device Mitutoyo SJ 400. The milling tool has been used with a diameter of 10 mm. Transverse roughness was measured at six locations of milled parts.

Introduction

At present, the technical industry forms the largest industrial production unit consisting of large and technologically demanding industries. Formed components that are increasingly used in both technical and non-technical practice are thin-walled components of various kinds. The definition of the thin-walled component has not yet been given. Based on studies by various authors, the thin-walled component could be characterized as a component formed by walls that are thin, are high and deform with small cutting forces. Thin-walled parts in the aerospace industry are very often large integral products that are produced by removing the material in the form of chips. Typically, 90–95% of the material is removed. The advantages of the

---

P. Michalik (✉) · J. Zajac · M. Hatala · D. Mital
Technical University of Kosice, Department of Automotive and Manufacturing Technologies,
Faculty of Manufacturing Technologies with seat in Presov, Presov, Slovak Republic
e-mail: peter.michalik@tuke.sk; jozef.zajac@tuke.sk; michal.hatala@tuke.sk;
dusan.mital@tuke.sk

Ł. Nowakovski
Politechnika Świętokrzyska, Katedra Technologii Mechanicznej i Metrologii, Kielce, Poland
e-mail: lukasn@tu.kielce.pl

thin-walled parts made in a chip-like manner compared to those made from a mould are being solid, lightweight, relatively inexpensive and precise but especially safe, which is particularly suitable for the aerospace industry. The number of components in the subsequent installation is reduced, thus shortening the total assembly time. Delivery times of finished products are reduced.

By control systems that include today's CNC machine tools, we are able to create programs with subroutine that can create different shaped parts. Current systems allow not only shop programming, programming in ISO code and STEP programming but also programming using CAD/CAM system [1].

The rapid development of machine production in recent years necessitates rapid implementation of modern engineering technologies and ever-higher demands on accuracy and quality of production as well as the function of the thin-walled components. Research on surface quality, especially dimensionally thin parts, is still in development, given the fact that the production of such components involves using new technologies, tools and instruments [2].

Thin-walled parts can mostly be used in the automotive, aerospace and energy industries. An important role of these components is to replace the original parts with lighter thin-walled components which have sufficient strength and functionality. The use of these components results in an overall weight reduction of structural units, thereby reducing a number of factors such as production costs, installation time and many more [3].

International ISO standards define the machined surface as the sum of the individual characteristics of the workpiece. Thus, it can be scratched that the machined surface is not only mechanical in nature. [4].

The method of manufacturing a CNC program for the manufacture of a particular component is influenced by its shape, experience and skill of the operator, CNC machine equipment, type of CNC machine control system and, last but not least, choice of hardware and software for component modelling and subsequent generation of G code of CNC program. [5].

Çolak et al. proposed optimization for multiaxis CNC machines. The quality of the machined surface was compared using artificial intelligent models. He made predictions of surface roughness using mathematical equations. They examined the prediction of surface roughness by means of gene expression. Parameters with the greatest influence were the tool rotation speed, the feed rate, the depth of cut, the shape and the layout of the tool's cutting edges. [6].

Seguy et al. investigated the connection between tool vibrations and surface roughness when milling thin-walled parts. They used the theory of linear stability of tool cutting edges to optimize milling of thin-walled parts. For a given phenomenon, they set an explicit mathematical model using a system of nonlinear equations. They set out a new description of the influence of vibrations of milling tools on the roughness of thin-walled parts. [7].

Jahan et al. evaluated the average surface roughness (Ra) for aluminium after measuring the ball-end milling. They set two ANFIS models for predicting values of mean arithmetic roughness (Ra). Experiments were performed for the various slopes of cutting edges, cutting speed of spindle rotation, tool feed rate and cutting depth. The ANFIS model was used to describe the roughness of the surface of a die using a polyline cutter. They determined an optimal combination of cutting parameters to achieve the desired roughness (Ra) finish [8].

Gulpak et al. predicted compensation strategies for the combination of thermal effects and residual stresses. This paper presents a new concept for the development of such compensation strategies for steel milling. They describe the effect of residual stresses caused by machining. Using finite element method, workpiece deformation was described by source stress [9].

Material, CAM Software for Produced CNC Program and Manufacturing

Dimensions of produced thin-walled component is seen in Fig. 26.1. The workpiece was made of duralumin ENAW2007. It has a good workability.

CAD model of part was saved in .ipt format [10]. Environments in CAM software solutions are showed on further figures. Zero-point selection is seen in Fig. 26.2; safety plane selection is seen in Fig. 26.3; tool selection is seen in Fig. 26.4.

Simulation of thin-walled component production is seen in Fig. 26.5.

For the experiment the following short-circuit admittance parameter was to be employed: cutting depth (DOC) 1 mm. For the manufacture of thin-walled parts, Hadfield steel may also be used [11].

The cutting tests were carried out on a three-axis milling centre Pinnacle VMC 650S with control system [12] Fanuc (Fig. 26.6), with a 4860 rpm spindle and feed of 860 mm min^{-1}. The tool was a CBN cylindrical end mill HM MG10 HX P 15358.300 with four cutting plates, 10 mm in diameter. Helix angle of inclination $\lambda = 55°$, rake angle $\gamma = 10°$ and length of the mill tool $L = 72$ mm.

Measurement of the Roughness of the Outer and Inner Surfaces of the Thin-Walled Component

Before the measurement traverse roughness on the thin-walled component, it is necessary to measure a calibration sample in Fig. 26.7. If the measured values are within the acceptable range of values, it is necessary for the roughness measurement using the device Mitutoyo SJ 400 and the calibrate control measurement to be performed again.

Surface roughness was measured on a Mitutoyo SJ400 measuring instrument at eight outer and inner locations of the periphery of the component and at four locations from a machined surface height of 20 mm (2.5, 7.5, 12.5, 17.5 mm). Measurement of traverse roughness on the inner place 1 is seen in Fig. 26.8.

Fig. 26.1 Drawing dimension of manufactured component in graphical environment Autodesk Inventor 2017

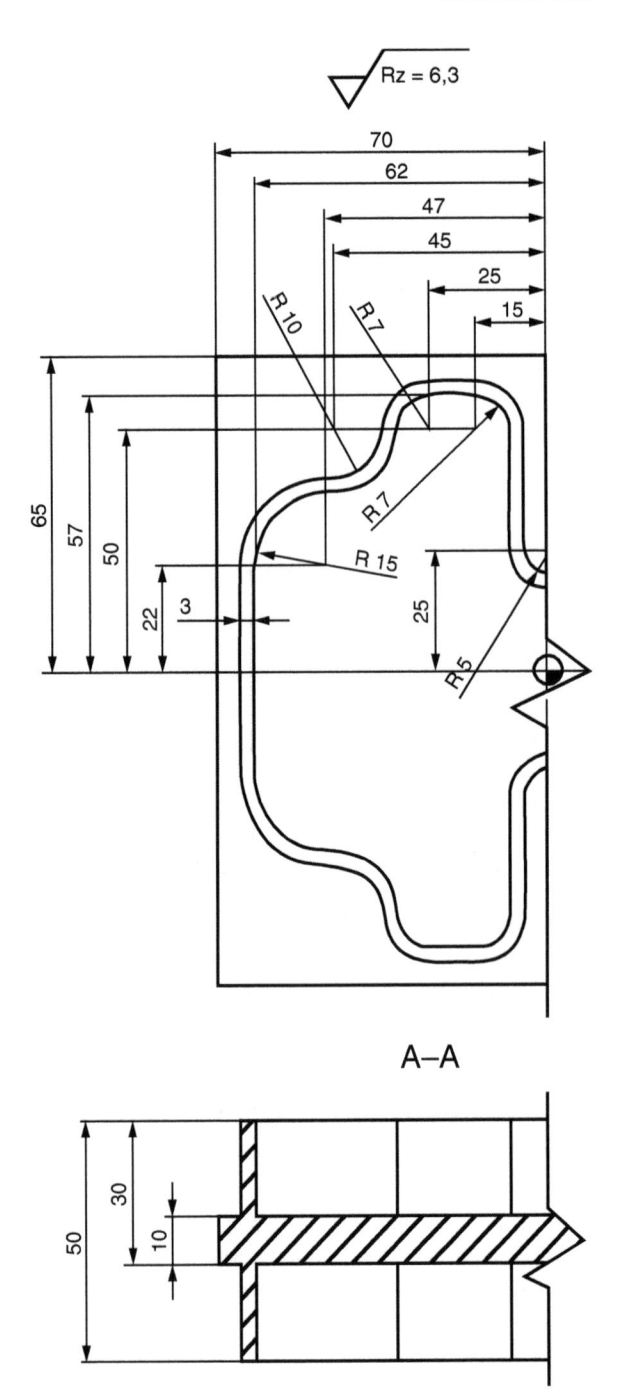

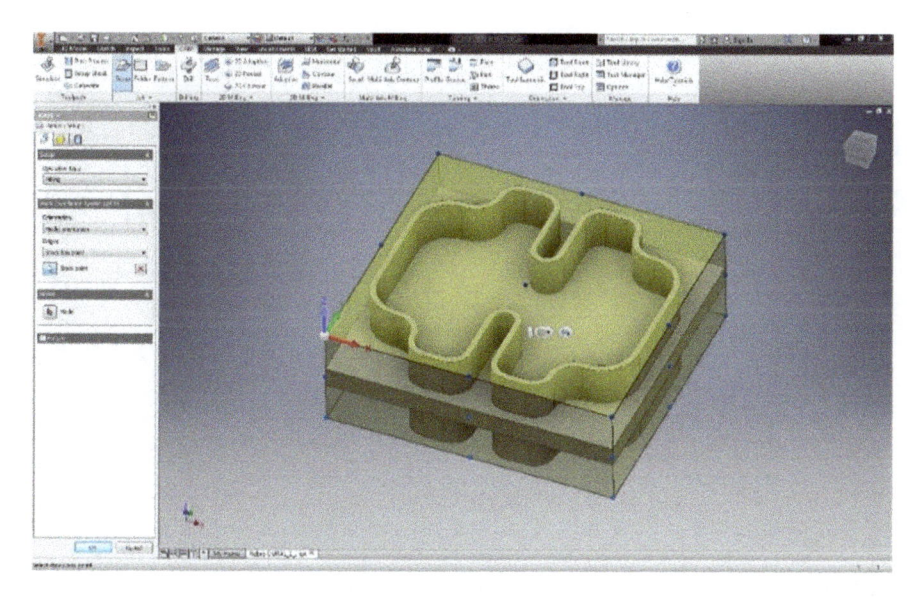

Fig. 26.2 Zero-point selection of thin-walled component

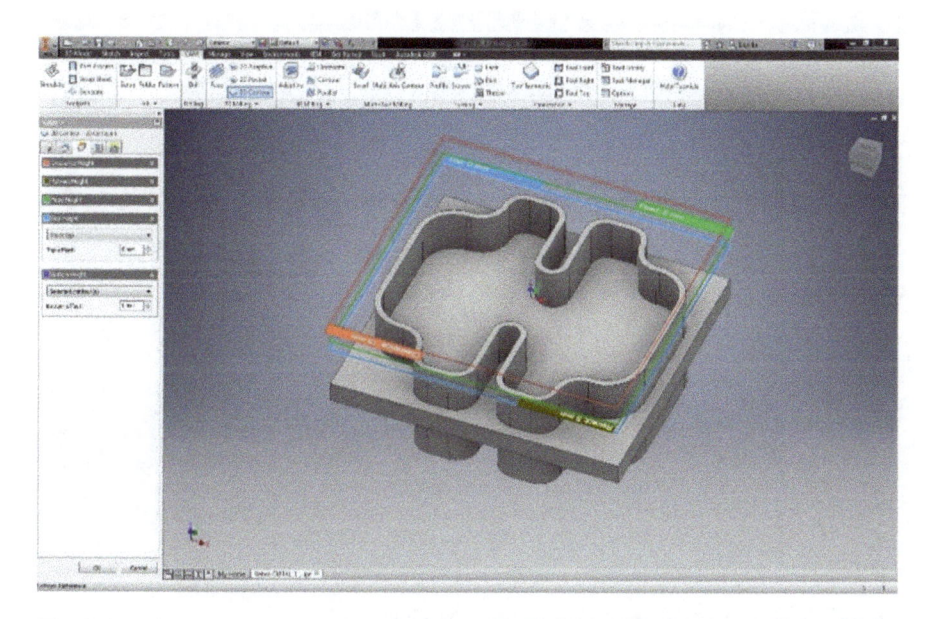

Fig. 26.3 Safety plane selection of thin-walled component for tool arrival

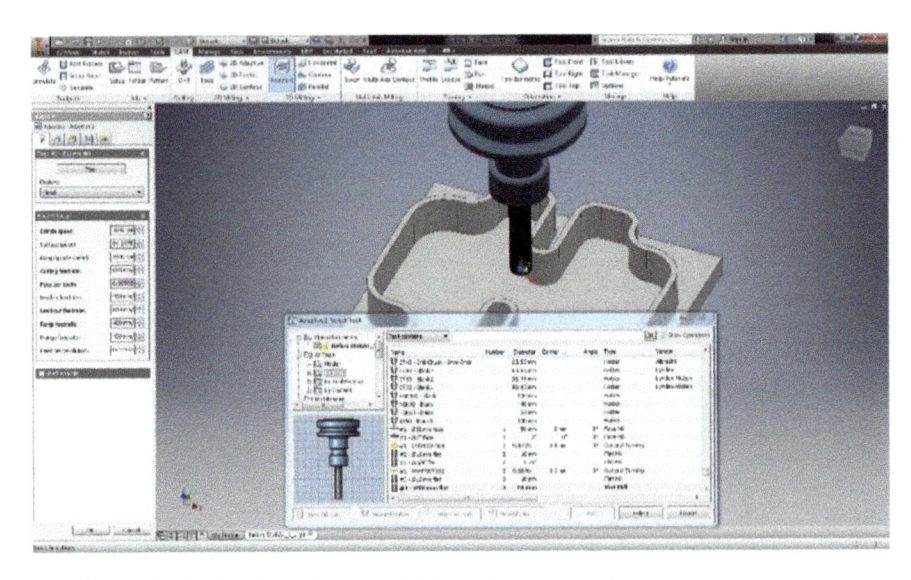

Fig. 26.4 Tool selection for production of thin-walled component

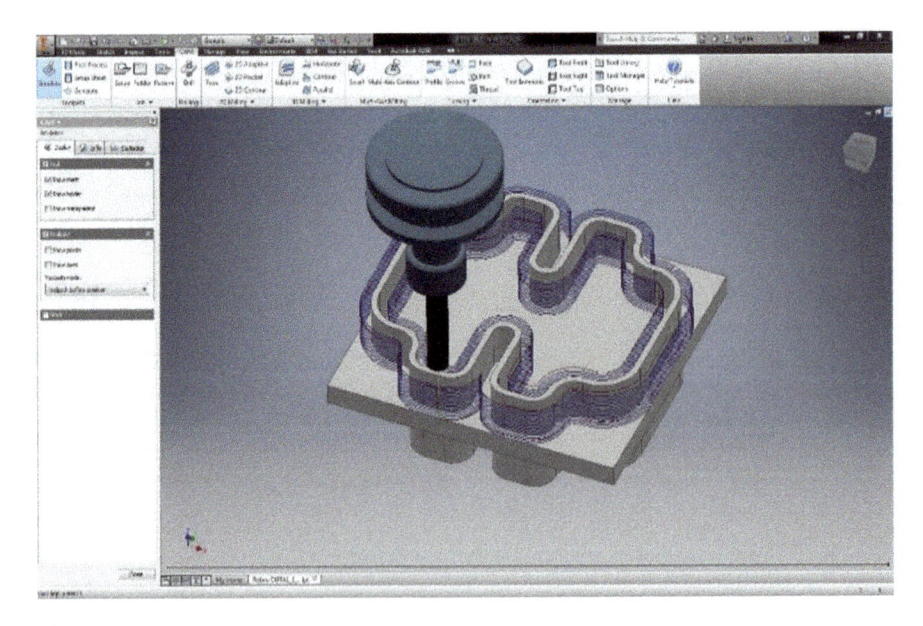

Fig. 26.5 Simulation of thin-walled component production

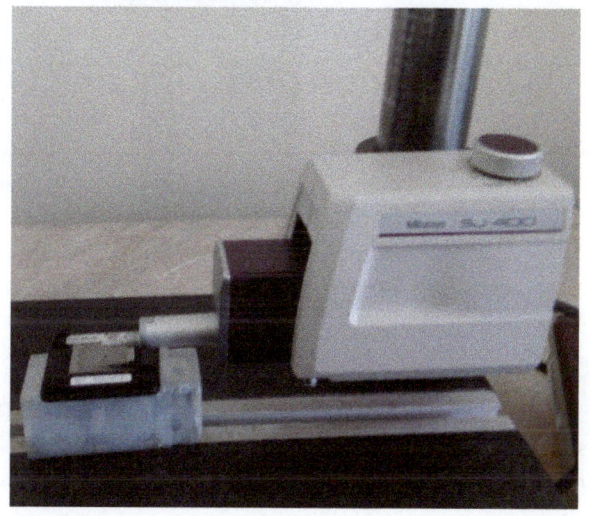

26 Evaluation of the Transverse Roughness of the Outer and Inner Surfaces...

Fig. 26.6 Three-axis milling centre Pinnacle VMC 650S with manufactured component and milling tool

Fig. 26.7 Measurement a calibration sample on the measuring device Mitutoyo SJ 400

Fig. 26.8 Measurement of transverse roughness on the inner place 1

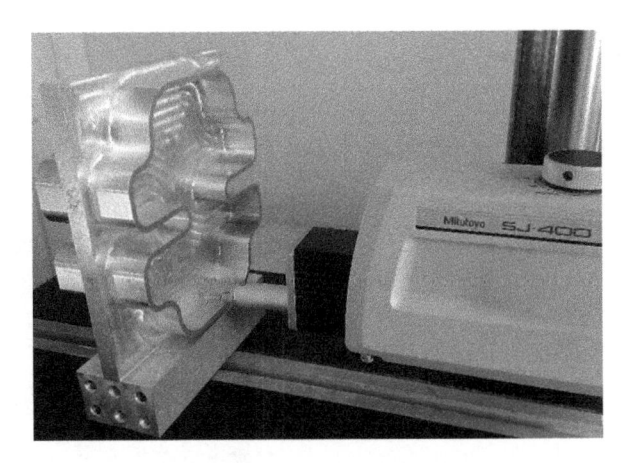

Fig. 26.9 Measurement of transverse roughness on the inner place 2

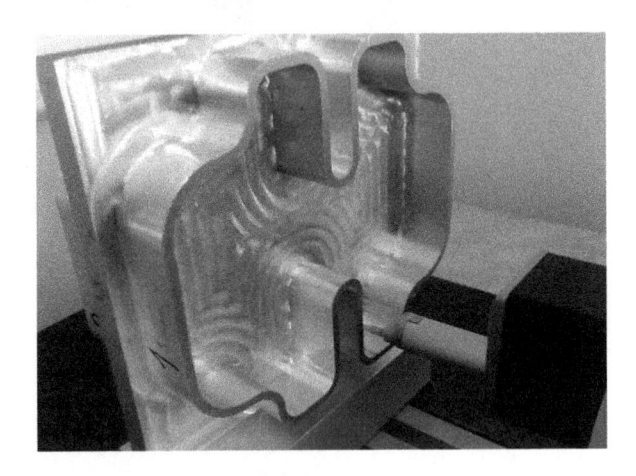

Measurement of traverse roughness on the inner place 2 is seen in Fig. 26.9. Measurement of traverse roughness on the inner place 3 is seen in Fig. 26.10. Measurement of traverse roughness on the outer place 1 is seen in Fig. 26.11.

Measurement of traverse roughness on the outer place 3 is seen in Fig. 26.12. Measurement of traverse roughness on the outer place 4 is seen in Fig. 26.13.

The following figures show the roughness values (Ra) of the outer and inner surfaces of the thin-walled component: depth 2.5 mm, Fig. 26.14; depth 7.5 mm, Fig. 26.15; depth 12.5 mm, Fig. 26.16; and depth 17.5 mm, Fig. 26.17.

Conclusion

Workshop programming for CNC machining is a midstage programming between manual programming and CAM automation. The greatest impact is the shape of the component, that is, the complexity and the fragmentation of the excavated surfaces. Next are

Fig. 26.10 Measurement of transverse roughness on the inner place 3

Fig. 26.11 Measurement of transverse roughness on the outer place 1

the skills and experience of the CNC machine operator as well as the technical level of in-use programming hardware and software. It can be stated, for practice, the way to create a CNC program selected so as to reduce as optionally eliminate the risk of a crisis situation in the production of components by production planning [13].

According to the values of transverse – in applications where roughness is measured, the lowest achieved value on the outer place was Ra = 0.21 µm in the place 8 in the depth 7.5 mm. On the inner place was measured value Ra = 0.2 µm, in the place 3 in the depth 12.5 mm. According to the values of transverse – in applications where roughness is measured, the highest achieved value on the outer place was Ra = 0.53 µm in the place 2 in the depth 7.5 mm. On the inner place was measured value Ra = 0.5 µm, in the place 3 in the depth 12.5 mm.

Fig. 26.12 Measurement of transverse roughness on the outer place 3

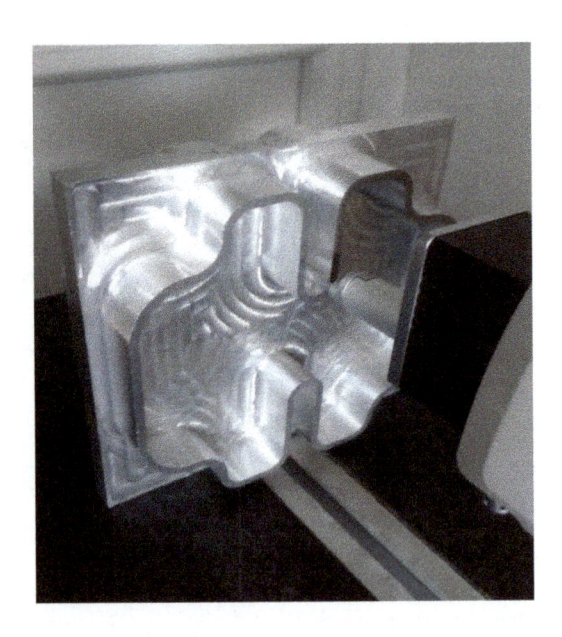

Fig. 26.13 Measurement of transverse roughness on the outer place 4

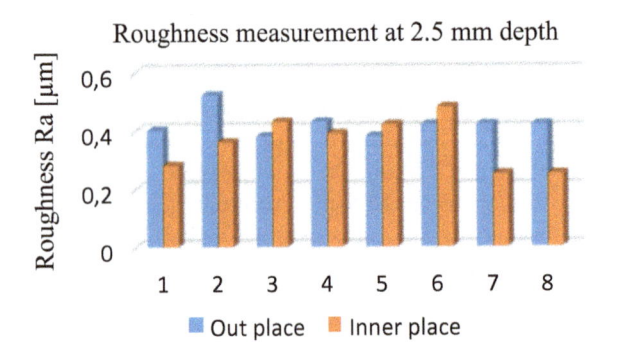

Fig. 26.14 Transverse roughness measurement at 2.5 depth of thin-walled component

Fig. 26.15 Transverse roughness measurement at 7.5 depth of thin-walled component

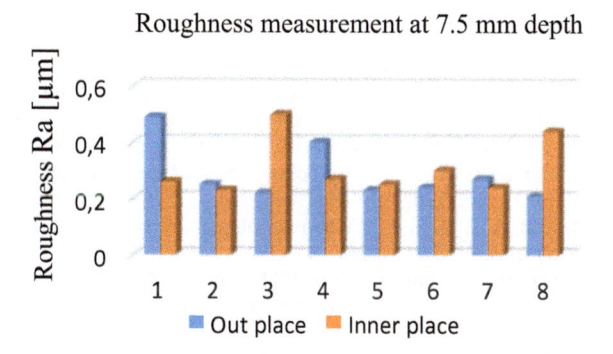

Fig. 26.16 Transverse roughness measurement at 12.5 depth of thin-walled component

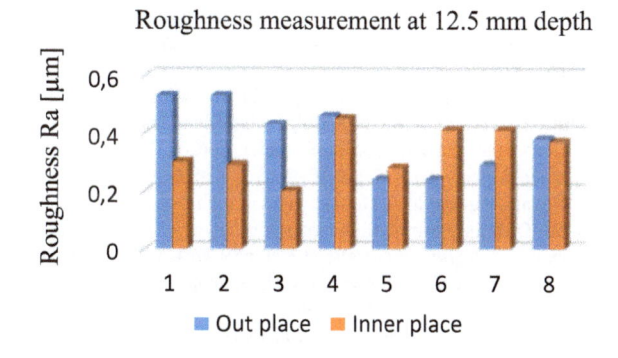

Fig. 26.17 Transverse roughness measurement at 17.5 depth of thin-walled component

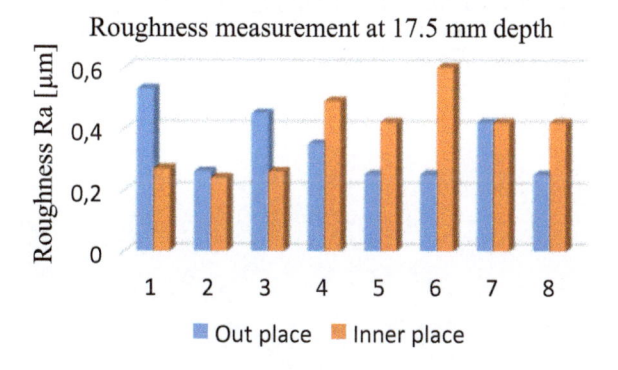

From the measured values, we can state that the roughness (Ra) with lower values was measured on the inner side of the component, which was due to a smaller machined area than the outer one.

Acknowledgement This work is a part of research projects VEGA 1/0045/18 and VEGA 1/0492/16.

References

1. Michalik, P., Zajac, J., Duplák, J., & Pivovarník, A. (2011). CAM software products for creation of programs for CNC machining. *Lecture Notes in Electrical Engineering, 141*, 421–425.
2. Novák-Marcinčin, J., Török, J., Janák, M., & Novakova-Marcincinova, L. (n.d.). Interactive monitoring of production process with use of augmented reality technology. *Applied Mechanics and Materials, 616*, 19–26.
3. Kral, J., et al. (n.d.). Creation of 3D parametric surfaces in CAD systems. *Acta Mechanica Slovaca, 2008*, 223–228.
4. Duplák, J., Panda, A., Kormoš, M., Pandová, I., & Jurko, S. (n.d.). Evaluation of T-vc dependence for the most commonly used cutting tools. *Key Engineering Materials, 663*, 278–285.
5. Michalik, P., Zajac, J., & Hatala, M. (2013). Programming CNC machines using computer-aided manufacturing software. *Advanced Science Letters, 19*, 369–373.
6. Çolak, O., Kurbanoğlu, C., & Kayacan, M. C. (2007). Milling surface roughness prediction using evolutionary programming methods. *Materials and Design, 28*, 657–666.
7. Seguy, S., & Dessein, G. (2008). Arnaud, :Surface roughness variation of thin wall milling, related to modal interactions. *International Journal of Machine Tools and Manufacture, 48*, 261–274.
8. Shahriar Jahan Hossain, N. A. (2012). Surface roughness prediction model for ball end milling operation using artificial intelligence. *Management Science and Engineering, 6*, 41–54.
9. Gulpak, M., Sölter, J., & Brinksmeier, E. (2013). Prediction of shape deviations in face milling of steel. *Procedia CIRP, 8*, 15–20.
10. Novak-Marcincin, J., Novakova-Marcincinova, L., Barna, J., & Janak, M. (2012). Application of FDM rapid prototyping technology in experimental gearbox development process. *Tehnički vjesnik – Technical Gazette, 19*(3), 689–694. ISSN 1330-3651 Spôsob prístupu: http://hrcak.srce.hr/index.php?show=clanak&id_clanak_jezik=129112.
11. Fedorko, G., Molnár, V., Pribulová, A., Futáš, P., & Baricová, D. (2011). *The influence of Ni and Cr-content on mechanical properties of Hadfield's steel*. In: Met. 2011, Tanger Ostrava, pp. 1–6.
12. Cep, R., et al. (2010). Ceramic cutting tool tests with interrupted cut simulator. In: P. of I.C. on I.T. Praha (Ed.), IN-TECH 2010, pp. 144–148.
13. Balog, M., Knapčiková, L., & Husár, J. (2016). *Plánovanie v strojárskej výrobe*. Brno: Tribun EU. ISBN 978-80-263-1078-5.

Part V
Smart Transportation Applications and Vehicle Data Processing System for Smart City Buses

Chapter 27
Designing Behavioral Changes in Smart Cities Using Interactive Smart Spaces

Predrag K. Nikolic and Adrian D. Cheok

Abstract This paper explores potentials of interactive media technology, public spaces, and interactive media art and design as potential drivers for behavioral changes and social innovations within future Smart City ecosystems. Design for Behavior Change as an approach is already accepted in several key areas such as ecology, safety, health, and well-being as well as widely adopted in social design. In this paper two interactive installations created by Predrag K. Nikolic, *InnerBody* and *Before & Beyond*, are used to explore the possibilities of employing interactive media art and design in the conceptualization of public spaces and the way it can affect and provoke user behavioral changes. Mostly considering importance of increasing user's consciousness related to sustainable community development.

Introduction

Over the past decade, design approaches have focused on human engagement capable of achieving social and public well-being. These tendencies are becoming increasingly significant for sustainable development strategies and widely reflect on a context of social innovation [1, 2]. The result of such processes and applied design models could come up with new smart services just like any innovation [3] but could also be an idea or social movement which encounters long-lasting human behavioral change [4, 5].

Design has potential to affect human behavior and as such has deep influence on our everyday life. Nevertheless, Design for Behavior Change as a design method is still under development and without clearly defined approaches and frameworks for

---

P. K. Nikolic (✉)
Cheung Kong School of Art & Design, Shantou University, Shantou, Guangdong, China
e-mail: predrag@stu.edu.cn

A. D. Cheok
Imagineering Institute, Nusajaya, Johor, Malaysia

effective implementation in some of key design areas such as ecological sustainability, safety, health, well-being, and social design [6]. In general, achieved changes in human behavior could end up as desirable or undesirable, but design efforts and strategies are always attempting to generate positive changes. Accordingly, Design for Behavior Change requires from a designer to understand people and predict how people behave in certain situation and to use design to encourage them to "do" or "not do" something [6].

This paper has intention to explore opportunities of using Design for Behavior Change as design approach and transform public spaces into interactive multisensory responsive environments capable to reflect on peoples' long-lasting behavioral changes. Hence, special attention is given to the aesthetical elements, metaphors, and meanings characteristic for users' perception of certain everyday forms and objects [7], to maximize user engagement toward meaningful experience which could elicit desirable changes and actions.

We have used activity theory as theoretical foundation and research methodologies from user-centered design process [8, 9] where focus is on the thing being designed (e.g., the object, communication, space, interface, service, etc.) to provoke user's behavioral changes. Besides that, the time-space aspect of user experience embedded into natural living environments could be considered important for future successful transformation of public spaces in Smart Cities into interactive multisensory environments where desired behavioral changes are reflecting on people. We believe that interactive media art and design can offer a research environment where people can interact through creative collaboration, aesthetic, and experiential level in a way that has immediate impact on the conscious and unconscious perception. That is the reason we decided to use *research through interactive media art* [10] and *research through design* as research approach, multisensory interactive responsive environments, and interactive everyday objects in the spatial design process of transforming Smart Cities' public spaces as a method to explore potentials for further development of Design for Behavior Change approaches and frameworks.

Background

Design in its different appearances such as for objects, services, environments, etc. has potential to influence human behavior and could create desirable as well as undesirable change [11–17]. Design has long history in its intentions to act upon positive changes in human perception and lifestyle. Hence, Designing for Behavior Change can be perceived through early understanding of behavior [18] where person's behavior is reflection of his or her personality or other "internal" factors and the physical and social environment. Clark [19] divides behavioral change approaches into those which are considering cognition from one side and context from another as the most important elements to shape behavior. Based on that we can use a framework for behavioral change design strategy which is derived from behavioral science but possible to apply in design context [6].

When referring to Design for Behavior Change models, we should go back to the 1980s and doctrines of design psychology or behavioral design, terms coined by Don Norman with respect to product design [16]. Respectively followed with the emotion design [20], persuasive technology [21] and design with intent [22] as design models have been considered more explicit in influencing human behavior.

In this study, interactive installations are used to investigate possibilities of transforming public spaces into smart environments where Design for Behavior Change can be applied and support implications on behavioral change. Interactive media art and recently HCI are intensively involved in investigating design approaches to utilize role of public spaces from merely playful, such as the BBC BigScreen Red Nose game [23] to diversity of aims and purposes [24] such as cultivating social values and sparking political discourse. Use of interactive multisensory responsive surrounding as a tool for experience design and better understanding of user interactions is revealing a plethora of possibilities in a new media design language where components are not only visual and verbal [25] but also experiential as participant can see and hear (and potentially feel) the response of the installation to his or her actions. Interactive projects presented in this paper are trying to provoke behaviorally change and design aesthetic and emotional users' experience, by allowing them to "escape the limitations of existing structures of meaning and expectation within a given practice" as with the Fictional Inquiry technique used by Iversen and Dindler [26] and to experiment with new ways of communication supported with interactive technology. In that sense Iversen and Dindler [26] describe the concept of aesthetic as "a profoundly meaningful transformation that provides a refreshed attitude towards the practices of everyday life, and as a change in our modes of perceiving and acting in the world." As such we are in position to use various aesthetical interventions and artistic dynamic forms in Design for Behavior Change approach which aims to change a way we act and react upon our or other's behavior. The use of artistically conceptualized public environments may result in better, more intensive reflection on user behavioral change and eventually to more innovative design approach that cannot be articulated on a purely conscious level [27].

Smart space is a responsive collaborative environment, which requires involvement of interdisciplinary fields such as computing, architecture, industrial design, interaction design, engineering, and cognitive psychology for the development. Most researchers are still using laboratory settings to demonstrate potentials of smart technologies, but they are missing from their consideration the importance of human dimensions such as emotional, perceptive, interpersonal, mindful, etc. Consequently, significant effort is needed to develop an interdisciplinary design framework that could express various viewpoints on smart computing technologies, while emphasizing the potential applications of smart spaces to transform our living environments [28].

Interactive installations *InnerBody* and *Before & Beyond* are artistic projects done by Predrag K. Nikolic with interaction design conceptual intention to trigger internal processes such as perceptive, mental, cognitive, and emotional and express them through external actions in a form of participants' behavioral changes. By placing them in a public space augmented with sensors and tactile and sensory technologies, our aim is to transform living surrounding into multisensory aesthetically composed environments, where artistic language and creative concept could contribute to a Design for Behavior Change methods and further development of design approaches within this field. We will describe the interactive installations' conceptual, interactive, and spatial elements of the created environments for the research experiments. Special focus will be on participants' observations and description of their experience within the installations. Accordingly, we will evaluate collected data and validate the proposed method as contribution to sustainable design choice within Smart Cities' living ecosystems.

Design for Behavior Change and Interactive Media Art

This study and the created art project have intention to reveal some of the art research potentials and to what extend it could contribute to traditional behavioral science methods. By putting people in an aesthetical environment, we can follow human patterns based on their perceptional, emotional, or intellectual experience, which usually is very hard to simulate in conventional laboratory environment.

Humans are likely to more intensively immerse and feel, react, express, or engage in the environment which uses metaphorical visual and verbal communication. Despite science (where facts and cognition are essential), art fundamentals are in experience and senses. In the case of this study, focus is on the relationship between art and behavioral science, more specific behavioral changes possible to induce through art. For this purpose, traditional arts do not offer enough space for research and conclusions as it considers viewer's passive physical behavior and inner contemplation. As such, it is an intimate experience we can share by describing it to others or keep only for ourselves. This, of course, does not mean that artists were not involved from the early beginning till today in scientific researches focused on the physical world, social design, and human needs. Examples include those of the Stonehenge designers, who helped further astronomical understanding, the Egyptian architects who created unparalleled construction technology, the early metal artists who helped discover new alloys and the chemistry of working with metal, the Renaissance artist-scientists who participated in the outpouring of scientific interest in everything from military technology to the shape of the universe, and the early twentieth-century artists who were among the first to grasp the revolutionary implications of theories such as relativity and quantum mechanics.

From the other side, science and engineering have grown increasingly self-confident and aggressive in the attempt to manipulate and control the physical world. Sometimes based on scientific research and sometimes pushing beyond those understandings, applied research has created new materials, products, and industries that have profoundly shaped everyday life and culture.

Historically, artists were inspired either with physical world like the sky, the seas, the earth, natural forces, or social, cultural, and philosophical phenomena. The synergy of scientific and social breakthroughs with nostalgia for romanticized antiquity brought about the formation of modern European civilization in the Renaissance. Science and engineering have been developed dramatically from that time onward and simply diverted artists to focus on interactions between individuals and society. They started to observe and artistically comment reflections on that relationship affected by technological development and new cultural interconnections. For artists expression and artwork embodiment became crucial the way visitors are engaging with others and the machine. Interactive art became artist's natural choice to establish new creative communication based on technological and sociocultural changes. The computer became a powerful tool suitable to imply on various ways as brain amplifier, number cruncher, and image manipulator.

Development and usage of digital technology led us to an unknown world of virtual reality and remote actions. As a result, modern research found new interesting space to explore where questions about our perception arise like: What aspects of our perception can we trust? How does virtual presence reflect on our physical bodies? What cultural forces drive interactive communication research? Why is it so important to be virtually present in distant places? Those crucial questions and many others drive frontier research agendas in digital technology fields such as new interfaces, artificial intelligence, and information visualization. In addition, artists/researchers are interested in cultural, social, and aesthetic implications in a new field of artistic experimentation in a variety of computer-related areas: digital video installation; interactive multimedia; virtual reality; installations that sense motion, gaze, facial expression, and touch; artificial intelligence; speech; surveillance; and information visualization. In our research, we used interactive installation titled InnerBody as the environment for the experiment. Details about the installation concept, experiential phases within interactive user journey, the installation storytelling line, and used research methodology will be described in the following chapters.

Interactive Project *InnerBody*

We developed *InnerBody* as an interactive installation where participants are invited to interact with the interface, shaped as model of the human heart, and take a preventive medical exam. The installation was exposed for the first time in 2014, and for that purpose public space of a medieval Belgrade Fort was transformed into improvised ambulance with hanging white sheets, MRI scans, and odors we are usually attaching to hospital environments. Before entering the interactive part of the installation, visitors were watching video showing MRI examination which the installation author has done on himself for the installation (Fig. 27.1). After entering the installation main part, we informed them about the type of examination they will be exposed and gave them instructions how to interact with the system. They did not know that medical exam was fake. The transformed public space into interactive multisensory environment base on tactile, olfactory, audio, and visual experience design has a critical role in provoking desired behavioral changes and supporting proposed design approach.

Participants were passing through several experiential phases within the installation experience design path. Firstly, they were exposed to the video showing MRI

Fig. 27.1 Entrance of the gun storage and the lobby which lead to the installation entrance

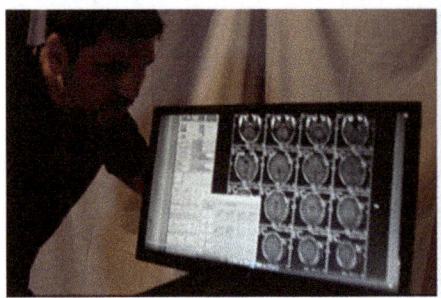

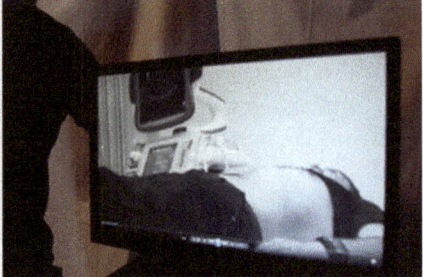

Fig. 27.2 Video which shows author's MRI examination

exam and engaging them with audiovisual sensations which referred to desired context (Fig. 27.2).

The next step in their experiential path has been entering the main installation space where visitors were exposed to human heart interface, smell suggestive of hospital environment, and projected audiovisual system outputs. In terms of design approach and artistic intervention applied on public space transformation, this was the part where aesthetical multisensory experience has been mixed with the pragmatic goals of achieving desired behavioral changes (Fig. 27.3).

Special attention was given to conceptualization and design of the human heart interface as we considered it as important interaction point in the process of visitors' experience design through multisensory perception, mainly because of the metaphors and meaning we are addressing to the heart, as symbolic representation of vitality, beginning and ending, living and dying, and health and sickness [7] and as such can reflect on us emotionally in that context. Visitors were asked to touch the heart to start the simulation of medical exam, which they did not know, and trigger the system outputs (Fig. 27.4).

Audiovisual outputs were controlled sound of heart beats and visual representation of several vital functions in human body. By establishing interactive multisensory dialog between users and the system outputs, even they have not been related to their physical state, we had an intention to explore the potential of using symbolic, spatial, and sensory language to affect their awareness and achieve desired health behavioral changes.

27 Designing Behavioral Changes in Smart Cities Using Interactive Smart Spaces 373

Fig. 27.3 The InnerBody heart interface. The output occurs when user grabs it

Fig. 27.4 The InnerBody model of a human heart and cardiovascular system interface

The data in the *InnerBody* experiment were collected from the following sources:

- Personal observation – the data was collected on the spot, while the users were interacting with the installation, before entering and after leaving the installation.
- User interview – this method was conducted after users' interaction with the installation.

Two groups of users participated in the *InnerBody* interface. The first group consisted of participants who were introduced to the installation narrative prior entering it. They were told about the idea, the purpose, and they knew what to expect. The second group included the participants who did not know anything about the purpose and the functioning of the installation prior entering and, hence, did not know what to expect as an outcome. A total number of 32 participants were personally observed while interacting with the installation – 6 of them were told about the purpose of the installation, while 26 were not. Our basic assumption was that the first group of participants, who were familiar with the installation narrative, would be more indifferent to the outcome of the installation procedure than the participants of the second group, who would become much more frightened and concerned about their health condition.

As mentioned before, the *InnerBody* interface was the stylized model of a human heart and cardiovascular system. The forms, objects, and its representations were something that the participants would easily recognize once they enter the installation. Nevertheless, the model of the human heart, the sound of a heartbeat, the scent of medicines, and the hospital-like white sheets were the elements with the purpose of triggering human sense of concern about health condition and the fear of dying. Participants were instructed to grab the model of the heart, which was slightly bigger than the normal human heart, to start "self-medical exam" and trigger the system. The first stage was observing participants while reading the instructions prior entering the installation. The only sound that could be heard while reading was the sound of MRI machine. In most of the cases, the expression on the faces of the participant would quickly turn from smiling to serious. During the upcoming interviews, the participants stated that the sound of the MRI machine, in combination with the instructions about the possible health condition that could be "found out" during the session, provoked a sense of urgency, alert, and even sudden fear inside of them. One of the participants, a 41-year-old male, said that suddenly he just felt he had to be serious about the "procedure." The second stage of the observation was watching participants grabbing the heart model to start the "examination." Many of the participants grabbed the heart model very gently to start the process. Later, during the interviews, we learned there were two reasons for such behavior: the first was the sense of discomfort provoked by reading the instructions and hearing MRI and the second was the sense of uncertainty about the upcoming "examination." The third stage of our observation was monitoring the effects of different stimuli of the installation on the four senses of the participants: sight, hearing, smell, and touch. As per environmental psychology, different stimuli from the outside world influence our senses [29], triggering the response of human organism (stimuli-organism-response (SOR) model). The responses range from dissatisfaction to satisfaction, from tension to relaxation, and from inferiority to superiority over a situation. It helps us understand the reasons behind human behavioral changes. Most of the behavioral change is driven by mental state which is usually affected by external stimuli. Touching the heart model was fun for most of the participants, but only for a while. In the case of three participants, a 42-year-old woman, an 18-year-old girl, and a 26-year-old man, it turned out to be an unpleasant experience. The paint of the heart model stained their hands. They instantly misinterpreted the red stains on their hands as blood. During our experiment, in most of the cases, auditory stimuli, such as the sound of MRI machine, caused tension and sense of concern for the participant's health state. On the other hand, the message at the end of the "examination," informing that the whole procedure was not a real examination, brought back the sense of relaxation and superiority over the situation to the participants. On the other hand, visual stimuli, such as watching monitor displaying fake heartbeats, in most of the cases made participants get serious expression on their faces. The fourth stage was observing participants in the final stage of their "examination" when getting informed about the "results." As previously stated, at the end of the session, the participants were informed about the purpose of the installation. It was interesting to observe changes of their facial expressions as well as the changes in their overall behavior. The tension was replaced by relaxation; the concern was replaced by happiness after they had got informed the whole "examination" was not real. Still, the final stage of the installation conveyed the sense of warning which effects were supposed to be interrogated through interviews with the participants.

User interviews were structured and done in the form of an informal, open conversation with participants. The goal of the interviews was to reveal the change in participants' behavior based on their experience. The questions were split in two phases:

Phase 1 included questioning conducted minutes after leaving the installation. The questions were the following:

(i) Describe your experience during interaction with the installation.
(ii) Describe your feelings related to different stimuli: sound, pictures, smell, and touch.
(iii) Can you say there is the difference in your feelings before entering the installation and after leaving it?

Since the purpose of the installation was to create the sense of moderate fear and concern about the participant's health state, it was important to have a follow-up on the installation experience. Phase 2 was the follow-up phase conducted 3 months after the installation experience. The questions posed were the following:

(iv) Since the installation, have you taken any kind of physical examination?
(v) If you have, can you say that the urge for physical examination was provoked by your installation experience?

The first phase of user interviews was conducted on 12 participants: 5 from the first group, who had been introduced with the installation narrative, and 7 from the second group, who had not been introduced with the installation. Although we had assumed that the participants from the first group would be more indifferent to the outcome of the experienced "examination" than participants from the second group, user interview phase 1 showed that there were participants from the first group who reacted dramatically to the stimuli they were exposed to. One of the users, a 42-year-old woman who had been introduced to the installation narrative, stated out that when she placed her hands on the heart model, her experience became so realistic that she was convinced that she could hear her heart started beating faster. Despite the fact she had been told that the sound of the heartbeat would come from the tape. Moreover, a 50-year-old woman stated that, although she had been told the installation story, during the "examination" she was trying to recall when she had her last appointment with her physician. We decided to follow-up the first group despite our first assumption that it would not be necessary. The second phase of the user interview was conducted on seven participants: three from the first group and four from the second group. Two participants from the first group, who had been familiar to the installation narrative, answered that they had developed serious concern about their health condition ever since the installation experience, while one of them even took a medical examination. From the second group, three participants made appointments with their physicians.

We concluded that our installation participants were frightened no matter they knew the whole "examination" was not real and that it was a part of a directed, synchronized performance. The participants understood the metaphors used in the installation, since the communication between them and the system was clear, easily understandable, and interactive. However, the interactive environment and suggestive objects used to design user experience created substantial amount of fear and managed to change the behavior of the participants who had been introduced with the installation narrative in the first place. We used deeply inherited fears in our consciousness we react on subconsciously to provoke desirable effects. We concluded that despite the fact our participants had been aware that it was fake examination, they started being afraid of their deepest fears. Based on that, we found interactive media art and multimodal storytelling as potential environments for the design of health behavioral change.

Interactive Project *Before & Beyond*

In the interactive installation, *Before & Beyond*, the author Predrag K. Nikolic tried to transform public space into playful environments where visitors will be engaging around physical interactions between each other. Hence, stimulate their internal processes such as motivate them to collaborate, bodily interact and communicate, having thoughts on their existence and natural surround. As result, to generate need for nearby social engagement and relationships development. The installation was exposed for the first time at the Maison Shanghai 2016 event as part of Final Fantasy exhibition. Space was responsive to visitors' body movements, direction of walking, and distance between participants. By entering the installation space, the visitors were attached to a one *string of energy* with characteristic color and sound as an abstract representation of their existence in a virtual world projected as system output on a display placed in front of them. To accurately track their location within transformed public space, we augmented it with sensory-based technologies, Kinect movement detection placed on the wall and Beacons integrated in the medallion around the neck of the participants (Fig. 27.5).

The initial idea was to track visitors only by using Kinect. During the prototyping phase, we concluded that the chosen solution is not accurate enough for some of

Fig. 27.5 After entering the installation space, every participant gets his personal *string of energy* projected on the screen

the most important types of visitors' behaviors we wanted to follow such as single-string tracking after they separate. Also, it was very crucial to provoke a variety of gestures and body interactions between participants which required usage of additional supportive technology to achieve our goals. We found Beacons in combination with Kinect as sufficient solution capable to fulfill our conceptual and interaction design needs. Personal *strings* (animated two-dimensional closed line in a shape of circle, enriched with a specific tone and color) which are projected on a screen are moving over the screen based on string owner (visitor) movements, walking directions and interacts with the other strings projected on the screen based on distances and interaction between visitors in physical space. When participants are close enough to each other, their strings are joining in one and vice versa; when they separate the joint string splits back to personal (single) strings (Fig. 27.5). The installation can host up to 11 persons which is directly connected with the initial inspiration for creating the installation and comes from string theory. According to this theory, every string of energy can exist in 11 dimensions [30]. Scientists who are supporting string theory believe that this theory can close the gap between quantum mechanics and gravity theory. Hence, the installation refers directly to questions related to the beginning of the world, beyond our perception and known reality. The installation tends to communicate all these ideas through art piece as important facts for understanding life, the universe, and mental and physical existence and affect visitors' social behavior. The metaphors of personal strings and what is behind the idea of attaching them to everybody who step in the installation were explained to the visitors by the assistants and printed on the installation separation wall. The ideas and meanings we wanted to communicate with the participants we found are very important for achieving our interactions' goals and desired engagement. Aesthetic experience has been accomplished with the audio-visual elements such as shapes of the

Fig. 27.6 Depending on distance and established relationships between visitors, visual and audio appearance of the strings changes

strings, background sound, sound of strings, etc. They are related to the installation concept, inspired with string theory and socio- emotional development, emotional management and the ability to establish positive and rewarding relationships with others [31] (Fig. 27.6).

The interaction concept of the Before & Beyond installation is to intrigue visitors to explore their own virtual existence and correlate it with others within a transformed public space. During that process, they are acquiring new personal properties such as color, shape, and sound attached to their *string*. After becoming an aware of it, to trigger social and cultural dimensions where they are becoming interested in collaboration with others and to develop new relationships within given context. In our case, physical as well as virtual space is becoming a place for body and social interactions and for strengthening our interpersonal relations as essential requirement for running social initiatives within communities with common goals and needs.

In case of the interactive installation *Before & Beyond*, we based our concept per idea that depending on content, context, and public space structure, we are in position to affect the social situation. For example, the degree of intimacy and how people will move and behave in the space influences aesthetic and interaction concept that will work well [32]. In case of the installation *Before & Beyond*, users are moving around in the responsive space in front of the display; the interaction space between the users and display is an agent which connects visitors' efforts and actions in physical space with virtual audiovisual outputs. By moving around and interacting between each other, participants are shaping social space the same way they are doing with their own everyday living surround. Public space configuration and context affect the social space, individual, and interpersonal interactions as the performer is not isolated but in a group.

The data in the *Beyond & Before* research experiment were collected from:

- Personal observation
- User interview

The goal of the interviews was to reveal the change in participants' behavior based on their experience. The questions were the following:

(i) Describe your experience during interaction with the installation.
(ii) Describe your relationship with other participants: before entering the installation, during a session, and after leaving the installation space.
(iii) Describe how spatial circumstances and environment affected your experience and behavior.

Through observations and informal interviews with the participants, we found out that the period they were alone in the installation space was not interesting and for some of them pointless. As per their answers, they tried to understand the relation between their movements and the *string* behavior, limitations, and the system responsiveness. After leaving the installation, they were not interested to enter it again. Even we instructed assistants how to explain the conceptual idea behind the installation, meanings of the visual and the audio outputs, as well as the way how to interact with the installation to avoid potential breakdown points [33], we did not increase significantly individual user engagement, movement, and gesture varieties. The major behavioral change happened when visitors intended to organize with each other and worked together. Their common goals were directed toward affecting the way audiovisual outputs are interacting between each other, which was the moment they were becoming active participants and contributors in the process of the installation creative development. The elements of collective efforts and needs to physically interact between each other changed their experience to more engaging and socio-tactile. Also, within the group they were learning faster how to control and manipulate the installation audiovisual outputs; as a result we lower down the possibility for breakdown points during users' interaction with the installation. Most of them describe their experience as: "We've been involved into playful game with total strangers. At the beginning, we tried to explore relation between our spatial positions and how they reflect on *Personal Strings* movements and system's audio-visual outputs. After that we were just enjoying in making something together with others who were involved. Touching and moving with others we considered in the beginning as necessity to achieve desired control over the audio-visual artifacts we were generating together. During that process, we end up developing our relationships in physical space too and somehow we spontaneously became connected."

From the moment, they realized they could join the strings by lowering down the distance between each other and getting in physical contact, started to stick together, bodily interact, and hug each other. Majority of the participants explained that before entering the installation, they did not have any intention to interact with other participants and then after figuring out the dependence between body position and *personal string*, they wanted to know what will happen if they touch with other strings. They started to shorten the distance between them and others and spontaneously got voluntarily involved in unexpected types of body interactions (hugs, holding hands, touching body parts, etc.). We also observed separately groups of users with some emotional connection (family, friends, partners), total strangers who met each other for the first time, and mixed groups made of these two. Then we tried to record the most frequent type of gestures and movement generated within the groups and find parameters we should consider in order to anticipate certain visitors' behavior. We used initially the following metrics [33].

- Time spent in the installation
- Number of repeated visits

The most frequent types of gestures within all the groups were hugs and holding arm to arm. According to participants' testimony, after being so close to each other, they changed their socio-emotional attitude related to people involved in the collective interaction, started to talk with them after leaving the installation, and wanted to enter the installation again.

Groups made of total strangers have had in average the longest time spent in the installation as well as number of repeated visits; second best were groups made of emotionally related participants (family, friends, partners). They were also the fastest learners as due to their relationships, they were starting to coordinate movements and gestures in very early phase of the interaction with the installation as well as to share knowledge between each other. Mixed groups had the poorest results considering both metrics as they became split as in the early phase of the interaction in two parts, related and non-related visitors. Expectedly, groups with different emotionally related members had the highest variety of gestures; the most engaged groups were those made of total strangers.

Conclusion

In the interactive project *InnerBody*, we concluded that despite the fact our participants had been aware that it was fake examination, they started being concerned for their health condition which motivates some of them to take real preventive medical exam. Furthermore, we found interactive media art with its aesthetical preferences and placed in public spaces could be used as potential design tool we can contribute to new approaches and frameworks within Design for Health Behavior Change field.

With same preferences, we can address findings from interactive project *Before & Beyond* as in this case we also provoked instant participants' behavioral changes and generated new types of socio-bodily interactions between total strangers as well as visitors who were emotionally related. Social situation, within transformed public space, joint participants into groups where all individual efforts were dedicated to common goals. As a result we manage to change their social behavior which resulted with new interpersonal and emotional relationships.

Furthermore, we are planning to expose *InnerBody* and Before & Beyond interactive installation under different sociocultural and spatial circumstances and conduct a systematic iterative data analysis after that, focusing on spatial and contextual aspects within and compare them with the observed social situation and behaviors [34]. In our future research directions, we want to better understand the potential of interactive media art and design and propose it as new Design for Behavior Change approaches as well as to develop a design model for using public places in Smart Cities to provoke new types of interactions for human behavioral change and social innovation.

References

1. Manzini, E. (2009). Viewpoint: New design knowledge. *Design Studies, 30*(1), 4–12.
2. Emilson, A., Seravalli, A., & Hillgren, P. (2011). Dealing with dilemmas: Participatory approaches in design for social innovation. *Swedish Design Research Journal, 1*(11), 23–29.
3. Murray, R., Caulier-Grice, J., & Mulgan, G. (2010). *The open book of social innovation.* London: The Young Foundation & NESTA.
4. Björgvinsson, E., Ehn, P., & Hillgren, P. (2012). Design things and design thinking: Contemporary participatory design challenges. *Design Issues, 28*(3), 101–116.
5. Dott Cornwall (2010). Design Council: What's Dott? http://www.dottcornwall.com/aboutdott/whats-dott. Accessed 27 Oct 2011.
6. Niedderer, K., MacKrill, J., Clune, S., Evans, M., Lockton, D., Ludden, G., Morris, A., Gutteridge, R., Gardiner, E., Cain, R., Hekkert, P. (2014). Joining Forces: Investigating the influence of design for behaviour change on sustainable innovation. In *NordDesign 2014: 10th Biannual conference on design and development* (pp. 620–630)
7. Nikolic, P. (2015). Multimodal Interactions: Embedding new meanings to known forms and objects, Lecture Notes of the Institute for Computer Sciences, Social Informatics and Telecommunications Engineering 2016. In B. Mandler et al. (Eds.), *Internet of things 360° 2015, part II, LNICST 170* (pp. 107–121). Rome: Springer International Publishing. https://doi.org/10.1007/978-3-319-47075-7.
8. Kuniavsky, M. (2003). *Observing the user experience: A practitioner's guide to user research (Morgan Kaufmann Series in Interactive Technologies) (the Morgan Kaufmann Series in Interactive Technologies).* San Francisco, CA: Morgan Kaufmann Publishers Inc..
9. Tullis, T., & Albert, W. (2008). *Measuring the user experience: Collecting, analyzing, and presenting usability metrics.* San Francisco, CA: Morgan Kaufmann Publishers Inc..
10. Sommerer, C., & Mignonneau, L. (2009). *Interactive art research.* Vienna: Springer.
11. Brown, T., & Wyatt, J. (2010). Design thinking for social innovation. *Stanford Social Innovation Review,* (Winter), 30–35.
12. Consolvo, S., McDonald, D. W., & Landay, J. A. (2009). Theory-driven design strategies for technologies that support behaviour change in everyday life. In *Proceedings of the CHI2009: Creative thought and self-improvement.* Boston, MA: ACM Press.
13. Fry, T. (2008). *Design futuring: Sustainability, ethics, and new practice.* Oxford: Berg.
14. Lockton, D. (2012). POSIWID and determinism in design for behaviour change. Working Paper Series, Brunel University, April 2012. http://bura.brunel.ac.uk/handle/2438/6394.
15. Niedderer, K. (2013). Mindful design as a driver for social behaviour change. In *Proceedings of the IASDR conference 2013.* Tokyo, Japan: IASDR.
16. Norman, D. (2002). *The design of everyday things.* New York: Basic Books.

17. Moggridge, B. (2008). Innovation through design. In International design culture conference, Korean Design Research Institute, Seoul National University, IDEO. Behavior and Technology Development, Springer, Berlin, pp. 222 (2006).
18. Lewin, K. (1935). *A dynamic theory of personality*. New York: McGraw Hill.
19. Clark, G. L. (2009). Human nature, the environment, and behavior. *SPACES Online, 7*(1).
20. Desmet, P. M. A., Overbeeke, C. J., & Tax, S. J. E. T. (2001). Designing products with added emotional value. *The Design Journal, 4*(1), 32–47.
21. Fogg, B. J. (2003). *Persuasive technology: Using computers to change what we think and do*. San Francisco, CA: Morgan Kaufman.
22. Lockton, D., Harrison, D. J., & Stanton, N. A. (2010). The design with intent method: A design tool for influencing user behaviour. *Applied Ergonomics, 41*(3), 382–392.
23. Fatah gen Schieck, A., Briones, C., & Mottram, C. (2008). The urban screen as a socialising platform: Exploring the role of place within the urban space. In *MEDIACITY: Situations, practices and encounters* (pp. 285–305). Berlin: Frank & Timme.
24. O'Hara, K., Glancy, M., & Robertshaw, S. (2008). Understanding collective play in an urban screen game. In *Proceedings of the 2008 ACM conference on computer supported cooperative work* (pp. 67–76). New York: ACM.
25. Sanders, E. B.-N. (2000). Generative tools for CoDesigning. In A. R. Scrivener, L. J. Ball, & A. Woodcock (Eds.), *Collaborative design*. London: Springer-Verlag.
26. Iversen, O.S., Dindler, C. (2008). Pursuing aesthetic inquiry in participatory design. Proceedings of the Tenth Conference on Participatory Design, Bloomington, Indiana.
27. Nelson, H., & Stolterman, E. (2003). *The design way – intentional change in an unpredictable world. Foundations and fundamentals of design competence*. Englewood Cliffs, NJ: Educational Technology Publications.
28. Hoffman, K. D., & Bateson, J. E. (2011). *Services marketing*. Boston, MA: Cengage Learning.
29. Witten, E. (1995). String theory dynamics in various dimensions. *Nuclear Physics B, 443*(1–2), 85.
30. Greene, B. (2010). *The elegant universe: Superstrings, hidden dimensions, and the quest for the ultimate theory*. New York: W. W. Norton & Company.
31. Cohen, J., Onunaku, N., Clothier, S., & Poppe, J. (2005). *Helping young children succeed: Strategies to promote early childhood social and emotional development*. Washington, DC: National Conference of State Legislatures and Zero to Three.
32. Martinovsky, B., Traum, D. R. (2003). Breakdown in human-machine interaction: The error is the clue. In Proceedings of the ISCA tutorial and research workshop on Error handling in dialogue systems.
33. Nikolic, P. K. (2015). Collective creativity: Utilizing the potentials of multimodal environments. In R. Giaffreda, D. Cagáňová, Y. Li, R. Riggio, & A. Voisard (Eds.), *Internet of things. IoT infrastructures. IoT360 2014. Lecture notes of the Institute for Computer Sciences, Social Informatics and Telecommunications Engineering* (Vol. 151). Cham: Springer.
34. Fischer, P. T., Zollner, C., Hoffmann, T., Piatza, S., & Hornecker, E. (2013). Beyond information and utility: Transforming public spaces with media facades. *IEEE Computer Graphics and Applications, 33*(2), 38–46. https://doi.org/10.1109/MCG.2012.126.

Chapter 28
Social Innovations in Context of Smart City

Richard Jurenka, Dagmar Cagáňová, Natália Horňáková, and Augustín Stareček

Abstract Increased dynamics of social processes imposes requirements on innovative approaches and solutions for resolving social problems in contemporary world that exceed the established ways of thinking and acting. A social innovation as a term covers a wide range of activities and tools. The widest definition specifies social innovation as any new strategies, concepts, and ideas. The paper contains theoretical description of smart city, social innovations, social enterprise, social entrepreneurship, social entrepreneur, definition of sustainable development in context of social innovations and best practice from social innovations represented by INNOVAT and YounGO projects. The first goal of the submitted article is to identify and define the key features of smart city and social innovations. The most important contribution of the article is to highlight the necessity of human capital in industrial enterprises. Research methods, consist of analysis, synthesis, deduction, induction, and comparison, are used in the article.

Introduction

Smart city is a community that is efficient, liveable, and sustainable and which increases quality of life. Smart city cannot exist and develop without smart human capital. Smart human capital is the fundamental element of each smart city. Only smart people could be an author of smart solutions and create smart ideas which

---

R. Jurenka (✉) · D. Cagáňová
Slovak University of Technology in Bratislava, Bratislava, Slovakia

Faculty of Materials Science and Technology in Trnava, Trnava, Slovakia
e-mail: richard.jurenka@stuba.sk; dagmar.caganova@stuba.sk

N. Horňáková · A. Stareček
Institute of Industrial Engineering and Management, Trnava, Slovakia

Slovak University of Technology in Bratislava, Bratislava, Slovakia
e-mail: natalia.hornakova@stuba.sk; augustin.starecek@stuba.sk

transform common city to smart city. Improvement of living conditions and working conditions and continual quality improvement of human capital can be achieved by social innovation. Social innovations in the context of a smart city allow achieving sustainable development, which in these days could ensure a higher quality of life. The main goal of social innovations is to support human capital. Involvement of government and insertion of these social goals in policy-making process and defining common strategy for each city should be fundamental.

Vision of Industry 4.0 will have a significant impact on employee qualification and on the labor market in general, whereby it will be also necessary to consider all social aspects and consequences. Impact of Industry 4.0 will lead to new work principles, change of employee's role, and changes in structure of job description of the most professions which will require new knowledge. The abovementioned impacts will affect the employee development and will also require the new settings of labor market and educational systems.

What Is a "Smart City"?

Examples of smart cities come in many variants, sizes, and types. This is because the idea of the smart city is relatively new and evolving, and the concept is very broad. Every city is unique, with its own historical development path, current characteristics, and future dynamic. The cities which call themselves smart, or are labeled as such by others, vary enormously. The evolution of the smart city concept is shaped by a complex mix of technologies, social and economic factors, governance arrangements, and policy and business drivers. The implementation of the smart city concept, therefore, follows very varied paths depending on each city's specific policies, objectives, funding, and scope [1].

Smart city (Fig. 28.1) is a city equipped with basic infrastructure which gives a decent and respectable quality of life and a clean and sustainable environment through the application of specific smart solutions [2].

The most effective definition of a smart city is a community that is efficient, liveable, and sustainable—and these three elements go hand in hand. A sustainable

Fig. 28.1 Smart city [3]

community is one which reduces the environmental consequences of urban life and is often an output of efforts to make the city more efficient and liveable [4].

The British Standards Institution (BSI) defines the term smart city as "the effective integration of human, physical and digital systems with the aim to built environment which can deliver sustainable and prosperous future for its citizens."

Citizen-focused definitions consider smart city as friendly, clean, and with well-developed transport infrastructure. Citizen-focused definitions are also familiar with words like "technology," "connected," "internet," and "modern" which extend the right meaning of smart city [5].

A smart city is also a learning city, which improves the competitiveness of urban contexts in the global knowledge economy. Learning cities are actively involved in building a skilled information economy workforce. Some authors established a typology of cities that are learning to be smart: individually proactive city, city cluster, one-to-one link between cities, and city network. A knowledge city is analogous to a learning city. It refers to "a city that was purposefully designed to encourage the nurturing of knowledge." Technopolis and ideapolis, early articulations of a knowledge city, have evolved into digital, intelligent, or smart city. The notion of knowledge city is interchangeable to a certain degree with similar evolving concepts such as intelligent city, educating city, or smart city. However, a knowledge city is heavily related to knowledge economy, and its distinction is stress on innovation. Knowledge-based urban development has become an important mechanism for the development of knowledge cities. The buzz concept of being clever, smart, skillful, creative, networked, connected, and competitive has become some of the key ingredients of knowledge-based urban development [7].

Despite the current wave of discussion and debate on the value, function, and future of smart cities, as a concept it resists easy definition. At its core, the idea of smart cities is rooted in the creation and connection of human capital, social capital, and information and communication technology (ICT) infrastructure in order to generate greater and more sustainable economic development and a better quality of life. Smart cities have been further defined along six axes or dimensions [1]:

- Smart economy
- Smart mobility
- Smart environment
- Smart people
- Smart living;
- Smart governance

But the concept of smart city is not static: there is no absolute definition of a smart city, no end point, but rather a process, or series of steps, by which cities become more "liveable" and resilient and, hence, able to respond quicker to new challenges. Thus, a smart city should enable every citizen to engage with all the services on offer, public as well as private, in a way best suited to his or her needs. It brings together hard infrastructure, social capital including local skills and community institutions, and (digital) technologies to fuel sustainable economic development and provide an attractive environment for all [8].

Fig. 28.2 Smart infrastructure [6]

Basic and Smart Infrastructure

Basic infrastructure (Fig. 28.2) includes assured water and electricity supply, efficient urban mobility and public transport, sanitation and solid waste management, e-governance and citizen participation, robust information technology connectivity, and safety and security of citizens [2].

Reaching of smart infrastructure is possible only in the way of applying information and communications technology (ICT) infrastructure. The availability and quality of the ICT infrastructure are important for smart cities. Indeed, smart object networks play a crucial role in building smart cities into a reality. ICT infrastructure includes wireless infrastructure (fiber-optic channels, Wi-Fi networks, wireless hotspots, kiosks) and service-oriented information systems. The implementation of an ICT infrastructure is fundamental to a smart city's development and depends on factors related to its availability and performance. There is a lack of literature that focuses on ICT infrastructure barriers of smart city initiatives [9].

Smart Solutions

Smart solutions consist of the following aspects: relevant public information, waste transformation to energy and fuel, waste transformation to compost, renewable source of energy, efficient energy and green building, smart pedestrian crossing, electronic service delivery, citizens' engagement, 100% treatment of wastewater, smart meters and management, monitoring water quality, smart parking, intelligent traffic management system, and others [2].

Now more than ever, cities need to provide public services more efficiently along with supporting sustainable and long-term economic growth. The latest researches suggest that the best way to do this is becoming "smart." This generally means to use new technologies (mainly information and communication technologies) and data to improve service delivery and address various economic, social, and environmental challenges. For example, smart energy meters can help cities manage energy demand, reduce cost, and safeguard the environment, while the move toward online health consultations can also reduce cost and improve the quality of services. Smart transport and mobility initiatives like traffic control center can also help the city manage traffic flows and reduce traffic jam, while making real-time bus arrival data publicly available can allow development of new mobile applications that make commuting in the city easier [9].

Connectivity in Smart City

Infographic (Fig. 28.3) outlines the key components of a smart city, especially components of top medium-sized smart cities [10].

Smart city consists of six key elements: smart people, smart mobility, smart economy, smart environment, smart government, and smart living. The abovementioned six elements are essential for achieving the smart city goals in real meaning.

These same six characteristics are deployed by a number of studies to develop indicators and smart city development strategies. This type of characterization framework is well justified and documented, and already used in practice by an increasing number of cities and policy makers. The framework aims to capture the

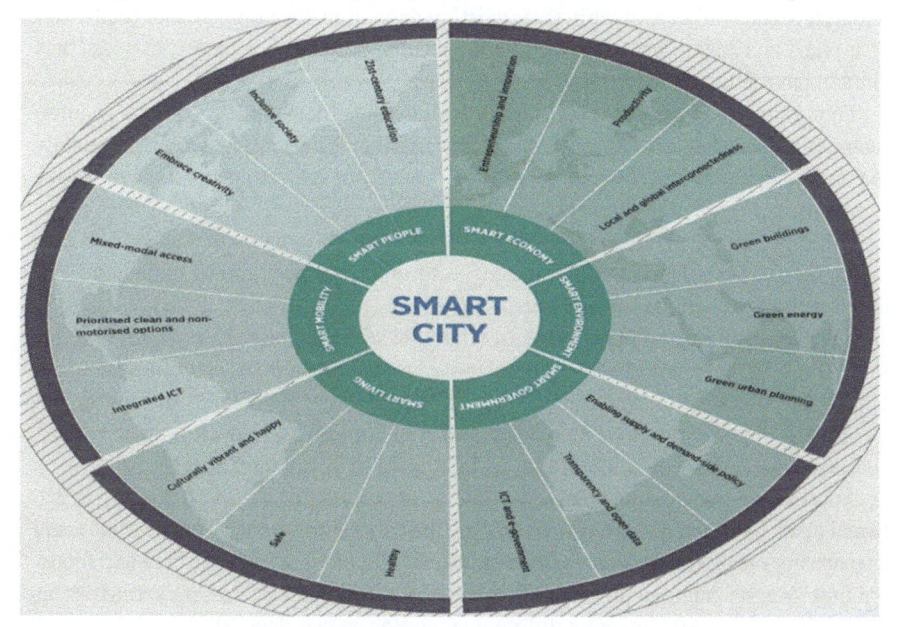

Fig. 28.3 Connectivity in smart cities [10]

key dimensions of European Smart Cities described above while retaining simplicity through specifying a relatively small number of characteristics which define these initiatives and cover the range of existing projects. When defining a smart city in the present study, at least one of the six characteristics must be present in a given smart city project or initiative. This is a baseline, however, and we must also keep in mind the smart city definitions and summary outlined above. These point to the deployment of multidimensional strategies, which consist of many components and projects designed to be synergistic and mutually supportive. Indeed, the most successful smart city strategies might be expected to adopt a multidimensional approach to maximize such synergy and minimize negative spill-over effects, as might happen, for example, if a smart economy strategy were prioritized which was detrimental to the environment. For this reason, we might expect to see more than one characteristic present in the most successful smart cities [1].

Basic Principles of Smart City

Cities that would like to take advantage of smart technologies need to set out their own vision for a smart city based on three basic principles: integration, pragmatism, and participation.

Smart cities should focus on integration rather than development of intelligent urban plans and projects that are isolated from other stakeholders and smart initiatives and thus can not achieve optimal use. City authorities also benefit most when they integrate smart initiatives within their existing economic development and public service plans and identify how new technologies can help them achieve the goals they already have [9].

The pragmatism principle recommends that cities focus on investments in intelligent projects and development programs that are financially viable, practically feasible and sustainable. Cities should also try to develop and pursue innovative initiatives from the other stakeholders.

Participation means that smart projects should be undertaken in partnership with businesses, stakeholders, the community, and other partners aim to make sure that they respond to local issues and needs [9].

Also, overcoming the barriers to growth of the smart technologies market requires joint working between cities, national government, businesses, stakeholders, and other users. These abovementioned subjects should work together on sharing capacity, identifying the required standards and regulations, and developing new risk-sharing models that will allow future technologies to be adopted at scale.

Approaches which incorporate these three principles and focus on already established goals have enabled cities to overcome the confusion associated with what being a smart city means. It has also enabled cities to strike the right balance between focusing on processes and outcomes and between top-down and bottom-up approaches, as well as how to use and integrate different technologies and data [9].

Having a clear vision and building partnerships are important prerequisites that cities need in order to progress their smart ambitions, but they are not sufficient. Many of the smart technologies and data sources—the enablers of smart cities—are relatively new and complex, and for the smart cities market to become successful, it needs the right conditions to grow and mature [9].

Barriers to Progress

Most smart initiatives involve the use of new and disruptive technologies that allow things to be done that were not possible before. As a result, smart technologies require the creation of new markets with new ways of working and new financial and governance models. These markets also need the right conditions to emerge: a new innovation and entrepreneurial ecosystem where stakeholders interact effectively and where new business models and ways of working can be created so that new technologies can be adapted. Without this ecosystem, the smart technologies industry is unlikely to grow and mature [9].

In particular, there appear seven barriers that need to be overcome in process of developing the smart city [9]:

- Cities find it difficult to work across departments and boundaries.
- Constrained demand from cities for smart initiatives.
- Cities lack technology-related skills and capacity.
- Business models for rolling out smart technologies are still underdeveloped.
- Increasing citizen participation is difficult.
- Cities have limited influence over some basic services.
- Concerns and risks about data privacy, security, and value.

Background of a Smart City

In the global profile of urban development, the smart city is emerging as an important basis for future city expansion. Europe's global competitors among the emerging economies are pursuing large smart city programs. India is planning to spend EUR 66 billion developing seven smart cities along the Delhi–Mumbai Industrial Corridor using a mixture of public–private partnerships (80%) and publicly funded trunk infrastructure investment (20%). China too is pursuing a smart city strategy as part of its efforts to stimulate economic development and eradicate poverty. As poverty in China is largely a rural phenomenon, the program seeks to attract rural workers to smart cities, which can then serve as giant urban employment hubs [1].

Europe does not face the problems of rural poverty or runaway mega-city development on the same scale as China or India, but the smart city idea is nonetheless highly relevant. It will be necessary to harness the power of smart cities in order to compete effectively with rival global economies. Moreover, experience with smart city development can help Europe to assist developing countries in managing mega-city development in ways that improve their welfare, reduce the risk of exported problems, and help them to become better trading partners for Europe. Most importantly, Europe has its own particular need for smart city thinking. The openness and connectivity of the European Single Market have allowed its cities to become hubs for the creative economy, technological and societal innovation, welfare enhancement, and sustainable development. They do this by drawing on resources (human or otherwise) throughout Europe and the globe and returning ideas, income, and other benefits. This complex ecosystem is robust and resilient, but it faces serious challenges, including economic and societal inequality, environmental change, and profound demographic transition. Other changes, including increased mobility and greater access to information, may both help and hinder this development. These developments directly affect the sustainability and the pan-European contributions of urban environments; they may be turned to advantage by smart city initiatives [1].

Social Innovations and Human Capital

Smart city is not only about technological approach, information revolution, smart mobility, or energy efficiency. Smart city is particularly about smart people, because without smart people nothing new and progressive will arise. The equally important aspect of smart city is its human capital and social innovation, which concern attention on the well-being of people.

Creativity is recognized as a key driver to smart city, and thus people, education, learning, and knowledge have central importance to smart city. The expansive notion of smart city includes creating a climate suitable for an emerging creative class. A creative city is one of smart city visions. Human infrastructure (i.e., creative occupations and workforce, knowledge networks, voluntary organizations, crime-free environments, after-dark entertainment economy) is a crucial axis for city development. Social infrastructure (intellectual capital and social capital) is indispensable endowment to smart cities. That infrastructure is about people and their relationship. Smart people generate and benefit from social capital. Smart city is about a mix of education/training, culture/arts, and business/commerce and a hybrid mix of social enterprise, cultural enterprise, and economic enterprise [7].

Future changes will become an opportunity for growth of qualification, flexibility, innovative ideas, and social innovations. Technical knowledge will be modified by Industry 4.0, and pure technical knowledge will retreat to skills which are capable to draft appropriate solutions for certain applications in praxis. In the future it will be crucial how and in what extend managers and technicians will be able to identify and utilize new opportunities of digital solutions. All of this requires completely new competencies.

Investment into the human capital through skills development and training is clearly vital for maintaining of production base in Europe. Availability of skilled labor is a key factor for growth and competitiveness of the automotive industry and will be essential to achieve leadership in breakthrough technologies. On the other hand, the industry already faces the lack of skilled labor and experience to clearly identify the skills that will be necessary in the future. According to the previous, it is necessary to ensure appropriate qualifications of the staff and their training and lifelong learning. Lifelong learning could be achieved by social innovations.

Lack of skills is a major concern. It is necessary to take urgent measures with long-term national targets to substantially modernize education and training in order to update the skills range through the new school curricula and provide education and training by using ICT and new forms of partnerships with employers. At European level, this problem is not only related to the automotive industry but also to other industries. The European Commission early in the statement on revaluation of investment in education and skills in order to achieve better socioeconomic outcomes sets the strategic priorities to address these issues. As a crosscutting issue, this topic is addressed within the framework of the European employment policy. In addition, following sector initiatives will be crucial. Due to changes in the necessary skills that can be observed on the EU labor market, the attraction of the new labor from the countries outside the EU is a possible supplementary solution of this problem.

Social Innovations

Social innovation as a concept and a summary of the tools includes a wide range of activities. The widest possible definition specifies social innovation as all new strategies, concepts, ideas, and organizations that enhance and support the improvement of conditions for the functioning of civil society.

Generally social innovations includes all activities that result in the qualitative changes of basic social structures and innovations, respectively, that have targeted significant social impact.

Authors from the University of Oxford perceived social innovation especially as new ideas and their practical applications that address dissatisfied social needs of people. Innovations are generators of processes of the desired social change [11].

Social innovation for the purpose of submitted article is understood as ideas to help individual actors to organize their activities and social interactions in order to successfully achieve common objectives. Social innovations also may include the development of new processes and procedures for structuring common work, the introduction of new social practices in the group, and the creation of new kinds of social institutions [12].

Social innovations are sustainable only if they are:

- Based on real social needs
- Addressed to specific community
- Initiated by leaders
- Realized as act for the common good
- Without personal interests

- Taking place in the context of existing systems and subsidizing each other

The notion of social innovation is not a new phenomenon; throughout the history people have always sought to find new solutions for pressing social needs [13]. Social innovations always copied the innovation of technology processes in the past. Nowadays they are developed individually and there is also institutional support for them.

Key definition criteria for social innovations are the following [12]:

- Focus on dissatisfied needs or social problems
- Novelty of the approach that brings positive social changes (changes in behavior, attitudes, social impacts)
- Production of added social value (social synergistic effect on the quality of social relationship)

Social innovation is characterized by the following specifications:

- Social innovation can be stated as a new combination or hybrid integration of existing elements (it is not the discovery of new elements, but the innovation can be found in a new connection, innovative application of known elements in non-traditional contexts).
- Their application in practice involves crossing the established borders.
- Produces new social relationships and links (networking) previously isolated individuals and groups.

Social innovation can be simply defined as new ideas, based on innovative combination of previously separated elements that work in practice in order to achieve social objectives. This distinguishes innovation from improvement, which only brings partial incremental changes, and also from creativity that is vital for innovation but does not include the application, or implementation into practice, which is the criterion of applicability of a new idea. In this sense it is clear that the term innovation is not only a new idea or the invention, but also it is a practical application in practice [11].

Social innovation is associated with the development of activities and services that are motivated by social objectives. This distinguishes them from business innovation, which is generally oriented to maximize the profit, which may also have positive social effects, but they are not their definition criterion [10].

Innovations have important social roles in society. Firstly, they are a tool for economic growth, and secondly they are the tool for social development. Linking these two tasks is a new challenge for creating social innovations. Social innovation as a term indicates a kind of cultural artifact which is a combination of previously separated elements. The main point of this is to apply individual element into different social context; then in this way, it is possible to create something new, which becomes a social innovation and also brings the improvement in social conditions of human being. Nowadays, the importance of social innovation is bigger than ever.

Social innovations, in terms of benefits, bring with themselves better and more effective fulfillment of the social needs of people. Resolving of human problems is very often the main goal of the social innovation. Sometimes social innovation can be more important for prosperity and successful business growth of company than technological innovation.

Technical and technological innovation requires handling new techniques and technical resources. Social innovation often requires changes in values, attitudes, and people's opinion, including learned ways of social action that increase the demand for their application. Current social innovations are also associated with the development of social partnership, state, and civil society, which places specific requirements on the ability of stakeholders to build and develop social capital.

Social innovation can bring social development, which consequently brings the improvement of opportunities and conditions for all employees. What is more, social innovation can ensure better working conditions, meet human needs, effectively motivate and encourage the staff, and provide continuous and quality education.

Social capital is gradually becoming a phenomenon for the functioning of the whole society, not only for the economic system. Social capital is the input and also output of the economic system. Social capital as a basic element of the social system is also a very important factor for successful functioning and prosperity of business companies. Social capital arises mainly in informal relationships, and for this strengthening is necessary to support informal groups of people.

Social innovation can also be considered as changes in the normative, cultural, and regulatory structures of individual society; it is because the contemporary society is characterized by a high level of dynamism that requires from the individual actors capability to initiate changes, capability to adapt to changes, and capability to learn from changes. Learning and flexible institutions, organizations, or groups of people are the fundamental for creating innovations.

The key role in the field of innovation has gotten social actors, who are capable to generate new ways of resolving social problems and are capable to effectively address these new ideas to the target group of people. These actors could be individuals, social groups, social movements, social organizations, or business companies. So in that way, we could talk about the individual, group, institutional, or business dynamics of social innovations. A very important prerequisite of individual or group dynamics is promoting the creativity and open space for new approaches in the area of social problems. Significant accelerator in these processes is the building of social partnerships that fulfill the role of the platform for the exchange of experience and also for unification of shared interests and creation of joint initiatives. At the institutional level, the concept of educational organizations has been strengthened together with their ability to adapt to environmental challenges, and all good experience and good practice are transferred between all organizations. For better realization of social innovation, it is important to create supportive and pro-innovative environment that motivates social actors to create innovative ideas and practices in solving social problems [12].

Social innovation is basically a social change in practical implementation. Application process of social innovation is in the wider meaning connected with the understanding of the introduced social change, its management, and the final control. New innovative thoughts and ideas have got a crucial innovative role only when a certain conditions are met. The main condition is supporting the change from the group of people.

The agenda of social innovator is to improve the conditions of human existence and life. Social innovator is making efforts to achieve social development, which is related to the following [12]:

- Improving of opportunities for all employees of the company
- Satisfying basic human needs
- Achieving the decent working conditions
- Offering higher standards of health services
- Continuous and quality education

Social Enterprise, Social Entrepreneurship, and Social Entrepreneur

In the area of social innovation, key terms have begun to appear that help to better understand the importance of the social sphere and its issues. These terms include social enterprise, social entrepreneurship, and social entrepreneur.

The term social enterprise in the form of an organizational actor, which is falling under social entrepreneurship, firstly appeared on the territory of Italy, in Europe. This term also found its expression in the form of specific legislation. The key feature of social enterprise is that it represents a hybrid form between market economic, civil society, and public policy [14].

The term of social entrepreneurship has been established in the second half of the 1990s. Under this concept fall a wide range of activities, such as voluntary activities, private individual activities, and corporate social responsibility. In the Anglo-American environment, the interconnection of different forms together with the creation of added value from social business has been highlighted. This concept of social entrepreneurship in European area has been closely linked to the third sector, which has a strong link with nonprofit sector [14].

Social entrepreneurship can be defined through two fundamental elements, namely, through the strategic focus on achieving business goals and through an innovative approach to achieving social benefits for its employees and their surroundings. This one combination creates a fundamental distinction between social business and other forms of business. Social actors symbolize agents of social changes, who through their efforts and initiatives exceed stereotypical ways of thinking and acting. Social business brings together three components: innovation, market orientation, and social mission.

The term social entrepreneur found its first link with initiatives that were based on American management and business schools in the mid-1990s. These schools have sought to identify and subsequently support new business activities. The main goal of these schools has been to resolve new social problems. In Europe they began to emphasize the collective nature of social entrepreneur, which is based on the activities of the organization acting in third sector [14].

Social innovation focuses attention on the unmet needs of people, which could have various forms. For example, social needs of people may relate to the overall quality of life, the quality of living conditions, the quality of infrastructure, the sufficiency of social services, the quality of health and public services, the balance between work and life, the quality of working life, and so on.

Social Innovations and Sustainable Development

Nowadays, it can be said that human development is not only about economic growth but is far more complex and has to lead to meeting social goals like improving quality of life, having access to quality in education, improving healthcare, reducing unemployment, reducing diseases, creating balance between work and personal life, and so on. All of this requires comprehensive approach to sustainable development, including versatile management of relationships between human, natural, sectoral, and structural aspects of sustainable development at all levels. It should be remembered that sustainable development meets the needs of current generation but at the same time does not jeopardize demands of the future generation.

Sustainable development is a targeted, long-lasting, complex, and synergetic process that affects all areas of life (spiritual, social, economic, environmental, and institutional), multilevel (local, regional, national, international), and directing through the application of practical instruments and tools to such a model of society, which satisfies the material, spiritual, and social needs and interests of the people, while always respecting the natural values [15].

Implementation of sustainable development principles can be achieved through different ways or instruments, but at the same time, these various forms of implementation should lead to sustainable models, practices, and changes in lifestyle of human being. Sustainable development can be divided into three basic aspects, namely, economic, environmental, and social. Social aspect strives for versatile development of man as a human being, society, equality of opportunity, and access to education for all without any discrimination.

Social aspect of sustainability is based on a number of principles, namely:

- The principle of respecting the needs and rights of future generations
- The principle of respect for human rights and freedoms
- The principle of cultural and social integrity
- The principle of tolerance
- The principle of emancipation and participation
- The principle of solidarity
- The principle of nondiscrimination
- The principle of equality of rights and opportunities
- The principle of subsidiarity

Goals of sustainable growth and development could be achieved through the application of innovations. Innovations could be applied in different spheres not only in technological, economic, and informational but especially in the social sphere, which affects the whole society life.

The concept of sustainable innovation is built on key factors of modern innovation, such as sustainable development, ecological thinking, participatory and continuous innovation, and innovative leadership. Sustainable innovation can be characterized as innovative activities that are based on ethical, economical, and environmentally sustainable principles. This approach combines opportunities related to sustainable development of practice together with new perspectives of innovative activities and management. The concept of sustainable innovation is based on five principles:

- Sustainable development
- Participatory innovation
- Continuous innovation
- Global innovation
- Innovative management

The abovementioned approach demonstrates the balance that arises between impact of the innovation process in the long terms on one side and impacts on society, economy, people, and external environment on the other side. This approach involves not only the production of new products and services but also social innovations. Social innovations, as part or as subcategory of society innovations, seek to meet social needs in more effectively way. Social innovations are very important for the prosperity and successful development of the countries, businesses, and human life.

Best Practice from Social Innovations: Social Innovation for Youth Social Entrepreneurship (INNOVAT) and Encouraging Social Entrepreneurship Among European Youth (YounGO)

The INNOVAT project, as the YounGO project is a part of the Erasmus+ grant scheme funded by the European Union. The main aim of the INNOVAT project within the Erasmus+ scheme projects is to strengthen the cooperation between eight organizations, five from the European Union and three from Latin America in innovation and mutual exchange of information and experience in the field of social work with youth. The main project partner is the organization INNOVAT JOVESOLIDES ESPAŇA. Project partners are Asociatiapentru Centrul European Integrare Socioprofesionala (ACTA), Romania; Interactive Media Knowledge Transfer (KT@INTERMEDI), Greece; The Innovation and Development Institute Principe Real (IDS), Portugal; Slovak University of Technology in Bratislava (STUBA), Slovakia; JEVESOLIDES (Colombia, Nicaragua a El Salvador) [16].

The projects YounGO and INNOVAT promote and support the creativity and innovation in social business. Both projects can also be considered as projects for the development of soft skills and competences that increase the employability of young people not only in the domestic labor market but also in the European Union [16].

The significance of the projects with the same or similar focus as YounGO and INNOVAT is not only in supporting the reduction of youth unemployment or graduates of higher education institutions/universities, but also in supporting the employment of other vulnerable groups of the population, such as disabled citizens and individuals dependent on a certain type of social support. An important aspect of the projects with elements of innovation and social focus is to create conditions for education and training in the given field in order to support the establishment of their own business, support and implementation of innovative and creative initiatives, and other projects aimed at young people [16].

The social entrepreneurship is currently an opportunity and a starting point for young people to develop and establish their own business activities, self-employment and financially independent activity, to integrate and apply to the labor market [16].

It follows that social entrepreneurship brings added value for all the abovementioned risk groups. Social entrepreneurship helps not only to reduce youth unemployment through self-employment/own business and enables to acquire new skills and abilities, but also reduces the unemployment of disadvantaged population groups by creating job opportunities [16].

Conclusion

Smart city, despite its great potential, deals with the human capital and social problems in the society and in industrial enterprises insufficiently. It is important that representatives or participants of smart city should pay more attention to promote this concept among the general public, along with stronger implementation into the sphere of social innovation in industrial enterprises at home and abroad.

The vision of smart cities gained importance in several years ago; smart city has perspective future in all its components. Smart city has a number of dimensions which are not only related to technology. Smart people and smart governance are fundamental for approaching the vision of smart city. It is very important to encouraging the citizen to become a more active and participative member of the community. It is necessary to meet new requirements from the labor market, and people will be exposed to new changes which will be caused by vision of Industry 4.0 in the future. These changes will be enormous and will require societal consensus and mainly support from the companies and government. This support could be given by social innovations. On this point the industrial companies should start to more intensively cooperate with school and educational institutions and mainly participate in constitution of new study programs. Well-established industrial companies should take a part in special education and create opportunities for gaining professional experience from industrial practice.

It is very important for young people to give them a chance to acquire an outlook and practical experience during their studies at university; this could be realized by compulsory internships. Internships could be organized in two forms, by summer internship and school internship. Summer internship would be only voluntary and should be a kind of extension of school internship. Internships should be realized in the range of 3 months with a very dynamic program that should be draft by a special mentor for young students. For the best students, there will be an option to realize internship in an international environment in some divisions which operate worldwide. Individual candidates could perform tasks in the field of manufacture, assembly, logistic, human resources, materials technology and information systems, and so on. By completing the internship, these students could have preferential right to processing the bachelor, master, or dissertation thesis. Students should during the internship gain a more realistic view on the work carried in industrial company. Students in this context could get practical insight into possible future career, and afterward they could motivate themselves to achieving better school results and identify themselves with future perspective job.

Another potential social innovation is applying the principles of age management in industrial practice within the European area, because age management solves problems of aging population, pressures of industrialization, and current migration policy. Age management principles can be applied as social innovation, for example, transferring of knowledge and skills of employees and then using knowledge capital of older generations and creating a cooperative corporate culture. Transfer of knowledge could be realized by mentoring.

Introduction of mentoring as a daily activity in industrial companies could have a lot of contribution. Mentoring opens the door to new experience, knowledge, information, skills, and competence. Mentoring among other things helps to more quickly adapt to the new environment as well as on colleagues. Mentoring allows to reach more qualified job performance, reduces mental effort, and eliminates potential concerns. Mentoring in that way represents the great social innovation for everyday use in industrial companies. Mentoring could be used for adaptation of the newly accepted employees, for talent development, and for development of supervisor's persons in specific departments. Mentoring could be also used for sharing the internal knowledge and know-how.

Social innovations should encourage the development of society and simultaneously count with future changes; for that reason it is important to find factors and relationships which facilitate the implementation of social innovation into the society and industrial companies.

Related to the abovementioned, there is a possibility to create a chance for the draw of the financial support from the European Social Fund, which can be used to support the development of social innovation within the concept of smart city. The main objective is to increase the quality of education, acquired skills, training, and development of society in all its aspects.

The main aim of the paper was to point out that any future changes will become an opportunity for social innovations to solve potential problems with qualification, flexibility, or knowledge which will be required from human capital. The contribution of the paper consists in analysis and summarization of social impacts which will bring future on human capital.

References

1. European Parliament. Mapping Smart Cities in the EU. (2014). [online]. Accessible at: http://www.europarl.europa.eu/RegData/etudes/etudes/join/2014/507480/IPOL-ITRE_ET(2014)507480_EN.pdf
2. THE TIMES OF INDIA. What is a 'smart city' and how it will work. [online]. Accessible at: http://timesofindia.indiatimes.com/What-is-a-smart-city-and-how-it-will-work/listshow/47128930.cms
3. Elets. Lucknow ties up with EU on smart cities [online]. Accessible at: http://smartcity.eletsonline.com/lucknow-ties-up-with-eu-on-smart-cities/
4. Charbel Aoun. The Smart City Cornerstone: Urban Efficiency. [online]. Accessible at: http://www.digital21.gov.hk/relatedDoc/download/2013/079%Electric%20%28Annex%29.pdf
5. Centre for cities. Smart Cities. [online]. Accessible at: http://www.centreforcities.org/wp-content/uploads/2014/08/14-05-29-Smart-Cities-briefing.pdf
6. Elets. Centre to guide cities for improving smart city plans. [online]. Accessible at: http://smartcity.eletsonline.com/centre-to-guide-cities-for-improving-smart-city-plans/
7. Nam, T., Pardo, T. Conceptualizing Smart City with Dimensions of Technology, People, and Institutions. New York: Center for Technology in Government, University of Albany. [online]. Accessible at: https://inta-aivn.org/images/cc/Urbanism/background%20documents/dgo_2011_smartcity.pdf
8. Department for Business Innovation & Skills. (2013). Smart cities: Background paper. [online]. Accessible at: https://assets.publishing.service.gov.uk/government/uploads/system/uploads/attachment_data/file/246019/bis-13-1209-smart-cities-background-paper-digital.pdf
9. Hawaii International Conference on System Sciences. Understanding Smart Cities: An Integratove framework. [online]. Accessible at:https://www.ctg.albany.edu/publications/journals/hicss_2012_smartcities/hicss_2012.pdf
10. RACOUNTER. Connectivity in smart cities. [online]. Accessible at: http://raconteur.net/infographics/connectivity-in-smart-cities
11. Mulgan, G., et al. (2007). *Social innovation: What i tis, why it matters and how it can be accelerated*. Oxford: Said Business School.
12. Lubelcová, G., et al. (2011). *Inovácie v sociálnych a verejných politikách: problémy konceptualizácie a nových nástrojov*. Bratislava: UK. ISBN 978-80-223-3043-5.
13. Cagáňová, D., et al. (2015). *Innovation in industrial enterprises and intercultural management*. Zielona Góra: University of Zielona Góra. ISBN 978-83-933843-4-1.
14. Defourny, J., & Nyssens, M. (2008). Social eterprise in Europe: Recent trends and developments. WP no. 08/01, EMES.
15. Jirásková, S. (2007). Inovácie a trvalo udržateľný rozvoj. In Manažment v teórii a praxi - online odborný časopis o nových trendoch v manažmente. ISSN 1336-7137, roč. 3. [online]. [cit. 2017-01-16]. Accessible at: http://casopisy.euke.sk/mtp/clanky/3-2007/5.%20jiraskova.pdf
16. Cagáňová, D. et al. (2017). *Managerial skills in industrial praxis*. Zielona Góra: Wydawnictwo Instytutu Informatiky i ZarządaniaProdukcją (IIZP). ISBN 978–83–65200-08-2.

# Chapter 29
Towards Creating Place Attachment and Social Communities in the Smart Cities

Matej Jaššo and Dagmar Petríková

Abstract Social communities, territorial identity and place attachment are examples of the important soft factors in smart city concepts. Meaningful network of links between physical environment, communication networks and social community is essential for every sustainable smart city. Smart cities and smart communities are also those ones that make more efficient use of physical infrastructure and engage effectively with local people in the process of citizen participation and building a place attachment towards given locality. Every concept of place attachment requires a particular work within the community, its effective transmission to all the members as well as to outward environment. In the sense of place attachment, there is the idea of urban gardening that generates uniqueness – specific character of places created by urban gardening contributes to calibration of unique place identity and develops emotional and social ties related to certain place. Urban gardening provides opportunities for social interactions that help residents develop their relationships in community, support community life and develop community and place attachment as well as enhance the quality of urban environment in the smart cities. Urban gardening is often viewed as one of the strategies which can improve urban sustainability and promote sustainable urban development in the smart cities.

Introduction

Cities provide the citizens with a number of services and functions to be used in the urban environment. Each of the functions – housing, employment, culture, sociability, leisure time activities and recreation – shows evidence of a characteristic structure and also of various needs of current population, with various impacts on the environment. In this regard smart cities and communities are also those ones that make more efficient use of physical infrastructure, engage effectively with

M. Jaššo (✉) · D. Petríková
Slovak University of Technology in Bratislava, Bratislava, Slovakia
e-mail: matej.jasso@stuba.sk; dagmar_petrikova@stuba.sk

© Springer International Publishing AG, part of Springer Nature 2019
D. Cagáňová et al. (eds.), *Smart Technology Trends in Industrial and Business Management*, EAI/Springer Innovations in Communication and Computing,
https://doi.org/10.1007/978-3-319-76998-1_29

local people in local governance and decision with emphasis placed on citizen participation and learn, adapt and innovate and thereby respond more effectively and promptly to changing circumstances by improving the intelligence of the city. Emphasis on the human factors related issues is reflected in many recent definition of smart city, e.g., and the main connotation and semiotics of the term "smart city" are nowadays leaning more towards the above-mentioned field. Current smart city concepts are based more upon the divergent than convergent creativity and are outgoing from initiatives from the 1990s, requesting "the creativity of being able to synthesize, to connect, to gauge impacts across different spheres of life, to see holistically, to understand how material changes affect our perceptions, to grasp the subtle ecologies of our systems of life and how to make them sustainable. We need skills of the broker" [15]. Strive for fulfilling the complexity and hierarchy has been replated by the urgent need to deliver uniqueness and specificity. Caragliu and Nijkamp [16] argue that "a city can be defined as 'smart' when investments in human and social capital and traditional (transport) and modern (ICT) communication infrastructure fuel sustainable development and a high quality of life, with a wise management of natural resources, through participatory action and engagement". Human and social capital is in this approach mentioned first, even before the classical core of most of the smart city approaches, namely, information and communication infrastructure. We have to bear in mind that each efficient smart city is rather mental and social (and not primarily technological) construct in its essence and philosophy of design.

Territorial identity and place attachment are prominent variables entering the process of designing smart city. They are very fragile mental structures which cannot be bought, emulated or stolen, but they are significantly contributing to the effectiveness of functioning of any social system based in certain territory. Social cohesion based on the highly profiled identification with the living space and deeply articulated place attachments are the fundamental preconditions of sustainability of any community or settlement structure. Systems based upon the mature and balanced links of various social relations and deeply integrated within certain territory display necessary portion of resilience also in case of outer/ inner threats and therefore stability in time and space.

Acceleration of spatial development has generated also the increased probability to face also the negative effects and created threats which were not acute even some years ago. Fragile spatial and societal structures have been exposed to huge pressure originated either from international markets, unfavourable demographic prognosis, environmental hazards or another sources of risk. Spatial planning faces the problem of increasingly higher uncertainty of the framework conditions of spatial development as well as necessity to react efficiently and flexibly to unpredictable external and internal shocks like floods, fires and economic disturbances confronted with unpredictable individual behaviour/decisions of multiple stakeholders. These factors represent risks not only for planning but first of all also for sustainability of spatial development. Spatial planning has been transformed and has become a process of permanent search without any warranty of outcome. Assessment and decision-making under uncertainty – it's the call of the day [1].

Urbanity and Its Interpretations

Urban and metropolitan milieu is an example of ultimate complexity on territorial level. This milieu displays manifold hierarchical and horizontal structures and registers including the contraversions and conflicts. Not many created systems do include so many variables and do involve so many involved actors. Moore [2] delivers the following overview of the approaches towards the city from the social perspective:

- The city as symbol and carrier of civilisation
- The city as land of economic and social opportunity
- The city as an initiating and controlling centre of a region or the nation
- The city as a melting pot versus the city as a mosaic of social worlds
- The city as heterogeneity, variety, diversity, the apex of culture and cosmopolitanism
- The city as a "feast" and the city as electronic stimulation
- The city as a place for transitory, second-hand, superficial contracts, as a place of reserve and indifference, of blasé or even predatory attitudes
- The city as depravity, the alienation of the person from the land and the subjugation of human values to the machines and commercialism

The common denominator of almost all current approaches is that the main scope of urbanity is shifting from "fulfilling the complexity and hierarchy" to "ability to deliver uniqueness and specificity". Schmeidler [17] introduced a manifold variety of approaches:

- Human ecological interpretation (size, density, heterogeneity of urban population)
- Functionalistic interpretation (quality of urban environment as the consequence of variability and completion of function)
- Socio-psychological interpretation (intimacy and emotionality of private life versus rational and cognitive framework of public life and work)
- Political interpretations (concept of so-called functional democracy, urbanity as a system rewarding wished behaviour)
- Civilisational interpretation (urbanity as an emancipation from nature)
- Postmodern interpretation (urbanity as an expression of "integrated chaos", space for alternative and non-legal activities, "the street is the last jungle where you can survive an adventure")

New urbanity is arising on the interface of various and manifold contexts (visual, symbolic, narrative, historical, political) etc. The tension "public-private" has been modified: it has been removed to the semipublic places. The role of the professionals (planners, architects) is to redefine the legibility of the place and its sense and the growing importance of the roles of other actors defining and redefining space is obvious. The place should be a point of meeting, interaction, exchange, transition, etc. Place is becoming a pattern in the "language" of people (Ch. Alexander) and generates specific metatext in the minds of people. Place and space are opportunities for projection and self-realisation – projection of values, ideas, principles, thoughts, etc. On the other hand, we are witnesses of certain controversial tendencies: commodification of spaces, privatisation of public spaces (shopping malls, corporate plazas...), fragmentarisation of spatial experiences and globalisation of local contexts (more in [3]). This shift has been reflected also in planning paradigm shift and modification of planning culture: from the system theories ("comprehensive planning") having their roots back in the 1950s towards the "incrementalism" of the 1970s and later to "cooperative planning" of the 1990s and 2000s [1]. The city of optimal infrastructural performance, social equity and rather normative regulation of spatial conflicts has been replaced by postmodern conceptual approaches enhancing uniqueness, imagery and soft assets [4]. Hierarchical planning cultures based upon the authoritarian decision-making proved to be inefficient and inappropriate when dealing with complex problems of high dynamics and multilateral impacts (see, e.g. [1, 5]). Change must be made by those living and acting outside the prevailing paradigm [6]. Planning has been transformed onto rather contingent nature [7] and has become a process of permanent searching without any warranty of outcome. Judgement under uncertainty – is the call of the day. Moreover, spatial planners, urban designers and architects are facing the ambiguity – lack of judgement criteria. Who knows what the stakeholders really want? Smooth, successful and genuine spatial development requires value compatibility and continuity. Integration of different values, basic assumptions and beliefs into a coherent spatial concept is a necessity and ultimate challenge for spatial planners (see [8]). Forester's concept "making sense together" has been completed by Healy's addition "while living differently" [9]. We are confronted with both positive (urban imagery, fun, celebration) and negative (urban anxiety, urban panic) connotations of urban environment. Current urban imagery is fragmented, deteritorialised, heterogeneous, diasporic, split apart, etc. Sense of a place is constantly changing, not necessarily held together, and the city is regarded as a partially connected multiplicity [3]. Archetypal perceptional patterns [10] appeared: the crowd as an ocean, skyscrapers as the mountains, the city as jungle, the cars as predators, etc. Revival of mythological contexts is represented by, e.g. "oceanic feeling" (term of Paul Tillich): the individual in the city is losing its freedom and is led by the crowd and the city itself (see [10]).

Territorial Identity and Place Attachment

Identity of any territorial subject (city, land, region) is derived from Latin word genius loci, encapsulating special rather intangible characteristics related to given place. Concept of place attachment is based on the psychological identification with certain place, space and community living in certain territory. Existential value provided by such relation is considered as meaningful. It generates the feeling of togetherness ("Wir – Gefuel"), saturates the needs of belonging and security and provides the milieu for everyday rituals, behavioural schemes and acting in various social roles. Phenomenon of territorial identity is neither exclusively architectural/geographical nor cultural/social issue and must be researched and treated by highly balanced interdisciplinary approach. Each territorial identity grows up from the combination of the natural characteristics of the living space (rivers, terrain, morphology) and the artificial interventions of the human being (settlements, infrastructure, culture, language). Karavathis and Ashworth [18] distinguish three ways how people make sense of places: artificial innovations of physical space (architecture, urban design), the ways how certain places are used (behavioural patterns) and various forms of place representation (movies, pictures). However, it has been shown in recent years that regional identity is no more the mere "track of the history", but it is a living organism, absorbing plethora of new impulses and influences. Territorial identity must be preserved but also actively promoted and fostered. Several new tools of strategic management (brand management, corporate identity of cities/regions) appeared in order to steer and communicate the regional identity in proper way.

It is generally supposed that highly profiled city/place identity and strong ties of place attachment are of utter importance for social cohesion within the territory [11]. Territorial identity is a crucial dimension in the concept of social identity and sense of belonging, and identity was one of the weakest points of the big modernist dreams (e.g. Brasilia). Place attachment saturates many psychological needs: the need for security, the need for self-realisation and the need for belonging and structuring the outer environment. Highly profiled identity contributes to the legibility of the place and space. The people are still generally territorial in their behavioural patterns. Slovak communities, mainly in smaller settlements (but even in urban milieu), always displayed rather strong and deep place attachment and deep identification with living place and environment. However, we can conclude from recent surveys (e.g. project Identity of River Basins; see more [12]) that both these phenomena (place attachment and territorial identification) are saturated more by emotional and social identification patterns ("I have grown up here", "my family lives her for decades") than by value-based identification patterns ("I am living here because I appreciated the value profile and behaviour of our municipality"). The territorial identification and sense of belonging is rather deep but in many cases rather monodimensional. During recent years in Slovakia, our housing and residential estates ceased to be the monolithic senseless places and have become chronicles of various stories and experiences which overcome sometimes the obsolete and uniform architectural language. It is obvious that the landmarks of identity are never only the physical (architectural) forms but rather the common experience, morals and stories. Identification with the living place goes far beyond the positive distinction (image) and should be based upon the common vision and values, which are present in given territory (environmental values, liberal values, etc.).

Urban gardening is in this dimension not only improvement of the physical structures but a unique platform for fostering the sense of community, reflection of place attachment and expression of the need of self-realisation. Although place attachment and processes of identification with place/territory are growing up from certain given predispositions, they are dynamic phenomena which should be effectively fostered and further developed. Every concept of place attachment requires a precise work within the community, its effective transmission to all the members as well as to outward environment. Direct participation of the inhabitants in this process is very important. The inhabitants are key players in this process – they are both creators of the place identity and also are the key target group in the process of its acceptation and evaluation. In order to ensure the highest quality and effectiveness of the process, it is necessary to approach the place in an interdisciplinary manner and with maximum emphasis on mutual functional and value compatibility of individual participants and the measures proposed. One of the most important conditions is authenticity of the concept. A very important category in terms of place identity and place attachment is an image of the place. Image is an abstract mental construction representing the subject in minds of audience. Positive image of a place/city means its goodwill, its good reputation or positive emotion appearing by thinking about the subject. The image is also the degree of affinity to subject manifested by significant groups of perceivers. Image of the place with significant presence of urban gardening structures goes far beyond pure visual appearance of green structures: it encapsulates also the values of solidarity, fairness, justice and advanced sense for quality of life.

Social Communities and Cohesion

Large parts of the Slovak cities and neighbourhoods are covered by residential areas of panel blocks of flats built in the 1970s and in the 1980s. These communities and settlements are often more than 30–40 years old and have become specific places with its own history, social climate and narratives. Unique and specific metatext of almost any Slovak city would remain unfinished without residential areas of panel blocks of flats. These areas have generated specific identity, social cohesion as well as social problems related to them. It is obvious that Slovak panel block housing areas failed to deliver the unique "tomorrow's quality of life" as once declared but on the other hand they never became the completely excluded localities without the vital contacts with the city' organism. Resilience of such neighbourhoods is dependent on the inner cohesion, an ability of the system to "stick together" or "withstand the external pressures" where peripheral location is always a risk factor. Unique and delicate combination of interaction of various subsystems (economic, social, cultural) within these territories makes their ability of resilience quite challenging task. We are witnessing efforts to increase their functional variability as well as initiatives aimed to actively form common values and civic culture (public initiatives against problematic developers, safeguarding the green areas on the Danube embankment, cycling routes, etc.). Humanisation of the public spaces and the refurbishment of the old panel blocks indicate common consensus and will to stay in the territory. Heterogeneity of the social milieu has generated both positive (social mix, different classes) and negative consequences (different identities competing somehow). Slovak panel housing estates have never got an appearance of classic hopeless ghettos. Underlying reasons are the manifold social mix and relatively open boundaries to the other parts of the city. These experiences have shown that social order and social control are better there where people are living together more than 30 years and went through all the phases of the life cycle of the housing estate. In these communities, there are some common experiences from problem-solving (refurbishment of the facades, gardening in the outyards, common sport activities, social life, etc.). Empirical evidence suggests that these communities are located predominantly in smaller houses with fewer flats, localised surprisingly outside from the places with good traffic connection or nodal placement. Slovak panel houses neighbourhoods never fulfilled the modernist dreams but, on the other hand, never became completely excluded locality cut-off from the city organism.

Towns and cities also provide an extensive supply of degraded or underused areas of brownfields that have the potential for increasing the quality of places and wait for more sustainable and sensitive redevelopment. Temporary use of the brownfield areas provides an opportunity to create a particular type of public space for urban gardening and the potential for community places that are open to the neighbourhood. This phenomenon helps change the current understanding of gardens, so that gardens are not just enhancing the life of gardeners and their immediate family or close friends but also serve as a tool to improve the life of local people and visitors. It is also of educational relevance, especially for children who spend their whole lives in a city. It gives them the possibility to learn more about gardening activities and recognise different kinds of flowers, fruit and vegetables. Neighbourhood spaces and courtyard garden settings in particular provide opportunities for social interaction that help residents develop their relationships within the community, support community life and develop community and place attachment as well as enhancing the quality of urban environment.

Urban Gardening

Urban gardening is more and more popular leisure time and recreational activity among city inhabitants that offer an opportunity for people from different backgrounds to participate in the activity of gardening and provide themselves with fresh fruit vegetables or herbs and at the same time to develop social relations among community members in the urban environment during the process of regular maintenance. Urban gardening contributes to increase green infrastructure in the city thus improving the quality of the urban environment, and at the same time, it is also a way of communicating within a city or its suburb. In this way urban gardening is a powerful tool for creating and building up community and fostering a deep place attachment. Urban gardening is an expression of positive values and attitudes towards environment and community. In relation to place attachment, urban gardening generates uniqueness – specific character of places created by urban gardening contributes to calibration of unique place identity and develops emotional and social ties related to certain place. It is a kind of "scene" yielding the stories and tales which secure inner composure of social community. In towns and cities, there is also endless quantity of degraded or underused areas of brownfields that are waiting for more sustainable and sensitive redevelopment. Temporary use of the brownfield areas is an opportunity for a particular type of public space for urban gardening and the potential for community places open to neighbourhood. This phenomenon helps change current understanding of gardens, when gardens are not just enhancing life of the gardeners and the immediate family or close friends but serve as a tool to improve the life of local people and visitors and it is also of educational character, especially for children, who spend their whole life in a city. It gives them possibility to learn more about gardening activities and recognise different kinds of flowers, fruit and vegetables. Neighbourhood spaces and courtyard garden places in particular provide opportunities for social interactions that help residents develop their relationships in community, support community life and develop community and place attachment as well as enhance the quality of urban environment. Donna Armstrong's survey of 63 community gardens grouped under 20 community garden programmes in upstate New York resulted in the description of numerous benefits of gardening:

- Improved social connections, raising awareness and activity of local policy
- Interactions between gardeners' groups through different programmes
- Identification of children with cultivated land
- Participation also of lower income households
- Stronger community cohesion – recognition of people on the streets
- Higher knowledge about local actors – easier action initiation process
- Social control of the neighbourhood
- Landscaping attempts not only on the community garden
- Establishment of neighbourhood organisations
- Establishment and maintenance of parks and playgrounds [13]

The quality of the urban environment has also become a crucial component of economic and social regeneration of abandoned and underused sites and brownfields in the cities. This creates not only economic revitalisation programmes but also programmes enhancing the quality of life of the urban population. Slovakia in May 2004 had become a member state of the European Union, which also brought many responsibilities and obligations of member states in the field of environmental protection and human health. Recently, there are a growing number of activities to promote sustainable urban development and the adoption of several documents and declarations in support of effective strategies that address the development of the urban environment towards meeting its quality for urban population. One of the approaches is the focus on ecosystem services that is part of the strategy of adaptation to the adverse effects of climate changes in cities. Bratislava as the capital of Slovakia has adopted such a strategy in 2015 and within that context supports creating community gardens on available plots of underused land or brownfields, with environmental and social benefits for the city. Community gardens are often viewed as one of the strategies, which can improve sustainability of urban environment as well as improve health and affect lifestyle of individuals.

Urban gardening in Bratislava is organised on the basis of voluntary work that has begun under the Pontis Foundation. The first attempts of community gardens have been connected with improving the courtyards in the residential areas, and the civic initiative "courtyard" has been established, with the aim to support motivation of the residents to improve public spaces in the community. Within this movement a specific project "Gaps" has started that mapped all the possible community places in the city that can be used for social activities. These community places are the plots that are underused and abandoned and as such are the holes in the urban fabric and can be designated as brownfields suitable for temporary use. Based on these available plots, the first activity "mobile gardens" has started where the main motivation for stakeholders to utilise these places has been the opportunity for gardening. Surprisingly, spin-off effect of this activity has brought rich social informal interactions that have been developed while spending time by urban gardening and sharing duties and experiences among local people of various age and nationality. Gradually people started to be involved in other après-gardening activities connected with consuming their own products together and having fun and socialise together up to becoming friends and spending time together in the afternoon and evenings. The next spin-off effect of this activity has been children education in becoming familiar with the type, colour and smell of flowers and vegetables as well as getting practical experience in helping with gardening. Last but not least, spin-off effect of this activity have been discussions about current situation on upgrading the outdoor environment in the community and creating semipublic spaces that are important for identification with local community and proactive behaviour of community members. Participation in urban gardening has generated synergy of the place attachment with the needs of the development of the community.

Now there are 4 types of "mobile gardens" in Bratislava. The first is in the old town in a gap between the block of flats; the second one is on a walkway under the building called Pyramid, and it is a combination of a community garden with a café; the third one is in the community close to nature under the slopes of the lesser Carpathians, and this one has extended its scope of gardening also to vineyards and tries to start with community winegrowing. Altogether there are about 276 people involved in these activities. Since these activities have only started 2 years ago, it is quite a success. Urban gardening on the underused plots is based on the lease of the plots for 3 years that have been guarded by the Pontis Foundation. Among the challenges and perspectives in Slovakia, the following ones seem to be most significant [14]:

- Limited research done yet (mapping of vacant spaces)
- Number of vacant/unused plots in cities
- Brownfields as potential space for urban gardening
- Learning from first successful examples: community garden Sasinkova, Bratislava, Old Town, and community vine yard and garden Pionierska, Bratislava – Nové Mesto
- Missing complex strategy for public spaces and legal support

- Missing support instruments for attracting gardeners (passportisation of available plots, clear rules)
- Promotion for land owners – usual fear of something new (gardeners will "stay forever", fear of plot degradation, administrational difficulties...)
- Transition of our cities
- Many others

Conclusions

Urban environment can have positive effects on creation and growth of communities as they have the opportunity to build a local identity and a sense of localism around a certain space. Place attachment is a significant factor influencing identification with local community and proactive behaviour of community members and generates territoriality based not only on routine but on the social commitment and value consensus among the members of community. In order to utilise the synergy of the place attachment with the needs of the development of the community and space/place overall, it is necessary to foster participative planning culture involving all the actors, making optimal mix between private, public and corporate elements. Special attention must be paid to non-formal tools: cooperation with the communities living in similar environment, introduction of best practice cases to public, building up clusters, non-formal cooperation with the municipality, city and region, etc. We have to face also other postcrisis paradigmatic changes – smart and learning municipalities/communities are those who are able to learn new, formerly unknown or unimportant competences in diffuse and rapidly changing world – ability to manage the network coordination, ability to learn and to forget, ability to deal with complexity and ambiguity and ability to deal with shifting and unstable alliances and partnerships.

Community gardens can have a huge impact on this process as well as on the quality of urban life beginning from producing fresh food to strengthening neighbourhood bonds. It can also have positive impacts on distressed neighbourhoods where vacant lots can be converted into community gardens or community green spaces, and these improvements can have an effect on residents' perception of safety outdoors, reduction of social problems and cultivation of social responsibility. Urban gardening becomes the unique signature of the given community, a largely understandable gesture of openness, solidarity, freedom and hospitality.

References

1. Jaššo, M. (2011). Plánovacie kultúra. In M. Finka (Ed.), *Priestorové plánovanie* (pp. 175–196). Bratislava: ROAD Spectra.
2. Moore, G.T. (1983). Knowing about Environmental Knowing: The Current State of Theory and Research on Environmental Cognition. In: Pipkin, J.S., La Gory, M. E., Blau, J., R.(Eds): *Remaking the City,* State University of New York Press, p.25.

3. Fahmi, W. S. (2006). The Urban incubator: (De) constructive interpretation of heterotopian spatiality and virtual image (ries). First Monday Online, Special Issues 4: Urban Screens: Discovering the Potential of Outdoor Screens for Urban Society.
4. Hain, S. (1997). Der Berliner Städtebaudiskurs als symbolisches Handeln und Ausdruck hegemonialer Interessen. *WeltTrends, 17*, 103–123.
5. Märker, O., & Schmidt-Belz, B. (2000). Online meditation for urban and regional planning. http://enviroinfo.isep.at/UI%20200/MaerkerO-12.07.2000.el.ath.pdf. Accessed 20 Jan 2008.
6. Kuhn, T. (1970). *The structure of scientific revolutions* (2nd ed.). Chicago: The University of Chicago Press.
7. Keim, K.-D., Jähnke, P., Kühn, M., & Liebmann, H. (2002). Transformation der Planungskultur? Ein Untersuchungsansatz im Spiegel stadt- und regionalplanerischer Praxisbeispiele in Berlin-Brandenburg. *Planungsrundschau – Zeitschrift für Planungstheorie und Planungspolitik, Ausgabe, 6*, 126–152.
8. Jaššo, M., & Kubo, L. (2015). *Urbánna sémiotika*. Bratislava: Spectra – ROAD.
9. Mäntysalo, R. (2005). Approaches to participation in urban planning theories. In I. Zetti & S. Brand (Eds.), *Rehabilitation of suburban areas – Brozzi and le Piagge Neighbourhoods* (pp. 23–38). Firenze: University of Florence.
10. Zlydneva, N. (2003). Urban fun. In: Sarapik, V.,Tuur, K.(Eds.) Proceedings of the Estonian Academy of Arts, Tallinn, 464 p. *Place and location. Studies in environmental aesthetics and semiotics III* (pp. 139–146).
11. Kearns, A., & Forrest, R. (2000). Social cohesion and multilevel urban governance. *Urban Studies, 37*(5–6), 995–1017.
12. Jaššo, M. (2005). Regional identity – Its background and management. In D. Petríková & I. Roch (Eds.), *Flusslandschaften ohne Grenzen - Mitteleuropäische Ansätze zu Management und Förderung landschaftsbezogener Identität* (pp. 171–179). Bratislava: ROAD-Spectra.
13. Armstrong, D. (2000). A survey of community gardens in upstate New York: Implications for health promotion and community development. *Health & Place, 6*, 319–327.
14. Petríková, D., & Szuhová, J. (2015). The role of networking, innovation and creativity in social responsibility to connect urban and rural environment. In *AESOP congress*. Definite Space – Fuzzy Responsibility, 13–16 Prague Czech Republic p.1222 ISBN 978-80-01-05782-7.
15. Landry, C., & Bianchini, F. (1995). *The Creative City*. London: Demos Publishing.
16. Caragliu, A., Del Bo, C., & Nijkamp, P. Smart cities in Europe. *Journal of Urban Technologies 18*(1). (1/2009), p. 52.
17. Schmeidler, K. (1997). Sociologie v architektonické a urbanistické tvorbě. Vyd. Ing. Zdeněk Novotný, Brno
18. Kavaratzis, M., & Ashworth, G. J. (2005). City branding. An effective assertion of identity or a transitory marketing trick? *Tijdschrift voor Economische en Sociale Geografie, 96*(5), 506–514.

Chapter 30
Awareness of Malicious Behavior as a Part of Smart Transportation in Taxi Services

Peter Pistek and Martin Polak

Abstract This paper is part of a wide work developed in the context of smart cities. Main objective is focused on taxi services for business class and its specific needs. We present a new comprehensive multimedia information system for taxi cabs. Competition points out a potential in this area; however, they primarily focus on advertising which discourages users rather than attracts them. In contrast to them, our system is also focused on protecting customers from being betrayed by the taxi service, providing them the possibility of continually monitoring the recommended route and also the real, traversed route on the map along with the number of actual kilometers traversed.

Introduction

Public transportation is indispensable part of every city. As mobility of citizens rises, the congestions and traffic collapses appear more often. As initiative of smart cities starts, the public transportation became one of the most important parts of it [1, 2].

Taxi services, as part of a public transportation, have not changed a lot in last few years. Nowadays, there is big competition among taxi companies, so they have to add new services to increase overall passenger satisfaction (i.e., in London 70% [3]). Because of permanent presence of taxi services, many cities include them as important stakeholder in public transportation in their process of conversion to smart cities [4]. For better imagination, in New York, the popular yellow taxis make more than 2 million kilometers each day [5].

In many countries, there are problems with overpriced taxi drives just because driver used longer road or because he used device which will measure longer distance traveled than the actual distance [6].

P. Pistek (✉) · M. Polak
Slovak University of Technology in Bratislava, Bratislava, Slovakia
e-mail: peter.pistek@stuba.sk

Related systems are usually just for playing commercials or just for watching TV, for example, in New York, there is a system called Verifone [7] inside cabs, and it can only be used for watching television and for paying of bill (this solution has a built-in feature for contactless payments). Another system called Cabadvertising [8] is just used for playing commercials. One of the best systems is sold in India and it is called Tabbie [9]. It is possible to play music, watch movies and see the position on the map on which the taxi is currently located. But there is no possibility to see whole route, so customer does not know if the driver is driving by optimal route or if the distance for which customer is paying is no longer than the actual traveled distance.

Other types of systems for taxi services, such as Hopin [10] and Uber [11], are focusing on the user when passenger is searching for cab. This area is wildly researched also for effective usage of cabs or evaluation of drivers [4, 12] when customer is not inside vehicle (i.e., process of booking of a cab). From our point of view, they all have one common disadvantage, they "do not care" about passenger when she/he is already in the cab. Usually, the only available additional service is free Wi-Fi.

Our system described in this paper is here to solve problems when user is inside a vehicle with the use of GPS inside tablet just for customer. As an additional feature, the system will be able to provide Internet access for passengers and play multimedia files – movies, television, music, and headlines which can be played directly into a built-in car audio system.

Solution

The system consists of two different programs which are in two different (customer's and driver's) tablets. As can be seen in Fig. 30.1, those programs communicate with each other. For this wireless network communication is used with direct connection to the Internet by a mobile network. TCP protocol is used for sending commands for activation and deactivation and sending ID of the first commercial. We also tried UDP protocol for commands, but during testing outside the laboratory, we found out that packet loss (due to many interferences) was beyond tolerance limits. This led to loss of some commands. The only use of UDP protocol is in position sending, because if sometimes packets get lost it is not a problem – the route is just less accurate as it could be.

We used tablet mounted on front seat so customer in back seat can control it (see Fig. 30.2). All features described later are implemented on this tablet. Tablet has the possibility to receive video from DVB-T server inside vehicle.

In each taxi, there is a "driver's" tablet, which can send commands to "customer's" tablet (activation, deactivation, sending ID of commercial, and sending location data). Driver's tablet activates "customer" tablet thanks to the command sent as data packet. This packet contains driver's name and locations of starting and end positions (this does not apply if customer order taxi without calling to dispatching/using an app) and also ID of first commercial, which will be played right after activation of tablet. Deactivation means that customer cannot do anything with her/his tablet – the application is disabled.

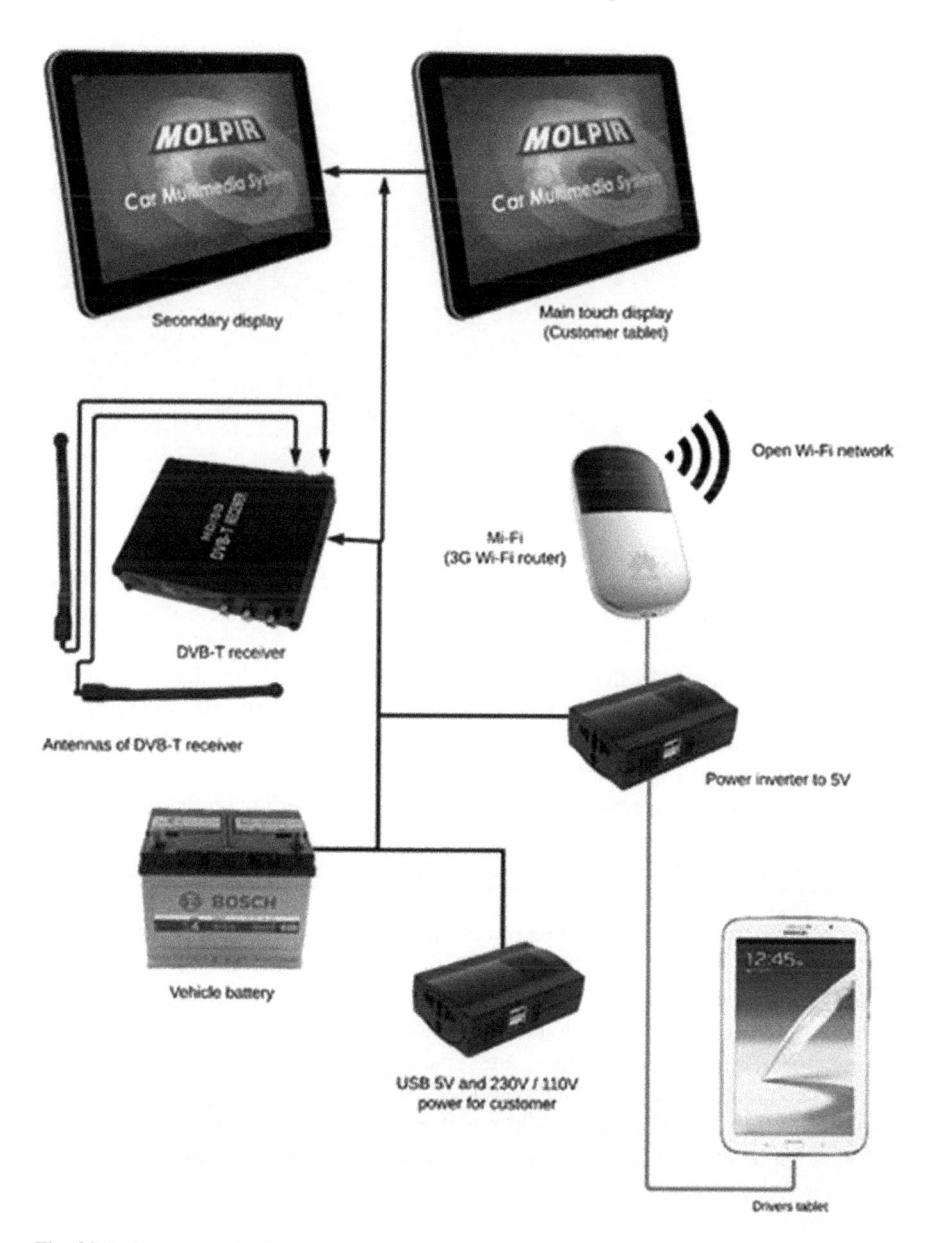

Fig. 30.1 System architecture

Fig. 30.2 Example of tablet attaching in the car

Applications

Our solution is based on the cooperation among several independent applications. This approach ensures minimal risk of compromising whole system when one of its parts is compromised.

For tablet maintenance, there are two applications – for synchronization and file browsing. Synchronization application is used for first installation, copying of the content (e.g., movies, music, settings), synchronization of data, and application's update. In file browser application, operator can only copy content from USB flash drive. Unlike synchronization, not every file has to be copied; operator can select a specific set of files for each tablet.

Main application (Fig. 30.3) is first application that passenger sees – right after commercial is played, but commercials are only optional. It depends on company policy if driver has to send this type of command. During the ride, a costumer can check traveled route and possible deviation from the optimal route on a map. It is obvious that this is not feature for long-time journeys. Customers have the possibility to use a daily newsletter downloaded by RSS and list of other applications for entertainment. Also on the top, there are some additional information about driver, start and endpoint of the route, estimated length of route according to the relevant source (Google Maps API), and actual traveled length of route so far.

Every application for multimedia and information has its own submenu and list of content (e.g., movies can be divided by genre, music by albums, and TV channels by its name, and Internet browser can have a quick access for the most used pages defined by an operator).

Synchronization

Multimedia data inside tablet are possible to synchronize via Wi-Fi or USB flash drive. At the beginning, we were proposing distributed synchronization so every cab could be used as server or client during Wi-Fi synchronization and part of the

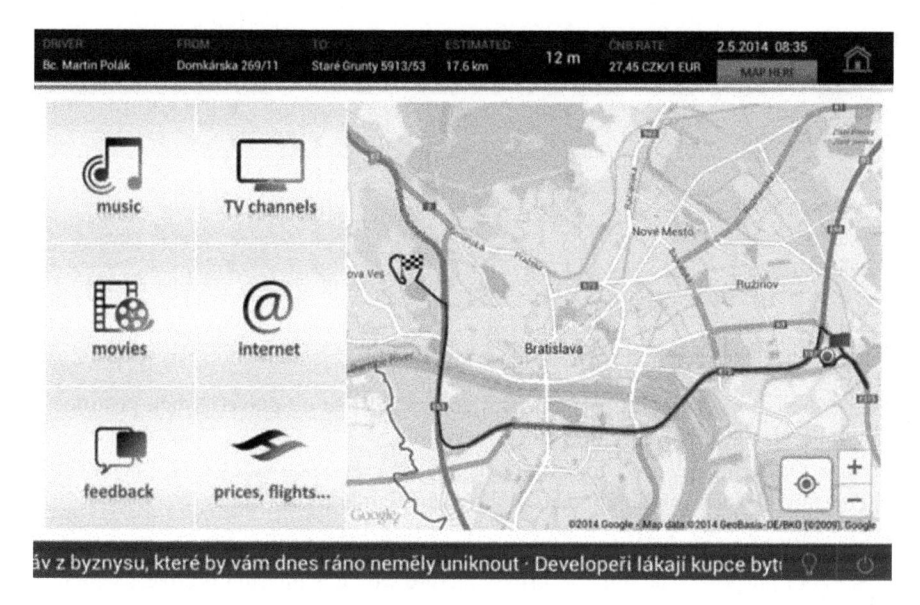

Fig. 30.3 Graphics interface of main application

synchronization would be managed at taxi positions throughout a city. Based on testing in real environment, we canceled this idea. In praxis there is not enough time and nearby taxi vehicles to process such synchronization (e.g., file size of movies); hence, it is possible to update multimedia content only in a taxi company service point.

Wireless synchronization is used for easier update of data (commercials, movies, music, settings, etc.) stored in a tablet. Wireless connection makes it possible to synchronize more vehicles at once by connecting to one access point. USB flash drive synchronization must to be done individually inside each vehicle.

Synchronization starts by sending a special type of control packet. Due to necessity to connect to different Wi-Fi networks, it is not allowed to implement synchronization as some kind of daemon service which would automatically perform an update. After receiving of the control packet, application searches for defined Wi-Fi networks. If successful, it performs an update and reconnects to the original Wi-Fi network.

Testing

System is deployed on approximately 150 vehicles of taxi company in Prague. Every vehicle is equipped with the same set of applications and the same configuration. If any crash happens, it appears on Crashlytics[1] site. During testing phase, the passengers used the system hundreds of times a day.

System is now fully functional. We add a comparison of related works in Table 30.1.

---

[1] https://crashlytics.com/

Table 30.1 Comparison of related works and TickTack solution

Features	Verifone	Cabadvertising	Tabbie	Hopin	Uber	Our solution
Commercials	✓	✓	✓	✗	✗	✓
Advertisement	✓	✓	✓	✗	✗	✗
Television	✓	✗	✓	✗	✗	✓
Map: current position	✗	✗	✓	✓✓	✓✓	✓
Map: traveled route	✗	✗	✗	✓	✓	✓
Internet browser	✗	✗	✗	✗	✗	✓

Difference between commercials and advertisement is understood as follows:

- Commercials – video clips with paid content. It can be played when customer will get inside cab or anytime during traveling.
- Advertisement – paid content visible at any time or when video or TV is played. They are usually small banners with dynamic content.

From Table 30.1, it can be seen that none of these solutions have as many features as our purpose. The only described feature which is missing is advertisement. The reason behind this is that advertisements were marked as annoying and are visible at any time. Paying customer usually does not want to be distracted by this kind of services or products' promoting.

The contribution to the smart transportation (as a part of smart cities) is increased reliance on taxi service thanks to the possibility of checking of the proper route which is needed mostly in cities with many tourists.

Conclusion

In the paper, we present a new solution for taxi services with the focus on smart transportation in cities and nearby area. Novel approach to smart transportation is also in tracking traveled route with comparison of optimal route according to well-known service. This way customers will not be lied to and can be sure that price is right and they are paying just for the route they really traveled.

Our solution is after testing phase and now is already deployed in a taxi company in Prague (Czech Republic) inside approximately 150 cabs thanks to our commercial partner. Based on every day usage, passengers like proposed features because they can trust this system and can be sure that taxi driver would not take more money than he should. Besides this, great advantage of system is to offer multimedia content (such as music video, television, headline news, and Internet) – the most popular choice seems to be playing of music of own choice.

There are plans to extend use of our system on other taxi companies also with better integration with existing reservation systems.

Acknowledgment This work was partially supported by the Slovak Science Grant Agency (VEGA 1/0616/14) and ITMS 26240220084.

References

1. Venezia, E., & Vergura, S. (2015). Transport issues and sustainable mobility in smart cities. In *Proceedings: International conference on clean electrical power (ICCEP)*. Taormina, Italy.
2. Soriano, F. R., et al. (2016). Smart cities technologies applied to sustainable transport. Open data management. In *Proceedings: 8th Euro American conference on telematics and information systems (EATIS)*. Cartagena, Colombia.
3. Department for Transport of United Kingdom. (2015). Taxi and private hire vehicle statistics: England.
4. Adewumi, A., Odunjo, V., & Mistra, S. (2015). Developing a mobile application for taxi service company in Nigeria. In *Proceedings: International conference on computing, communication and security (ICCCS)*. Pamplemousses, Mauritius.
5. Bloomberg, M. R, Yassky, D. (2014). *New York City Taxi & limousine commission: Taxicab Factbook*. New York.
6. National Transport Authority of Ireland. (2016). Statistical bulletin number 01/2016, Taxi statistics for Ireland. Ireland.
7. V. media, VNET | Taxi TV. (2015). [Online]. Available: http://www.verifonemedia.com/digital-ooh/taxi/. [Cit. 16.5.2015].
8. Cabadvertising, Cabadvertising. (2015). [Online]. Available: http://cabadvertise.co.uk/. [Cit. 16.5.2015].
9. L. Technologies. (2014). Tabbie (Tablet in a Cab), [Online]. Available: http://luminositydigital.com/products.html. [Cit. 15.2.2015].
10. Hopin s.r.o., Hopin, (2015). [Online]. Available: http://www.hopin.sk. [Cit. 16.2.2015].
11. Uber, Uber. (2015). [Online]. Available: https://www.uber.com. [Cit. 20.2.2015].
12. Ch. Zhang et al. (2016). A social-network-optimized taxi-sharing service. In *IT professional*, Vol. 18, 4, 34–40, casopis (t.j. nema lokaciu - http://ieeexplore.ieee.org/document/7535094/).

Chapter 31
Alternative Lights for Public Transport in Smart Cities

Michal Cehlár and Dušan Kudelas

Abstract It is estimated that in the year 2050, almost 70% of people will live in urban areas. Cities need to go through evolution and change into smart cities. They are energy efficient, save resources, produce low emissions, and provide citizens a better quality of life. Also in Slovakia public opinion and idea about what smart city means in practice is formed. It is a complex innovation in all spheres of life of the people and infrastructure of the city, using new materials and the latest information and communication technologies. The aim of the presented paper is to show chosen developed technologies at the Institute of Earth Sources, which can be used for smart cities.

Introduction

Residents expect from their city real innovation, not only small improvements. They want single point of contact and access to information at the level of the twenty-first century, without unnecessary forms, stamps, and office visit. They want to know if their residence is not on the maps of crime or flood maps in the critical region. They want to keep track of traffic conditions and have information about free parking spaces [10].

Together with the concept of smart city, we get another term – placemaking. This deals with life and spaces between objects. In the foreground are the people and their needs. People who are from the city, from their living space, expect some elementary functionality and require decent infrastructure and smooth transport, sufficient amenities and services, lots of greenery, low emissions, and a safe environment for themselves and their children. In each group of people, each sector (e.g., entrepreneurs in trade and services, cultural, community and nonprofit organizations), however, has specific requirements [10].

M. Cehlár (✉) · D. Kudelas
Ústav zemských zdrojov, Fakulta BERG TU v Košiciach, Košice, Slovakia
e-mail: michal.cehlar@tuke.sk; dusan.kudelas@tuke.sk

© Springer International Publishing AG, part of Springer Nature 2019
D. Cagáňová et al. (eds.), *Smart Technology Trends in Industrial and Business Management*, EAI/Springer Innovations in Communication and Computing,
https://doi.org/10.1007/978-3-319-76998-1_31

Buildings are responsible for 36–40% of global energy consumption. Therefore, the concept of smart city is often associated with energy savings, finding alternative sources, and reducing carbon footprint. Thanks to smart energy solutions, the concept has reached 5% of the potential savings in smart cities. One of the solutions is a smart-connected infrastructure, which is able to increase its value tenfold. For example, data from energy metering and lighting (public and private) that is collected and evaluated contribute to better energy balance [10].

The concept of "smart cities" is characterized by the interconnection of information and communication technologies (ICT) with already extensive and costly critical infrastructure such as energy networks, transport networks, waste management, health, etc. and thanks to precise monitoring and control system allows increased efficiency and amount of savings. In addition to the deployment of intelligent sensors and universal access to smartphones (and their complementarity), "smart cities have the potential to empower communities and realize significant benefits in terms of efficiency, creativity, and even participation of local citizens in the democratic process. Therefore, the concept of smart cities naturally becomes an interesting opportunity for the city government, and often so there is a rapid adoption of "smart" solution. However, uncontrolled, rapid development of smart initiatives and technology complexity for integrating existing infrastructure with ICT opens the door to potential cyber threats, which in turn can have a negative impact on the community [6].

Mobility and Transportation in Smart City

Mobility and transportation are essential parts of the urban infrastructure. Smart city should be easily accessible to visitors and its inhabitants. Travel through the city should be smooth and comfortable but also environmentally friendly. The aim is to provide a versatile, efficient, safe, and comfortable transportation systems that are connected to the infrastructure of information and communication technologies and open data [9].

Within the concept of smart cities are the trends in transport as follows [9]:

- Operation monitoring system
- Sharing of transport
- Intelligent traffic management
- Intelligent traffic lights
- Intelligent traffic information
- Sharing experiences of citizens with transportation
- Intelligent parking spaces
- Bike sharing
- Electric vehicles
- Optimization and popularization of public transport

In most large cities in the Czech Republic and Slovakia, there are similar problems of transport as in Brno [9]:

- Incomplete main ring road
- The lack of a complete system of superior communications
- State railway junction
- Unfinished part of the city sewer and unresolved flood protection
- Collision of parking and traffic (mainly pedestrian)
- Stagnation in the development of infrastructure for rail transportation
- Preference of individual car transport to the detriment of pedestrians, cyclists, and public transport
- Lack of bicycle paths and bicycle lanes
- Lack of parking and garage space in the city center and housing estates
- The lack of car parks
- Lack of public transport attractiveness

In terms of transport, urban mobility is often used. Mobility is defined here as the ability to move as efficiently as possible from place to place and is related to expanding road capacities and faster means of transport. With regard to smart cities, however, the authors of this article prefer to move away from mobility toward availability. With clever urban planning and with the use of modern technology, for example, homeworking, demand for transport can be greatly reduced [7]. Another approach with the future is today's dynamically growing car sharing [8]. There are two different ways for car sharing. Car owners and their cars to customers through Car Sharing companies and either private companies rent cars directly to customers or individuals. Drivers who do not own a car and drive a few km a year, are more likely to use the car sharing method. Companies also provide mobile apps to manage bookings, track mileage and total costs, or even unlock and lock the car. There are already many companies in the world like the Czech Republic who are trying to gain a share in the growing market.

Car parking in big cities is getting heavier and it is a relatively big problem. According to various studies, drivers spend an average of about 20 min searching for a parking space. This results in an increase in CO_2 emissions; cars getting into places unnecessarily, thereby increasing the likelihood of traffic jams; and, the most important of all, people wasting time. Of course, any future road construction or its reconstruction should be an Intelligent Transport System (ITS) that integrates information and telecommunication technologies with transport engineering with the support of other related fields (economics, transport theory, system engineering, etc.) to ensure the management of transport and shipping processes for existing infrastructure [10]. Their aim is above all to provide better and safer traffic management and more effective support for the transport of persons or things. The main benefit of introducing intelligent transport systems and services in terms of societal benefits is to increase traffic safety. The important goals of ITS are, of course, the sustainability of the city and the reduction of emissions. With this in mind, the concept of Green ITS is also used – transport management and management with regard to urban sustainability [10]. If there is an unexpected event in traffic, such as a traffic accident or traffic jam, traffic can be regulated, decelerated/accelerated, deflected, and so on.

Another part of smart transport is, for example, urban public transport (public transport). It can be assumed that the entrance will be limited to the center of most large cities or a fee will be charged for each entry. The city's goal is clear, to minimize city traffic in order to reduce emissions and maximize passenger transport via public transport. It is necessary to construct parking garages on the outskirts of the city, where the drivers can leave their vehicles and then continue with public transport. With regard to smart cities, these aspects must be interconnected, so that the driver gets information about the free parking space on the parking lot, can book it (or pay), navigate to the place, and can buy a public transport ticket and get information to his or her destination.

Autonomous vehicles are also a huge topic, for example, in logistics, they are nothing new and have been used for many years in modern storage and production plants where they provide full or partial automated delivery of goods, parts, or materials. Of course, road and real traffic is quite different, but from the point of view of the difficulty of autonomous driving, it is probably the easiest ride on the highway.

Proposed Public Lighting

Public lighting is not only part of the local color of each town or village but is also an essential part of the transport system. One of the important aspects of smart cities is not only lighting in motorized traffic but also lighting of walkways. It is considerable part of the expenses of municipal government. With the growth of prices for consumed energy by lighting system, grows requirement of self-sufficiency in the acquisition. Currently it is taking place in the whole territory of Slovakia reconstruction of public lighting, which is addressed by implementing European standards. There are several types of street lamps, which on its operation used renewable energy sources. Our proposal is an intelligent LED solar lamp of public lighting with wind aggregate. In general, the lighting system consists of a set of features and technical means to ensure the correct lighting such as lamps, luminaires, poles with electric equipment, power cables, switchgear, etc [5].

Intelligent LED street lighting combines a camera system, wireless networks, and sensors. In addition to significant energy savings and energy supply management, it contributes to a better awareness of the city traffic conditions and parking, while it increase safety on the streets. Linkable also for applications in smartphones, with street lamps, it serves as a "beacon" for navigation of tourists. They can also be transmitters of marketing messages and important announcements of the city management. Ecologic production and moderate energy consumption are only one side of the coin [3]. It is important to be able to predict the state of freedom of the regulatory power and the unused energy office building over the weekend to divert to residential block. Smart city's vision is to achieve complete energy self-sufficiency city, which will not produce and "export" waste, by respecting local conditions and individual character of the city [10].

Public lighting in Slovakia is made up of several types of lighting columns with the most commonly used as concrete and steel. Lights are placed on these types using the boom arm with different angles or directly [1]. The lamps currently used are incandescent, fluorescent lamps, compact fluorescent lamps, LEDs, and other species.

As a source of electricity, we propose two alternatives: Savonius rotor and technology "wind belt," respectively. Savonius rotor belongs to the device with a vertical axis of rotation (VAWT). Rotor blades with a vertical axis of rotation are long and rounded at both ends and fixed. Their main advantage over systems with the horizontal axis of rotation is that it is not necessary to capture the wind flow in different directions.

Savonius rotor is low-speed wind machine. It is a type of vertical-axial wind turbine. It uses the pressure difference in the flow of air and the round hollow semicircular scoop which, when viewed from above, has the form of letter "S". The efficiency of the device is moved only by the $\eta = 0.2$ but is structurally simple with good start-up characteristics even at low speeds of wind and is characterized by a high starting torque [2].

At present, there are large structures with two or three wings because the higher number reduces efficiency. Semicircular blades are curved and mounted in the middle of the opposite. The direction of rotation can be from the right to the left. The speed of the rotor depends on its diameter, the load, and the wind speed. Our proposed road lamp with wind aggregate consists of a steel mast with a height of 8 m and a length of boom 1.5 m.

Rotor area is 4 m2. It is essential that the blades are made of a light material; it is proposed to construct the plastic with addition of glass fiber. The rotor is mounted through the cage surrounding the shaft that transmits torque to the transmission. Transmission with gears is used for converting the input torque for the rotational movement. It allows to set up several gears and transfers energy to the flywheel, which converts mechanical energy from the rotor into electrical energy which produces a direct current. In order to ensure business continuity in times of operation lamp, it is necessary to provide switching between battery and distribution networks which support element having a switching function. This element battery charged automatically connects the lamp to the network which distributes electricity. The battery is charged continuously for 24 hours. The final appearance of such a lamp in 3D can be seen in Fig. 31.1.

Savonius rotor does not cover the overall demand for electricity, and the lighting should be equipped with PV panel.

Another alternative is to use technology of wind belt. The aim of our research is the applicability of technologies to generate electricity from the airstream as part of an autonomous light source. Lack of equipment for the use of low potential winds prompted the present investigation; one of the objectives is to optimize technology for air flow rate which is low at the same time the production of sufficient power for lighting and meet the safety parameters.

The principle of wind belt is as follows: tensioned membrane (tape), under the influence of wind flow vibrates (see Fig. 31.2). Flapping causes movements of the

Fig. 31.1 View of a wind accumulation lamp

Fig. 31.2 Wind belt technology [8]

permanent magnet, which is attached to the membrane. Oscillating magnet moving between these coils induces a voltage coils. Since the Windbelt and number of other "flutter" wind harvester devices have been designed, almost all have efficiencies below turbine machines. But the main advantage is that it can produce electricity at wind speed lower as conventional wind devices. One-meter version that could be used to power LED lights can generate 10 W average.

At present, it remains the most effective way of using solar energy, so-called photovoltaic panels (see Fig. 31.3). Scientists are already quite a long period con-

Fig. 31.3 Innovative transparent PV panels [4]

sidering the idea, as one would use photovoltaic panels in practice even more effectively [7].

Innovative photovoltaic panels are the first fully transparent, which is considered a key point of this innovation. Upgrading these panels is the fact that during the day, the sun's rays impinge on transparent photovoltaic panel that performs two functions. The first function is that the space required by the sun's rays is released, and the second function is that some of these rays are directed to the perimeter of the panel. The great advantage of this innovation is also that all the so-called glass building to become producers of their own electricity would also save costs and reduce the pressure on the ecology. In the past there were a number of attempts to package glass building by photovoltaic panels; however, the problem is that the panels change the visibility of these windows while reducing the amount of natural light in a given area, which was simply undesirable. The problem with many developed technologies is still low efficiency [4].

Conclusion

The use of renewable sources for smart cities is essential, but useful technologies are particularly photovoltaic panels and wind energy. In implementing the concept of smart cities, however, it must also have in mind not only trendy and marketing issues but also consider the potential risks, for example, cooling of server stations for managing the infrastructure and with the possible increase of energy consumption, cyber threats, and possible energy outages. Therefore, autonomous lighting sources and new technology could be beneficial solutions, but these risks need to be solved before general application.

References

1. Verejné osvetlenie [online]. (2005). [cit. 2012-04-26]. Dostupné z: http://www.verejneosvetlenie.sk/master/Goinggreendiy.com [online]. 2009 [cit. 2010-04-01]. How to build a solar panel. Dostupné z WWW: http://goinggreendiy.com/wind-turbine-informatio/
2. Schulz, H. (2005). Savoniův rotor – návod na stavbu. Ostrava: HEL. 80 s. ISBN 808-616-72-67.
3. Rybár, R., Kudelas, D., Rybárová, J., & Beer, M. (2013). Parallel manifold header on foam material basis for vacuum tube solar collectors/Radim Rybár ... [et al.] – 2013. In: Advance science letters. Vol. 19, no. 2, pp 591–594. ISSN 1936-6612.
4. Radimak, E. Priehľadné fotovoltaické panely ako nový zdroj energie. http://www.setri.sk/priehladne-fotovoltaicke-panely-ako-novy-zdroj-energie/
5. Lom, M., & Přibyl, O. Smart Cities aneb města budoucnosti II. http://elektro.tzb-info.cz/informacni-a-telekomunikacni-technologie/14209-smart-cities-aneb-mesta-budoucnosti-ii
6. Tobiáš, L. Smart cities a kybernetická bezpečnost. SCmagazín 02/16. http://www.scmagazine.cz/casopis/02-16/smart-cities-a-kyberneticka-bezpecnost?locale=cs
7. http://www.smart-magazine.com/en/tomorrows-sustainable-city-smart-electric-drive/
8. http://www.p-stadtkultur.de/made-in-darmstadt-wolt
9. https://cs.wikipedia.org/wiki/Smart_City#Chytr.C3.A1_doprava_.28Smart_Mobility.29
10. Digitálna transformácia inteligentné mesto. Atos e-book.

Chapter 32
Smart Wristband System for Improving Quality of Life for Users in Traffic Environment

Dragan Peraković, Marko Periša, Rosana Elizabeta Sente, Petra Zorić, Boris Bucak, Andrej Ignjatić, Vlatka Mišić, Matea Vuletić, Nada Bijelica, Luka Brletić, and Ana Papac

Abstract The aim of this research is to develop a smart wristband system architecture that provides real-time information to users in traffic environment. Users in traffic environment are persons with visual impairment, hearing impairment, and locomotor impairment, elderly, children, and persons without disabilities. The purpose of service is to provide users with accurate and real-time information and raising the level of quality of life. Previous research and solutions have provided information about user needs and demands. Conducted survey defined functionalities of system based on Cloud Computing for the Blind concept which provides 24/7 support for the delivery of services and safety of users. Architecture of the service is designed according to universal design and Ambient Assisted Living concept. With simulation testing and experimental methods based on the Arduino and Raspberry Pi platforms, the work of proposed system is tested in laboratory and real-world

---

D. Peraković (✉) · M. Periša · R. E. Sente · P. Zorić
Department of Information and Communication Traffic, University of Zagreb, Zagreb, Republic of Croatia
e-mail: dperakovic@fpz.hr; mperisa@fpz.hr; rsente@fpz.hr; pzoric@fpz.hr

B. Bucak
Ericsson Nikola Tesla, Zagreb, Republic of Croatia
e-mail: boris.bucak@ericsson.com

A. Ignjatić
Alca Zagreb d.o.o., Zagreb, Republic of Croatia
e-mail: andrej.ignjatic@alca.eu

V. Mišić
CARNET – National Cert, Zagreb, Republic of Croatia

M. Vuletić
Republic of Croatia Ministry of Interior, Zagreb, Republic of Croatia

N. Bijelica · L. Brletić · A. Papac
Laboratory of Development and Research of Information and Communication Assistive Technology, University of Zagreb, Zagreb, Republic of Croatia

environment which proved real-time information delivery to users. Having the ability to notify users in real time will increase the level of autonomy and safety while moving through traffic network.

Introduction

With the development of information and communication technology (ICT) comes the offer of many services that are implemented in various terminal devices (TD). In order to use currently available services, market offers a wide variety of TD to users. The fact is that a large number of people have a problem with adjusting to the use of functionality based on ICT. For this reason, it is necessary to develop a system which will deliver services to users in a way that they can use it separately and independently. This research proposes system architecture for informing users in traffic environment and everyday needs – Smart Assist for All (SAforA). This research analyzes currently available solutions in the field of assistive technologies, their advantages, and disadvantages. User demands are defined with conducted surveys. Existing hardware and software solutions are still not satisfying in terms of developing quality forms of assistive technologies.

Parents with children, the elderly, persons with disabilities, and persons without disabilities make a targeted group of users (TGU) of research. The analysis of the currently available assistive technology solutions is demonstrated, their advantages and disadvantages and survey that defines the user requirements. By applying the proposed system for informing users in traffic environment, it is expected to increase the level of quality of user's life due to possibilities to customize available information and increase their mobility.

Previous Research

Currently available studies provide information about how many people are familiar with modern technology and the possibilities that they offer. As a result, it is possible to define some of the functionalities that proposed service should provide. According to a study conducted in the United Kingdom which involved 920 people (parents with younger children), the results gave how many people are interested to use location services to know where their children are. In the United States of America, 72% of parents wants to have a real-time information about where their children are. Thirty-six percent of them want their children to use a smart watch with a built-in GPS, while thirty-nine percent of them would like to receive a warning from a child's smart watch if they call an emergency number. At the same time, they find a physical activity of children important; therefore, 22% of parents want to have a fitness services at children's devices to follow a number of their steps [1].

The research supported by Yeungnam University Research Grant got the results that a large number of respondents were not familiar with the concept of smart clothes. This technology may have a wide application in practice of locating children and elderly and to encourage self-reliance and independence [2].

Jawbone UP fitness bracelet collects information about user's body activity. It works with built-in Smart Coach application which provides information about user's activity and tips for improvement of the user results. The information that is being collected is heart rate, temperature and breathing which allows monitoring of user's sleeping time. Disadvantages are lack of display and synchronization with MTD to obtain needed information.

Garmin Vivosmart is another fitness bracelet which has a display. It provides information about movement, burned calories, heart rate, time of movement, and the number of taken steps. There is no built-in GPS module, and therefore it cannot accurately calculate the distance.

Some of the smart bracelets have built-in GPS so that the users can see how many kilometers have they walked or ran. Some of the smart bracelets have the functionality to detect obstacles on the road using variety of sensors. Users get information through vibration signals. Such example is Sunu smartband that is based on beacon technology. Another example of smart bracelet is a bracelet that can locate users. They can be used to improve children's safety. As the number of kidnapping of children increases sharply, this functionality can be considered good for trying to reduce that number. Some of the smart bracelets have the functionality to make an emergency call by pressing the button on it. In that case, parents can get an information about children's location. The disadvantage is that a kidnapper can take off the bracelet without triggering its alarm. Locating smart bracelets can be used for locating elderly that have dementia problems. Such example is Hemayati smart bracelet, Vega GPS bracelet, etc. [3–6]. Another example of assistive technology that serves to improve the quality of life of all users consists of an application, wearable device, and wireless connection to which it connects to the information gathering system. The wearable device has sensors for monitoring the heart rate, temperature, accelerometer, gyroscope, compass, and GPS. It provides the possibility of preventing various illnesses, reducing healthcare costs, and monitoring elderly people [7]. Smart bracelet, Bluetooth, and MTD can measure vital signs and behavior patterns of user in order to predict health problems. The common problem is the position of the smart bracelet on the user's hand that affects poor detection of vital signs. Due to the large amount of collected data, there is a problem of complexity of information processing [8]. Another solution based on the same devices can detect falls. Using 3G and 4G networks, information is sent to contact persons in the event of detecting the fall. To obtain more accurate and better data on fall detection, it is necessary to detect the fall of the bracelet and MTD [9].

Abbi smart bracelet is intended for blind persons, primarily children, with the aim to encourage them to explore the environment. It is consisted of Bluetooth communication, motion sensors, and sound synthesis system. The bracelet connects to iBeacons that allow estimation of distance from the point of interest. With sounds it impels users to explore the surrounding environment. It is used for independent movement and recognition of persons, and it sends voice alert messages in unsuitable situations. The main problem of this solution is in audio sounds because they need to be very well-defined sound so that the user can hear the difference between bracelet sounds and sounds in the environment [10]. MTD Sony Xperia Z and Sony Smartwatch are proposed for visually impaired persons to obtain better location information. In combination with added algorithms for tracking the steps, it is possible to get better information about location of a user on geographic map than using GPS [11].

Previous solutions to increase the level of quality of life of individual user groups are offered in form of smart bracelet. After analyzing existing solutions, it can be concluded that there are a lot of similar smart bracelets with limited functionalities and that those solutions are not suitable for all user groups. Manufactures do not think about their upgrade and thus limit the possibility of their use for other user groups.

Methodology

The main goal of this study is to propose a system architecture to deliver information services to users in traffic environment and everyday needs based on ICT. Currently available services do not satisfy needs of all user groups. Because of it, functionalities must be interoperable so that all users can use them by their own choice. The hypothesis of this study is that the application of ICT can increase the level of quality of life and improve the user's movement through transport network.

One of the goals of this study is to define service functionalities that will meet the needs of end users with purpose to increase the level of quality of life. Information about the health, social and cultural life, traffic environment, and everyday needs are considered. For the analysis of user requirements, it is necessary to define characteristics of TGU. Persons with disabilities include those who have any restriction or lack of ability to perform certain activities which prevents their full and effective participation in society on an equal basis with others [12].

Materials and Methods

Needs of TGU were investigated in order to define user requirements and assure that the functionalities of services are fully customized to the TGU. They are related to a fully functioning in daily activities and the availability and knowledge of new solutions and services based on modern ICTs. To obtain relevant information, research was carried out, one in 2014 and one in 2015. The first survey was conducted in nine homes for the elderly in Zagreb.

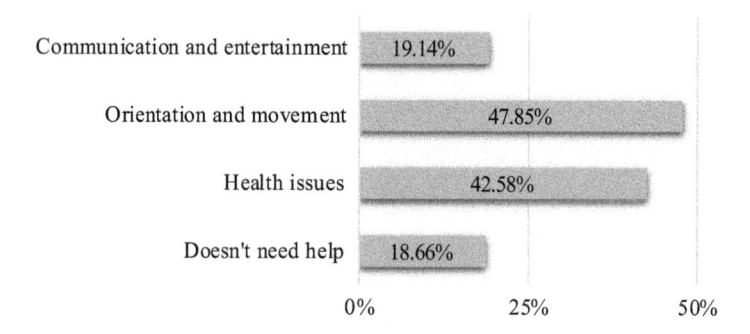

Fig. 32.1 Type of needed help

- Communication and entertainment: 19.14%
- Orientation and movement: 47.85%
- Health issues: 42.58%
- Doesn't need help: 18.66%

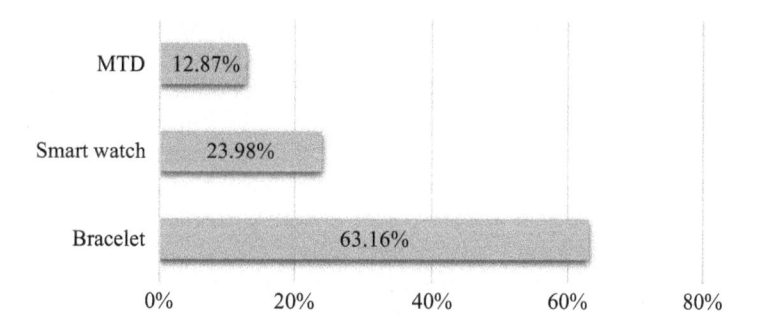

Fig. 32.2 Form of the service

- MTD: 12.87%
- Smart watch: 23.98%
- Bracelet: 63.16%

Method of survey and interview research was carried out on the possibilities of the use of modern ICT by the elderly to obtain information related to their use of services based on these technologies [13]. From the total number of 209 respondents, 53.58% were older than 80 years, while 5.26% were younger than 60 years.

The respondents were asked about the type of their disability, and 49.28% have visual impairment, which is the highest one, while 44.98% have locomotor impairment. Figure 32.1 shows the reasons for the use of services based on new technologies. It was surprisingly that 73.68% respondents use MTD, where 14.19% use smartphone, while the rest of them use dumbphone. The respondents were interested (67.94%) in the service that could help them in their everyday activities and that is customized to their needs. Figure 32.2 shows in what form would they use the service. It was shown that respondents have problems with orientation and movement, so they were asked about using services that would identify them and inform them while using public transport, and 53.11% were interested in that type of service.

The second survey was conducted in collaboration with UP2DATE organization in the premises of Laboratory of Development and Research of Information and Communication Assistive Technology (ICATLab) at Faculty of Transport and Traffic Sciences in Zagreb. Research on user needs while moving traffic network was conducted by survey method. The aim of this research was to collect all relevant information about the difficulties that users face when moving through transport network and the ways and possibilities of using modern ICT in their daily

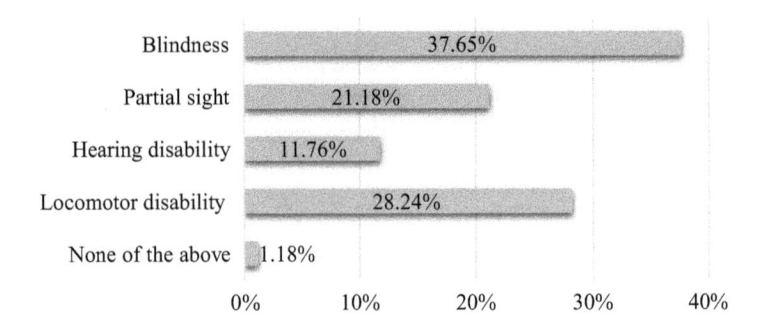

```
Blindness          37.65%
Partial sight      21.18%
Hearing disability 11.76%
Locomotor disability 28.24%
None of the above   1.18%
         0%   10%   20%   30%   40%
```

Fig. 32.3 User disabilities

activities. From total number of 112 respondents, 28.57% were in age between 21 and 24 years, while 6.25% were older than 60 years. Most of them were students, 51.79%, and employees, 31.25% of them. When they were asked if they have some kind of disability, 72.32% said yes. The highest disability was vision impairment where 37.65% of respondents said that they are blind, while locomotor impairment was the second one highest, 28.24%. Other disabilities are shown in Fig. 32.3.

They were asked if they are satisfied with current ways of informing them about public transport, indoors, intersections, streets and squares. In all these cases, most users are not satisfied with the current way of informing. Respondents were most dissatisfied with the way of informing at intersections, 72.73% of them. Other cases are not so far behind for such many dissatisfied users, and 68.18% of them stated that they were not satisfied with information on the squares.

A large number of respondents, 93.75%, use MTD of which 73.33% use Android Operating System (OS). The most common use of MTD was for calls and messages (33.67%), but only 1.6% used them to locate objects. MTD are used for reminders (16.00%) and navigation (15.0%). According to the result obtained from the survey, the two most appropriate ways to inform users are sound information using MTD (30.00%) and visual information (27.39%). Of all respondents, 95.54% said they would use service for informing them about their environment and whichcan to adapt to their needs.

Due to increase the safety of users and allow them aid in dangerous situations, respondents were asked if they would like to have a service that would transmit an SOS message. Most of them replied that they would like to use this type of service (73.64%) and would like the message to be sent to the responsible or legal person (31.21%), emergency services (25.48%), or spouse (24.20%).

In the survey, there were 21.62% parents that were asked if they would use locating services to know where their children are, SOS paging if their child is in some kind of threat and detection of removal of smart wristband. The entire TGU replied that they would use SOS paging service, but 90.91% also would use other two services. Respondents said they would be willing to share information about their experience in health institutes (83.36%), points of interest (88.07%), and appropriate and verified movement routes (90.74%).

To determine functionalities of service to inform users in traffic, they were asked about the importance of certain information, and the results are shown in Table 32.1. Importance of certain information that concerned information about objects, current location (CL), locating items like wallets and keys (LI), directing users with voice guidance and automatic return route (DUWV), health status (HS), open area like markets and squares (OA), barriers, automatic SOS call (SOS), and fall detection (FD) was evaluated according to their importance on a scale from 1 (very irrelevant) to 5 (very relevant).

With given results, it is possible to determine the level of importance of service functionalities. In automatic transmission of SOS call, 33.94% of users consider to be very important, and 36.19% also find information about barriers very important. Navigation route with voice guidance 43.12% of respondents find important and 47.71% of them find information about current location also important. Table 32.2 shows the importance of traffic information for users. On a scale from 1 to 5, respondents evaluated information on traffic problems (TP), station information (SI), information about intersections (INT), information about public transport (PT), traffic environment (TE) information related to street names (SN) and in-formation about traffic lights (TL).

The results gave the possibility to define information which are needed to users while moving through traffic network. As additional functionalities, the respondents (81.82%) stated that they would use contactless payment, while the remote locking and unlocking of MTD would use 75.23% of respondents. Other additional services

Table 32.1 Importance of information for users

	1	2	3	4	5
Object (%)	7.34	5.50	13.76	37.61	35.78
CL (%)	2.75	3.67	11.01	47.71	34.86
LI (%)	6.48	18.52	20.37	36.11	18.52
DUWV (%)	5.50	11.01	16.51	43.12	23.85
HS (%)	9.09	12.73	29.09	30.00	19.09
OA (%)	5.50	5.50	29.36	31.19	28.44
Barrier (%)	2.86	5.71	15.24	40.00	36.19
SOS (%)	10.09	10.09	18.35	27.52	33.94
FD (%)	9.17	6.42	24.77	32.11	27.52

Table 32.2 Importance of traffic information

	1	2	3	4	5
TP (%)	3.70	5.56	12.96	43.52	34.26
SI (%)	1.87	4.67	11.21	41.12	41.12
PT (%)	0.91	3.64	6.36	44.55	44.55
TE (%)	2.68	0.89	10.71	49.11	36.61
INT (%)	3.67	3.67	17.43	44.95	30.28
SN (%)	3.67	0.92	10.09	56.88	28.44
TL (%)	3.64	1.82	1.82	44.55	48.18

that the respondents want to use are medical services, alarm reminder, flashlight, time weather information, clock, and customized car service options.

Relevant parameters can be divided into system components and services. The parameters for the design of the proposed system are hardware and network components. Hardware component consists of MTD and smart wristband. Suitable ways for informing users are audio and visual informing. Network component consists of Bluetooth system, Cloud Computing for the Blind (CCfB) and Internet of Things (IoT). Bluetooth technology is used as a linking element in the system while CCfB is used for system requirements and the needs of users for real-time information. IoT is used because of the need to connect the entire system to a global network and access real-time information. The conducted research has provided indicators for designing new IC service. This service is shown through value chain that shows all system members.

Value Chain of Customer Information Service

According to ITU-T, the IoT ecosystem consists of five components: the manufacturer of equipment and devices, the network service provider, the platform provider, the application provider, and the user interface [14]. User represents the demand, while other four components represent the offer and are shown in Fig. 32.4.

The manufacturers of equipment and devices are all parties responsible for supplying the network service provider and the application provider with the required equipment. The network service provider performs the task of accessing and integrating resources for other stakeholders. In the domain of network service provider, there is also a communication network for data transfer. Data transfer of SAforA service is between smart wristband and MTD, MTD and CCfB, and CCfB and

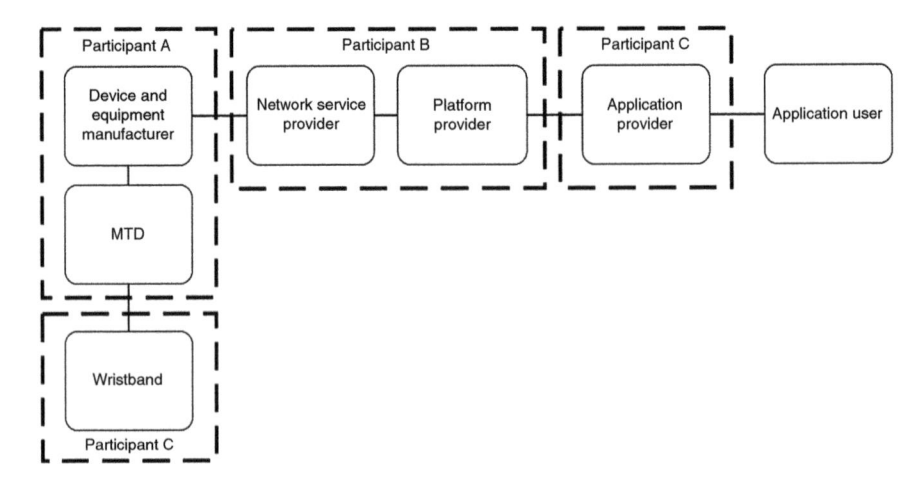

Fig. 32.4 Value Chain

stakeholders. The platform provider provides integration and open interfaces, and its purpose is to link the network service provider and the application provider. The application provider ensures IoT application to users, but it uses resources which are provided by device manufacturer, network service providers, and platform providers.

Service Accessibility

Systemmust be designed to be accessible to all users. The concept of universal design is an approach of designing products, services, and the environment so that they can be usable to all people, without the need for adjustments or special design. Universal design can increase the utilization of the environment or products without significant increase in its cost while reducing the need for design changes later when capacity circumstances change [15].

The proposed system for informing in the form of smart wristband has an ergonomically friendly design. Equitable use is achieved by designing services according to customer requirements. Users will get required information in the fewest possible steps which achieves flexibility of service and low physical effort.

Content and information will be clear and easy to see or hear. Using and understanding of the entire system for all users will be easy [16]. The service, in addition to the application of universal design, should also have designed architecture services to Ambient Assisted Living (AAL) concept [17].

Technology Analysis

With ICT it is possible to increase the level of quality of life of users if they are implemented in ways that are simple to use for all users. When designing a system, it is necessary to check the compatibility of technologies and their capabilities in order to provide real-time and accurate information.

Bluetooth and Near-Field Communication Technology

Near-field communication (NFC) and Bluetooth technologies are used in proposed model of assistive technology. They link users, enabling them to communicate, other transport entities, and the overall traffic environment into a unified whole by applying the principles of IoT. Bluetooth technology (BLE v.4.0.) is an open standard that can be implemented in IoT and AAL concepts [18]. NFC technology is designed for use on devices that are located close to each other [19]. The big advantage over other technologies is that it doesn't require pairing devices which facilitates the use of technology.

Arduino and Raspberry Pi Platforms

For the development of a prototype smart wristband, it is possible to use Arduino platform. Arduino platform is an open-source family of microcontrollers that can be paired with various sensors. Microcontroller has modest possibilities when compared to the performances of today's computers and even MTD. Arduino UNO is small and has low power consumptions. More advanced modules that can wirelessly connect to Arduino platform are called shields. Except them, there are a lot of programs created by Arduino community that are shared on the Internet.

With a relatively low-cost price and additional modules and shields, Arduino platform is a great choice for testing and developing new concepts and ideas [16].

Raspberry Pi is also a platform that can be used as an IoT device because of its small size and possibilities of connection to the Internet and to other devices. In relation to Arduino, Raspberry Pi models have great processing power. It has an integrate Ethernet port that allows easier to connect to the Internet. Disadvantages of Raspberry Pi are high-energy consumption and inability to process analog signals and interface that requires access via keyboard, mouse, and computer monitor.

For those reason, Raspberry Pi is often used as a central processor unit that collects data from other devices and sensors and then forwards them to the CC for storage and processing [20].

Solution Proposal

Based on research done in the field of application of CC technology, it is proposed to extend the functionalities of CCfB concept through conceptual architecture of the system [21, 22]. The functionalities are focused on efficient delivery of information to users by using sensor technology.

As it's shown in Fig. 32.5, system architecture consists of eight elements: sensors, smart bracelet, MTD, terminal device (TD), mobile application, web application, user database and database.

Sensors located in the architecture collect data from environment of user. Data is collected in the proposed informing system, in the form of smart wristband, and is stored in the database. All data from sensor located on the smart wristband are stored in the database. Database contains data that require further processing and adapting into information comprehensible to the end user. User database, which contains information of accounts and their settings, is a part of CCfB architecture of service provider. It provides necessary data for verification and adaptation of content to the user priority. Adaptation of content is performed based on user characteristics and functionalities that each user chooses.

Users can access their data through mobile applications on MTD or web applications using a web browser. These applications are integral part of CCfB architecture (service provider part), and through these applications, users demand information collected from smart wristband. Data is processed and adapted to the needs of users

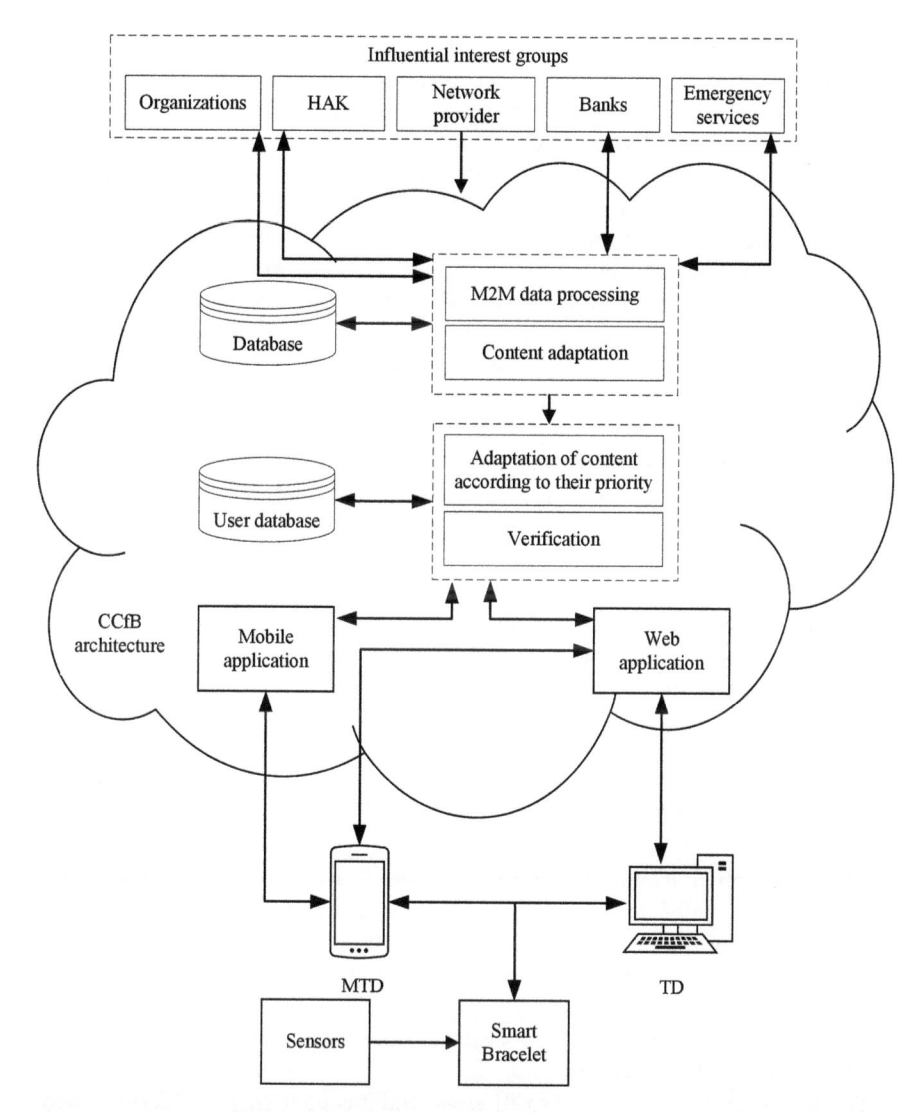

Fig. 32.5 System architecture and influential groups

and is presented as a relevant information in the application. Users can configure smart wristband using MTD and Bluetooth connection or computer and IP/GPRC communication channel. All data from the sensors are sent to central processor unit for processing that is in smart wristband, where they transform into information that will be sent over the network to the database. Individual sensors located on smart wristband can be turned on or off when setting smart wristband. In that case, users will get only those information from sensors that are turned on. As noted above, users can configure smart wristband in two ways: by using MTD or computer. If the smart wristband is configured with MTD, Bluetooth connection is used. After identification and authentication, smart wristband connects with MTD and receives configuration information that apply automatically. Configuration via computer goes through web application. User logs into his account and chooses settings that he wants smart wristband to do. User accounts are stored in CCfB database of provider, who during the logging in identifies and authorizes user. The user then changes the settings in smart wristband, and after saving the changes are send via IP/GPRS connectivity to smart wristband. In the case where settings were changed for more smart wristbands, settings will be sent to all smart wristbands. It is possible to access web application by MTD using web browser. Application solution is able to guide the user through voice, visual, and tactile information which is not possible using the web browser. For example, if user wants to get direction to some place, he needs to set a route on his MTD. After the data is processed, it's sent to smart wristband that presents them as a visual, audio, or tactile information.

Important interest groups (IIG) communicate with CCfB service provider using IP protocol and available Internet infrastructure. This kind of communication is two-way, and through it system can inform IIG about location and user condition, and also, they can request information about a specific user if they have permission to do so. Users can allow specific IIG to have information about them while configuring smart wristband. This method allows IIG to have access only to that information that users allow them. Service provider is in charged for this functionality because he processes information in CCfB architecture.

Determining the Functionalities of SAforA Service

Delivery of proposed IC services based on modern ICT includes users whose needs effect on the design of the service. Stakeholders help users of the service during certain situations. Information service consists of basic and additional functionality that provide all relevant information. Use case diagram is shown in Fig. 32.6. Additional functionalities are watch, flashlight, weather forecast/temperature, alarm reminder, fitness services, contactless payment, locking/unlocking MTD, and automatic adjustment of car functions.

The functionalities are offered to all users, and it is up to them to choose which functionality they want to use. There are recommendations for each user on what kind of functionalities they should use. For example, a person with visual impairment is recommended to use fall detection functionality, Bluetooth locating items, SOS calls, route creating with an automatic return route, and pulse detection.

Fall detection is a functionality that is recommended for the elderly, visually impaired persons, and persons with mobility impairments who live alone and can be found in situations where they need help. After falling, user can be unconscious or can't get up and retrieve a MTD to call for help. For this purpose, a smart wristband must be able to detect the fall and automatically send a message to responsible persons or relevant departments. There are three cases how system that detects the fall decides how to handle received information: if the user gets up, the system will

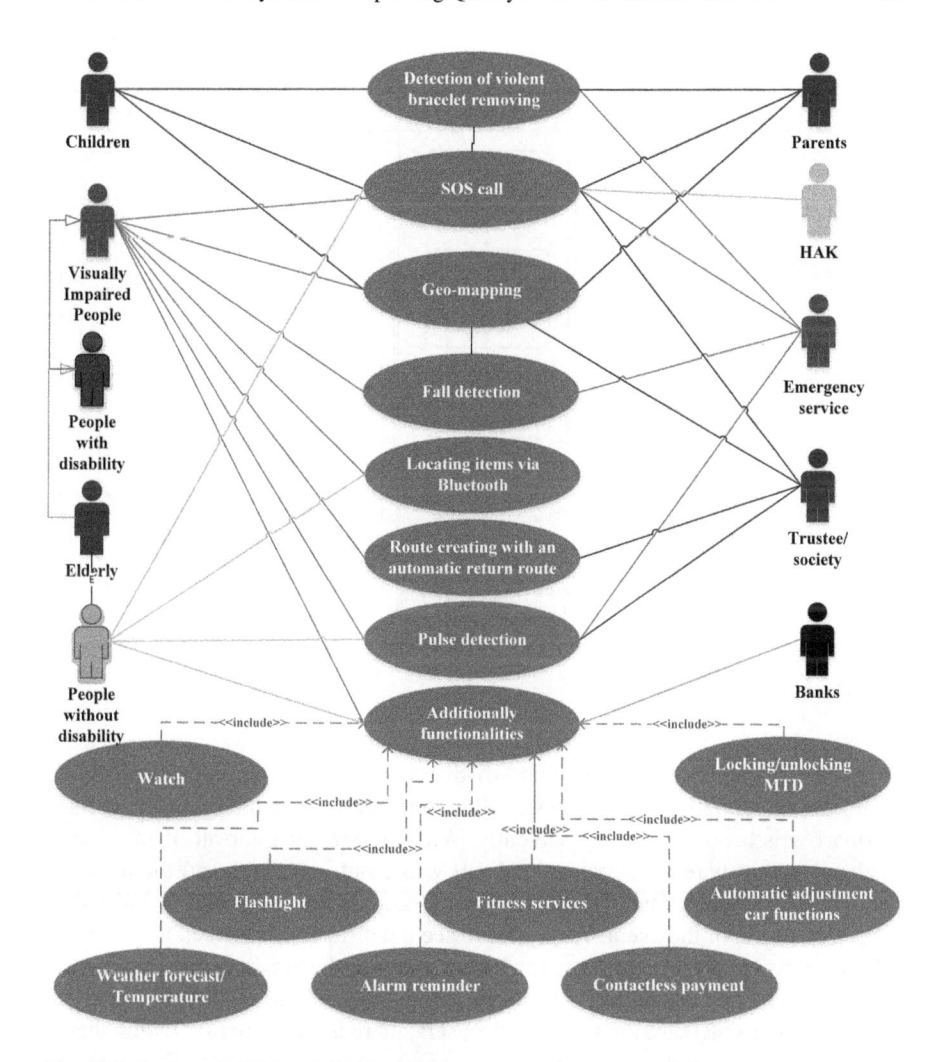

Fig. 32.6 Basic and additional functionalities

receive information through smart wristband and won't take any emergency actions; if the user after the fall remains lying on the floor and presses the SOS button, the process of sending an emergency message will start; and if the user after the fall remains unconscious, there is a time frame of few seconds in which the system should detect if the user got up. Figure 32.7 shows UML diagram for fall detection.

By using accelerometer and gyroscope, smart wristband can determine whether the user stood up, and, if not, it will send the SOS message to the predetermined destination. When receiving a message via GPRS network using a GSP/GPRS (Global System for Mobile Communications) module on smart wristband, the responsible person and/or emergency institutions decide on how to proceed with the injured person. User decides which persons will be notified in case of the emergency.

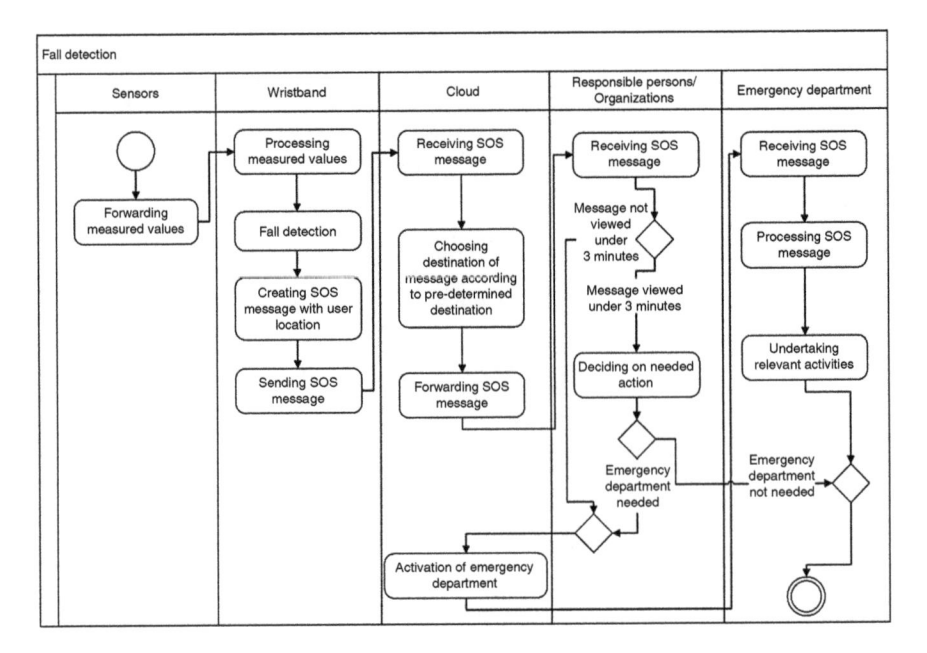

Fig. 32.7 Fall detection

In case of violent removal of smart wristband, user is not able to press the SOS button. In such situation, it is necessary to automatically send the information that the smart wristband is removed violently. With the violent removal of the smart wristband, a thin wire that circles the smart wristband is broken, and the distress message is sent to responsible persons. The message is automatic and includes the location and time of the event. To be able to see the difference between normal and violent removal of the device, there is a button that must be pressed when the device wants to be removed normally. By pressing and holding the button, distress message won't be sent. To increase the safety of the users, there is a time period within there must be no removal of the device even if you press the security button. This reduces the risk of unauthorized removal of the smart wristband. In Fig. 32.8, UML diagram shows the functionality of detecting the violent removal of the wristband.

GPS module can be used to define the range of movement and locate the user. Studies conducted in ICATLab (the Laboratory of Development and Research of Information and Communication Assistive Technology) tested the reliability of GPS application solutions on MTD, and the accuracy interval ranged from 1 to 40 [m]. Different kinds of MTD were used: Sony Xperia Z, Z1, and Z3, Samsung Galaxy S4 and S6, and Nokia Lumia 925. For the children and elderly that have caregivers, it is possible to set the geographic area within they can move. In case of leaving the area within a set period of time, the responsible person or association will receive alerts and take further actions. User can also choose via the mobile application a destination after which he will get the information about the fastest route to destination. The route will be selected based on the characteristic of impairment that user has.

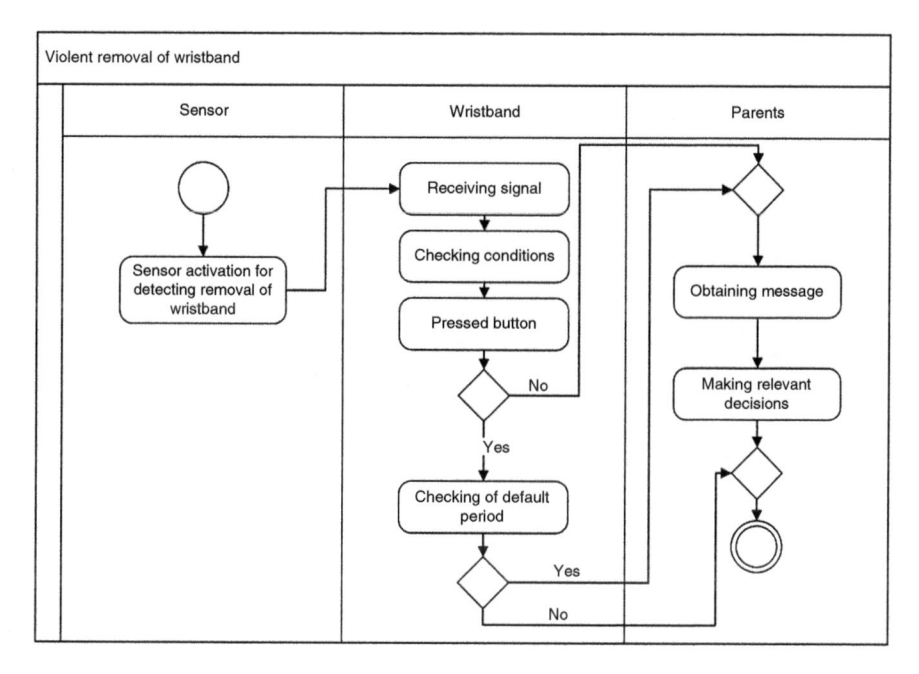

32 Smart Wristband System for Improving Quality of Life for Users in Traffic…

Fig. 32.8 Violent removal of wristband

All persons, regardless of whether they have disability or not, want to receive the necessary assistance as soon as possible. In situations where the user has no time to make a call or is unable to do it, a simple action, such as pressing a button on a smart wristband is enough to alert other persons. For this reason, sending the SOS message in case of a danger is also one of the main functionalities of the proposed system. The message is forwarded to a specific destination, such as to a person of trust or emergency service of the country where the person is located. Figure 32.9 shows the functionality of sending the SOS message.

The message content includes GPS coordinates, the name of the user in danger, and the information that the user is in danger and that the message was automatically generated and sent. The message is sent in SMS format using built-in GSM/GPRS module with the SIM card. A smart wristband with a GSM/GPRS module will have a built-in GPS module that will determine the user's coordinates. With this, it is possible to find users in danger more quickly and easily. For easier usage of the smart wristband, it is suggested to use a virtual SIM card that takes up less physical space on smart wristband.

For example, if the user has visual impairment, presented route will be the one that has Beacon devices set on intersections. This allows better guiding and locating the user because user is guided by tactile information which he gets through vibrating motors on the smart wristband and/or through audio signal that produces the smart wristband and is activated only in the presence of the Beacon device that is set on the intersection. In addition to external guidance, it is possible to use the functionality of the inner guidance that also uses Bluetooth Beacon technology.

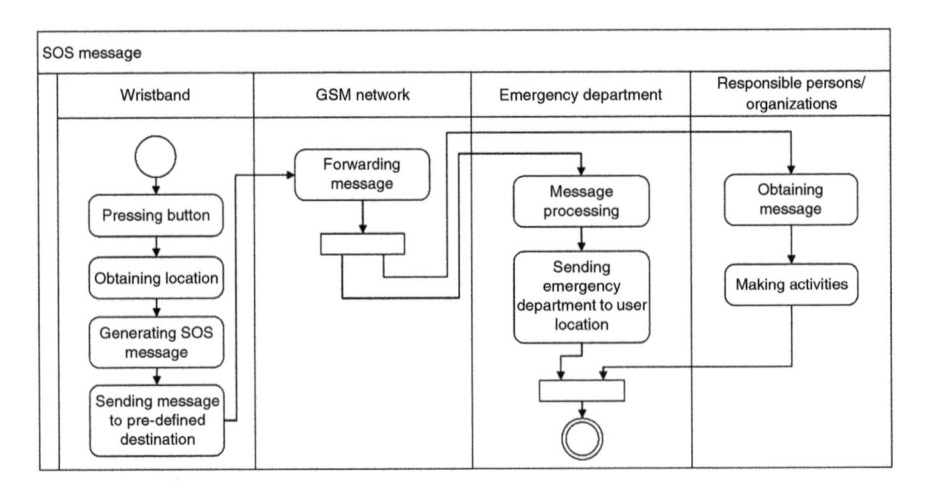

Fig. 32.9 SOS message

Bluetooth technology is used to locate items. By gathering the signal strength, the distance to another device can be determined. Using Beacon mobile device, the Bluetooth signal is transmitted at certain time intervals, and it contains basic information such as numeric identifiers. The searching process starts when the users in mobile application choose the item that they want to find. Before the item can be found, the usermust mark it and register it in the application. To find specific item, the users can receive sound or tactile information, depending on how they want to receive the information. This mode of communication is suitable for people with hearing and visual impairment. Visually impaired users will use sound signal, while users with hearing impairment will use tactile signal. The moment when the object is found, pressing the key on the smart wristband stops the search for the item, and the vibration and audio information is turned off. This is shown in Fig. 32.10.

Pulse monitoring can be interesting to a wider range of users regardless of whether they are impaired persons, the elderly, or people with health problems. Users can track their pulse for medical purposes or simply to see how sports affects the heart. The functionality is implemented by using modules for heart measuring rate. The sensor is placed on the smart wristband that processes the received data and afterward sends them to a computer cloud where the information is being stored.

Additional functionalities of the system user can define in its user profile, and they will be shown on their screen of the smart wristband. Smart wristband should have functionality to provide information about time because it reduces the need for using MTD. Because of it, smart wristband will have a built-in clock that displays time on the smart wristband screen. Furthermore, a LED light will be installed that will allow users to navigate in dark areas or under low or no light conditions.

Weather and temperature provides information about the temperature of the place where the user is located, and the weather forecast of the area. User with health issues can have the need to receive information about temperature changes and humidity. They can receive the information in two ways.

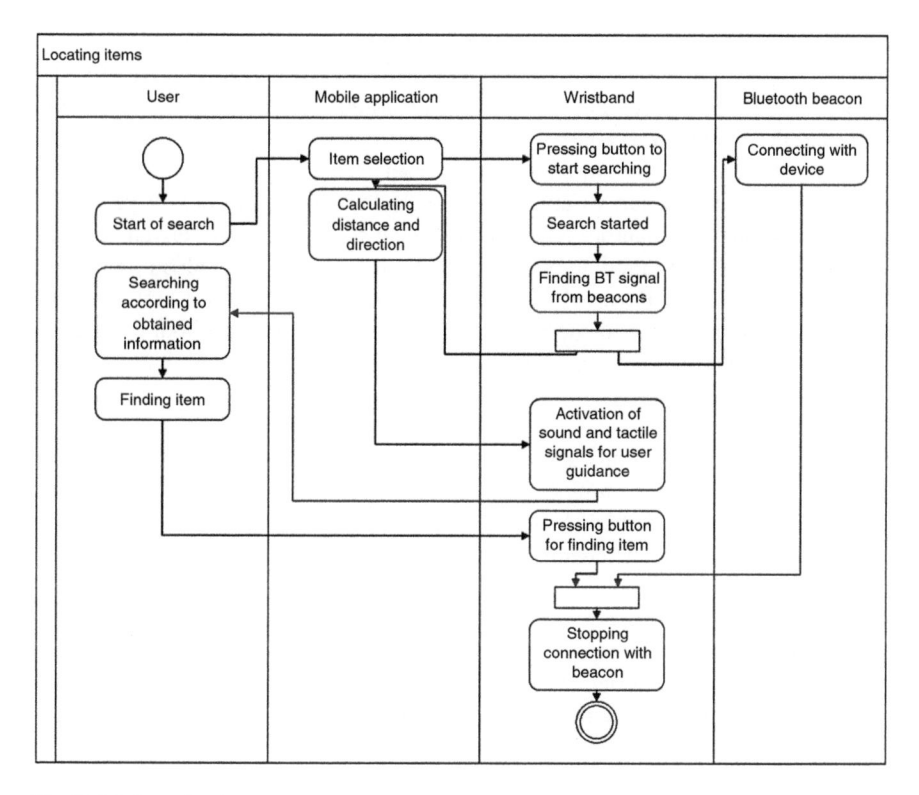

Fig. 32.10 Locating items

The first option to provide information is on the wristband screen where user can see current temperature, wind speed, precipitation probability, and air humidity levels. Second option to provide information is achieved through sound module because visually impaired persons cannot get information from the display. All information displayed on the screen are transmitted to audio channel as well. Since weather information is collected through MTD, this functionality can only be used in situations where there is an active Bluetooth connection with smart wristband. Information is collected from services based on CCfB solution.

Smart wristband supports MTD locking through NFC module. To unlock MTD user needs to touch the MTD with smart wristband. MTD detects NFC module in smart wristband and unlocks itself. To lock the MTD, all the user needs to do is touch it again with wristband. One of the technologies used by the wristband to communicate with modern city automobile is Bluetooth technology. Bluetooth signal is sent in short time period after the button is pressed which gives the user a desired time frame in which he can get close to the automobile and by doing that activates all of its functionalities. When Bluetooth signal is detected, which is unique for each user, automobile starts adjusting itself based on current user preferences. This includes adjusting the seat height, adjusting rear view mirrors, and adjusting the steering wheel position. User preferences are set upon first connection with the automobile. It is possible to define minimum distance required for user to connect smart wristband with the automobile. This service depends on automobile maker which installs Bluetooth technology in automobiles and controls additional functionalities which are available.

Contactless payment in stores makes it easier to use payment services for all users. NFC module in smart wristband will be used to achieve contactless payment. To protect private information, NFC module needs to be manually activated. After short time period, NFC module is automatically deactivated. It is also possible to deactivate it manually. Payment transaction is executed when smart wristband is close to POS (point of sale) device. Contactless payment is done without PIN entry, and it does not require credit card. This makes it easier for people with disabilities to use payment transactions.

Alarm reminder service is defined as an additional functionality of a system for informing users. In synchronization with a mobile application, the user can set the reminder. Since an alarm reminder requires synchronization with MTD, Bluetooth connection is used to link them. When MTD and smart wristband are no longer connected, the alarm reminder remains active.

Fitness service also belongs to additional functionalities. User defines his/her height, weight, age, and gender so that the service can be tailored to each user. While sporting, it is not necessary to carry MTD for data collection, but all user and physical activity data are stored on the outer memory of the smart wristband. Since there is a heart rate detector in the smart wristband, it is possible to monitor the pulse when performing various sports exercises. All information is available on a mobile application where there are recommendations about how much should the user be moving.

Proposal for the Development of the Smart Wristband

For the development of smart wristband, it is proposed to use Arduino platform. To determine whether it is possible to collect the necessary data from the sensors and by their reading provides defined functionalities of service, experimental methods were conducted in ICATLab. The smart wristband needs to be designed in a way so that it can be used by all user groups. The proposal of the smart wristband is given in Fig. 32.11. The modules are located on the smart wristband in a way to reduce the danger to the users and allow them to use all the functionalities according to their needs. Smart wristband can be used with a minimum effort and situations as accidentally key pressing are avoided.

To properly calibrate the acceleration and vibration sensors, a test has been conducted in a laboratory environment. Arduino UNO was connected with a module for measuring acceleration and vibration detection module by using USB cable that is connected to a laptop. By this, the whole platform becomes mobile, and it is possible to conduct tests of fall detecting. The procedure was repeated until they got the approximate sensor value that reduces the number of false-positive results.

Fig. 32.11 Conceptual overview of the smart wristband

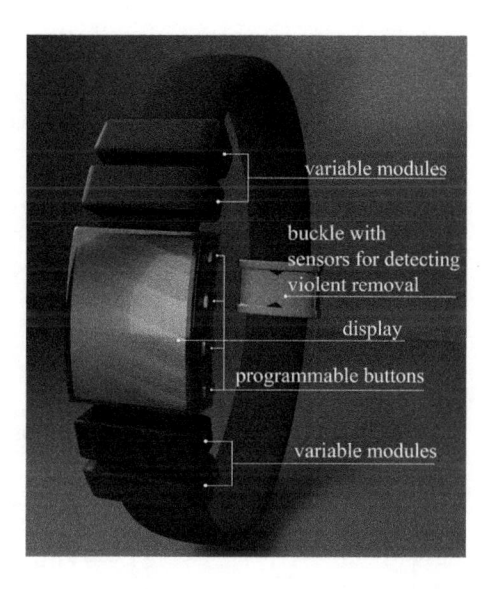

Raspberry Pi platform was used to simulate CCfB because of his ability to simulate a large computer network. Testing was conducted on the Raspberry Pi Model B (RPi3), a miniature computer with low energy consumption that uses operating system based on the Linux kernel [23]. It can use Ncat program that can send and receive TCP and UDP packages on open ports. The command syntax is simple and can redirect packages to other ports even when the received packed is in the different protocol than it was during the sending.

The system which sends out an SOS message consists of assistive device, communication channel, MTD, servers, and sensors. Assistive device is based on the Arduino platform; communication channel is based on IP protocol, and in laboratory environment it represents a local network while in real environment represents GSM/GPRS; MTD receives a text message, and CCfB represents a server. RPi3 is used to represent mentioned server, runs a Ncat program and waits on receiving packages from the network. The sensors are located on the assistive device, and all readings are processed on the Arduino platform. The procedure is shown on Fig. 32.12. The first scenario is that the person presses the SOS button on the assistive device. By pressing the SOS button, data packet is sent via GSM/GPRS mobile network to a server that is located in the CCfB. According to the user needs, a close person can be defined as a person that will receive an SOS message. That part of communication is simulated by SMS to a given number of MTD. If the message was sent, the IC system will receive an information. In case that the information wasn't set, the system will forward the message to emergency services. Message forwarding is tested by using the e-mail service and can be send automatically by using RPi3 [23].

In the second scenario, the example of fall detection is tested to see if the SOS message will be send. Sensors on Arduino UNO platform detect vibrations and acceleration toward the Earth's surface. Arduino UNO, based on the program that is installed on it, recognizes whether the person is going down the stairs or moving in the elevator or if the person has fallen. When collecting data from the sensor detects

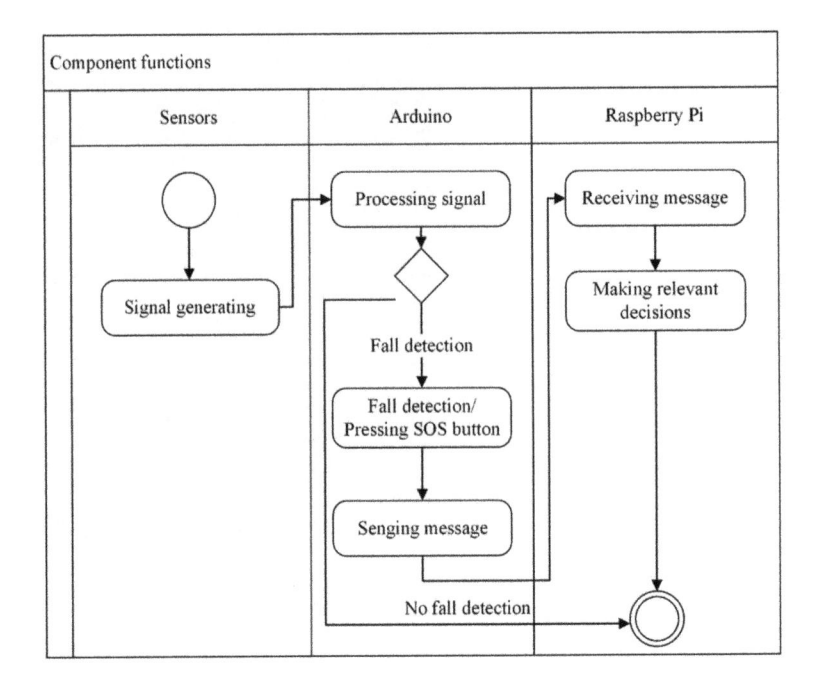

Fig. 32.12 Functions of the prototype components

a fall, SOS packet is sent using Wi-Fi module and local IP network to RPi3. After receiving the packet, system is being informed that the user has fallen and that within a defined set of time did not get up. The simulation in the laboratory stops at this step.

Conclusion

Different characteristics, interests, skills, and the level of disability make users different from each other. Universal design, as an approach in designing products, service, and environment, allows the integration of user characteristics in one system without the need for adaption. Currently implemented solutions on the market don't provide the possibility of equal use of experience to all consumers.

Data about daily needs of users was collected by using the methods of surveying and interviewing. According to the results of questionnaires, relevant parameters were defined. These parameters were used in the design of the system intended for informing users in traffic environment. Research results also gave the information about what type of functionalities should the system provide, and they were divided into basic and additional functionalities. Based on the study and the experimental method, it has been established that the optimal solution to customer needs comes in a form of smart wristband in combination with an application to MTD.

Conceptual architecture of the system for informing users – SAforA is proposed based on defined functionalities. Given the currently available solutions on the market and existing modern technology, the proposed architecture is IC system based on the CCfB concept. The above architecture is currently one of the better choices for connecting accurate and real-time information in one common system that is available to the user. System with multiple stakeholders involved in the formation of the information available to the users is based on the mentioned architecture.

Real-time communication between user and the system was simulated using experimental method with the use of laboratory equipment in laboratory and real-world environment. After the conducted simulation, it was proved that it is possible to make the data transfer from a terminal device to the system that is based on the CCfB concept. Safety of user information and the 24/7 delivery is provided. The hypothesis that using ICT can increase the level of quality of life and improve the user's movement through the transport network was proved through conducted researches and simulation verification. By providing accurate and real-time information, it is possible to increase autonomy and independence of the user while moving through a part of the transport network.

In future research, it is proposed to explore the possibilities of new technologies to design new modules for expanding and adjusting the service. Implementation of new modules that add new functionalities within the service is simple because of the modularity of the proposed smart wristband.

Acknowledgments This research has been carried out as part of the project "System of automatic identification and informing of mobile entities in the traffic environment," Faculty of Transport and Traffic Sciences, University of Zagreb, 2016. The research was awarded with Rector's award at University of Zagreb, Zagreb in 2016.

References

1. BusinessWire. (2014). Parents want technology to track their children. http://www.businesswire.com/news/home/20141111005176/en/Parents-Technology-Track-Children
2. Park, S., Harden, A. J., Nam, J., Saiki, D., Hall, S. S., & Kandiah, J. (2012). Attitudes and acceptability of smart wear technology: Qualitative analysis from the perspective of caregivers. *International Journal of Human Ecology, 13*(2), 87–100.
3. Angulo, I., Onieva, E., Perallos, A., Salaberria, I., Bahillo, A., Azpilicueta, L., Falcone, F., Astrain, J. J., & Villadangos, J. (2015). Low cost real time location system based in radio frequency identification for the provision of social and safety services. *Wireless Personal Communications, 84*(4), 2797–2814. https://doi.org/10.1007/s11277-015-2767-6.
4. Deshmukh, A., Mishra, A. K., Patil, V., & Saraf, K. (2014). Wireless personal safety bracelet. *International Journal of Computer Appliations, 107*(23), 11–13.
5. Huang, J. C. S., Lin, Y. T., Yu, J. K. L., Liu, K., & Kuo, Y. H. (2015). *A wearable NFC wristband to locate dementia patients through a participatory sensing system*. Dallas: International Conference on Healthcare Informatics. 21–23 October 2015.
6. Zhang, Y., & Rau, P. L. P. (2015). Playing with multiple wearable devices: Exploring the influence of display, motion and gender. *Computers in Human Beahviour, 50*, 148–158.
7. Noh, C.-B., & Na, W. (2016). Portable health monitoring system using wearable devices. *Indian Journal of Science and Technology, 9*(36), 1–5.
8. Postolache, O., Silva Girão, P., Santiago, F. (2011.) Enabling telecare assessment with pervasive sensing and Android OS smartphone. *Proceedings of the 2011 IEEE International*

Workshop on Medical Measurement and Applications Proceedings (MeMeA), Bari, Italy, 30–31 May 2011.
9. Ahanathapillai, V., Amor, J., Goodwin, Z., & James, C. J. (2015). Preliminary study on activity monitoring using an android smart-watch. *Healthcare Technology Letters, 2*(1), 34–39.
10. Freeman, E., and Brewster, S. (2016). Using sound to help visually impaired children play independently. *Proceedings of the 2016 CHI Conference Extended Abstracts on Human Factors in Computing Systems – CHI EA '16, San Jose, California*, USA, 7–12 May 2016.
11. Chippendale, T., D'Alto, et al. (2015). Personal shopping assistance and navigator system for visually impaired people. In *Lecture notes in computer science* (pp. 357–390). Cham: Springer.
12. Law on Croatian Register of Persons with Disabilities. (2001). *NN 64/01*. Zagreb: Narodne novine.
13. Black, R. D., Weinberg, L. A., & Brodwin, M. G. (2015). Universal design for learning and instruction: Perspectives of students with disabilities in higher education. *Exceptionality Education International, 25*(2), 1–26.
14. Mazhelis, O., Luoma, E., & Warma, H. (2012). Defining and internet-of-things ecosystem. *Lecture Notes in Computer Science, 7469*, 1–14.
15. Periša, M., Jovović, I., & Peraković, D. (2014a). *Recommendations for the development of information and communication services for increasing mobility of visually impaired persons. Proceedings of the Conference Universal Learning Design, Masaryk University*. Paris, France, 9–11 July 2014.
16. Mann, W. C. (2005). *Smart technology for aging. disability and independence*. Hoboken: Wiley.
17. Garcia, N. M., & Rodrigues, J. J. (2015). *Ambient assisted living*. Boca Raton: CRC Press.
18. Gomez, C., Oller, J., & Paradells, J. (2012). Overview and evaluation of bluetooth low energy: An emerging low-power wireless technology. *Sensors, 12*(9), 11734–11753.
19. Al-Ofeishat, H., & Al Rababah, M. (2012). Near field communication. *International Journal of Computer Science and Network Security, 12*(2), 93–99.
20. Periša, M., Sente, R. E., & Brletić, L. (2016). *Proposal of information communication technology architecture for people with disability*. The 4th Online Scientific Conference – ScieConf., Žilina, Slovakia, 6–10 June 2016.
21. Periša, M., Peraković, D., & Šarić, S. (2014b). Conceptual model of providing traffic navigation services to visually impaired persons. *Promet - Traffic & Transportation, 26*(3), 209–218. https://doi.org/10.7307/ptt.v26i3.1492.
22. Peraković, D., Periša, M., Sente, R. E., Bijelica, N., Brletić, L., Bucak, B., Ignjatić, A., Mišić, V., Papac, A., Vuletić, M., & Zorić, P. (2016). *Information and Communication System for informing Users in Traffic Environment – SAforA*. EAI International Conference on Management of Manufacturing Systems, Bratislava, Slovakia, 22–24 November 2016.
23. Jain, S., Vaibhav, A., & Goyal, L. (2014). *Raspberry Pi based interactive home automation system through E-mail*. Manav: 2014 International Conference on Optimization, Reliability, and Information Technology – ICROIT, Rachna International University. 6–8 February 2014.

Chapter 33
Smart Transportation Applications and Vehicle Data Processing System for Smart City Buses

Serkan Mezarcıöz, Enis Aytar, Murat Demizdüzen, Mert Özkaynak, and Kadir Aydın

Abstract With the application of a data processing and reporting system, the vehicle had capability of processing, sending, and reporting data in the vehicle as requested by the driver, owner, and/or fleet management center via Internet. This data can be saved and monitored online by the fleet management center via Internet. By processing this data, performance of drivers of the fleet can also be evaluated.

With the application of this system, vehicle can be diagnosed without any need of special diagnose tool, and all error codes for engine, transmission, brake, or any other system in the bus can be monitored on fully programmable, touchable driver information display of the bus. Another advantage of employing this system to the city bus is to see the condition of the vehicle, situation of the systems, error codes, and special parameters like pressure in the brake lines, throttle pedal position, and park brake status just before any accident. And this information can help to find out the cause of the accident.

Passenger counting systems can help to fleet managers to plan route of city buses effectively by supplying the information of number of passenger inside a bus. By combining passenger counting system with a GPS system, location of the bus and number of passenger in the vehicle can be monitored online, and an effective plan of city transportation can be done. Application of IP (Internet Protocol) cameras will provide online information about the vehicle inside and surrounding to the fleet management center. If the vehicle deviates from its standard route or panic buttons located near the driver and inside the passenger cabin are pressed, an emergency signal can be sent to the fleet management center, and vehicle can be monitored simultaneously via IP cameras.

S. Mezarcıöz (✉) · E. Aytar · M. Demizdüzen · M. Özkaynak
Temsa Ulaşım Araçları San. Ve Tic. A.Ş, Sarıhamzalı mahallesi, Turhan Cemal Beriker Bulvarı, Seyhan, Adana, Turkey
e-mail: serkan.mezarcioz@temsa.com; enis.aytar@temsa.com; murat.demirduzen@temsa.com; mert.ozkaynak@temsa.com

K. Aydın
Çukurova University, Automotive Engineering Department, Balcalı, Adana, Turkey
e-mail: kdraydin@cu.edu.tr

© Springer International Publishing AG, part of Springer Nature 2019
D. Cagáňová et al. (eds.), *Smart Technology Trends in Industrial and Business Management*, EAI/Springer Innovations in Communication and Computing,
https://doi.org/10.1007/978-3-319-76998-1_33

By checking the position of the vehicle via GPS system, advertisements or touristic information near the region can be displayed in special LCD monitors located between the glazes of side glasses.

A special mobile application was designed for the use of drivers, fleet managers, and passengers of a smart city bus.

In the driver section of the application, some properties of smart city bus can be controlled via mobile application. Driver can start/stop the vehicle and control the A/C system, preheater, lights, doors, horn, etc. of the bus by using this mobile application.

In the fleet manager section, some of important vehicle parameters can be monitored by fleet manager via mobile application.

In the passenger section of the application, passengers can see the estimated time of arrival for their station with the rate of fullness and real-time position of the bus supplied by GPS.

Smart City Concept

Smart city is a recent topic, but it is spreading very fast, as it is perceived like a winning strategy to cope with some severe urban problems such as traffic, pollution, energy consumption, and waste treatment. A smart city is a complex, long-term vision of a better urban area, aiming at reducing its environmental footprint and at creating better quality of life for citizens. Mobility is one of the most difficult topics to face in metropolitan large areas. It involves both environmental and economic aspects and needs both high technologies and virtuous people behaviors. Smart mobility is largely permeated by Information and Communication Technologies (ICT), used in both backward and forward applications, to support the optimization of traffic fluxes and also to collect citizens' opinions about liveability in cities or quality of local public transport services [1].

Smart city is considered like a winning urban strategy using technology to increase the quality of life in urban space, both improving the environmental quality and delivering better services to the citizens [2].

Intelligent Transport Systems

Information technology (IT) has transformed many industries, from education to health care and to government, and is now in the early stages of transforming transportation systems. While many think improving a country's transportation system solely means building new roads or repairing aging infrastructures, the future of transportation lies not only in concrete and steel but also increasingly in using IT. IT enables elements within the transportation system—vehicles, roads, traffic lights, message signs, etc.—to become intelligent by embedding them with microchips and sensors and empowering them to communicate with each other through wireless technologies. In the leading nations in the world, ITS bring significant improvement in transportation system performance, including reduced congestion and increased safety and traveler convenience [3].

Intelligent transport systems (ITS) are advanced applications to collect, store, and process data, information, and knowledge aiming at planning, implementing, and evaluating integrated initiatives and policies of smart mobility [1].

Fleet Management

Fleet management is an administrative approach that allows companies to organize and coordinate work vehicles with the aim to improve efficiency, reduce costs, and provide compliance with government regulations. While most commonly used for vehicle tracking, fleet management includes following and recording mechanical diagnostics and driver behavior [4].

Vehicle Data Processing and Reporting System Overview

With the application of data processing and reporting system to a smart city bus, the vehicle has the capability of processing, sending, and reporting data in the vehicle as requested by the driver, municipality, owner, and/or fleet management center via the Internet.

Vehicle data processing and reporting system include driver information display, telemetry device, 3G communication, big data infrastructure, web application, application database, smart city bus, and end user as shown below in Fig. 33.1. Telemetry device is connected to vehicle CANBUS and obtains CAN messages via its transceiver. Obtained messages are evaluated with its Linux-based processor according to SAE J1939-71 standard. Then telemetry device sends CAN messages to big data infrastructure via 3G communication and driver information display via Ethernet. After gathering CAN messages on server, we can make big data analysis with designed web application. Also, we can define different user and authorization groups. Some of user groups have limited access and only can monitor allowed messages and design their own dashboards if needed, while some of groups have access to see all messages, making big data analysis and prepare reports for municipalities or customers about their vehicles and drivers. Allowed user groups also can work on error estimation and warn the driver and vehicle owner before the error occurs by analyzing all messages which are located on application database for a long time.

Municipalities, owner, or fleet manager can obtain location, engine speed, vehicle speed, throttle pedal position, brake pedal position, engine coolant level, engine

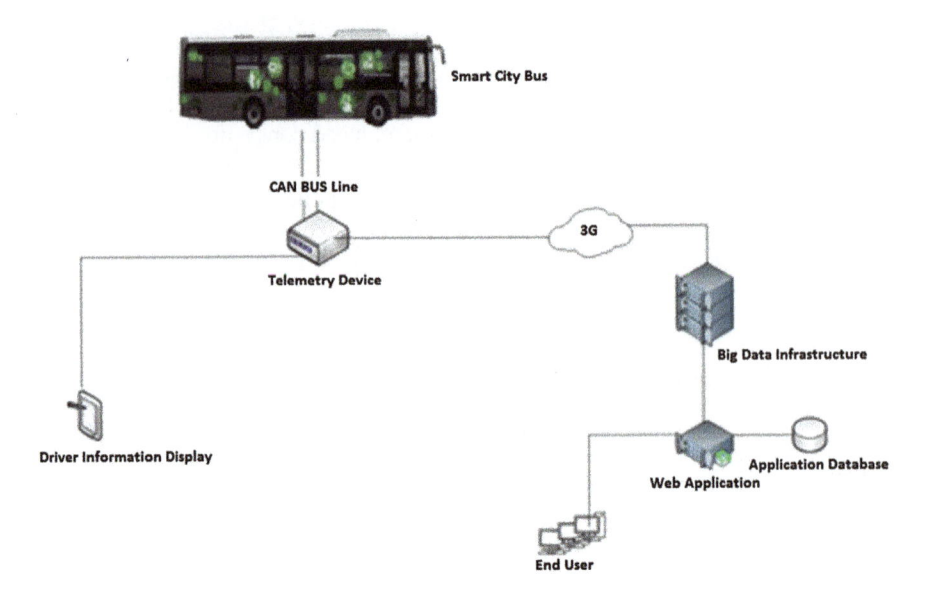

Fig. 33.1 General overview of the system

coolant temperature, engine oil pressure, transmission oil pressure, engine oil level, fuel level, instantaneous/average fuel consumption, total distance, trip distance, fuel consumption at idle speed, AdBlue level, AdBlue consumption, total engine working hour, total engine working hour at idle speed, average engine speed, average throttle pedal position, average vehicle speed, average fuel consumption, remaining km for maintenance, park brake status, brake lining thickness, door status, battery voltage, alternator status, front/rear brake tank pressure, current gear, retarder status, engine oil temperature, transmission oil temperature, compressor status, time/date information, kickdown status, failure status for engine, transmission, brake and suspension systems, etc. These messages can be customized for customer demands. Some sample views of web application are shown below such as map view (Fig. 33.2) and CAN BUS messages (Fig. 33.3).

By this study, municipalities, owner, or fleet manager can limit desired messages, and the system warns them if the limited values are exceeded. For example, they can limit vehicle speed to 80 km/h and if the driver exceeded this limit value, the system sends an e-mail and warns the responsible people. By processing gathered data, performance of drivers of the fleet can also be evaluated. Drivers can be rated according to their fuel consumption, engine speed, vehicle speed, current gear, throttle pedal position, brake pedal position, etc. values.

With the application of this system, vehicle can be diagnosed without any need of special diagnose tool, and all error codes for engine, transmission, brake, suspension, and any other system in the bus can be monitored on fully programmable, touchable driver information display. Important messages can be monitored by drivers

Fig. 33.2 A sample view from map view of the web application

via touchable driver information display (Figs. 33.4 and 33.5). Messages on touchable driver information display can be customized according to customer demands. If any error occurs with engine, transmission, brake, or suspension system, corresponding text color returns to red and when the driver touches this text, related fault page screen opens automatically (Fig. 33.6). The driver can realize the problem without any need of special diagnose tool. Also, if the driver exceeded limited values, corresponding value color returns to red to warn the driver.

Another advantage of applying this system to the city bus is to see the condition of the vehicle, situation of the systems, error codes, and special parameters like front/rear brake tank pressure, throttle pedal position, brake pedal position, and park brake status just before any accident. These messages can help to find out the cause of the accident.

Auxiliary Components Supporting System

Smart transportation application for city buses consists of location-based passenger counting system, IP cameras and panic buttons, and location-based advertisement system with special LCD monitors located between the glazes of side glasses.

Time Date	Instantenous Fuel Cons.	Rear Right Brake Lining	Throttle Pedal Position	Coolant Temp.	Vehicle Speed	Engine Speed	Total Distance
10.02.2016 06:14	1 km/lt	92%	100%	45 C°	49 km/h	1690 rpm	10330 km
10.02.2016 06:14	127 km/lt	92%	0%	47 C°	18 km/h	1175 rpm	10330 km
10.02.2016 06:14	1 km/lt	-	100%	48 C°	42 km/h	1486 rpm	10331 km
10.02.2016 06:14	1 km/lt	-	77%	49 C°	38 km/h	1318 rpm	10331 km
10.02.2016 06:15	127 km/lt	-	0%	51 C°	42 km/h	1480 rpm	10331 km
10.02.2016 06:15	1 km/lt	92%	100%	51 C°	45 km/h	1550 rpm	10331 km
10.02.2016 06:15	1 km/lt	92%	100%	53 C°	65 km/h	1630 rpm	10331 km
10.02.2016 06:15	1 km/lt	92%	100%	54 C°	71 km/h	1820 rpm	10331 km
10.02.2016 06:16	2 km/lt	92%	100%	55 C°	81 km/h	1735 rpm	10332 km
10.02.2016 06:16	3 km/lt	-	100%	57 C°	79 km/h	1705 rpm	10332 km
10.02.2016 06:16	14 km/lt	92%	100%	57 C°	80 km/h	1714 rpm	10332 km
10.02.2016 06:16	3 km/lt	-	100%	58 C°	79 km/h	1700 rpm	10333 km
10.02.2016 06:17	2 km/lt	92%	100%	58 C°	78 km/h	1675 rpm	10333 km

Time Date	Park Brake Status	Brake Status	Vehicle Speed	Front Right Brake Lining	Transmission Oil Temp.	Engine Oil Pres.	Engine Oil Level
10.02.2016 06:14	●	○	49 km/h	-	44 C°	82 B	72%
10.02.2016 06:14	●	○	18 km/h	94%	44 C°	82 B	72%
10.02.2016 06:14	●	●	42 km/h	94%	45 C°	74 B	72%
10.02.2016 06:14	●	●	38 km/h	-	45 C°	82 B	72%
10.02.2016 06:15	●	●	42 km/h	-	45 C°	82 B	72%
10.02.2016 06:15	●	○	45 km/h	-	46 C°	92 B	72%
10.02.2016 06:15	●	○	65 km/h	94%	46 C°	97 B	72%
10.02.2016 06:15	●	○	71 km/h	94%	46 C°	90 B	72%
10.02.2016 06:16	●	●	81 km/h	94%	46 C°	82 B	72%
10.02.2016 06:16	●	○	79 km/h	-	47 C°	82 B	72%
10.02.2016 06:16	●	○	80 km/h	94%	47 C°	80 B	72%
10.02.2016 06:16	●	○	79 km/h	-	48 C°	79 B	72%
10.02.2016 06:17	●	○	78 km/h	94%	48 C°	97 B	72%

Fig. 33.3 A sample view from CAN BUS messages of the web application

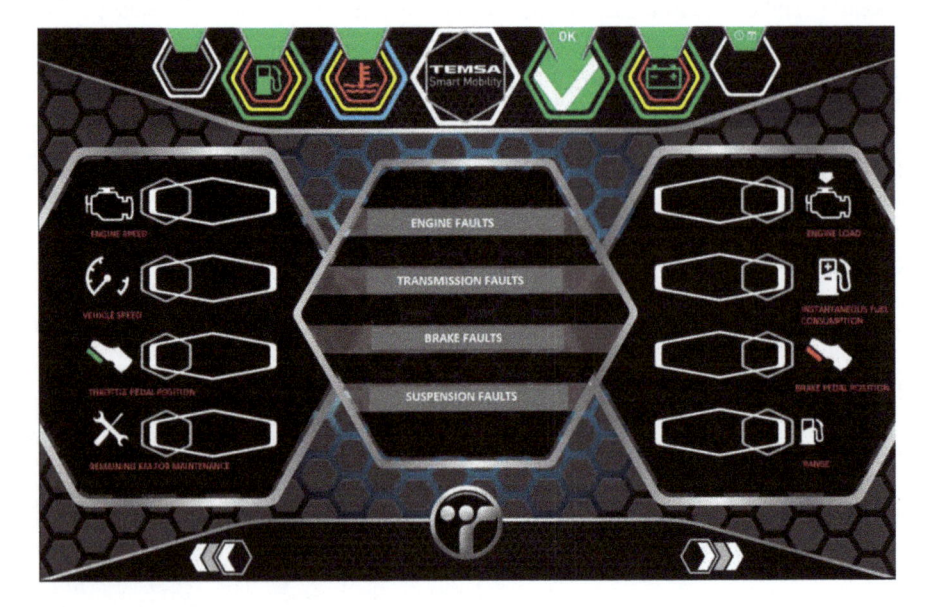

Fig. 33.4 Diagnose screen of the vehicle

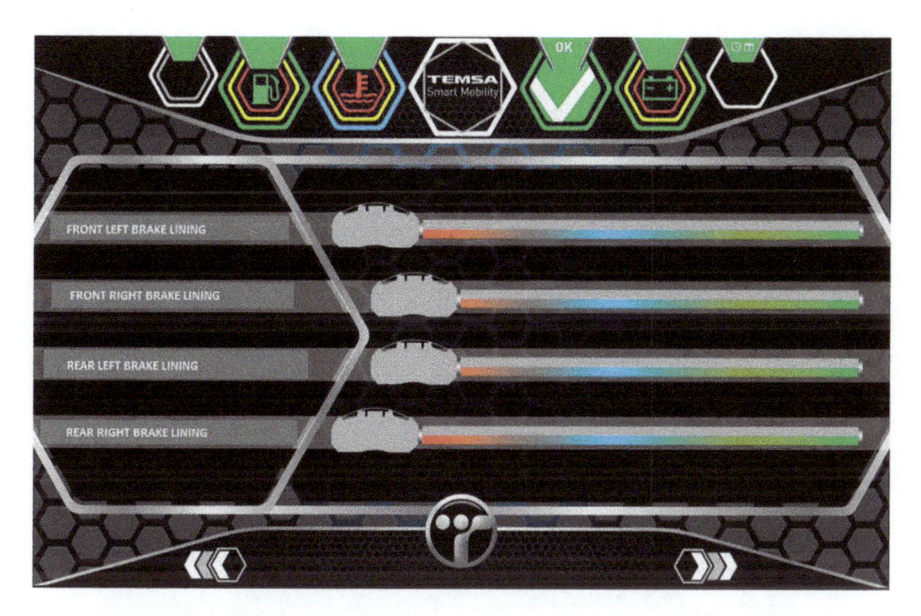

Fig. 33.5 Brake lining status of the vehicle

Global Positioning System (GPS)

A tracking system employing Global Positioning System (GPS) satellites provides extremely accurate position, velocity, and time information for vehicles or any other animate or inanimate object within any mobile radio communication system or information system, including those operating in high-rise urban areas. The tracking system includes a sensor mounted on each object, a communication link, a workstation, and a GPS reference receiver [5].

Embedded GPS receivers in vehicles' onboard units (OBUs, a common term for telematics devices) receive signals from several different satellites to calculate the device's (and thus the vehicle's) position [3].

Passenger Counting System

Using electronic infrared beams or mechanical treadle mats, automatic passenger counters (APCs) have the ability to count transit passengers as they board and a light transit vehicles at individual stops. When coupled with stop location information, archived APC data can be post-processed to generate disaggregate data in both time and space [6].

Passenger counting systems can help to fleet managers to plan route of city buses effectively by supplying the information of number of passenger inside a bus. By combining passenger counting system with a GPS system, location of the bus and

Fig. 33.6 A sample view from engine and transmission fault codes of the vehicle

number of passenger in the vehicle can be monitored online, and an effective plan of city transportation can be done. With this system estimated time of arrival for the next station with the rate of fullness can be calculated, and this information can be supplied to the fleet management center.

Passenger counting system consists of passenger counting unit, 8 infrared sensors, GPS antenna, GSM antenna, Ethernet connection, and vehicle parameters as shown below in Fig. 33.7. The vehicle has a handrail in the middle of front door, so two infrared sensors are enough for the front door to obtain high accuracy, while three infrared sensors are used at the middle and rear door. All sensors are connected to the passenger counting unit. PCU works with 24 V DC power, and all sensors take power from PCU. GPS antenna is connected for location information; GSM

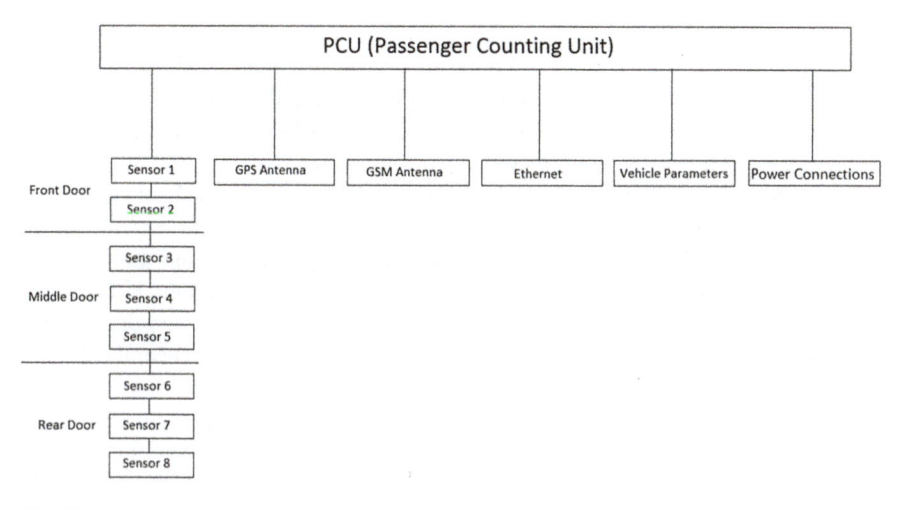

Fig. 33.7 General overview of passenger counting system

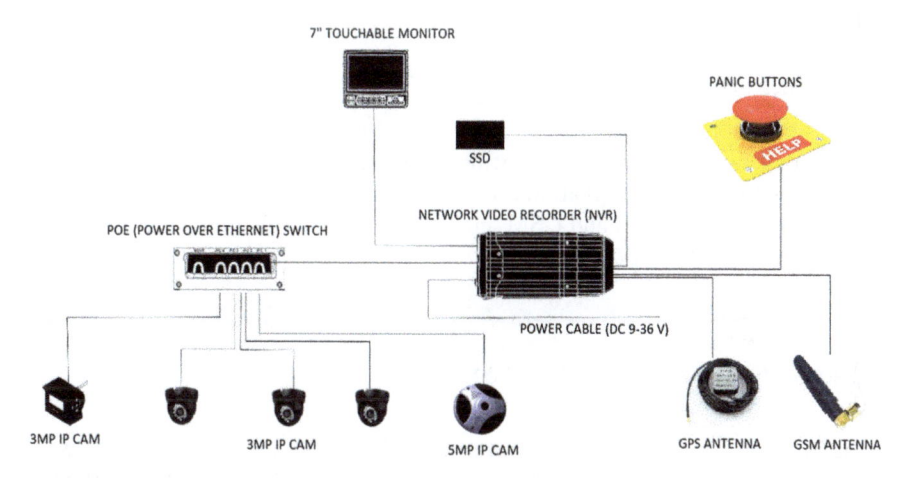

Fig. 33.8 General overview of safety package

antenna and Ethernet connection are for data transfer. PCU also needs some vehicle parameters such as ignition status, door status for each door separately, and vehicle speed information. We make necessary adjustment on PCU such as FTP server address, port, user name, password, etc. Then, PCU starts to send data to the server.

IP Cameras

IP camera and panic button system consists of IP cameras, POE switch, network video recorder, solid-state drive (SSD), 7″ touchable monitor, GPS antenna, GSM antenna, panic buttons, and power connections as shown below in Fig. 33.8.

3MP IP camera with its cover is used for rear view; other 3MP IP cameras are used for front view, driver area, and third door area. 5MP IP camera which can record 3600 is used for the inside of the vehicle and mounted middle of the vehicle. All cameras are directly connected to the POE switch via RJ45 cables. POE switch is connected to the network video recorder via RJ45 cable. Network video recorder saves recordings in its hard disk, displays for the driver via a 7" touchable monitor, and sends to server via GSM antenna.

Application of IP (Internet Protocol) cameras will provide online information about the vehicle inside and surrounding to the fleet management center. If the vehicle deviates from its standard route or if any insecure situation occurs during the travel, driver or passengers can press panic buttons which are located near the driver area and inside passenger cabin. Then, an emergency signal and location information can be sent to the fleet management center, and vehicle can be monitored simultaneously via IP cameras. With the application of IP cameras, drivers know that every behavior of them is recorded, so they push themselves to obey traffic rules and passenger safety strictly.

Panic Buttons

Panic buttons directly connected to the network video recorder and network video recorder send emergency signal and location information whenever one of panic buttons is pressed via GPS and GSM antennas. Network video recorder can work with a power of 9–36 V DC.

Location-Based Advertisement System

Location-based advertisement system consists of Linux PC, GPS, GPRS, and power source as shown below in Fig. 33.9. Linux PC decides to which advertisement will be displayed at screens by comparing location information which comes from GPS

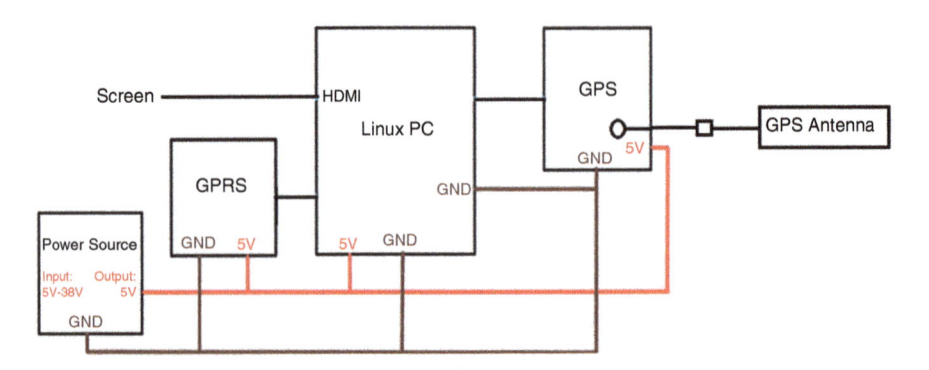

Fig. 33.9 General overview of location-based advertisement system

and previously defined location information in its software. Any advertisement, video, or picture can be displayed. GPRS module is used to be able to update Linux PC's software and advertisement. Power source is used to supply power to all system. Its input voltage range is between 5 V and 38 V. Its output voltage is 5 V and supplies power to all other components.

We present to have a chance to be able to display different advertisements with only one system to municipalities, owners, or fleet managers instead of a constant and unique advertisement. They can display advertisements about nearest hotels, restaurants, shopping centers, touristic areas, and historical places according to vehicles' location information. They can display information about local government if desired. Also, we present them to have a chance to update their advertisement via USB or Internet quickly.

Another feature of this application is LCD monitors located between the glazes of side glasses. Municipalities, owners, or fleet managers can be able to select LCD monitors located between the glazes of side glasses instead of standard 19″ monitor in vehicles. The risk of crashing to monitors by passengers and breakdown of monitors by external effects disappear by LCD monitors located between the glazes of side glasses. The system is compatible with both LCD monitors located between the glazes of side glasses and standard 19″ monitors.

Smart City Bus Mobile Application

The system consists of mobile application, Wi-Fi communication, telemetry device, GPS, vehicle CAN BUS line, body controller, multiplexer, and smart city bus as shown below in Fig. 33.10. Telemetry device is connected to vehicle CAN BUS line and obtain CAN messages via its transreceiver. Obtained messages are evaluated with its Linux-based processor according to SAE J1939-71 standard. Then telemetry device sends CAN messages to mobile application via Wi-Fi communication. Telemetry device has GPS connection for location information and Wi-Fi feature to communicate with mobile application.

Mobile application sends a command by its application buttons or voice control feature; telemetry device processes this command and sends a CAN message to body controller of the smart city bus. Body controller evaluates this message and sends command to activate or deactivate multiplexer outputs for vehicle functions. The driver can turn on/off park lamps, low beams, high beams, fog lamps, interior lamps, left signals, right signals; open/close front, middle and rear doors; activate/deactivate kneeling, lifting, and side kneeling; control horn, air conditioner, and preheater; start/stop the engine, etc. by using this mobile application. These functions can be customized according to customer demands. Some sample screenshots from driver section of the mobile application can be seen in Fig. 33.11.

In the fleet manager section, some of important vehicle parameters, like engine speed, vehicle speed, throttle pedal position, engine coolant temperature, front/rear

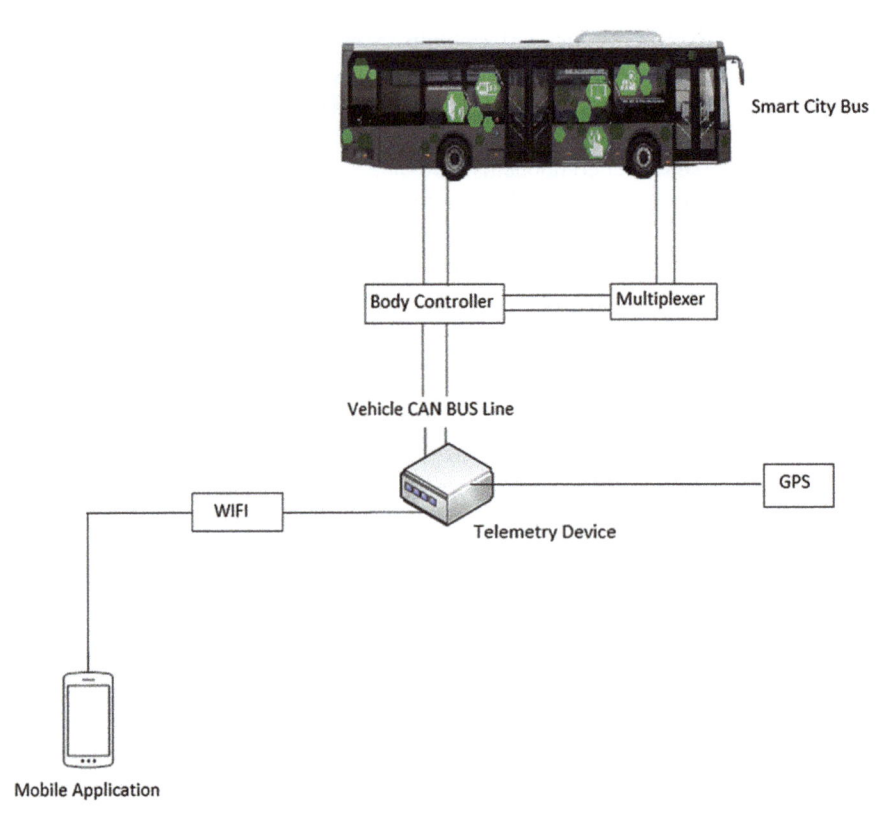

Fig. 33.10 General overview of the mobile application system

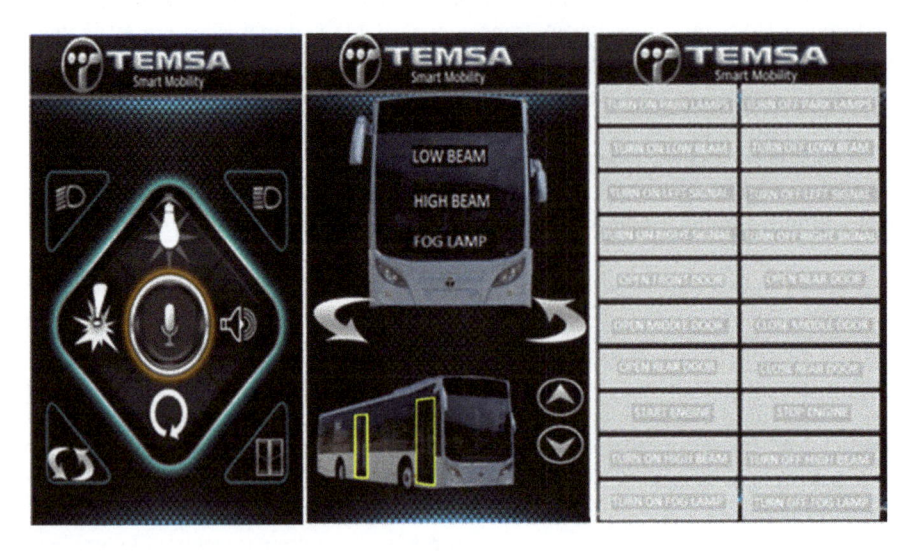

Fig. 33.11 Sample screenshots from driver section of the mobile application

brake tank pressure, park brake status, brake lining thickness, fuel level, instantaneous/average fuel consumption, total distance, battery voltage, AdBlue level, engine oil level, engine coolant level, remaining km for maintenance, failure status for engine, transmission, brake and suspension systems, etc. can be monitored by fleet manager via mobile application. These messages can be customized according to customer demands.

Fleet managers or garage technicians don't have to go to each vehicle to check fuel level, AdBlue level, engine coolant level, engine oil level, battery voltage, etc. They can control all these parameters in their offices via mobile application. They also don't have to go to each vehicle before its route to check if the vehicle has any error or not. They can connect to the vehicle and check the errors. Intercity drivers can also use this feature during their break. They can control fuel level, AdBlue level, engine oil level, engine coolant level, etc. via mobile application. Mobile dashboard and vehicle status pages of the application can be seen in Fig. 33.12.

In the passenger section of the application, passengers can see estimated time of arrival for their station with the rate of fullness and real-time position of the bus supplied by GPS. Passengers can organize their travel according to these data and minimize the time lost during city transportation. If passenger is inside the vehicle, passenger can connect to the media archive of the vehicle with Wi-Fi and watch videos or listen to music, which are already stored in the vehicle computer.

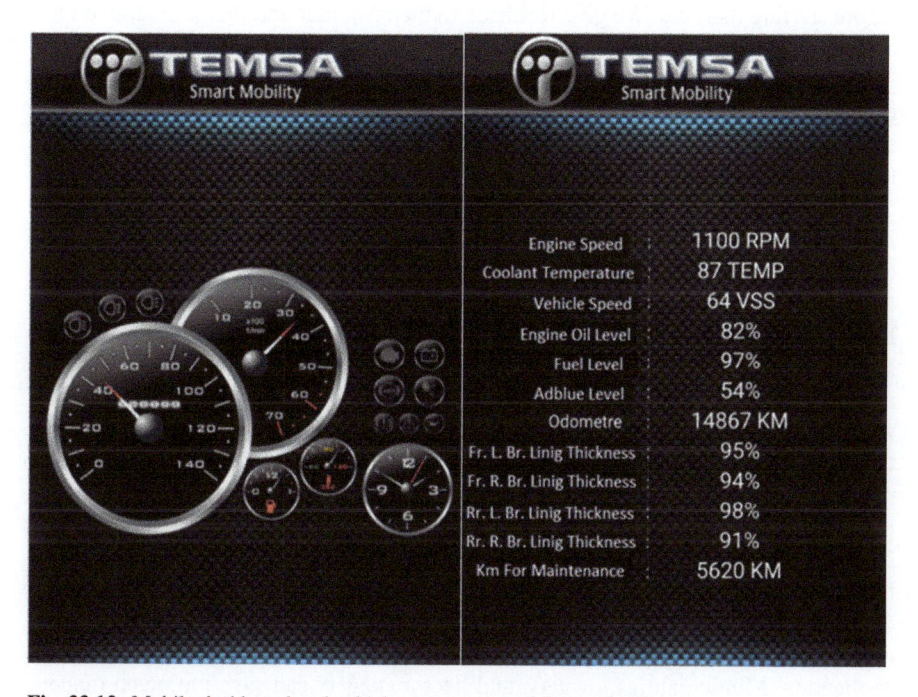

Fig. 33.12 Mobile dashboard and vehicle status pages of the application

Conclusion

According to (ENEA, 2016) experiences made so far in the EU countries, USA and Japan show that the introduction of ITS technologies has significantly contributed to improve the efficiency, safety, environmental impact, and overall productivity of the transportation system. These applications, as pointed out by the European Commission, are an attractive solution to many of the problems of the transport sector: in the road sector, it is possible to record reductions in journey times (15–20%), in energy consumption (12%), and in emissions of pollutants (10%), as well as increases in network capacity (5–10%) and decreases in the number of accidents (10–15%). Significant results have also been achieved in the fleet management and logistics processes of goods and in the exercise of public passenger transport [7].

By this we expect to reduce maintenance cost and fuel consumption of vehicles by analyzing related messages which are located on application database. For example, if a customer complains about lifetime of brake linings, we analyze the vehicle messages and maybe we find out that the driver doesn't use retarder (auxiliary brake system for buses) and so lifetime of brake linings decrease. We inform the customer and driver about the relationship of retarder and brake linings to increase lifetime of brake linings. If a customer complains about fuel consumption, we analyze the vehicle messages, and maybe we find out that the driver usually uses the vehicle at high engine speed, high throttle pedal percentage, unsuitable gear position, and kickdown situation, or the driver starts the engine and wait for a long time at idle speed before starting its route. We give feedback to the customer and driver to reduce fuel consumption.

We expect to present an effective routing plan to municipalities, owners, and fleet managers by location-based passenger counting system. Municipalities, owners, and fleet managers can make an effective plan of city transportation according to location, time, and fullness rate information. City transportation will be safer with the application of safety package which includes IP cameras and panic buttons. Drivers know that every behavior of them is recorded, so they push themselves to obey traffic rules and passenger safety strictly. We present to municipalities, owners, and fleet managers to have a chance to update their advertisement via USB or Internet quickly and without any additional cost by location-based advertisement system. Municipalities, owners, and fleet managers don't have to display constant and unique advertisement at exterior surface of the vehicle as it is up to now. In this way, their advertisement income increases significantly.

We present to control desired vehicle functions remotely to drivers and fleet managers. They can start the engine and turn on A/C or preheater without going to the vehicle. Responsible people can activate necessary functions and can control from outside to see if it works or not by mobile application such as low beam, high beam, left/right signals, exterior lights, etc. Fleet managers or garage technicians don't have to go each vehicle to check vehicle liquids such as fuel, AdBlue, oil, etc. or errors before its route. They can make all these controls remotely by this mobile application. Municipalities, owners, and fleet managers can save too much money and time by using this mobile application. Intercity drivers can also use this feature during their break. They can control fuel level, AdBlue level, engine oil level, engine coolant level, etc. via mobile application. Passengers can see estimated time of arrival for their station with the rate of fullness, and they can organize their travel according to these data. The most important feature of this mobile application is customization property according to customer demands.

References

1. Benevolo, C., Dameri, R. P., & D'Auria, B. (2016). *Smart mobility in smart city action taxonomy, ICT intensity and public benefits*. Lecture notes in information systems and organisation, p. 17.
2. Hall, P. (2000). Creative cities and economic development. *Urban Studies, 37*, 633–649.
3. Ezell, S. (2010). Intelligent Transportation Systems booklet, Information Technology and Innovation Foundation, Washington, USA. p.58.
4. http://whatis.techtarget.com/definition/fleet-management
5. https://www.google.com/patents/US5225842
6. CUTR. (2010). *A guidebook for using automatic passenger counter data for National Transit Database (NTD) reporting*. National Center for Transit Research, University of South Florida, p. 58.
7. ENEA. (2016). http://old.enea.it/produzione_scientifica/pdf_brief/Valenti_ITStrasporti.pdf

Chapter 34
A Model Approach for the Formation of Synergy Effects in the Automotive Industry with Big Data Solutions: Application for Distribution and Transport Service Strategy

Martin Holubčík, Gabriel Koman, Michal Varmus, and Milan Kubina

Abstract The automotive industry is currently the leading industry sector in Slovakia. Innovative solutions and sustainability play key roles in the continuous progress of companies. Among these, Big Data solutions have a significant impact. This chapter explores the use of Big Data solutions in the automotive industry and their connection to synergy effects in strategic management. Of these novel approaches, Big Data solutions and emerging synergy effects are necessary to develop recommendations on successful collaborative solutions. For example, a variety of services are required in the manufacture of large-scale electric vehicles – from shipping to the delivery of parts. These services closely connect the customer and the company that meets his or her needs. The use of Big Data technology for electric cars and a greener automobile industry with regard to distribution and transport service are some of the most competitive strategies in this industry.

Introduction

The current business environment is creating new challenges in the marketplace, especially in the automotive sector. Companies are facing low-cost and targeted competition, cooperation risk, customer requirements, new technological developments, progress in science and research, and other factors – not only in the domestic market, but increasingly in the global market. Thus, innovations in managerial thinking are necessary. A shift in managerial thinking is essential for a proper understanding of cooperative organizations. Managers need information (data) to

---

M. Holubčík (✉) · G. Koman · M. Varmus · M. Kubina
Department of Management Theories, University of Žilina, Žilina, Slovak Republic
e-mail: martin.holubcik@fri.uniza.sk; gabriel.koman@fri.uniza.sk; michal.varmus@fri.uniza.sk; milan.kubina@fri.uniza.sk

© Springer International Publishing AG, part of Springer Nature 2019
D. Cagáňová et al. (eds.), *Smart Technology Trends in Industrial and Business Management*, EAI/Springer Innovations in Communication and Computing,
https://doi.org/10.1007/978-3-319-76998-1_34

gain a clear picture of the market and its players, which form a causal relationship with quality information. In the dynamic and chaotic processes of a company's internal and external environments, interactive and unstructured data are the basis for the formation of synergy effects (evolution).

The starting points of synergy can be observed in many, if not all, research areas. Synergy is a core discipline that changes the perception of the added value generated by objects such as individuals, companies, clusters, and organizations. Managers should analyze and examine various synergetic phenomena, the conditions of potential impacts, and the resulting values. In this way, a new approach to synergy can be achieved as a link, connection, or engagement in cooperation within a specific environment that is changing, developing, and reacting. A synergy effect is based on synergy in a particular environment that creates value, expected results, and unexpected effects.

Second, Big Data solutions present processing of large amounts of heterogeneous data, and these can be the bearer of critical information value. There are mainly unstructured and semi-structured data that are generated by man, machine and their combination. This phenomenon observed in recent years is a great potential, mainly due to the rapid development in information and communication technologies. People and machines currently generate large volumes of disparate data that can be potential carriers of significant information value for a company if it can capture and process it. Timely, relevant and reliable information is mined from available data, becoming an inherent input with background and a component in decision-making of a company. It is possible to consider using Big Data solutions in almost every area of the business process. A uniform integrated data base in a company, when there are diverse data from across the company and its environment, becomes available for the needs of strategy management. Through these new approaches, Big Data solutions and emerging synergy effects, it is necessary to propose recommendations in order to create successful collaborative solutions.

The strategic management of companies within identified starting points provides a basis for modelling and validation of these results in practise and their further investigation. It represents a potential way (strategy management) and place (cooperative organizational forms) to create synergy effects. It is thus where Big Data solutions represent a source of values (data base) that can provide relevant and valuable information obtained by analysing various occurences of synergy phenomena and data. Detecting patterns and key factors affect potential value and the impact of opportunities.

For theory and practice it creates a significantly new methodological approach as a way to appropriately manage cooperative organizational forms (group of cooperative enterprises) and form synergy effects with the use of Big Data technology.

The purpose of this article is to use this new approach in the automotive industry, which is currently one of the leading industries in the Slovak market.

According to the Statistical Office of the Slovak Republic, on January 1, 2016, a total of 665 big companies (over 250 employees) were registered in Slovakia with 279 businesses in industrial production, which also included the automotive industry (types of economic activity: CL – manufacture of transport equipment, CK –

machinery and equipment, etc.), which represented approximately 41.2% of the companies.

The problem is continually expanding amounts of diverse data that have potentially important information value.

The objective is, based on theoretical knowledge of the issue, to identify the potential of Big Data solutions for purposes of synergy effects in strategic management of collaborative relations in the automotive industry. These findings present a case study of the results in theoretical and practical comprehensive research which has been processed in the following text. The value of this information is enrichment of knowledge in the automotive industry and potential applications in industry. Methods that are used in this article are: content analysis, induction and deduction of information, logical reasoning and a qualitative approach in information processing.

The research intention of this article is to introduce and demonstrate the potential of Big Data technology and its solutions in the automotive industry through the use of synergy as an approach in strategy and cooperation management.

The article follows our previous publications: Cooperation as base for synergy [18]; Application of Big Data technology in the knowledge transfer process between business and academia [22]; Evaluation of the innovative business performance [61]; Possibility of improving efficiency within business intelligence systems in companies [24].

Identification of the Main Theoretical Approaches

Synergy is one of the processes in the cooperative environment in which there are an increasing number of interactions with important individual links between them. If a group of companies want to achieve a synergistic effect it is necessary that the group is properly managed and that management is based on data.

The elements of cooperation in the interactive relationship (setting goals, establishing a cooperative equilibrium, consolidating mutual trust, filtering members' cooperation, rules and standards of cooperation, maintaining mutual information and communications environment, complying with reciprocity, creating a positive experience, the combination of resources, setting of the control mechanism) are incomplete as a set of recommendations for management members in cooperation. At present, it is progress and development of cooperative management that addresses application management in cooperation. In a global environment, with its growing diversity of members, interactions, and distances, it is necessary to apply cooperative management. Soviar et al. [53] disclose cooperative management as efficient and effective relationship management in the sense of cooperation between relatively independent organizations or individuals in order to increase their competitiveness.

Linking cooperation and synergies needs to be supported with the following baseline factors:

- Cooperation in the context of evolution: According to Nowak [35], this is one of the approaches of evolution—cooperation and its different forms—while adding, "cooperation is necessary for the evolution to build new levels of organizations"
- The diversity of the members of cooperation: The global environment creates the diversity of members, as stated by Axelrod [4], and co-operation is common among members of the same species and also between the members of different species; it is the central idea of reciprocity
- Interactions within cooperation: Fehr and Schmidt [16] point to determine in which one member affects the whole group cooperating with the changes in her behavior, which can be beneficial or prejudicial within the group.

To better understand the issue of synergies is within the survey carried out on comprehensive analysis on real cases (secondary data). The findings represent the following status synergies within the business environment.

Cases of cooperation are stronger than ever, not only in the past but mainly looking towards the future. Strategic alliances are setting an example of such cooperation, namely, an example of synergy in management (control example of cooperation in management; organizational forms of cooperation). In 2003, the Austrian Institute "Hernstein International Management Institute" conducted a study of 450 central European businesses, which found a high proportion (51–67%) of companies that used cooperative management style. However, individual interviews with managers showed that they used mostly consensual, or patriarchal and authoritarian methods of management [52]. Currently, management style results in increased operational capacity and aggressiveness to strengthen the company's competitiveness by creating alliances and networks. However, it depends on environmental conditions and companies situated there. Accordingly, different conditions should be used for different management styles.

Authors Ireland, Hitt and Vaidyanath [21] stated that most alliances end in failure, but highlighted the importance of alliance attributes. They stated that alliances are a source of competitive advantage and create bigger value on the market. Businesses usually seek access to relevant sources through the alliances. Management efficiency of alliances should be based on the benefits of their implementation. The whole process starts with choosing the right partner, followed by the building of social capital, knowledge and maximizing cooperation on the basis of confidence-building.

Authors Hamel, Doz and Prahalan [17], during their five-year content analysis of 15 strategic alliances, focused on how companies use competitive cooperation to increase (increase, improve) their internal skills and technologies, while preventing the transfer of competitive advantages to ambitious partners. Where large investments are needed to develop and promote products to new markets, only a few companies can go separately in every situation. Cooperation is competition in a different form. Occasional conflicts are evidence of the mutual beneficial cooperation in which only a few co-operations can maintain a constant win-win strategy. The authors also argue that cooperation has its limits, and the company must defend against competitive compromise. For example, strategic alliances include constantly evolving negotiations which go beyond the realistic conditions of agreements or goals of top management. Learning from partners is paramount, but each alliance is actually a window to the wide partnership skills, building new knowledge through cooperating organizations.

Kang and Sakai [65], for the study of international strategic alliances, identified providing a synergistic effect. These strategic alliances, mergers and acquisitions are expected to provide a synergistic effect in the long term by reducing the overlap between trading partners and reducing costs (critical level for the market, adding new lines of business, providing financial support for ailing businesses). Output (product) for alliance latecomers (up to several years) is limited (it is characterized by a one-time collaboration in development activities). Strategic alliances prefer small companies with unique technological advantages to cooperate with larger companies with financial resources that enable them to increase core (core, unique and their own skills and knowledge of business) competencies while maintaining their independence. Initially, transaction costs input by acquisitions and mergers are much higher than for strategic alliances.

The Big Data Technology

Manager's decision-making currently depends on timely, available and relevant information within the context of problem solving.

The continuous development of information and communication technologies (ICTs) gives rise to new systems and platforms that can potentially carry important information. Generated data is typically stored in different database structures. Currently the most comprehensive database structure is represent by the data warehouse.

The ICT provides tools for managers to obtain information they need for the decision-making process. This ensures effective decisions through innovative solutions and tools, which can meet the needs of managers in dealing with specific situations, i.e. office automation, etc. [2]. Businesses generate large volumes of data. Manual and mechanical systems for converting this data into information cannot handle this large amount with respect to time, costs and complexity of processes.

For this reason the transformation of data into information nowadays provides complex information and communication technology as information systems with distribution of information to specific needs (manager implementing decision) [3, 26, 59, 60].

The impact of ICT on decision making of managers is crucial in terms of global exposure of the company, especially for certain needs [2]: defining the number of variants for global markets and the Internet; reducing uncertainty; saving time and finances; timely communication with various consultants, work teams or experts, who appear in the decision-making process.

Based on these findings, it is possible to assume that information and communication technologies emerge in the whole decision-making process as a supporting tool.

Information and communication technology and specific information systems allow managers to handle amounts of data and transform them into relevant information through specific tools and techniques, i.e. provide information support in different phases of the decision-making process. The influence of 'informatization' of the whole society [1, 36] can be assumed to follow the trend of using information and communication technology in the decision-making processes of managers that will be constantly deepened and specified according to current needs and requirements, i.e. the need for processing unstructured data to obtain important information [10].

The amount of data is growing, but businesses trying to handle this data through traditional systems cannot store, process and use it. This phenomenon is mainly due to different structured data, i.e. unstructured and semi-structured data that make processing challenging [6]. Large volumes of unstructured data have different characteristics than data generated in businesses, while data warehousing and data management tools cannot efficiently process and analyse these amounts in terms of time and cost [15, 55]. For this reason Big Data technology was established.

The impact of large amounts of data was recorded in the 1970s, when scientists from Berkeley identified that by 1999 there would be over 1.5 trillion bits of information generated. The very basis for definition of large data (Big Data) was introduced in 2001 by Doug Laney, who called constantly-increasing amounts of data "3 V", i.e. volume, variety and velocity. The problem of large amounts of data was first mentioned in 1997, when NASA scientists tagged problems with visualization, since graphic data reached such a volume that the data file could not be placed in the main memory or on a computer hard drive [5, 31].

Large data represent a defined term that describes a large number of complex data sets as well as advanced technologies to collect and store data [54]. It is possible to talk about large amounts of data in two basic senses. First, a large volume of data has to be processed in a reasonably short time; second, there is unstructured data or information which we need in real time.

The Big Data term refers to large amounts of information coming from different sources such as transaction records, boot files, social media, sensors, web applications, etc. Big Data is not only a large amount of data, but also extremely diverse data types distributed at different speeds and frequencies (Stanimirovič and Miškovic). Big Data implies that the amount of data is difficult to process in conventional, commercially available means [57]. Big Data is the next generation of data warehousing and business analysis that is ready to provide the highest support for saving costs and increasing business efficiency [33].

Big Data is data that require an excessive amount of time and space for storage, transmission, processing and use of available resources [64]. The emergence of new technologies in the field of Big Data was supported by the assumption that the amount of data generated is increasing at a rate that is at odds with the speed of development of current technologies for the processing of the data, i.e. current technology cannot save to memory the amount of data generated. Precisely for this reason arose technology for the processing of data, MapReduce and Hadoop open platforms, from Google and Yahoo. Without having to store them in the process technologies, the data is arranged in rows, as is the case with conventional database tables; this means that the technology can work with data without the need for a single hierarchy or homogeneity. Law firms, which have a lot of data generated in the context of the Internet, have an interest in analysing these data for different reasons, in particular, economic reasons. Therefore, these businesses are creating an initiative for the creation of new trends in the processing of the quantity of diverse data. All generated data contain some information value, which can bring different benefits to the enterprise [27].

The foundations of architecture for Big Data analysis tools, infrastructure and applications serve the purpose of implementation of the various operations of the available data. The current database systems are not sufficient in terms of performance and infrastructure for the processing of large volumes of disparate data. To resolve this issue new technologies have been created in the context of Big Data infrastructure, which include Hadoop, NoSQL, and MPP (Massively Parallel Processing) [29, 30].

Hadoop represents a tool aimed at storing, processing and analyzing large amounts of structured and unstructured data, i.e. it can work with diverse data types, such as technology, analysis tools, software solutions, etc. It is possible in the architecture of Big Data to expand or integrate according to the requirements of the enterprise. An example of that is the integration architecture of Big Data with Business Intelligence.

The principle of working with the data in the context of Big Data is consistent with the principle of traditional Business Intelligence, which also applies in the case of architecture. The essential difference, however, is the type of data processed. In the framework of the enlarged Hadoop architecture (Fig. 34.1), it is evident that the architecture of the Big Data may contain a layer of a data warehouse that stores structured data in multidimensional databases for the purposes of business generated in the advanced analysis. Beyond the data warehouse are Business Intelligence layer works, or analytical tools and principles of BI, which handle structured data. Thus, Big Data technology represents a certain connecting cell among analytical tools for processing various data and, at the same time, the system of Business Intelligence tools. Among the key layers are especially those that work over a lot of architecture of unstructured data that are generated in the vicinity of the enterprise, or in the real world. These include the following layers: a layer for the data warehouse, data processing, layer for layer, a layer of data management and data access, and a layer of data connection [46].

Similarly, in the context of Big Data system architecture's integrated Business Intelligence platform, IBM's Big Data (Fig. 34.1), there are structured and unstructured data merging into a single database structure, or into a single architecture [51].

From the above it is clear that the providers of technology integration in Big Data are trying to integrate not only the analytical tools of Business Intelligence, but to

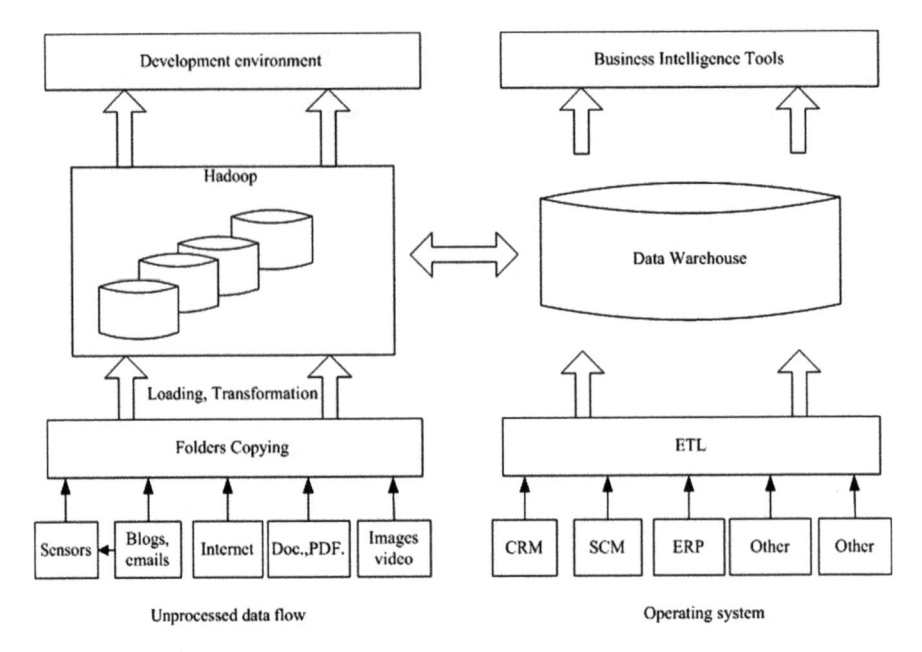

Fig. 34.1 IBM's Big Data architecture. (Own elaboration edited by Big Data: From great expectations to practical use)

link the entire infrastructure of the enterprise into a single unit with the architecture of the Big Data.

Integration of Big Data solutions within a company with the possibility of processing structured data in combination with other disparate data that are generated in a company and its external environment can provide significant information value in terms of gaining competitive advantages, increasing profit or other various business areas, for example [7–9, 13, 25, 39, 41, 44, 45, 49]:

- Effective work and evaluation of data through modern technologies and advanced analytical tools
- Increasing the efficiency of marketing activities
- Reducing the risk of loss by prediction of consumer purchasing behavior
- Influencing and shaping consumers purchasing behavior on the basis of available information obtained from data gained from social networks, etc.
- Adapting products according to needs and requirements of customers (i.e. through data from the self-service cashier, camera records in store, etc.)
- Targeting of marketing campaigns (offer specific products and services to a particular customer)
- More effective workflows and processes (i.e. in a call center through the analysis of voice calls)
- Improving and extending service (e.g. in the health sector by evaluating various patient data that is generated, for example, by Smart and fitness equipment)
- Customer identification, prediction and detection of fraud

- Support of logistics processes in terms of reducing product delivery, saving shipping costs and others through analysis of thousands of delivery variants in real time
- Optimization for supply and use of energy
- The possibility to transfer management on a data-driven organization

Obviously, Big Data solutions can be used in almost any area of business. It is all about data, i.e. Big Data solution deployment is conditional on a number of various generated and used data. Data are the carriers of information value. The advanced technology and analytical tools of different platforms of Big Data solutions offer companies the opportunity to evaluate a variety of data and to mine valuable information for strategy management of the company. In the process of strategic management of the company, information is used from across the whole company and also the external environment. Data are generated from the lowest level of management to top levels and other data sources, from which the information value allows the company to inspect and plan its position and future activity with regard to a continually changing and turbulent market environment. In this sense, it is also possible to consider detection of factors, or in-depth analysis of available data, which give rise to synergies and subsequently synergy effects. The capturing and evaluating of these phenomena enable companies to uncover additional value or risk factors that may occur in implementation of decision-making on a strategic level and their impact on the lower levels of company management, processes, people and overall positions of the market, i.e. detection of synergies in the process of strategic management provide companies the opportunity to harness them or prepare for possible negative impacts from implementing decisions.

The most frequent applications are observing deployment of Big Data solutions in sectors of product sales (wholesale), logistics centers (warehouse management, orders and deliveries), production (managing production program), health service (patients' health monitoring), sport (predicting matches, planning teams, investment, etc.) and so on. With powerful analytical tools this solution is deployed to support decision making in more and more areas of the business. Big Data find significant applications also in the area of the automotive industry.

Case Study: Solutions for the Automotive Industry

The purpose of deployment of Big Data solutions in the automotive industry is particularly stimulated by fast digitalizing of the entire industry and the difficulty in collecting and processing the amount of available data. The consequence of this phenomenon is gradual or resolute transition of traditional industries (as well as online) to digitalize with the support of implementation processes through Internet and cloud-based solutions [14, 50].

On the basis of this solution, future decision-making in the automotive industry is expected to use data and information from five dimensions [50, 58]:

- Electronics
- Autonomy
- Linking
- Mobile services
- Information

Big Data solutions are currently used in almost every aspect of the automotive industry, such as safety, design and so on. This intervention has an impact on total production in the sector and also on specific products (cars) especially in the following areas [11, 12, 19, 20, 23, 28, 32, 34, 37, 40, 42–45, 47, 48, 50, 56, 62, 63]:

- **The design**: assessed on the basis of available collected and processed data throughout the lifecycle of vehicles, i.e. analysis of data in real time with deployment of vehicles can identify opportunities to improve safety, aerodynamics, placing a product on the market, etc.
- **The acquisition**: optimization of supply chain management for improving efficiency of the process by analysing the amount of data from different suppliers.
- **The production**: through the Big Data solutions and advanced analytics it is possible to predict the state of a production line or product quality. The results of simulation can subsequently detect a hot spot in the production process and improve production planning, repairing, human capital and eventually the entire production process.
- **The marketing**: and development of products based on customer feedback and in time marketing campaigns (knowledge of customers and their needs).
- **The finance**: obtaining relevant information from customer behaviour provides a base for better understanding and developing effective funding programs across all market segments. There is a presumption that it is possible to identify opportunities for new sources of revenue for a company.
- **The performance**: in terms of gathering, analysing and evaluating information systems of a vehicle and its surroundings in real-time for purposes of identifying characteristics of future generations of vehicles.
- **The services and technical support**: will have available information about current status of a vehicle. This makes it possible to predict and identify potential failure or dangerous situations and create action and solutions for their elimination.
- **The aftermarket services**: based on information and knowledge of drivers combined with unstructured data needs for development of partnerships.

These examples of deployment and use of Big Data solutions are just some of the number of possibilities for use in the automotive industry. The digitalization of the automotive industry and processing amounts of diverse data is the basis for effective management and planning of all processes in the automotive industry. This is confirmed by General Motors' investment into human potential with 10,000 employees, with a focus on information technology [44, 45]. It is expected that by 2020 about 90% of cars will be connected to the network in terms of M2M (machine-to-machine) communication, which represents a generation and sharing of large amounts of diverse data available in the entire automotive industry [38].

Use of Big Data Technology in Cases of Electric Cars and a Greener Automobile Industry in Distribution and Transport Service Strategy

There are many opportunities to use Big Data technology in this new direction of the automobile industry. There is widespread technology at present in optical barcode systems and frequency technologies such as radio frequency identification, electronic product codes and others. This is a baseline to use Big Data technology.

For example, Tesla Motors is famous for their electric cars. Nevertheless, other brands are not well known for environmentally friendly car models that do not need any kinds of fuels. But they have started and continue with this trend. This has the effect that society is talking about a greener automobile industry as a sustainable solution. A new model of an automobile could be run by electric, hybrid or any other renewable energy with use of Big Data technology in their manufacturing, distribution, sales and customer services.

The environmental issues are important for many car brands. They are paying more attention towards sustainable development that could contribute to the society and future. The current automobile industry should carry out new projects focusing on inventing pollution-free car models which are not harmful to the environment.

Market readiness for electric mobility sector growth is shown in the current trend in the growing sales of electric cars. This trend is not only a customer's need for both passenger transport and freight transport. In 2016, more than 221,000 of these vehicles were sold in Europe, which represents a 13% increase over the previous year. Surprising growth was recorded in the Chinese market, which grew 85% to 351,000 vehicles [66, 68]. Green automotive industry research and development is the current direction of the automotive industry. Electromobility in Europe is a starting point for the future, but also uncertain and changing transport development.

We also expect a wider connection with new technologies such as Industry 4.0, Big Data technology, and Internet of Everything that will make these changes. In general, the goal of electromobility in the European market is to contribute to many changes, such as green car initiatives, smart energy, the future of the environment, intelligent infrastructure and many others [67]. The International Energy Agency talks about a significant increase in the number of clean electric cars, plug-in hybrid cars, and hydrogen-powered cars, with a slight retreat from conventional hybrid vehicles without charging and almost a complete retreat of diesel and petrol engines [66, 68].

Electromobility trends are focused on increasing battery charging (battery swapping, wiring charging, wall boxing), common use of electric vehicles for the benefit of the company (car sharing, taxi and distribution services, car cell elderly, etc.), improving infrastructure, as well as application and information support for electric cars in various campaigns that bring the benefits of electric cars to people and businesses. Application support includes Car Control (APPM), Charger Reservation [69], Open charging GPS map (Open charge map), and applications with dedicated electric cars (Parknow), among others [71, 72].

The main issue in terms of market development is the state of the charging infrastructure. States and governments have started to promote this trend and to finance it from the state budget. Most European countries are involved in the development of electric cars, helping to develop this market, and actively supporting potential customers in the form of purchasing, users in the form of various concessions, and in the form of investment in infrastructure [70, 73].

The Distribution and Transport Service Strategy with the Use of Electric Vehicles

The provision of a variety of services, from shipment to delivery of parts, is a significant area of use for large-scale electric vehicles. These services most closely connect the customer and the company that meet its needs. The company needs to keep the customer, not lose it; therefore, it is increasingly connected to it through the Internet, different marketing communication, and especially fast, quality, timely and reliable distribution. They need their transport to be cheap, fast, and trouble-free. Green transport adds to the reputation. The solution of electromobiles is that they are supposed to bring about a new revolution not only in lower transport costs, fewer services performed, and more satisfied customers, but also the recognition of the company and a better reputation in the given business sector for the use of a green transport solution.

One solution would be to build a distribution link between carriers. By combining their forces, they build up the recharging stations in the territory and jointly divide the distribution, transport, and import using the roads in both directions as well as the filled capacity of the vehicle. Another solution is to dedicate certain finances to science and research for solar panels in vehicle glass for automatic charging as well as increasing battery capacity. IBM is developing a lithium-oxygen battery, which should not only have more capacity but also a much lower weight, which would enormously improve performance in electric cars [83]. Also, in the development of solar cell technology in glass with 95% visibility, one can find suitable cheap technology for use in glass cars for automatic battery recharging [82].

Integral to the whole of the infrastructure and support of it is the unified Big Data analytics platform. This platform is able to initiate, process, and distribute evaluated data for decision-making in real time. We are already seeing major changes in the automotive industry by deploying and using Big Data in businesses. This solution was used within the automotive industry at Mercedes AMG. Mercedes-AMG is a Mercedes-Benz performance division. The main activity of the company is the production of Mercedes performance vehicles. As the Daimler-Benz company is part of the automotive industry, it is possible to anticipate an interest in participating in the future in the field of electric vehicles. As Mercedes-AMG is focused on the production of high performance vehicles, these features are also required in the field of information technology, particularly in the field of test management, quality and optimization. The emphasis on the performance of information and communication technology has led Mercedes-AMG to turn to the Big Data solution. The benefits of implementing the Big Data solution at Mercedes AMG include [84–86]:

- The ability to work with a variety of real-time data
- Provision of information support to company management within other company structures (e.g. project management)
- Production of three new car models, based on customer needs and requirements
- More precise specification of customer needs and requirements
- The company has had the most successful year in terms of product sales by introducing and using Big Data
- Prediction of errors
- Intuitive detection of variations in the testing process due to visualization (GUI) of real-time data processing results
- The rapid identification of errors through comprehensive analytical tools
- Reduction of the time of testing the wrong product by up to 94%
- Sending real-time results of tests to mobile or desktop computers
- Gaining one day for testing per week
- Increasing product testing capability
- Increasing the time to implement corrective actions and incorporating customer requirements
- Reduction of operating costs

The area of electric vehicles is used in transport and distribution. In this case, it is also important to keep in mind the rapidly available information that is obtained from the available data. Like in Mercedes-AMG, it is also appropriate to use the Big Data platform for this purpose. The suitability of the Big Data platform within the transport and distribution industry has been verified by UPS. The company is committed to providing package delivery services and specialized shipping and logistics services at a global level with over 200 countries worldwide. The basic services of the company are [74]:

- Logistics, distribution and forwarding services to over 195 countries
- Freight transport (air, sea, railway)
- Supply chain creation, planning and others

UPS is currently monitoring more than 16.3 million packages for approximately 8.8 million customers worldwide. Every day, an average of 39.5 million requests from customers for tracking goods are processed in the company. More than 16 petabytes of heterogeneous sensor data in more than 46,000 vehicles are generated within business stores. This includes data such as: package status, vehicle speed, direction, braking, power of propulsion equipment, etc.

UPS's intention to implement Big Data was to enable the use of real-time information from these data to reduce costs, accelerate deliveries, or optimize the route of vehicles in the field [7]. The benefits of implementing Big Data at UPS include [7, 75–77]:

- The cost saving associated with 10 million gallons of natural gas
- Saving of emissions in the amount of 100,000 units
- Saving of 30 million $ per vehicle mile
- Accelerate delivery times by removing routes with left-hand turns
- Saving 85 million miles thanks to real-time vehicle optimization
- The ability to track the whole process of real-time delivery processes online
- The possibility of processing different data from available sources (sensors, weather, situation, etc.) and combining them with various analyses

The above-mentioned examples of Big Data in the automotive industry are based on current technology solutions for automotive production but similar effects can be achieved by deploying Big Data as an analytical platform in the automotive industry focusing on electric cars. Achieving the benefits of introducing the Big Data solution for the automotive industry with a focus on electromobiles can also be achieved in other areas of application of the development of electric vehicles, such as [78–81]:

- Public transport as a mode of transportation where it is possible to use electric cars in terms of new types of passenger transport. At the same time, it would be possible to apply data analysis in the context of monitoring data on transport (speed, allowable mass, sensory data on plant propulsion, safety data). By analysing these data in real time, it is possible to provide the required, real-time service at the event of a malfunction or other event. At the same time, it is also possible to apply advanced predictive models, for example, in the case of planning repairs, equipment upgrades or, in the case of support for the manufacturing process (e.g. testing) of bulk transport machines in the city. By introducing the Big Data solution in public transport, especially in the automotive industry, it will achieve similar benefits to deploying this solution at Mercedes-AMG and UPS in cost savings, removing bottlenecks, supporting science and research, reducing emissions, fuel economy, faster passenger transport, real-time capacity planning, promoting environmental access to business, increasing customer satisfaction and living standards, etc.
- Long distance transport continuously provides research, development and battery production, enabling long routes to be managed with a minimum number of charging stops. At the same time, it is possible to monitor and evaluate vehicle data, to ensure the required passenger comfort, to plan maintenance or to decommission for safety reasons. By evaluating available data, UPS-like benefits can be achieved, e.g. cost savings, emissions, and contributing to an environmentally-friendly business.
- Rail transport, for the expansion and support of science and research in the field of electric trains powered by solar panels. In this way, it is possible to provide regular transport to areas where the rail corridors are not connected to electricity, and therefore the train traffic can only be implemented in a limited way, irregularly and with the aid of trains based on combustion engines. Similarly, to Mercedes-AMG, it is possible to see the opportunity to apply the Big Data solution in this area of the automotive industry, i.e. research and development of

electric motors and supportive devices to ensure cleaner and more regular transport by trains powered by electric motors and solar energy.
- Enterprise warehouse logistics is a part of the automotive industry, which produces forklift trucks, cranes, handling machines, etc., which provide the necessary processes within large warehouse complexes of large companies. From a manufacturing perspective in the automotive industry, it is possible to apply the Big Data solution similarly to Mercedes-AMG, for example, for testing engines and other components of these storage machines. From the point of view of warehouse owners, it is possible to make better decisions based on data obtained from warehouse machines in combination with the warehouse information system. In this way, it is possible to more efficiently manage the warehouse economy of the enterprise or automate certain processes by using, for example, drones moving depot items unmanned. At the same time, it is possible to reduce downtime, save costs, and optimize material flows within warehouses based on the analysis of available data. The company can also define the needs and requirements of a supplier of warehouse machines (electric vehicles) that can be dealt with immediately by the supplier in the event of their occurrence. In this way, it is possible to support strategic co-operative links between companies of different industries to promote production and efficient use of electric cars in practice.
- Electric motors in combination with combustion engines can already be used nowadays mainly due to insufficient battery power. This means that there is a combination of a short-long route where a short-distance electric motor and a long-distance combustion engine can be used. By introducing and using the Big Data solution within this area, it is possible to support science and research in the field of electromotor and combustion engine interconnection. At the same time it is possible to shave ineffectiveness and deficiencies in the production process or in the actual use of the vehicle, the elimination of which would make it possible to produce more efficient, more economical and better engines with greater reach.
- Taxi service in the city is another area of application of electric cars. In the taxi, it is necessary to ensure that there are enough electric vehicles on a single charge so that there is no discharge of the electric car during the customer's transport. Due to the technical characteristics of the vehicle, Big Data technology can be used similarly to Mercedes-AMG. Because of the actual use of a vehicle for business purposes, it is possible to apply the Big Data solution, for example, to analyse and evaluate real-time traffic data and to propose the most optimal route (as in the case of UPS). In this way it is possible to ensure customer satisfaction while saving the costs associated with combustion engines (emissions, petrol consumption, etc.). At the same time, it is possible to monitor and evaluate in real time the technical state of the vehicle (e.g. battery capacity) and to connect to SmartCity to plan servicing processes (e.g. navigate the car to the nearest free charging station, display the occupancy status of the charging stations and predict their release according to the charging status of the connected electric vehicles, etc.).

- Delivery of products in the city (e.g. meals) can be considered in this respect as in the case of a taxi service to shorten the delivery time, select the most optimal route, plan and predict service, etc.
- Due to the size of the motor, shipping is characterized by high fuel consumption; therefore, replacing larger electric motors will help reduce fuel consumption and deliver more power, with short-range single-buoyancy routes, and be long-lasting with solar cells through solar energy. The use of Big Data is similar in this area to the production and testing of engines at Mercedes-AMG. Even with shipping, data processing capabilities can be applied in this case, as in the automotive industry (real-time data analysis, optimal route selection, technical status monitoring and service planning, etc.), as well as the production of electric motors and batteries. The cooperation of the automotive and shipping industries in terms of the production and use of electric motors in the shipping industry could have a significant impact on the ecological transport. It is possible to prevent ecological disasters in the event of a ship's accident and the spillage of gallons of fuel into the oceans.
- Air transport can, like the ship, cooperatively interact with the automotive industry for the development of electric motors and batteries that can serve to support or fully replace existing combustion aircraft engines. Application of Big Data is similar to that of Mercedes-AMG for speedy engine failure and engine reliability testing, which is a key issue in aviation. In this way, it is possible to ensure the ecology of transport or to achieve (in a combination of electric and aviation) an increase of flight time without the necessity of re-fueling.
- The state and public spheres contain police and rescue components where it is also possible to use electromobiles and analyse processing of a variety of real-time data. Already at present, some state police have electric cars for short distances. This allows saving emissions and fuel costs and contributes to the state's eco-friendly environment. The use of the Big Data solution is similar in this case to UPS: real-time track optimization, if combined with SmartCity, for example, can provide a green wave of traffic lights in the event of an ambulance vehicle (ambulance, fire brigade, etc.).

This is an emerging strategy where the various competitors share capacities, transport costs, the two-way use of routes to draw from this partnership, and greater customer satisfaction. There is a higher value strategy for customers – a better reputation in the eyes of customers by committing the company to driving on electric vehicles for a better living standard in the city. They also gain customers who are supporters of a healthy lifestyle, environmentalists, and innovators. The customer may feel comfortable using the services of a company that drives electric cars and even be thankful to it for reduced air pollution. The customer often discourages the lack of recharging stations, as well as the ownership of fuel cars, which he or she has to sell or give up over time and with enough capital.

These give customers a kind of comfort such as:

- Transport of orders directly in front of the door
- Distribution of food and hot meals

- Transport of persons to different places
- Transport of packages and goods

Common characteristics (by introducing electromobility):

- The lower fuel cost vs. the price of electricity
- The charging stations: the expansion and the necessary infrastructure within the global interconnection of goods and services

The factors supporting implementation:

- High fuel prices
- Abrasion of combustion engines (more frequent service)
- Frequency of transport
- The increasing consumer consumption and dependence on carriers
- A large number of cars and others forms of transportation

Distribution and transportation services are nowadays an everyday affair. That is why we see the potential use of electric cars in this area very high, both for individual companies and for their current or potential customers. In spite of the disadvantages that the use of electric cars currently carries with it, it is a matter of fact that the current development of electric cars is one of the most important trends in the future. This would lead to a higher electric vehicle awareness, which would most likely influence the overall use of Big Data technology in the automotive industry. Apart from that, they can increase their level of development in many areas. On the topic of electric cars there will be much future interest, because as compared to conventional vehicles, there are "no" limitations.

In case of the strategy of building green cars and sustainable manufacturing plants, they reach the goal of attracting consumers that are interested in sustainability, which is a future trend. Electric cars should be specified according to the geographic allocation and the territory they wish to operate in the future. One must identify and find the right customer who will bring into the community a new unrecognized thing that causes others to change their attitude towards the use of electric cars. The future of electric cars therefore depends on customer preferences, so they need to be gradually absorbed into this problem. Start with customers gradually to communicate and address their demands and for the needs of electric vehicles in transport.

Conclusion

These statements have been prepared and formulated not only from analysis of available information that has been included in this article, but also from the knowledge and experience of the authors.

Based on the above findings, in the context of forming synergy effects in support of Big Data solutions in the automotive industry, we define the following statements:

1. Specific solutions in the form of informational systems are required to obtain necessary and relevant information, which is now a vital input for decision-making at strategic management level.
2. The development and use of Big Data solutions in the automotive industry are important in terms of generated disparate data that may contain potentially important information.
3. The Big Data solution can be used in almost every field of the automotive industry as strategic management with needs to streamline the entire industry.
4. The opportunity for use of Big Data solutions is created by the existence of ever-expanding disparate data. Information obtained in these solutions may significantly affect forming synergy effects in strategy management of these processes.
5. The decision making in strategy management is influenced by uncovering synergy effects created with intensive reliance on data and information.

On the basis of the previous facts it is necessary to select the following elements for the formation of proposals and applying the synergy approach for Big Data solutions used in the automotive industry:

– Cooperative management style. The current capacity for action and aggression in an environment for strengthening the competitiveness is manifested by creating mutually beneficial interactions (such as alliances, networks, etc.).
– The environment of companies must be prepared to adapt to the use of new approaches and ways of working with data – through them the whole environment changes, develops and responds.
– The value of a company or group of companies is now considered in the context of access to resources and the benefits it can provide and create. These data represent a new source of the current companies' environment, the utilization rate and proportionally increasing importance to the competitiveness of companies.
– Identifying partners for cooperation is dependent on data.
– Mutually beneficial cooperation makes goals and creates win-win strategies.
– Cooperation is a window to a broad range of skills and the development of new knowledge.

Through these changes the intensity level of synergy increases in the cooperative environment and thus the occurrence of synergistic effects increases. Strategy (policy) of the company has been categorized into different types by the current authors and are discussed in detail.

The potential of processing amounts of disparate data and obtaining information lead to an improvement in many ways (driving performance of cars, avoidance of accidents, increased customer satisfaction, etc.). This can ultimately cause a synergy effect from the level of strategy management in production to efficiency in the automotive industry.

One of these strategies is just cooperation, namely, cooperation with automotive companies in a certain area within defined conditions and competitive competition. The product of such interactions can also be synergy.

If cooperation is necessary for the creation of synergies, then through the strategic management of cooperative organizational forms, synergistic effects in the automotive company environment can be obtained. This effect is a new critical factor of success of the present, more and more globalized environment.

Digitalization of the automotive industry and processing amounts of diverse data are the bases for effective management and planning of all processes in the automotive industry. With powerful analytical tools this solution can be deployed to support decision making in more and more areas of the business. Big Data find significant applications in the area of the automotive industry.

We put higher importance on the focus of new successful collaborative solutions. This case study suggests using Big Data solutions in the automotive industry to create synergy effects based on a cooperative environment which is then the starting point to realize more intense research in the previous context in certain automotive areas.

Acknowledgements This work was supported by the Slovak Republic scientific grants VEGA 1/0617/16 and APVV-15-0511.

References

1. Best practice guideline. Big Data. http://datascienceassn.org/sites/default/files/Big%20Data%20Best%20Practice%20Guideline.pdf
2. ICTs for decision making problems and prospects. http://www.gu.edu.pk/New/GUJR/PDF/Dec-2009/5%20Qammar%20Afaq%20Paper%201.pdf
3. Aldea, C. C., Popescu, A. D., Draghici, A., & Draghici, G. (2012). ICT tools functionalities analysis for the decision making process of their implementation in virtual engineering teams. *Procedia Technology, 5,* 649–658.
4. Axelrod, R., & Hamilton, W. D. (1981). The evolution of cooperation. *Science, 211,* 1390–1396.
5. The impact of information technology in facilitating communication and collaboration in Libyan public sector organisations. http://itc.scix.net/data/works/att/w78-2009-1-61.pdf
6. Buday, T. (2011). Unconventional sources of data for conventional BI systems (in Slovak). http://beta.itnews.sk/temy/5-2011/2011-05-12/139761-nekonvencne-zdroje-dat-pre-konvencne-systemy-bi
7. How UPS uses Big Data with every delivery. http://businessintelligence.com/big-data-case-studies/ups-uses-big-data-every-delivery/
8. Macy's gets a leg up on competition with business intelligence. http://businessintelligence.com/big-data-case-studies/macys-gets-leg-competition-business-intelligence/
9. Tesco's legendary Big Data benefits. http://businessintelligence.com/big-data-case-studies/tescos-legendary-big-data-benefits/
10. Černý, M. (2013). Eight technology trends that transform libraries in the information society (in: Czech). *ITlib, 2,* 30–36.
11. Big Data in the automotive industry. http://www.csc.com/auto/insights/103241-big_data_in_the_automotive_industry
12. Connected car fleet management with big data analytics. https://www.datameer.com/company/datameer-blog/connected-car-big-data-automotive-industry-future/
13. Big Data in big companies. International Institute for Analytics. http://www.sas.com/resources/asset/Big-Data-in-Big-Companies.pdf

14. Big Data and analytics in the automotive industry. https://www2.deloitte.com/us/en/pages/manufacturing/articles/big-data-and-analytics-in-the-automotive-industry.html
15. Big Data. New ways of processing and analyzing large volumes of data. https://www.system-online.cz/clanky/big-data.htm
16. Fehr, E., & Schmidt, K. M. (1999). A theory of fairness, competition, and cooperation. Harvard college and the Massachusetts Institute of Technology. *The Quarterly Journal of Economics, 114*, 817–868.
17. Hamel, G., Doz, Y., & Prahalad, C. K. (1989). Collaborate with your competitors – and win. Harward business review.
18. Holubčík, M. (2015). Cooperation as base for synergy. *eXclusive e-Journal*, ISSN: 1339-4509
19. Fast-track automotive innovation. https://www.hpe.com/h20195/v2/GetPDF.aspx/4AA6-5167ENW.pdf
20. Digital disruption and the future of the automotive industry. https://www-935.ibm.com/services/multimedia/IBMCAI-Digital-disruption-in-automotive.pdf
21. Ireland, R. D., Hitt, M. A., & Vaidyanath, D. (2002). Alliance management as a source of competitive advantage. *Journal of Management, 28*, 413–446.
22. Koman, G., & Kundrikova, J. (2016). Application of Big Data technology in knowledge transfer process between business and academia. *Procedia Economics and Finance, 39*, 605–611.
23. Global Automotive Executive Survey. (2016). https://home.kpmg.com/xx/en/home/insights/2015/12/kpmg-global-automotive-executive-survey-2016.html
24. Kubina, M., Varmus, M., & Kubinova, I. (2015). Possibility of improving efficiency within business intelligence systems in companies. *Procedia – Economics and Finance, 26*, 300–305.
25. 7 Reasons why Big Data easier for enterprises to life. http://www.zive.sk/clanok/66175/7-dovodov-preco-big-data-ulahcia-firmam-zivot
26. Marielle, D. H., & Sol, G. H. (2001). The impact of information and communication technology on interorganizational coordination: Guidelines from theory. *Informing Science – Special Series on Information Exchange in Electronic Markets, 4*, 129–138.
27. Mayer-Schonberg, V., & Cukier, K. (2013). *Big data: A revolution that will transform how we live, work, and think*. London: John Murray Publishers.
28. Car data: paving the way to value-creating mobility. Perspectives on a new automotive business model. https://www.mckinsey.de/files/mckinsey_car_data_march_2016.pdf
29. Mcnulty, E. (2014). Understand big data: Infrastructure. http://dataconomy.com/2014/06/understanding-big-data-infrastructure/
30. Understand big data: The ecosystem. http://dataconomy.com/2014/06/understanding-big-data-ecosystem/
31. What is ´big data,´ anyway?. http://www.strategy-business.com/blog/What-Is-Big-Data-Anyway?gko=28596
32. The automotive industry as a digital business. http://www.ntti3.com/blog/the-automotive-industry-as-a-digital-business/
33. Minelli, M., Chambers, M., Dhiraj, A. (2013). Big data, big analytics. Emerging business intelligence and analytic trends for today's businesses.
34. Nedelcu, B. (2013). About Big Data and its challenges and benefits in manaufacturing. *Database Systems Journal, 4*, 10–19.
35. Nowak, M. A. (2006). Five rules for the evolution of cooperation. *Science, 8*, 1560–1563.
36. Ogbomo, O. M., Ogbomo, F. E. (2008). Importance of information and communication technologies (ICTs) in making a healthy information society: A case study of Ethiope east local government area of delta state, Nigeria. *Library Philosophy and Practice*, ISSN: 1522-0222.
37. Model of handling Big Data and knowledge management in automotive industry. https://ideas.repec.org/h/tkp/mklp16/731-740.html
38. Connectivity in the Automotive Sector. http://www.pinsentmasons.com/en/media/publications/connectivity-in-the-automotive-sector/
39. Big Data case study: Tesco. http://www.speakersconnect.com/robert-plant-big-data-case-study-tesco/

40. How to build a secure connected car. http://www.forbes.com/sites/ibm/2014/11/10/how-to-build-a-secure-connected-car/#56891ee44e40
41. 12 Big Data definitions: What's yours?. http://www.forbes.com/sites/gilpress/2014/09/03/12-big-data-definitions-whats-yours/#41a8246e21a9
42. RampelliI, S., Vadlamani, S. K., & Nukala, S. P. K. (2014). Big Data and data analytics: An action platform for the science of prediction. *Journal of Engineering Research and Applications, 4*, 24–27.
43. Big Data is driving your car. https://www.linkedin.com/pulse/iot-big-data-preventative-predictions-matthew-reaney?trk=mp-reader-card
44. Walmart makes Big Data part of its DNA. https://datafloq.com/read/walmart-making-big-data-part-dna/509
45. Self-driving cars will create 2 petabytes of data, What are the Big Data opportunities for the car industry? https://datafloq.com/read/self-driving-cars-create-2-petabytes-data-annually/172
46. Big Data is scaling BI and analytics. http://www.information-management.com/issues/21_5/big-data-is-scaling-bi-and-analytics-10021093-1.html
47. How is Big Data impacting sales in the automotive industry? http://docplayer.net/10728874-Big-data-how-is-it-impacting-sales-in-the-automotive-industry.html
48. The connected vehicle: Big Data, big opportunities. http://www.sas.com/content/dam/SAS/en_us/doc/whitepaper1/connected-vehicle-107832.pdf
49. "Big" with a company that won the big data (in: Slovak). http://businessworld.cz/analyzy/big-firmy-to-s-big-daty-vyhraly-11472
50. The automotive industry's Big Data callenge (Part 2). https://corporate-innovation.co/2016/02/11/the-automotive-industrys-big-data-challenge-part-2/
51. Big Data: From great expectations to practical use (in Slovak). https://www.systemonline.cz/business-intelligence/big-data-od-velkych-ocekavani-k-praktickemu-vyuziti.htm
52. Souček, Z. (2005). *The company of the 21st century* (in Slovak). Professional Publishing: Praha. ISBN 80-86419-88-6.
53. Soviar, J., Lendel, V., Kocifaj, M., & Čavošová, E. (2013). *Cooperation management (in Slovak)*. Žilina: EDIS.
54. Hu, W., Kaabouch, N. (2014). Big Data management, technologies, and applications. In *Information Science Reference* (pp. 222–269).
55. Statistical Office of the Slovak Republic. Businesses: by legal forms of economic activities (NACE Rev. 2) the size and number of employees. https://slovak.statistics.sk
56. Stricker, K., Wegener, R., & Anding, M. (2014). *Big Data revolutioniert die Automobilindustrie*. Bain & Company.
57. Synek, T. (2014). How to cut IT costs by using big data technologies. (in Slovak) INFOWARE 10, pp. 23–24.
58. The automotive industry's Big Data callenge (Part 1). https://corporate-innovation.co/2016/01/25/the-automotive-industrys-big-data-challenge-part-1/
59. Ericsson. (2014). Unhabitat for a better urban future. The role of ICT in the new urban agenda.
60. Ujunju, O. M., Wanyembi, G., & Wabwoba, F. (2012). Evaluating the role of information and communication technology (ICT) support towards process of management in institutions of higher learning. *International Journal of Advanced Computer Science and Applications., 3*, 55–58.
61. Varmus, M., & Lendel, V. (2014). Evaluation of the innovative business performance. *Procedia – Social and Behavioural Science, 129*, 504–511.
62. Three ways Big Data is helping to build better cars. http://www.forbes.com/sites/ibm/2014/12/18/3-ways-big-data-is-helping-to-build-better-cars/#ae4cc4826e30
63. Wyman, O. Implementing Big Data is the hardest part. http://www.oliverwyman.de/insights/publications/2015/jul/implementing-big-data-is-the-hardest-part.html#.WHyD91PhCM8
64. Yildirim, A., Özdoğan, C., & Watson, D. (2014). Parallel data reduction techniques for big datasets. Big data management, technologies, and applications. In *Information Science Reference* (pp. 72–93).

65. Kang, N. H., & Sakai, K.(2000). International strategic alliances. *OECD Science, Technology and Industry*, ISSN: 1815-1965 .
66. EVvolumes.: Global Plug-in Sales for 2016. The electric vehicle world sales database. http://www.ev-volumes.com/country/total-world-plug-in-vehicle-volumes/
67. EurActiv.: Electromobility – return to the future. https://euractiv.sk/fokus/energetika/elektromobilita-navrat-do-buducnosti-000304/
68. International Energy Agency. Global EV Outlook 2016, Beyond one million electric cars. https://www.iea.org/publications/freepublications/publication/Global_EV_Outlook_2016.pdf
69. Collins, T.. Airbnb-style electric car site lets you charge up your vehicle on someone else's driveway. http://www.dailymail.co.uk/sciencetech/article-4469050/Airbnb-style-booking-service-charging-electric-cars.html
70. APPM Management Consultants en Policy Research Corporation. Nederland inductieland?! Een verkennende studie naar de mogelijkheden en potentieel voor inductieladen. http://www.rvo.nl/sites/default/files/2015/01/Nederland%20Inductieland%20-%20Een%20verkennende%20studie%20naar%20mogelijkheden%20en%20potentieel%20voor%20inductieladen.pdf
71. Open Charge Map. https://openchargemap.org/site
72. ParkNow. https://eu.park-now.com/.
73. Amsterdam Roundtable Foundation and McKinsey & Company The Netherlands. (2014) Evolution, Electric vehicles in Europe: gearing up for a new phase?
74. UPS. http://www.ups.com
75. Bessis, N., Dobre, C. (2014). *Big Data and internet of things: A roadmap for smart* environments (470p). Springer. ISBN 978-3-319-05028-7.
76. Westberg, T. The road to optimization. http://www.slideshare.net/UtahBroadband/2015-broadband-tech-summit-todd-westberg-ups-presentation
77. Davenport, T., & Dyché, J.. Big Data in big companies. *International Institute for Analytics*. http://www.sas.com/resources/asset/Big-Data-in-Big-Companies.pdf
78. ABB.:TOSA electrical bus charging infrastructure. http://new.abb.com/substations/railway-and-urban-transport-electrification/tosa-electrical-bus-charging-infrastructure
79. Zverková, S. Electromobiles have dominated the planet, today they report a return (in Slovak). https://autobild.cas.sk/clanok/185635/elektromobily-ovladali-planetu-dnes-hlasia-navrat/
80. Kmáč, D. TEST: Volkswagen e-Golf – Electricity, the best drive for the best hatchback? (in Slovak). https://automix.atlas.sk/tlac/?IdText=854011
81. Kollár, M. Even the cops will go to the electricity. Thanks to Tesla (in Slovak). https://medialne.etrend.sk/internet/uz-aj-polisi-pojdu-na-elektrinu-vdaka-tesle.html
82. Williams, M. Why are mobile phones batteries still so crap? http://www.techradar.com/news/phone-and-communications/mobile-phones/why-are-mobile-phone-batteries-still-so-crap--1162779
83. Pavuk, J. Mobile chargers will soon end up in the dumpster (in Slovak). https://www.etrend.sk/firmy/nabijate-mobil-dvakrat-denne-nabijacky-uz-o-chvilu-skoncia-v-kosi.html
84. Donato, C. (2014). Mercedes-AMG gibt mit Echtzeitanalysen noch mehr Gas. http://news.sap.com/germany/mercedes-amg-gibt-mit-echtzeitanalysen-noch-mehr-gas/sthash.L1SCdp6J.dpuf
85. Overby, S. (2014). Mercedes-AMG: A showcase for real-time business decisions. http://clients.23k.com/SAP/7910-SoH_Ebook/Daimler_AG_AMG/SAP-Mercedes-AMG-REPORT.pdf
86. Mercedes-AMG. (2017). Mercedes AMG: About company. https://www.mercedes-amg.com/about_company2.php?lang=eng

Index

A
Abbi smart bracelet, 431
Acceleration, 55, 340–342
Acceleration change factor π_T, 140
Age management, 398
Agent-based systems, 166
Aggregated complexity index (CI), 74
Aluminium
    average surface roughness evaluation, 354
Analytical hierarchy process (AHP)
    advantages, 7
    application, 10
    disadvantages, 7, 8
    innovation management, 9, 10, 12, 13
    mathematical principles, 6
    principles, 8
    scientific analysis, 7
    structured hierarchy, 7
    usage, 9
Application server
    backend, 93
    database management system, 93
    real-time verification, 95
    requirements, 88
    Ruby on Rails framework, 92
    SNMP traps processing, 93–94
    web-based user interface, 92
Application support includes Car Control (APPM), 477
Approximate calculation of definite integrals, 62
Approximate calculation of nonlinear equations, 62
Arduino platform, 438, 446

Assistive technology
    Android OS, 434
    Arduino platform, 438
    Bluetooth, 437
    MTD, 434, 436
    NFC, 437
    Raspberry Pi, 438
    service accessibility, 437
    service functionalities, 435
    smart wristband, 436
    SOS paging service, 434
    traffic information, 435
    UP2DATE organization, 433
    user disability, 433–434
    value chain, 436
Automated guided vehicle (AGV), 195, 197
Automatic passenger counters (APCs), 457
Automotive Cluster Slovakia (ACS), 304, 306, 307
Automotive industry
    acquisition, 476
    aftermarket services, 476
    decision-making dimensions, 475
    design, 476
    and distribution and transport service strategy, 477–483
    electric vehicles, use of, 478–483
    General Motors' investment, 476
    Industry 4.0, 218
    internet and cloud-based solutions, 475
    marketing, 476
    production, 476
    quality assurance, 223, 224

Automotive industry (cont.)
   services and technical support, 476
   supply chain, 219, 220
Automotive sector, 304–306, 311, 315
Automotive testing, 85, 99
Autonomous lighting, 425, 427
Auxiliary components supporting system
   GPS, 457
   passenger counting systems, 457–459
   smart transportation application, 455
Average value, 74

B
Big data, 222
Big Data solutions
   applications, 475
   automotive industry, 468, 469
      (see also Automotive industry)
   business analysis, 472
   business process, 468
   companies, 467
   competitive advantages, 474
   data warehousing and management tools, 472
   decision-making, strategic level, 475
   Hadoop architecture, 473
   IBM's, 474
   ICTs, 471, 472
   infrastructure, 473
   law firms, 473
   managerial thinking, 467
   managers, 468
   MapReduce and Hadoop open platforms, 473
   NASA scientists, 472
   patterns detection and key factors, 468
   synergy effects formation, 468
   unstructured and semi-structured data, 468
Bluetooth Beacon technology, 443
Bluetooth technologies, 437, 439
British Standards Institution (BSI), 385
Business
   decision-making, 257
   flexibility, 257
   management methods, 256
   management of organizations, 257
   in market economy, 256

C
CAM software
   graphical environment Autodesk Inventor 2017, 356
   thin-walled component
      safety plane selection, 357
      simulation, 358
      three-axis milling centre Pinnacle VMC 650S, 359
      tool selection, 358
      zero-point selection, 357
C-E diagnosis, 339
CEIT Intelligent Logistic Management, 193–196
Central European Institute of Technology (CEIT), 167
Charger Reservation, 477
City transportation, 464
Climate change, 103, 108, 115–117
Climate mitigation
   effects, 116
   semi-public space, 124, 125, 127
   smart community, 122, 124
Cloud computing, 222
Cloud Computing for the Blind (CCfB), 436, 438, 440, 445, 447, 449
Clusters
   ACS and MAC, 304
   activities, 305, 306
   automotive sector, 304
   cost savings of scale, 305
   defined, 303
   efficiency and productivity, national and global level, 304
   enterprises, 311
   firms cooperation, 311–312
   implementation risks, 313–315
   innovation performance, 304
   marketing and management activities, 308, 311
   Porter's diamond, components, 309–314
   regional economies, 304
   stakeholders, 308–309
   subcontractors development, 305
Code of Ethics for Advertising Practice, 284
Collaborative solutions, 36, 468, 485
Common-pool resources (CPRs), 126, 127
Community, 122, 124
Competitiveness, 231, 240, 245
Complexity indices, 77, 80
Complex systems theory, 111
Composite manufacturing process, 320, 322
Computer-aided manufacturing (CAM) software, 158, 159
Computer-aided process planning (CAPP) system
   CA systems, 328
   computer assistance, 328

corporate planning and organizational structure, 330
East European (EE) post communism countries, 328
enterprise unit's structure, 151
EU's competitiveness, 327
group technology, 154
individual approach, 152
individual companies view, 151
innovation strategies, countries, 327
IS (*see* Information systems (IS))
marketing and sales, 330 (*see also* Multi-variant process planning)
newly designed computer aided systems, 152
plant, principle scheme, 330
process planning systems, 330
procurement, 330
production system control, 331
production system design, 330, 332
purchasing all parts of software from different vendors, 153
quality assurance and accounting systems, 330
renewed momentum, 327
software delivered completely by general contractor, 153
software development by company itself, 154
technological documentation, 152
technological processes typification, 154
type technological process, 154, 155
user requirements, 152
Computer aided (CA) systems, 328
Computer-aided virtual environment (CAVE), 192, 193
Computer numerical control (CNC), 157
Consumer behaviour, 284, 287, 288
Consumer education, 289
Cooperation management, 36
 ability to rapidly react to changes, 271
 actual respondents, 264
 administratively undemanding methods of management, 273
 building of relationships, attributes, 265
 comprehensive mapping, 264
 creative approach, 272
 decentralized management, 272
 definition, 265
 direct evaluation and testing, new ideas, 272
 dynamic cooperation organizational structures, 276
 economic dimension, 276
 geographical proximity, partners, 266
 groups and individuals, 272

high added value, 273
higher degree of uncertainty and risk, 272
impressions/gut feelings, 263
informal team work, 273
information background and knowledge creation, 266
interorganizational activities, 265
interpretation variability, 265
lower requirements, management system, 273
management levels, 273
matrix of cooperation organizational structures, 274
matrix organizational structure, 275, 276
multiple employees, roles of, 271
mutual trust, 266
network organizational structure, 274
ongoing market development, 263
organization, 271
organizational elements and connections, 273
organizational factors, 266
organizational structures, 264
research methods, 264
role of innovation, 266
social dimension, 276
stakeholders, 264
wider, regional environment, 266
Cooperative relations, 308, 309
Corporate social responsibility (CSR)
 company environmental records, 298
 Fisher exact test, 298
 stakeholder, 292, 295
 transparency, 297, 299
CPRs, *see* Common-pool resources (CPRs)
Crisis, 117
Critical values
 bisection method, 63, 64
 F-distribution, 66–68
 normal and Student's t-distribution, 62
 Simpson's rule, 63
 standard numerical methods, 62
 statistical methods, 62
 tables/specialized statistical software, 62
 trapezoidal rule, 64
 χ^2 distribution, 65–66
Customer behaviour models, 282
Cutter location (CL), 158
Cutting zone
 after cold cutting technologies, 331
Cyber-physical system (CPS), 188
 description, 188
 digital factory, 188
 smart factory, 188
 virtual factory, 188

Cybersecurity, 105
Cycle counting, 86
Cycle index, 73

D
Decision algorithm, 6
Decision-making
  AHP (*see* Analytical hierarchy process (AHP))
  algorithm, 6
  content and steps, 5
  final decision, 4
  management, 4
  problem, 4, 5
  situations, 4, 5
  variations, 5
Decision points (DS) index, 73
Decision support systems (DSS), 309
Density index, 73
Design for behavior change
  activity theory, 368
  aesthetical interventions, 369
  interactive media art, 370, 371
  physical and social environment, 368
  transform public spaces, 368
Digital factory concept, 188, 189, 192, 193
Digital humanism, 165
Distribution redundancy index (RD), 73
Dynamic complexity, 72
Dynamic environment, 4

E
East European (EE) post communism countries, 328
Echo boomers, 166
Electronic counter IP module
  photography of, 90
  requirements, 88
  signal processing diagram, 90
  wall-mounted panel, 91, 92
Energy system, 24
Energy transition
  macro level, 25, 26
  meso level, 25
  micro level, 24
  regime level, 25
  sociotechnical systems, 24–26
  spatial planning, 23, 28
  sustainable energy systems, 104
  systemic change, 26
  TM, 112
Energy value chain, 104

Engagement of employees in innovation
  corporate strategy, 231
  decision-making, 231
  economic growth, 239
  everybody is fully engaged, 231
  motivation, 238, 241
  regional structure, 237
  stratification criteria, 237
  troubleshooting, 231
  unconscious engagement, 230
Environmental policy
  business (*see* Business)
  business sector, 250
  company management requirements, 258
  company's profitability, 249
  competitive advantages, 250
  construction activities, 255
  controller, role of, 258
  financial management, 257
  high-quality controlling, 256
  vs. indicators of profitability
    (*see* Indicators of profitability)
  management process, 256
  material resource use (DMC) per capita, 254, 255
  national policy priorities and responses, 254
  regression analysis, 250
  resource productivity, 255, 256
  transformation and stability development, 258
European Alliance for Innovation (EAI), 36
European Factories of the Future Research Association (EFFRA), 184
European Recovery Plan, 184
Expert Choice, 9, 10, 14

F
Factories of the Future (FoF), 194, 199
  CEIT and ZIMS, 167
  clusters, 168
  collaborations with enterprises, 175
  EFFRA, 167, 169
  good practice, 170
  human resources officers, 167
  knowledge-based economies and societies, 169
  learning of university model, 167
  national project, 172–174
  perspective of orientation, study programme, 175
  project partner organizations, 169
  Ranking of Collaborations, 175

real salary development, 166
regional positive areas, 170
regional problematic areas, 170
Slovak Republic practice, 170–172
students collaborations, 175
study programmes, 174
system of learning, 168
young talent programmes, 170
ZIP, 169
Failure mode and effect analysis (FMEA), 339, 340
Fall detection, 440, 441
Fatigue testing machines
    classification, 86
    cycle counting, 86
    range counting method, 86
    VBA language, 86
Fault intensity value, 135
F-distribution, 66–68
Feed rate, 347, 350
Fictional Inquiry technique, 369
Finishing mill, 341, 342
Fisher exact test, 298
Fleet management, 453
Fleet managers
    drivers performance, 454
    garage technicians, 463
    IP cameras, 460
    LCD monitors, 461
    mobile application, 463
    municipalities/owner, 453
    passenger counting systems, 457
Flow-shop system layout, 79–81
Food industry, 294, 295
Friedman test, 296
Future data analytics, 98

G
Generation Y marketing, 166
Generative process planning, 330
Global Positioning System (GPS), 457
Global Reporting Initiative (GRI), 294
Green markets
    core customer, 287
    customer attitudes and behaviour, 286
    inside-sustainability, 287
    mid-level, 287
    outside-sustainability, 286
    periphery customers, 287
    sustainability, 282, 286
Green purchases, 286
Greener automobile industry, 477
Group technology, 154

H
Haptic Virtual Collaborative Development Environment (HVACDE), 192
Heat islands, 116, 120–122, 126
Holonic concept, 189
Hot rolling
    checkout, 339
    FMEA, 340
    input control, 338
    interoperational control, 338
    laminar strip cooling, 338
    processes, 338
    risk number, 339
    slab warehouse, 338
Human capital, 239, 383, 390, 391, 397, 402

I
Identity, 405, 406, 410
I-Government, 117–119
Important interest groups (IIG), 440
Incrementalism, 404
Indicators of profitability
    competitive advantages, 251
    competitiveness, Slovak companies, 252
    description, 250
    internal and external indicators, 251
    'Operating return on sales' indicator, 253
    'Return on assets' indicator, 253
    'Return on equity' indicator, 253
    'Share of EBITDA in sales' indicator, 253
    'Share of value added in sales' indicator, 253
    SWOT analysis, 251
    transformation process, 251
Inductive line
    cost, 209
    scheme, 209
Industry 4.0
    Big data, 222
    cloud computing, 222
    digitalization, 103
    educational system, 222
    fragments of philosophy, 221
    functionality, 220
    ICT, 221
    intelligent and flexible processes, 104
    IoT, 104, 105, 222
    IT/ICT, 112
    OEMs, 221
    quality, 220, 221
    risk analysis, 222
    security/safety, 221

Industry 4.0 (*cont.*)
   SK and AT, DE and CH
      average assessment comparison, 108
      data and collection, 106, 107
      IT/ICT, 108
      percentage comparison, 108, 109, 111
      sustainable manufacturing, 104
      technologies, 104, 105, 111
Information and communication technology (ICT), 120, 122, 386, 471, 472
   quality of life, 432
   TGU, 432
Information systems (IS)
   business units, European Union, 332
   defined, 329
   entrepreneurial surroundings, 329
   manufacturing information systems, 329
   optimization and variation rules, 331
   process planners, 329
   production software selection, 329
   research, European market, 332
   small and medium-sized companies, 331
   structural concept, 331
Information technology (IT)
   description, 452
   software, 107, 109
   transportation system, 452, 453
Innovation management
   criteria, 10
   Expert Choice, 9
   individual criteria
      activities and variants, 11
      final evaluation, 12
      paired comparison, 12
      weights to variants, 11, 12
Innovations, 14, 107, 109, 111, 112, 392, 393
   employees (*see* Engagement of employees in innovation)
   model with high engagement level, 240, 242
   organisations, 230
   potential of human resources, 230
Intelligent manufacturing systems (IMS), 185, 194
Intelligent transport systems (ITS), 423, 453
Interaction, *see* Interactive installations
Interactive installations, 369, 370
   Before & Beyond, 376, 377, 379, 380
   InnerBody
      data, 373
      external stimuli, 374
      heart interface, 373, 374
      MRI examination, 371, 372, 374
      relaxation, 375
      stimuli monitoring, 374
      user interviews, 375, 376
      users participation, 374
Interactive media art, 368, 369
Interactive media design, 368, 370, 371
Intercity drivers, 465
Interface, 371, 372, 374
International Energy Agency, 477
Internet of Things (IoT), 222, 223, 436–438
   application server, 88
   automotive industry plant, 89
   cheap operating system and database server, 89
   electronic counter IP module, 88, 90–91
   manager's objective, 87
   new information system, 87
   SNMP, 88, 91
   system requirements, 87–88
Internships, 398
IP cameras, 459–460

J
Job-shop system layout, 75–78

K
Kanban principle, 200, 201, 207, 208, 210, 211, 214
Knowledge environment
   system of learning, 168

L
Labels test
   Alien 9540 Squiggle, 138
   inaccurate/alternate orientation, 138
   installation, 138
   nominal pressure load, 140
   types, 138
Learning from the process, 168
   approximative (gross) production management, 191
   knowledge-based systems, 190
   physical production system, 190
   system of learning, 190
Learning of university model, 167
Location-based advertisement system
   constant and unique, 461
   LCD monitors, 461
   overview of, 460

Index 495

M

Magnitude redundancy index (RM), 74
Management of Manufacturing Systems (MMS) Conference, 36
Manager's Excel VBA-based dashboard, 97–98
Market readiness, electric mobility sector growth, 477
Marketing research, 295, 300
Mass customization (MC), 71, 183
Material flows
  AutoCAD software, 200
  Kanban system, 201
  process diagram, 203
  resources, 202
  Sankey diagram, 202, 213
  supermarket, 207
Mean time between failures (MTBF), 135
Mean value$_{-\lambda s}$, 135
Mechanical cycles counting
  accuracy verification, 94
  comparison method, 95–96
  IP module counter with prescaler, 94–95
  simple-range counting method, 86
Mechanical engineering industry, 331
Millennium generation, 166
Mobility, 223, 423, 430, 440
Model–View–Controller (MVC) architecture framework, 92
Moravian-Silesian Automotive Cluster (MAC), 304, 307
MTD Sony Xperia Z, 432
Multi-agent system (MAS), 189
Multilevel governance, 117, 118, 122
Multimedia
  file browser application, 416
  graphics interface, 417
  synchronization application, 416, 417
Multisensory environments, 368, 369
Multi-variant process planning
  basis of group representative, 160
  CAM software, 158, 159
  CNC, 157
  complex-group representatives, 159
  designers ambitions, 155
  heterogeneous information system, 156
  manufacturing sequence approach, 156
  NC program, 156
  new component within group technology, 159
  object assignment, group representative, 161
  process plants, 156
  production segment, 156, 157
  tested information system, 161
  theory of, 155
  3D model, NC program verification, 161

N

Natural fibers, 327
Near-field communication (NFC), 437
Nemenyi test, 296
Nepal irrigation systems, 119
Net generation, 166
Niche management, 23
  area-based conditions, 29, 30
  internal/external processes, 29
  local-regional interface, 28
  macro level, 25, 26
  meso level, 25
  micro level, 24
  niche-space relationship, 29
  regime level, 25, 28
  RES, 28
  SE, 28, 29
  sociotechnical systems, 25, 26, 28
  systemic change, 26
  transition management, 26, 27

O

Onboard units (OBUs), 457
Open charging GPS map (Open charge map), 477
Operating return on sales' indicator, 253
Operational complexity, 72
Organization
  in clusters, 306
  educational institutions, 306
  marketing activities, 311
  stakeholders, 308
Over-Kanban production board, 211

P

Panic buttons, 460
Pareto analysis, 340–343
Passenger counting system
  APCs, 457
  fleet managers, 457
  overview of, 459
Path index, 73
Personal strings, 379, 380
Photovoltaic panels (PV), 427
Place attachment, 402, 404–408, 410
Place-based approach, 109
Placemaking, 421
Plan for Every Part (PFEP), 201
Pneumatic artificial muscles (PAMs)
  acceleration, 341, 342
  applications, 338
  benefits, 338
  control system, 339, 340

Pneumatic artificial muscles (PAMs) (cont.)
   electropneumatic valve, 341, 342
   experimental device, 53, 339, 340
   operating conditions, 340
   output force, 338, 339
   position control, 54, 338
   PWM, 341
Potential of human resources, 230
Probability distributions, 62, 68
Process diagram, 203
Process optimization
   future data analytics, 98
   information system administrator, 96
   manager's Excel VBA-based dashboard, 97–98
   software applications, 99
   software tools, 99
   test engineer, 96–97
Product λ_s disturbances determination, 142–143
Production systems
   analysis, 205
   AutoCAD software, 200
   benefits, 215
   cooling aggregate, process diagram, 205
   current and future state comparison, 215
   enterprises manufacture, 199
   FoF, 199
   implementation advantages, 214
   implementation plans, 206–213
   improvement criteria, 206
   improvement tasks, 200
   input storehouse layout, 204
   Kanban system, 214
   material flows, 200
   PFEP, 201
   process diagram, 203
   rack storehouse, 204
   R&D, 199
   return on investment, 215
   Sankey diagram, 202
   technological procedure, 202
   transport devices utilization, 205, 206
   universities and development centres, 199
Public lighting
   intelligent LED solar lamp, 424
   lighting columns, 425
   PV panels, 427
   rotor, 425
   VAWT, 425
   wind belt, 425
Pulse-width modulation, 339, 340, 342
Purchase decision-making, 287

Q
Quality assurance, 219, 223, 224

R
Rack storehouse, 204
Range counting method, 86
Raspberry Pi, 438, 447
Renewable energy sources (RES), 24, 28, 104, 105, 109, 111
Reporting system, see Vehicle data processing
Research institutes, 185, 197
Responsive public spaces, 368, 378
Return on assets' indicator, 253
Return on equity' indicator, 253
RFID labels
   acceleration tests, 136
   Arrhenius theory, 136
   bar code/manual entry, 134
   control, quantitative and qualitative indicators, 133
   faults categorization, 135
   intensity of component/device faults, 136
   no-failure operation, 134
   product's technical specifications, 135
   reliability, 134
   sustainability, 134
   tag unit, 137–138
   technical life expectancy, 135
   testing of reliability, 134
   time accelerated, test functionality, 135
   unambiguous attribution, 134
Roll mill, 341, 342
Roughing mill, 340–342

S
Sankey diagram, 202
Self-governance, 125
Semi-public spaces
   access-controlled environment, 124
   CPRs, 119
   institutional, 125
   physical environment, 124
   private space, 124
   property relations, 125
   qualities, 125
   self-governance, 126, 127
Shannon's information entropy calculation, 75
Share of EBITDA in sales' indicator, 253
Share of value added in sales' indicator, 253
Shewhart control charts, 339
Simple Network Management Protocol (SNMP), 88, 91, 93–95

Simulation software
  composite, 320, 322, 323
  industry, 320, 323
  materials input, 323
  panel element, 320
Slovak enterprises
  benefits resulting from cooperation, 269
  building up close cooperation, 270
  cooperating with other organizations, 268
  cooperation in near future, 268
  description, 266
  developed cooperation areas, 267
  improvement areas, cooperation, 269
  level of use, cooperation management, 270–271
  personal interviews, 267
  research goals, 267
  suppliers and respondents, 267
  technical cooperation, 267
Slovak Republic practice, 170–172
Smart Assist for All (SAforA), 430
  fall detection, 440, 441
  functionalities, 440
Smart cities, 116, 368, 370, 381
  changes, 397
  citizen-focused definitions, 385
  climate change, 115
  connectivity, 387
  definition, 385
  energy savings, 422
  human capital, 383
  ICT infrastructure, 386
  ICTs, 422
  ITC, 116
  mobility and transportation, 422–424
  quality of life, 106, 384
  smart infrastructure, 386
  smart people and governance, 397
  smart solutions, 386, 387
  spatial planning, 107
  sustainable community, 384–385
  taxi services, 413
  TM, 107
Smart city bus mobile application
  body controller evaluation, 461
  dashboard and vehicle status pages, 463
  fleet managers, 461, 463
  Linux-based processor, 461
  overview of, 461, 462
  passengers, 463
  sample screenshots, driver section, 462
  telemetry device, 461
Smart city concept
  ICT, 452
  mobility, 452
  urban problems, 452
  winning urban strategy, 452
Smart community, 122, 123
Smart energy transition, 108
Smart factory system, 188
Smart governance
  climate mitigation, 118
  CPRs, 126, 127
  decision-making system, 117
  definition, 118
  ICTs, 119, 120, 124
  ITC, 116
  local governance, 119
  multilevel governance, 117, 118
  Nepal irrigation systems, 119
  self-governing, 125
  semi-public, 119, 124–126
  smart community, 123
  social networks, 123
  stakeholders, 121
Smart Industry, 104, 112
Smart spaces, 369
Smart transport, 424
Smart transportation, 418
Smart water management
  big data techniques, 116
  cross-scale theoretical model, 123
  cultural approach, 118, 121–122
  decision-making, 122
  engineering approach, 118–120
  landscape-oriented, 118, 120, 121
  regional scale, 128
  research question, 118
  resilience framework
    hierarchical scales, 124
    landscape heterogeneity, 125–127
    principles, 122, 123
    scales, 124, 125
    spatial and scale mismatches, 127
  smart city, 116, 118
  spatial planning, 117
  vertical and horizontal axes, 118, 119
  water crises, 119
Smart wristband system
  Abbi smart bracelet, 431
  alarm reminder service, 446
  Arduino platform, 446
  Arduino UNO platform, 447
  Bluetooth technology, 445
  built-in clock, 444
  built-in GPS, 430, 431
  CCfB database, 440
  component functions, 447

Smart wristband system (cont.)
  configuration, 440
  emergency calling, 430
  fitness service, 446
  GPS module, 442, 443
  GSM/GPRS module, 443
  locating items, 444
  MTD, 440, 442
  MTD Sony Xperia Z, 432
  NFC module, 445, 446
  pulse monitoring, 444
  quality of life, 431
  Raspberry Pi, 447
  Sony Smartwatch, 432
  SOS message, 441, 443, 447
  violent removal, 442
  visual impairment, 443
  Vivosmart, 431
  weather and temperature, 444, 445
Social capital, 393, 402
Social cohesion, 402, 406, 407
Social communities, 406, 407
Social enterprise, 394
Social entrepreneur, 394
Social entrepreneurship, 394
Social innovation
  age management, 398
  benefits, 392
  changes, 391, 393
  characterization, 392
  criteria, 392
  human capital, 390, 391
  social development, 393, 394
  sustainable development, 384, 392, 395, 396
Social networking, 123
Sociotechnical systems, 24, 29
Solar wind lamp, 424
Sony Smartwatch, 432
SOS message, 443, 447
SOS paging service, 434
Spatial planning, 117, 402
  decision-making, 27
  RE, 108
  and spatial development, 29
  sustainable frameworks, 107
  TM, 108
Stakeholders
  classification of companies, 295
  CSR, 292, 293 (see also Corporate social responsibility (CSR))
  Friedman test, 296
  involvement advantages, 292

marketing activities, 295, 296
  smart governance, 121
Stimuli-organism-response (SOR), 374
String theory, 377, 378
Structural complexity
  flow-shop system layout, 79–81
  individual complexity coefficients, 72
  job-shop system layout, 75–78
  MC product strategy, 71
  mutual comparison, 82
  mutual paths, 82
  negative aspect, activities types, 71
  operational and dynamic, 72
  production system assessment, 72–75
  research question (RQ), 72, 82
Sunu smartband, 431
Supermarket
  attachable device cost, truck, 210
  costs of new racks, 208
  design, 207
  material flows improvement, 201, 207
  over-Kanban production board, 211
  pre-assembly relocation, 212
  preparing factors, 207
Supply chain, 219, 220
Surface roughness, 348, 350
Sustainability
  audit, 283
  green market, 282
  green purchases, 286
  inside-sustainability, 287
  organizations, 286
  outside-sustainability, 286
  reporting, 293
Sustainable development, 284, 367
  social innovation, 395, 396
Sustainable energy (SE), 104
  decentralized and community-driven, 104
  energy transition, 112
  place-based approach, 107
  RES, 30, 109
  smart city, 24, 109
  sociotechnical transition, 105
  spatial development, 105
  spatial planning, 105
  technical/technocratic, 111
  TM, 27
Sustainable marketing management, 281
  consumers awareness, 288
  consumption, 282
    barriers, 285
    code of ethics, advertising practice, 284
    eco-label award scheme, 283

Index

energy-using products, 283
European Environment Agency, 284
industrial policy action plan, 284
organizational aspect, 285
Slovak Environment Agency, 284
World Wildlife Fund, 284
customer behaviour, 282
purchase decision-making, 286, 287
qualitative research, 282
rate availability, 288
sustainability audit, 283
Sustainable reporting, 297, 300
SWOT analysis
competitiveness of Slovak enterprises, 251, 252
Synergy effects
automotive industry, 484
Big Data solutions, 468
company's internal and external environments, 468
description, 468, 469
linking cooperation, 469
strategic alliances, 470
strategy and cooperation management, 469
value, expected results and unexpected effects, 468
Systemic change/system innovation, 26

T

Targeted group of users (TGU), 430, 432
Taxi service
cabadvertising, 414
commercials and advertisement, 418
customer tablet, 414
driver's tablet, 414
internet access, 414
multimedia (*see* Multimedia)
Tabbie, 414
Verifone, 414
Terminal devices (TD), 430
Territorial identity, 402, 404, 405
Tesla Motors, electric cars, 477
Tested products and abbreviated test, 140–142
Testing of reliability, 134
Thin-walled component
accuracy and quality of production, 354
calibration sample measurement, 359
safety plane selection, 357
simulation, 358
technical and non-technical practice, 353
tool selection, 358

transverse roughness measurement
depth of, 362, 363
inner place, 360, 361
outer place, 361, 362
zero-point selection, 357
Transect coding, 118, 119, 125, 126
Transition management (TM), 26, 27, 30, 31, 107, 112
Transportation
car parking, 423
car sharing, 423
ITS, 423
problems, 423
urban mobility, 423
Transverse roughness evaluation
ANFIS model, 355
(*see also* CAM software)
CNC machine tools, 354
CNC program manufacturing, 354
industrial production unit, 353
international ISO standards, 354
multiaxis CNC machines, 354
nonlinear equations, 354
surface quality research, 354 (*see also* Thin-walled component)
thin-walled parts, aerospace industry, 353
Turning, 345–347
Type technological process, 154

U

University of Žilina
Industrial Engineering department, 179–180
mechanical engineering faculty, 176–178
Urban gardening
benefits, 408
Bratislava, 409
mobile gardens, 409
place attachment, 405, 406
place identity, 406
Slovakia, 409
social relations, 407, 409
sustainability, 408
Urbanity, 403, 404
Urbanized landscape
accessibility, 127
definition, 125
heterogeneous, 126
monocentric transect pattern, 127
polycentric spatial pattern, 127
transect coding, 125

V

Value chain, 436
Variant process planning, 329
Vector method, 74
Vehicle data processing
   advantages, 455
   brake lining status, vehicle, 457
   diagnose screen, vehicle, 456
   drivers, 454
   engine and transmission fault codes, 458
   system overview, 453, 454
   telemetry device, 453
   user and authorization groups, 453
   web application, 454–456
Vertical axis of rotation (VAWT), 425
Visual Basic for Application (VBA), 86
Vivosmart, 431

W

Water crisis, 117, 119, 127
Water governance, 119, 120, 122–125, 128
Water resilience framework
   adaptive water governance, 123
   landscape heterogeneity, 125–127
   principles, 122, 123
   scales
      complex systems, 124
      hierarchical, 124
      large, 124
      meso, 124
      micro, 125
   spatial and scale mismatches, 127
Water-resilient city, 122
Web generation, 166
Wind belt, 425
Wood-plastic composites (WPCs)
   additives, 330
   adjustable clamping, 331
   biocides, 330
   cellulose-based natural fibers and thermoplastics/thermosets, 327
   construction industry, 328
   cutting samples, 331
   cutting speed, 346, 348, 350, 351
   environmentally friendly/green materials, 328
   experimental material, 346
   experimental samples, 332
   feed rate, 350
   heterogeneity of material, 328
   lignin derivates, 329
   machining process, 347
   Mitutoyo SJ 400, 331
   optical profilometer, 346, 348
   outdoor applications, 346
   plasticizers, 330
   polyethylene (PE), 329
   polymer modifications, 330
   polypropylene/polypropene, 330
   polyvinyl chloride (PVC), 330
   production technologies, 327
   radiographic method, 328
   stabilizers, 330
   surface roughness, 332, 348, 350
   technical parameters, 332
   test samples, 346
   thermoplastic matrixes, 329
   turning, 345–347
   volume ratio, plastic matrix (HDPE), 330
   Water jet 3020b–1Z, 331
   wood fibers, 329
   wood filler and polymer matrix, 328
   wood flour density, 328, 329

Urbanized landscape (*cont.*)
   transect segments, 125
   water system (*see* Smart water management)
   zoning codes, 126

X

χ^2 distribution, 65–66

Z

Žilina Innovation Policy (ZIP), 169
Žilina Intelligent Manufacturing System (ZIMS), 167
   cooperation model, 185
   CPS, 188–189
   disposition, 186
   EFFRA, 184
   European Recovery Plan, 184
   Holonic concept, 189
   innovative solutions, 186
   joint researches and innovation projects, 183–184
   laboratory sections, 187
   lean and flexible production systems, 185
   MC, 183
   partnerships and mutual cooperation, 184
   research and innovation priorities, 184
   research institutes and universities, 185
   technologies and methods, 186
   technology base, 191, 192
   University of Zilina, 185